CONTACT CATALYSIS

Volume 1

Contributions by the members of the
Catalysis Club of the Hungarian Academy of Sciences

Authors

M. Bakos
P. Fejes
G. Gergely
S. Holly
K. Ibrányi-Árkosi
D. Kalló
J. Király
J. Ladik
J. Petró
L. Radics
K. Sasvári
F. Solymosi
A. Tungler

CONTACT CATALYSIS

Editor-in-chief

Prof. Z. G. SZABÓ

Member of the Hungarian Academy of Sciences

Assistant editor

Dr. D. KALLÓ

Volume 1

ELSEVIER SCIENTIFIC PUBLISHING COMPANY

Amsterdam — Oxford — New York 1976

This is a completely revised edition of the original Hungarian *Kontakt Katalízis*, published by Akadémiai Kiadó, Budapest

Manuscript revised by

G. SCHAY

Member of the Hungarian Academy of Sciences

Translated by

GY. GALAMBOS, MRS. E. JAKAB, GY. JALSOVSZKY, T. MOHÁCSI, MRS. F. SOÓS and MRS. P. SZŐKE

English translation supervised by

D. A. DURHAM

The distribution of this work is being handled by the following publishers:
for the U.S.A. and Canada

Elsevier/North Holland, Inc.
52 Vanderbilt Avenue
New York, New York 10017

for the East European countries, China, Cuba, Korean People's Republic, Mongolia and People's Republic of Viet-Nam

Akadémiai Kiadó, The Publishing House of the Hungarian Academy of Sciences, Budapest

for all remaining areas

Elsevier Scientific Publishing Company
335 Jan van Galenstraat
P. O. Box 211, Amsterdam, The Netherlands

Library of Congress Cataloging in Publication Data

Szabó, Zoltán G. 1908 – Contact catalysis.
Rev. ed. of the work originally published in Hungarian 1966 under title: Kontakt katalízis.
Bibliography: p. –
Includes index.
1. Catalysis. I. Kalló, Dénes. II. Bakos, Miklós. III. Title.

QD505.S94 541′.395 75-34394

ISBN 0-444-99853-5 (v. 1)

Set. Nr. ISBN 0-444-99852-7

Coedition published by Elsevier Scientific Publishing Company, Amsterdam, The Netherlands and Akadémiai Kiadó, The Publishing House of the Hungarian Academy of Sciences, Budapest, Hungary

Printed in Hungary

CONTENTS

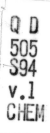

PREFACE

The first half of the 20th century often used to be called the century of chemistry. In view of the recent rapid development of the chemical industry, however, I feel that this expression, and numerous others, may refer even more aptly to the second half of our century. The modern chemical industry is superior to past industry not only in the dimension of its operation but also, and perhaps even more essentially, in a qualitative sense. Whereas, in the past, most of the individual technological steps were developed empirically, by "optimalization", our recent knowledge of the nature of certain processes and chemical reactions is essentially more sophisticated. Catalysis is involved in the majority of these processes, even when the catalytic effect is caused involuntarily. Although the role of homogeneous catalysis in industrial processes has recently increased, the application of contact catalysis is still of fundamental importance.

For the above reasons the Catalysis Club of the Hungarian Academy of Sciences decided to compile a monograph on contact catalysis, such a monograph which would not only be based on a scientific background but one which would also provide detailed practical instructions for the application of catalysis. With this aim in mind a team of authors, including also non-members of the Club, was organized under the supervision of an Editorial Board. This decision proved to be correct because it would have been a task beyond the capabilities of one or two authors to elaborate the projected wide subject range of contact catalysis which spans from solid-state physics to reaction kinetics and reactor design, and also because in the topics included in this monograph we have had the benefit of specialists who have graduated since the Second World War. Of course, as with every book written by a team of authors, much effort was required to ensure a suitable balance of the various chapters and a unified approach, and to avoid repetition as far as possible; but any difficulties or failings arising are, it is to be hoped, more than compensated for by the advantage that each chapter has been written by an authority in the field concerned. In certain places the reader may find repetition, or brief summaries of material already discussed at length. These summaries, referring to previous chapters have been included to provide some background and possibly facilitate subsequent treatment.

The special problems and editorial activities concerning this book were always on the agenda of the sessions of the Catalysis Club; thus this whole book is the result of a great deal of collaborative effort. No such standard handbook similar to this in content and size has ever been published anywhere in the world.

This monograph can be used as a textbook for university students, and also as a reference book for scientists and industrial experts working in the field of catalysis. The arrangement of the chapters follows the principle that contact catalysis is the result of elementary reaction steps occurring on the surface, and thus resembling, in several respects, ordinary chemical reactions. Accordingly, the discussion proceeds from a theoretical background towards industrial problems; first the theory of the structure of matter is discussed, this being necessary for understanding the chemistry of catalyzed reactions and the mechanisms of the catalytic effect. The process of contact catalysis is discussed in the order of our recent knowledge, and not in an historical order. The authors have striven to give, beyond the theoretical facts, useful practical advice on the selection, preparation and investigation of catalysts, and on the investigation techniques of the catalytic processes. A large number of references can be found at the end of each chapter, enabling the reader to go into further details of the problems. Stress has been laid upon illustrating the applications of the theory with particular catalytic reactions, including the discussion of typical and industrially important processes.

With regard to the method of the treatment, it should be noted that the authors endeavoured to discuss the subject in a rigorous way, by means of mathematical formalism, wherever possible. Nowadays not only theoretical, but also practical research work on reaction kinetics, for instance, the design of reactors, involves a mathematical treatment. Nevertheless, care was taken to ensure that the mathematical formalism did not become too detailed and conceal the physical or chemical background for readers less familiar with this subject. Thus mathematical formulae play rather the role of signposts, by means of which also the practically oriented chemists can proceed when solving their particular problems.

The present English version of the book is not simply the translation of the original Hungarian text; (i) corrections and amendments have been made in the Hungarian text wherever this appeared to be necessary, (ii) the most recent results published in the literature of catalysis and their application have been included; (iii) due to the limitations imposed on the size of the English edition, the original text was condensed in several places, and some parts of lesser importance have been omitted, thereby enabling the inclusion of the new parts without increasing the projected size of the book.

We hope that the carefully selected contents and the chosen way of treatment will enable both theoretical and practical experts to use this book to their advantage.

Budapest,
February, 1975

Zoltán G. Szabó

PART I

ELECTRONIC STRUCTURE OF METALS AND SEMICONDUCTORS

INTRODUCTION

J. Ladik

In order to understand the processes taking place during heterogeneous catalysis it is necessary to know, among other things, the properties and structures of the solids used as catalysts. Catalytically active solid substances are primarily metals and semiconductors. Accordingly, this Part will deal with the elements of the quantum mechanical theory of the electronic structures of metals and semiconductors, and it will be shown how the electrical, magnetic and surface properties of these materials, which are most important from the point of view of the catalytic activities of metals and semiconductors, can be interpreted on the basis of the electronic structures. The discussion, however, will be elementary throughout, and in the case of mathematically more involved derivations only the results will be given. In several places only a qualitative treatment will be presented. Thus readers having less practice in theoretical physics and mathematics will have no difficulty in grasping the general treatment necessary for an understanding of the subsequent chapters of the book.

In Chapter 1 of this Part the basic principles of solid-state physics, indispensable for later discussions concerning the geometry of crystals, the concept of Brillouin zones, the different types of bonding in crystals, etc., will be summarized. In Chapter 2, which deals with the electronic structures of metals, an analysis of the free electron model and the band model for the description of the electronic structure will be given. In connection with the latter the relation between crystal symmetry and the structure of the energy band system is indicated, and methods for the approximation of crystal orbitals having the symmetry of the crystal (the Bloch crystal orbitals) are discussed.

In Chapter 3, which deals with the electronic structures of semiconducting materials, the band model of semiconductors is described, and this is followed by a qualitative survey of the most important features of the mechanisms of intrinsic and impurity (n- and p-type) conduction of semiconductors. Thereafter the Fermi levels of different types of semiconductors are determined according to Fermi-Dirac statistics. Finally, the different types of defects in solids are discussed and relationships are pointed out between the conduction properties and defect structures of various n- and p-type conducting, intrinsic and amphoteric oxides.

In Chapter 4 of this Part, which summarises electrical and magnetic properties of metals and semiconductors, mainly a semi-quantitative discussion of the electrical conductivities of metals and semiconductors is presented, on the basis of the energy band model. This is followed by a description of how the magnitude of the Hall effect and the paramagnetic and diamagnetic susceptibilities of metals and semiconductors can be deduced from their electronic structures.

In Chapter 5, which deals with the surface electronic states and other surface properties of solids, the band model is used to give the necessary conditions for the occurrence of different surface states (of the Tamm and Shockley type), utilizing the results of calculations made on various simple models. Finally, from the calculations reported in the literature, results on the surface states of some solids are discussed*.

* It was considered necessary to include this chapter since the more detailed calculations concerning the surface states of solids provide a possibility for the future construction of a more exact quantum mechanical theory of catalysis.

CHAPTER 1. BASIC CONCEPTS IN SOLID-STATE PHYSICS

J. Ladik

SECTION 1.1. ELEMENTS OF THE GEOMETRICAL STRUCTURES OF CRYSTALS

Thermodynamically stable solids have a crystalline structure. The atoms, ions or molecules constituting the crystal are contained in a three-dimensional lattice of translational symmetry. In other words, the geometrically equivalent building units of the crystal are located in lattice points described by the lattice vectors:

$$\mathbf{R}_{n_1,n_2,n_3} = n_1\mathbf{a}_1 + n_2\mathbf{a}_2 + n_3\mathbf{a}_3 \tag{I.1}$$

[Eq. (I.1) describes the other equivalent lattice points if the origin of the coordinate system is taken to coincide with a lattice point.] In the above equation n_1, n_2 and n_3 are arbitrary positive or negative integers, and the elementary translation vectors $\mathbf{a}_1$, $\mathbf{a}_2$ and $\mathbf{a}_3$ are, in general, not orthogonal and their absolute values are not necessarily identical. Due to the three-dimensional translation symmetry of the crystal any two points within the crystal are physically equivalent if their position vectors differ only by a lattice vector:

$$\mathbf{r}_2 = \mathbf{r}_1 + \mathbf{R}_{n_1,n_2,n_3} \tag{I. 2}$$

On the basis of the angles α, β, γ, defined by the elementary translation vectors (see Fig. I.1), and the relative values a_1/a_2, and a_3/a_2, it is possible to classify crystals into seven crystal systems. These are given in Table I.1.

The macroscopic crystals classified in crystal systems can be assigned to 32 point groups according to the symmetry operations* characterizing their symmetry in a more detailed way. The macroscopic symmetry operations (elements) are the following: rotation axis, plane of symmetry, centre of inversion and rotation-inversion axis. It can be shown [1] that the rotation and rotation-inversion axes of crystals can only be one-, two-, three-, four-, and six-fold. The macroscopic geometrical structures of perfect crystals (where

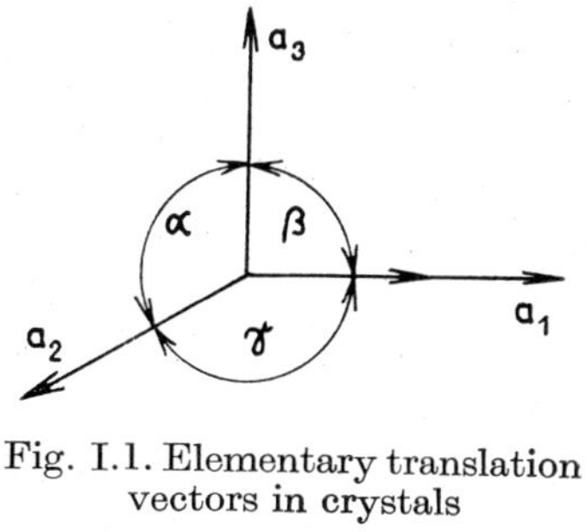

Fig. I.1. Elementary translation vectors in crystals

* A symmetry operation is an operation which, when acting on a system, transforms it into another configuration, indistinguishable from the original system. For example, when a benzene ring is rotated through 60°, 120° or 180° about the axis passing perpendicularly through the plane of the molecule at its centre, the new configurations of the molecule are indistinguishable from the original one.

TABLE I.1

The crystal systems

Triclinic	$\alpha \neq 90°,\ \beta \neq 90°,\ \gamma \neq 90°;\ \mathbf{a}_3 \leq \mathbf{a}_2 \leq \mathbf{a}_1$
Monoclinic	$\alpha = \gamma = 90°,\ \beta \neq 90°;\ \mathbf{a}_3 \leq \mathbf{a}_1;\ \mathbf{a}_2$ arbitrary
Orthorhombic	$\alpha = \beta = \gamma = 90°;\ \mathbf{a}_3 < \mathbf{a}_1 < \mathbf{a}_2$
Trigonal	$\alpha = \beta = \gamma \neq 90°;\ \mathbf{a}_1 = \mathbf{a}_2 = \mathbf{a}_3$
Tetragonal	$\alpha = \beta = \gamma = 90°;\ \mathbf{a}_1 = \mathbf{a}_2 \neq \mathbf{a}_3$
Hexagonal	$\alpha = \beta = 90°,\ \gamma = 120°;\ \mathbf{a}_1 = \mathbf{a}_2;\ \mathbf{a}_3$ arbitrary
Cubic	$\alpha = \beta = \gamma = 90°;\ \mathbf{a}_1 = \mathbf{a}_2 = \mathbf{a}_3$

the equivalent planes are uniformly developed) are exactly characterized by the symmetry properties of the given crystal point group, and even the geometry of real crystals more or less follows this symmetry. The symmetry operations characteristic of the individual crystal point groups are not given here, as they can be found in several textbooks [2].

The positions of the lattice planes with respect to the crystal axes can be most conveniently characterized by the Miller indices. These can be obtained by determining the intersections of the three crystal axes with the lattice plane in question (the orientations of the crystal axes are identical to those of the elementary translation vectors), and by expressing the coordinates of the intersection points in terms of the corresponding lattice constant (the length of the elementary translation vector). Next the reciprocals of these numbers are formed, and the values obtained are multiplied by the least common multiple of the numbers in the denominator. For example, if the lattice plane intersects the crystal axes in points

$$(2n\mathbf{a}_1, 0, 0); \quad (0, n\mathbf{a}_2, 0); \quad (0, 0, 3n\mathbf{a}_3)$$

the position of the plane is described in place of the numbers $2n$, n, $3n$ by the appropriate Miller indices:

$$(hkl) = \left(\frac{1}{2n}, \frac{1}{n}, \frac{1}{3n}\right) 6n = (3\ \ 6\ \ 2)$$

The unit cell of a crystal can be determined if the orientations and magnitudes of the elementary translation vectors are known. The unit cell is a parallelepiped which, if subjected to any translation operation (except for the identity operation $n_1 = n_2 = n_3 = 0$) allowed by the crystal symmetry, will never produce a parallelepiped congruent with the original one, and the set of parallelepipeds obtained by applying all the possible translation operations contains all points of the space.

Using the above definition of the unit cell there are still a number of possibilities for its construction in the case of a given crystal symmetry, and therefore the unit cell is usually chosen so that its apices should be lattice points. Furthermore, in order to have the best correspondence between the symmetry of the unit cell and that of the macroscopic crystal, the unit cell is chosen in such a way that the centres of planes, or the centre of the parallelepiped should also coincide with a lattice point. The three-dimensional lattices containing these types of unit cell are called face-centred and body-centred lattices, respectively, while if only the apices coincide with the lattice points, the lattice is primitive. It can be shown [3] that the number of unit cells and, therefore, lattice types constructed in the above manner is 14. These are the following: (1) primitive triclinic, (2) primitive monoclinic, (3) base-centred monoclinic, (4) primitive orthorhombic, (5) base-centred orthorhombic, (6) face-centred orthorhombic, (7) body-centred orthorhombic, (8) primitive tetragonal, (9) body-centred tetragonal, (10) base-centred hexagonal, (11) primitive trigonal, (12) primitive cubic, (13) body-centred cubic, and (14) face-centred cubic lattices. Illustrations of the different types of unit cells can be found in other textbooks [4].

SECTION 1.2. THE RECIPROCAL LATTICE AND THE BRILLOUIN ZONES

If the elementary translation vectors $\mathbf{a}_1$, $\mathbf{a}_2$, and $\mathbf{a}_3$ of a given three-dimensional lattice are known, the reciprocal vectors $\mathbf{b}_1$, $\mathbf{b}_2$ and $\mathbf{b}_3$, obeying the relationships

$$\mathbf{a}_i \mathbf{b}_j = 2\pi\delta_{ij}^* \ (i, j = 1, 2, 3) \tag{I.3}$$

can also be determined (δ_{ij} is the Kronecker symbol: $\delta_{ij} = 1$ if $i = j$; $\delta_{ij} = 0$ if $i \neq j$). It can easily be seen that the vectors described by the equations

$$\begin{aligned} \mathbf{b}_1 &= \frac{2\pi(\mathbf{a}_2 \times \mathbf{a}_3)}{\mathbf{a}_1(\mathbf{a}_2 \times \mathbf{a}_3)} \\ \mathbf{b}_2 &= \frac{2\pi(\mathbf{a}_3 \times \mathbf{a}_1)}{\mathbf{a}_1(\mathbf{a}_2 \times \mathbf{a}_3)} \\ \mathbf{b}_3 &= \frac{2\pi(\mathbf{a}_1 \times \mathbf{a}_2)}{\mathbf{a}_1(\mathbf{a}_2 \times \mathbf{a}_3)} \end{aligned} \tag{I.4}$$

satisfy just the relationships given by Eq. (I.3).

Taking the vectors $\mathbf{b}_1$, $\mathbf{b}_2$ and $\mathbf{b}_3$ thus obtained as the primitive translation vectors of another lattice, the so-called reciprocal lattice, the lattice points of the latter can be described in a manner analogous to Eq. (I.1), by the lattice vectors

$$\mathbf{K_m} = m_1\mathbf{b}_1 + m_2\mathbf{b}_2 + m_3\mathbf{b}_3 \tag{I.5}$$

Here $\mathbf{m}$ refers to the set of positive or negative integers m_1, m_2, m_3. Considering Eq. (I.4) it can be seen that the dimensions of the vectors $\mathbf{b}_i$ ($i =$

*The coefficient 2π in Eq. (I.3) defining the reciprocal vectors $\mathbf{b}_i$ ($i=1, 2, 3$) simplifies the form of the exponential factor of the wave function.

$= 1, 2, 3$) and of the lattice vectors $\mathbf{K_m}$ of the reciprocal lattice are reciprocal length, l^{-1}.

On the other hand the De Broglie relationship

$$\lambda = \frac{h}{p} \tag{I.6}$$

assigns a wave of wavelength λ to each electron (h is the Planck constant, and p is the momentum of the electron). Instead of by wavelength, electron waves can also be characterized by their wavenumber, which is the reciprocal of the former. Corresponding to the vector character of momentum, according to Eq. (I.6) the wavenumber can also be taken as a vector, and thus via multiplication by 2π the wavenumber vector of the electron wave can be obtained:*

$$\mathbf{k} = \frac{2\pi}{h}\mathbf{p} = \frac{2\pi}{\lambda}\mathbf{e}_n \tag{I.7}$$

(where $\mathbf{e}_n$ is the unit vector in the direction of the momentum). The array of vectors $\mathbf{k}$ constitutes the wavenumber space (or briefly, $\mathbf{k}$-space). As can be seen from Eqs (I.6) and (I.7), the dimensions of the wavenumber vector are also cm^{-1}, and hence for practical reasons, the $\mathbf{k}$-space can be identified with the space of the reciprocal lattice. This identification is advantageous in several respects as far as the study of the electronic structures of solids is concerned. It will be shown in Section I.1.5 that if we want to utilize the three-dimensional translation symmetry of the electric potential acting on the electrons in the crystal,

$$V(\mathbf{r} + \mathbf{R}_n) = V(\mathbf{r}) \tag{I.8}$$

($\mathbf{R}_n$ is an arbitrary lattice vector), in the construction of the wave functions providing the solution of the Schrödinger equation, the wave functions $\psi_\mathbf{k}(\mathbf{r})$ should be written in the form

$$\psi_\mathbf{k}(\mathbf{r}) = e^{i\mathbf{kr}}\, u_\mathbf{k}(\mathbf{r}) \quad (i = \sqrt{-1}) \tag{I.9}$$

(Bloch crystal orbitals). The function $u_\mathbf{k}(\mathbf{r})$ has the same three-dimensional periodicity as the function $V(\mathbf{r})$

$$u_\mathbf{k}(\mathbf{r}) = u_\mathbf{k}(\mathbf{r} + \mathbf{R}_n) \tag{I.10}$$

and both factors of the wave function $\psi_\mathbf{k}(\mathbf{r})$ depend on the wavenumber vector $\mathbf{k}$.

Provided that an arbitrary wavenumber vector $\mathbf{k}$ can be written in the form

$$\mathbf{k} = \mathbf{k}' + \mathbf{K}_m \tag{I.11}$$

* The coefficient 2π in Eq. (I.7) again serves to simplify the exponential factor of the wave function.

the corresponding Bloch function can also be written as

$$\psi_{\mathbf{k}}(\mathbf{r}) = e^{i\mathbf{k}'\mathbf{r}} e^{i\mathbf{K}_m\mathbf{r}} u_{\mathbf{k}}(\mathbf{r}) \tag{I.12}$$

It can easily be shown, however, that the term $e^{i\mathbf{K}_m\mathbf{r}}$ in Eq. (I.12), has the property

$$e^{i\mathbf{K}_m(\mathbf{r}+\mathbf{R}_n)} = e^{i\mathbf{K}_m\mathbf{r}} \tag{I.13}$$

and hence it can be included into the function $u_{\mathbf{k}}(\mathbf{r})$ having the periodicity of the lattice. For this reason it is more practical to regard the wave functions as functions only of $\mathbf{k}'$, instead of $\mathbf{k}$. The individual functions $\psi_{\mathbf{k}'}(\mathbf{r})$ will thus be multiple-valued functions of vectors $\mathbf{k}'$ terminating within the unit cell of the $\mathbf{k}$-space. The unit cell of the $\mathbf{k}$-space determined by the symmetry of the crystal is called the first Brillouin zone of the crystal.

The unit cell of the $\mathbf{k}$-space, i.e. the first Brillouin zone, can be defined in a number of ways, similarly to the unit cell of the crystal as defined in the usual geometrical space. To avoid ambiguity, the generally accepted convention is that the unit cell of the $\mathbf{k}$-space has a symmetrical position with respect to the point $\mathbf{k} = \mathbf{0}$, and that only the shortest vectors $\mathbf{k}'$ terminate within or at the surface of this unit cell. (If an arbitrary vector $\mathbf{K}_m$ of the reciprocal lattice is added to any vector $\mathbf{k}'$ terminating within the Brillouin zone, no shorter vector may be obtained than the longest vector $\mathbf{k}'$ terminating within the first Brillouin zone.) This requirement can be shown to be satisfied by constructing the first Brillouin zone in the following way. A lattice point of the reciprocal lattice is taken as the origin ($\mathbf{k} = \mathbf{0}$), and vectors are drawn from this point to the other points of the reciprocal lattice. Each of the obtained lattice vectors of the reciprocal lattice is bisected by a perpendicular plane. Among the polyhedra of the $\mathbf{k}$-space formed by these planes, the one having the smallest volume is the first Brillouin zone.

The second, third, etc. Brillouin zones can be defined in a similar way. The second Brillouin zone is the polyhedron having the second smallest volume in the $\mathbf{k}$-space defined by the planes bisecting vectors $\mathbf{K}_m$ and subsequent Brillouin zones can be defined in an analogous manner. For the solution of most of the problems relating to electronic structure it is sufficient to determine the first Brillouin zone of the crystal lattice.

Since the first Brillouin zone is the unit cell of the reciprocal lattice, its volume is $\mathbf{b}_1(\mathbf{b}_2 \times \mathbf{b}_3)$. If the primitive translation vectors of the reciprocal lattice are expressed by means of Eq. (I.4) in terms of the primitive translation vectors $\mathbf{a}_1$, $\mathbf{a}_2$, $\mathbf{a}_3$ of the geometrical lattice, it is found that

$$V_{1.\text{B.z.}} = \mathbf{b}_1(\mathbf{b}_2 \times \mathbf{b}_3) = \frac{8\pi^3}{\mathbf{a}_1(\mathbf{a}_2 \times \mathbf{a}_3)} = \frac{8\pi^3}{\Omega} \tag{I.14}$$

where Ω is the volume of the unit cell of the crystal lattice (and not of the reciprocal lattice!). For a more detailed treatment of Brillouin zones see, for example, the paper of Reitz [5].

SECTION 1.3. DIFFERENT TYPES OF BONDS IN CRYSTALS

Crystals can be classified into four main groups according to the type of the bonds between their building units. These are the following:

(1) molecular crystals,
(2) covalent crystals,
(3) ionic crystals,
(4) metals.

The building units of *molecular crystals* are molecules (e.g. Cl_2, I_2, noble gases, CH_4), and there are only weak van der Waals-type forces acting between the molecules (accordingly, the melting points of molecular crystals are low). As these forces are approximately independent of the relative orientation of the molecules*, the molecules are packed in the closest possible manner. Several molecular crystals (Ne, Ar, Kr, CH_4 lattices, etc.) therefore form closely packed cubic lattices.

Due to the weakness of the intermolecular interactions, the physical properties of the molecules in molecular crystals are very similar to the gas-phase properties, even in the solid phase. Thus, the bond length in the I_2 molecule is 2.65 Å in the gaseous phase, and 2.70 Å in the solid phase (on the other hand, the minimum distance between the atoms of neighbouring molecules in the I_2 lattice is 3.54 Å). Another example is given by the vibrational frequencies of benzene; the crystal frequencies corresponding to the gas phase values of 3099 and 3045 cm^{-1} are 3089 and 3034 cm^{-1}, respectively. Consequently, the molecules retain their individual characteristics in the molecular crystals.

In *covalent* (valence) *crystals* the atoms situated in the lattice points are bonded to one another by covalent chemical bonds. The electrons forming the bonds can be regarded as localized between the two atoms bonded by them (just as in the case of saturated compounds containing only single bonds). A characteristic example of a covalent lattice is the diamond lattice, in which carbon atoms in the sp^3 hybrid state form strong covalent bonds with their four neighbours, and the value of the bond angle is 109°, characteristic of the tetrahedral arrangement. Silicon, germanium and grey tin have similar crystal lattices. In the tetrahedral lattice the volume occupied by an atom is fairly large. This suggests that the forces bonding the constituents are of an oriented nature and therefore have a covalent character. A high melting point, due to strong bonding, and great hardness are further characteristics of covalent lattices.

* There are three types of van der Waals forces: the orientation force acting between molecules with permanent dipoles; the induction force acting between a molecule with a permanent dipole and another non-polar, but polarizable molecule; and finally the dispersion forces acting between molecules having no permanent dipoles (for more details see, for example, Ref. [6]). Of the three types of forces it is only the first that depends strongly on the relative orientation of the molecules. However, since the molecules building up molecular crystals generally have no, or only a small, dipole moment, apart from special cases, it is the dispersion forces that play the decisive role in molecular crystals, and these forces scarcely depend on the orientation.

Ionic crystals contain atoms of very different electronegativities (electron-attracting power). One partner is usually an electropositive metal atom, and the other an atom of high electronegativity. These are arranged in the crystal lattice in such a manner that the number of electronegative atoms surrounding the electropositive one, and the number of electropositive atoms surrounding the electronegative one, is a maximum. As a consequence, the electropositive atom loses almost all of its outermost, loosely bound electrons, which are taken up by the electronegative atoms. This results in the lattice points being occupied by ions having closed shells. For example, in the face-centred cubic lattice of sodium chloride each Na^+ ion is surrounded by six Cl^- ions, whereas each Cl^- ion is surrounded by six Na^+ ions.

In the calculation of the energy of the ionic lattice the first step is to sum the electrostatic energies corresponding to the interactions between the ions. For example, for any Na^+ ion in the sodium chloride lattice it is necessary to take into account the attraction of six Cl^- ions at a distance $R_{Na^+-Cl^-} = R$, the repulsion of the next twelve Na^+ ions at a distance $\sqrt{2}R$, the attraction of the eight Cl^- ions at a distance $\sqrt{3}R$, etc. Accordingly, the electrostatic energy acting on the Na^+ ion can be given by the following expression (elementary charge is denoted by e):

$$E_{Na^+} = -e^2\left(\frac{6}{R} - \frac{12}{\sqrt{2}R} + \frac{8}{\sqrt{3}R} - \ldots\right) \qquad \text{(I.15)}$$

A similar expression can be derived for any Cl^- ion too. Now when the total electrostatic energy is calculated by summing the energies of type (I.15), each interaction is taken twice. Eq. (I.15) therefore, also gives the electrostatic energy of a single NaCl molecule.

For other ionic lattices the expression for the electrostatic energy is different from that of Eq. (I.15), being a function of the geometrical structure of the crystal lattice, but for every case the energy can be given as the sum of terms depending on e^2/R. Several methods have been elaborated for the summation of such expressions. The results can be given in the form

$$E_1 = E_{\text{electrost./ion pair}} = -Ae^2/R \qquad \text{(I.16)}$$

where A is the Madelung constant whose value depends on the geometrical structure of the crystal lattice. Thus, for example, in the case of a face-centred cubic lattice (e.g. NaCl) $A = 1.75$, whereas for a body-centred cubic lattice (e.g. CsCl) $A = 1.76$. For other cases widely different A values are obtained.

According to Eq. (I.16), provided that other energy terms did not make a significant contribution, the crystal lattice would have minimum energy if all the ions were collected in one point ($R = 0$). In reality, however, when the distance between the ions is reduced, their electron orbitals penetrate one another, and so repulsive energy terms appear. These terms rapidly decrease with increasing R, and according to detailed calculations they can be given in the form

$$E_2 = E_{\text{repulsive/ion pair}} = c_1 e^{-R/c_2} \qquad \text{(I. 17)}$$

where c_1 and c_2 are constants which can be calculated from the experimentally determined values of certain physical constants of the crystal (lattice constant, compressibility). In the NaCl lattice, at the equilibrium internuclear distance $E_1 = -8.92$ eV and $E_2 = +1.03$ eV [7], i.e. $E_1 + E_2 = -7.89$ eV/ion pair, while the experimental value is -7.86 eV/ion pair [7].

The two most important characteristics of *metal lattices* are that (1) they have good electrical conductivity, and (2) the atomic coordination number (i.e. the number of neighbours for any chosen atom) is always high.

These properties are due to the fact that metals most frequently crystallize in body-centred or face-centred cubic or in close-packed hexagonal lattices. In the first type of lattice every atom has fourteeen immediate neighbours (eight atoms at a distance R, and a further six at a distance $2R/\sqrt{3} = 1.15R$). In the other two lattice types every atom is surrounded by twelve equidistant neighbours.

The good electrical conductivities of metal lattices containing atoms o low electronegativity result from the fact that their valence electrons may move relatively freely within the crystal, and therefore can move along the field even if the electrical field strength is low (for more details see section I.4.1). The high coordination number also suggests that the electrons forming metallic bonds are strongly delocalized. A detailed discussion of the electronic structure of metals follows in Chapter I.2.

Finally it must be noted that several solid substances represent intermediate types between the above-mentioned four main categories. Thus, for instance, intrinsic (impurity-free) semiconductors represent an intermediate between covalent crystals and metals.

SECTION 1.4. PRELIMINARY SURVEY OF METHODS FOR APPROXIMATING THE ELECTRONIC STRUCTURE OF METALS AND SEMICONDUCTORS

The simplest model for approximating the electronic structure of metals and semiconductors is the free electron model, where the potential influencing the freely moving electrons within the crystal is constant, i.e. $V(\mathbf{r}) = -W$. The total energy of the electron in this case is given by the simple expression

$$E = \frac{1}{2m_e}\mathbf{p}^2 - W \tag{I.18}$$

where $\mathbf{p}$ is the momentum vector of the electron, and m_e is its mass. (It should be noted that even if the potential is not constant within the crystal, it can be shown that Eq. (I.18) still holds if the mass of an electron m_e is replaced by m_e^*, the effective mass.) For the sake of generality m_e^* will be used throughout this Part in place of m_e.

In building up the free electron model the next step is to determine the number of levels falling between E and $E + dE$. For this purpose, however, instead of solving the Schrödinger equation with potential W, a simplified method is applied. By this method the six coordinates defining the position of each electron in the phase space, the spatial coordinates x, y, z and the momentum components p_x, p_y, p_z, can generally be given for each electron. (The phase space is the space defined by the product of the coordinate space and the space determined by the momentum vectors, the momentum space. The volume element of the phase space is therefore $d\chi = dVdP$, where dV and dP are the volume elements for the coordinate space and the momentum space, respectively.) The phase space is next divided into cells of size h^3, and to each cell of the phase space two electron states are assigned which can be occupied by two electrons of opposite spins [8]. (It can be shown that the state density obtained by such a construction of the phase space agrees with the true state density derived by quantum mechanical means.)

Since all electrons may travel freely in the total volume V of the crystal according to the basic assumption of the model, each cell of the phase space contains the volume V of the coordinate space. Hence, the individual cells are completely determined by the appropriate volume element of the momentum space. Thus, the number of cells belonging to the parallelepiped in the phase space whose edges span from p_x to $p_x + \Delta p_x$, from p_y to $p_y + \Delta p_y$, and from p_z to $p_z + \Delta p_z$, is

$$G = \frac{2}{h^3} V \, \Delta p_x \, \Delta p_y \, \Delta p_z \tag{I.19}$$

Since the volume of the macroscopic crystals is usually larger than 10^{-12} cm^3, the quantities Δp_x, Δp_y, Δp_z for a single cell are infinitesimally small. (It follows from the Heisenberg uncertainty principle that if $V = 10^{-12}$ cm^3, then, for example, $\Delta p_x \approx \dfrac{10^{-27}}{10^{-4}} = 10^{-23}$ g·cm·sec^{-1}, which can be regarded as infinitesimally small compared to the pertinent interval of the momentum changes.) Eq. (I.19) can, therefore, also be written in differential form:

$$dG = \frac{2}{h^3} V dp_x \, dp_y \, dp_z \tag{I.20}$$

The number of levels in the energy interval $E + dE$ can now be determined. Rearranging Eq. (I. 18) and substituting m_e^* in place of m_e, it can be seen that the levels of energy lower than E are found within a sphere of the momentum space of radius $p = \sqrt{(E + W)2\, m_e^*}$. The volume of the spherical shell between p and $p + dp$ of the momentum space is

$$dP = 4\pi p^2 \, dp \tag{I.21}$$

Introducing the notation

$$\varepsilon = E + W \tag{I.22}$$

one obtains $p^2 = 2m_e^*\varepsilon$ and $pdp = m_e^* d\varepsilon$. Substituting these expressions into Eq. (I.21):

$$dP = 2\pi(2m_e^*)^{3/2} \sqrt{\varepsilon} \, d\varepsilon \tag{I.23}$$

According to Eqs (I.20) and (I.21), therefore, the number of states belonging to the element $\mathrm{d}P$ of the momentum space or to the energy interval $\mathrm{d}E = \mathrm{d}\varepsilon$, respectively, is

$$g(\varepsilon)\,\mathrm{d}\varepsilon = \frac{4\pi V}{h^3}\,(2m_e^*)^{3/2}\,\sqrt{\varepsilon}\,\mathrm{d}\varepsilon = C\,\sqrt{\varepsilon}\,\mathrm{d}\varepsilon \tag{I.24}$$

where

$$C = \frac{4\pi V(2m_e^*)^{3/2}}{h^3} \tag{I.25}$$

For the determination of the number of the electrons (occupied states) in the energy interval $\mathrm{d}\varepsilon$ it is necessary to know the function determining the probability w_i that the level of energy ε_i is occupied. This distribution function is to be obtained from statistical mechanics. Electrons, with half spins, obey Fermi-Dirac statistics; this takes into account that according to the Pauli principle only one electron can occupy a state also characterized by the spin quantum number, and therefore the occupation number is either 0 or 1. The Fermi-Dirac distribution function is

$$w_i = \{\exp[(\varepsilon_i - \varepsilon')/kT] + 1\}^{-1} \tag{I.26}$$

where k is the Boltzmann constant, T the absolute temperature and ε' a parameter.

If the number of levels in the energy interval $\mathrm{d}\varepsilon$ is multiplied by the function (I.26) taken as continuous (since the number of electrons is large, the levels ε_i are very densely spaced), the number of electrons having energy in the interval $\mathrm{d}\varepsilon$ can be obtained:

$$\mathrm{d}n = \frac{C\,\sqrt{\varepsilon}\,\mathrm{d}\varepsilon}{\mathrm{e}^{\frac{\varepsilon-\varepsilon'}{kT}} + 1} \tag{I.27}$$

The parameter ε' can be determined from Eq. (I.27) by means of the equation

$$N = \int_0^N \mathrm{d}n = C \int_0^\infty \frac{\sqrt{\varepsilon}\,\mathrm{d}\varepsilon}{\mathrm{e}^{\frac{\varepsilon-\varepsilon'}{kT}} + 1} \tag{I.28}$$

where N is the number of electrons. In general, these calculations are fairly complicated, but they become greatly simplified when $T = 0$, and are still reasonably simple at low temperatures. An illustration of this, and a description of the application of the free electron model to the calculation of the specific heat of metals, follows in Chapter I.2.

The electronic structure of metals and semiconductors can certainly be best approximated by the crystal orbital method. One of the basic assumptions of this method is that the valence electrons may move freely throughout the entire volume of the crystal, and their motion should therefore be described by wave functions delocalized over the whole crystal, the so-called crystal orbitals. It is not assumed, however, that the potential is constant within the crystal; instead the potential acting on the electrons and arising from the Coulomb interaction with other electrons and nuclei, as in the real situation, is considered to follow the three-dimensional periodicity of the lattice [see Eq. (I.8)]. At the same time, the individual energy levels are also determined more accurately by the approximate solution of the corresponding Schrödinger equation. As we shall see in Section I.1.5, consideration of the spatial translation symmetry of the potential involves certain restrictions for the form of the crystal orbitals, which simplifies the approximate solution of the Schrödinger equation.

The valence electron levels of metals or semiconductors are split as a result of interactions, and the number of the split levels is given by the number of atoms in the crystal. Since the number of atoms in the macroscopic crystal is extremely large, the number of levels originating from a single atomic level is also very high, and the distance between the component levels becomes very small. For this reason the multitude of levels originating from an atomic level can be taken to a good approximation as a continuous range of energy values accessible to the electrons, in other words, an *energy band* (see Fig. I.2).

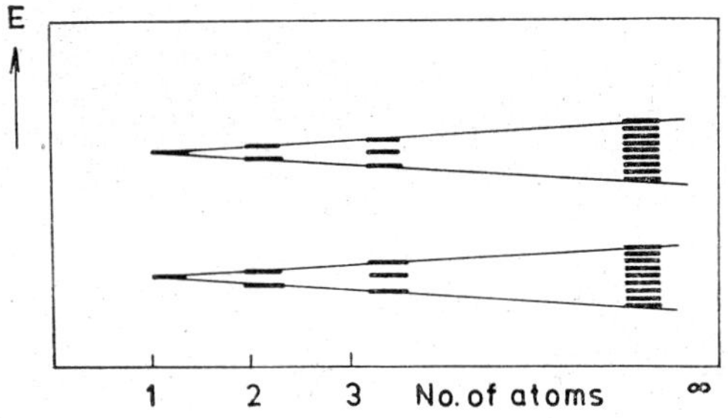

Fig. I.2. The splitting of atomic levels to energy bands in crystals

The extent of the splitting, i.e. the width of the energy bands, also depends on the magnitude of the interaction between the atomic wave functions (atomic orbitals) in addition to the number of atoms constituting the crystal. The extent of the interaction can be well characterized by the overlap integral between wave functions ψ_i and ψ_j of the valence electrons belonging to the neighbouring atoms i and j:

$$S_{ij} = \int \psi_i \, \psi_j \, \mathrm{d}V \tag{I.29}$$

The value of the overlap integral for the 2s valence orbitals of Li is 0.50 [9] if the distance between the Li atoms is 3.03 Å (the value found for the Li lattice), which is rather high. (Between two neighbouring π orbitals of the benzene ring S_{ij} is 0.26.)

When the atoms constituting the metal lattice have more interacting valence electrons, there are more energy bands. If the corresponding atomic levels are closely spaced and the bands are also wide, some bands may overlap, and in certain cases all energy bands may merge into a single, very broad band. This is the case, for example, with the face-centred cubic crystal lattice of nickel [10]. In such cases the number of levels in the energy interval dE may be very high.

In Chapter I.2 dealing with the electronic structures of metals we shall return to the discussion of the crystal orbital method.

The third main method for approximating the electronic structure of metals is the valence bond method, in which the different possible valence states are superimposed. A valence state (or limiting structure) is a given pairing scheme of the valence electrons of the atoms. For instance, the covalent valence states of 4 Li atoms are

$$
\begin{array}{ccc}
\text{Li--Li} & \text{Li\ Li} & \text{Li}\diagdown\quad\diagup\text{Li} \\
& |\quad | & \diagup\quad\diagdown \\
\text{Li--Li} & \text{Li\ Li} & \text{Li}\quad\text{Li}
\end{array}
\quad\text{and}
$$

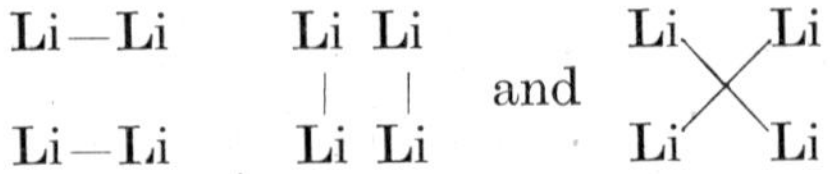

while *one* of the possible ionic valence states is

$$
\begin{array}{c}
\text{Li--Li}^{(-)} \\
|\quad| \\
\text{Li}^{(+)}\text{Li}
\end{array}
$$

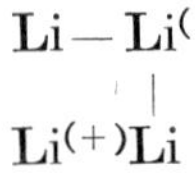

Great difficulties are caused, however, by the fact that the number of possible valence states is extremely large for metals, and therefore the description becomes very complicated. Pauling [11], for instance, has shown that the number of covalent valence states of a Li crystal containing $2N$ atoms is π^N, even if only the interactions with the closest eight neighbours are taken into account, and the interactions with the six next closest neighbours at a not much larger distance away are neglected. For a description of metallic conduction, however, the ionic states should also be taken into account. Their number can be shown [11] to be even larger: $(2.32\pi)^N$.

It should be noted with regard to the ionic valence states of metals, that valence structures containing negative ions may be formed because the metal atoms always contain empty levels close to the highest occupied level, and thus easily gain a further electron. For example, the first empty level of the Li atom, namely the level of the 2p electrons, is higher than the 2s level by only 1.84 eV [12]. In the case of alkali metal atoms the highest occupied level is usually an ns level, and the lowest empty level an np level (where n is the radial quantum number, $n = 2, 3, 4, \ldots$). The situation is similar with the alkaline earth metal atoms of configuration $(ns)^2$. Transition metals (such as Fe, Co, Ni) have partially filled 3d sub-shells, so they may gain further electron(s) even more easily.

The small energy difference between the occupied and empty levels of metal atoms very often causes the hybridization (mixing) of these orbitals. Thus, the Li^- ion*, or the negative ions of alkali metals are in the digonal sp hybrid state, and the negative ions of alkaline earth metals are in the trigonal sp^2 hybrid state. In the case of transition metals various hybrid states may occur, in which 3d, 4s and 4p electrons may take part. Consequently, the hybridization of atomic orbitals should be taken into account in the more accurate calculation of the electronic structure of metals, and this can be carried out in both the crystal orbital and the valence bond methods.

---

* The two outer electrons of the Li^- ion might occupy a $(2s)^2$ configuration. It can be proved, however, that the $(sp)^2$ configuration is more stable.

The hybridization of valence electrons generally increases the number of unpaired electrons and, accordingly, the number of bonding possibilities in the metals. For the calculation of the partial bonding number (the number n which, when multiplied by 2, gives the number of valence electrons forming a given bond; in a molecule containing only single bonds, $n = 1$), which plays a major role in the determination of bond distances in a metal lattice, however, the number of unpaired electrons compared to the number of atoms, i.e. the "electron concentration", should be known.

In order to clarify the concept of the partial bonding number let us again take the simple example of the body-centred Li lattice. Let us denote the partial bonding number of the bonds formed with the first 8 neighbours by n_8, and the value corresponding to the bonds formed with the further 6 neighbours by n_6. Since there is only one valence electron in the Li atom:

$$8n_8 + 6n_6 = 1 \tag{I.30}$$

At the same time, using the atomic radius equation developed by Pauling [13]:

$$R(1) - R(n) = 0.300 \lg n \tag{I.31}$$

it follows that

$$R(1) - R(n_8) = 0.300 \lg n_8 \tag{I.32a}$$

and

$$R(1) - R(n_6) = 0.300 \lg n_6 \tag{I.32b}$$

Using the values $R(n_8) = 1.516$ Å and $R(n_6) = 1.751$ Å determined from the interatomic distances measured in the Li lattice, the set of equations (I.30), (I.32a), and (I.32b) yield the solutions $R(1) = 1.230$ Å, $n_8 = 1/9$, and $n_6 = 1/54$.

A comparison of the results of the valence bond and crystal orbital methods shows that the valence bond method is more advantageous for the determination of bond distances in metals and for the classification of the bond types encountered. For the discussion of phenomena arising from the delocalization of valence electrons (mainly metallic conductivity) the crystal orbital method is much more favourable, and for this reason recent theoretical calculations on metals have been carried out using this method. In this book, therefore, the elementary discussion of the electronic structure of metals will be based on the crystal orbital method. As a first necessary step it will be shown in the following section why Bloch crystal orbitals must have the form given by Eq. (I.9).

SECTION 1.5. BLOCH CRYSTAL ORBITALS

The Schrödinger equation pertaining to a delocalized electron moving within a crystal lattice is

$$\frac{-\hbar^2}{2m_e} \Delta\psi_{\mathbf{k}}(\mathbf{r}) + V(\mathbf{r})\,\psi_{\mathbf{k}}(\mathbf{r}) = E_{\mathbf{k}}\,\psi_{\mathbf{k}}(\mathbf{r}) \tag{I.33}$$

where $\hbar = h/2\pi$, Δ is the Laplace operator defined as

$$\Delta = \frac{\partial^2}{\partial x^2} + \frac{\partial^2}{\partial y^2} + \frac{\partial^2}{\partial z^2}$$

and the potential function $V(\mathbf{r})$ acting on the electron shows the lattice periodicity, as was mentioned in Section I.1.2 [cf. Eq. (I.8)].

It can be shown [9] that due to the lattice periodicity of the potential function the following equation holds for wave functions $\psi_{\mathbf{k}}(\mathbf{r})$ satisfying Eq. (I.33):

$$\psi_{\mathbf{k}}(\mathbf{r} + \mathbf{R}_j) = \sigma_j \, \psi_{\mathbf{k}}(\mathbf{r}) \tag{I.34}$$

where σ_j is a complex constant. It is convenient to write the latter in the form

$$\sigma_j = e^{i\mathbf{k}\mathbf{R}_j} \tag{I.35}$$

(where $\mathbf{k}$ is the wavenumber vector and $\mathbf{R}_j$ is a lattice vector).

Introducing the function $u_{\mathbf{k}}(\mathbf{r})$ defined by the relationship

$$u_{\mathbf{k}}(\mathbf{r}) = e^{-i\mathbf{k}\mathbf{r}} \, \psi_{\mathbf{k}}(\mathbf{r}) \tag{I.36}$$

and taking into account Eqs (I.34) and (I.35), it follows that

$$\begin{aligned} u_{\mathbf{k}}(\mathbf{r} + \mathbf{R}_j) &= e^{-i\mathbf{k}(\mathbf{r}+\mathbf{R}_j)} \, \psi_{\mathbf{k}}(\mathbf{r} + \mathbf{R}_j) = e^{-i\mathbf{k}(\mathbf{r}+\mathbf{R}_j)} \, e^{i\mathbf{k}\mathbf{R}_j} \, \psi_{\mathbf{k}}(\mathbf{r}) = \\ &= e^{-i\mathbf{k}\mathbf{r}} \, \psi_{\mathbf{k}}(\mathbf{r}) = u_{\mathbf{k}}(\mathbf{r}) \end{aligned} \tag{I.37}$$

Hence, it can be seen that the function $u_{\mathbf{k}}(\mathbf{r})$ defined by Eq. (I.36) also has the periodicity of the lattice, as already mentioned in Section I.1.2. In order to utilize the periodicity of the potential, the wave functions $\psi_{\mathbf{k}}(\mathbf{r})$ of the electrons delocalized within the crystal lattice are advantageously written in the form

$$\psi_{\mathbf{k}}(\mathbf{r}) = e^{i\mathbf{k}\mathbf{r}} \, u_{\mathbf{k}}(\mathbf{r}) \tag{I.38}$$

Wave functions of the above form are called Bloch crystal orbitals.

CHAPTER 2. ELECTRONIC STRUCTURE OF METALS

J. Ladik

SECTION 2.1. FREE ELECTRON MODEL OF METALS

For the development of the free electron model of metals presented in Section I.1.4, the parameter ε' of the Fermi-Dirac distribution function is first determined for the case $T = 0$. This parameter is of great physical significance. If the energy levels ε_i are very close to each other, Eq. (I.26) can be written in the continuous form

$$w(\varepsilon) = \frac{1}{e^{\frac{\varepsilon - \varepsilon_0'}{kT}} + 1} \tag{I.39}$$

(ε_0' is the ε' value corresponding to $T = 0$). It follows from both Eqs (I.26) and (I.39) that for $T = 0$, $w = 1$ if $\varepsilon \leq \varepsilon_0'$, and $w = 0$ if $\varepsilon > \varepsilon_0'$. In other words, at the absolute zero all levels of energy lower than ε_0' are populated, and the remaining levels are vacant. Taking this into account, it follows from Eq. (I.27) that

$$\mathrm{d}n = \begin{cases} C\sqrt{\varepsilon}\,\mathrm{d}\varepsilon & 0 \leq \varepsilon \leq \varepsilon_0' \quad (T = 0) \\ 0 & \varepsilon > \varepsilon_0' \end{cases} \tag{I.40}$$

In the theory of solids the concept of the Fermi level plays an important role. By definition this is the level filled at the given temperature with a probability of 0.5. It follows from the above discussion and from this definition that at $T = 0$ the Fermi level is equal to ε_0', i.e.

$$\varepsilon_F = \varepsilon_0' \; (T = 0) \tag{I.41}$$

If Eq. (I.40) is taken into consideration at the absolute zero of temperature, Eq. (I.28) can be simplified to

$$N = C \int_0^{\varepsilon_0'} \sqrt{\varepsilon}\,\mathrm{d}\varepsilon = \frac{2}{3} C\, \varepsilon_0'^{3/2} \tag{I.42}$$

Expressing ε_0' from the above expression and substituting C from Eq. (I.25):

$$\varepsilon_0' = \frac{h^2}{2m_e^*}\left(\frac{3n_0}{8\pi}\right)^{2/3}, \quad n_0 = \frac{N}{V} \tag{I.43}$$

If ε_0' is known, the energy of the electron gas averaged for a single electron ($\bar{\varepsilon}_0$) can also be determined at $T = 0$:

$$\bar{\varepsilon}_0 = \frac{1}{N}\int_0^{\varepsilon_0'} C\sqrt{\varepsilon}\,\mathrm{d}\varepsilon = \frac{C}{N}\,\frac{2}{5}\,\varepsilon_0'^{5/2} = \frac{3}{5}\,\varepsilon_0' = \frac{3}{10}\,\frac{h^2}{m_e^*}\left(\frac{3\,n_0}{8\,\pi}\right)^{2/3} \qquad \text{(I.44)}$$

$\left(\frac{3}{5}\varepsilon_0'\right.$ can be obtained by performing the integration, resubstituting the expression for C from Eq. (I.25), and making the appropriate simplifications.$\left.\right)$

When $T \neq 0$, ε' can be determined on the basis of Eq. (I.28). This rather complicated task can be somewhat simplified if $kT \ll \varepsilon_0'$. (This condition is still applicable at the melting points of most metals.) Sommerfeld and Bethe developed a procedure for this case, whereby the integration can be performed analytically. The somewhat lengthy derivation is not reproduced here (for the details see, for example, Ref. [15]), and only the final result is quoted. Accordingly:

$$\varepsilon' = \varepsilon_0'\left[1 - \frac{\pi^2}{12}\left(\frac{kT}{\varepsilon_0'}\right)^2\right] \qquad (kT \ll \varepsilon_0') \qquad \text{(I.45a)}$$

It can be seen that the constant ε' now contains a small correction term depending on T, in addition to the constant ε_0'.

On the other hand, when the level density function is not taken on the basis of the free electron model [i.e. $\mathrm{d}n \neq C\sqrt{\varepsilon}\,\mathrm{d}\varepsilon$, see Eq. (I.40)], instead of Eq. (I.45a) the expression

$$\varepsilon' = \varepsilon_0' - \frac{1}{6}\pi^2 k^2 T^2\left(\frac{\mathrm{d}\ln n(\varepsilon)}{\mathrm{d}\varepsilon}\right)_{\varepsilon=\varepsilon_0'} \qquad (kT \ll \varepsilon_0') \qquad \text{(I.45b)}$$

can be derived.

The average energy, $\bar{\varepsilon}$, for $T \neq 0$, but $kT \ll \varepsilon_0'$, has also been determined according to the Sommerfeld-Bethe method, yielding the expression

$$\bar{\varepsilon} = \frac{1}{N}\int_0^{\infty} \frac{1}{e^{\frac{\varepsilon-\varepsilon'}{kT}} + 1}\, C\sqrt{\varepsilon}\,\mathrm{d}\varepsilon \qquad \text{(I.46)}$$

Substituting Eq. (I.45a) into the above formula and integrating:

$$\bar{\varepsilon} = \bar{\varepsilon}_0\left[1 + \frac{5}{12}\pi^2\left(\frac{kT}{\varepsilon_0'}\right)^2\right] \qquad (kT \ll \varepsilon_0') \qquad \text{(I.47)}$$

$\varepsilon_0 = \frac{3}{5}\varepsilon_0'$, see Eq. (I.44), showing that the expression has a form similar to Eq. (I.45a) for ε'. It is to be noted that when $T \neq 0$, but $kT \ll \varepsilon_0'$, since the form of the distribution function $w(\varepsilon)$ is only slightly different from the form valid for $T = 0$ (see Fig. I.3), the value of ε for which $w(\varepsilon) = 1/2$, that is, the Fermi level will again be ε_0' to a good approximation. When the crystal orbital method (the band model) is used, the Fermi level can again be taken to a good approximation, at temperatures which occur in practice, as the uppermost level of the partially filled band.

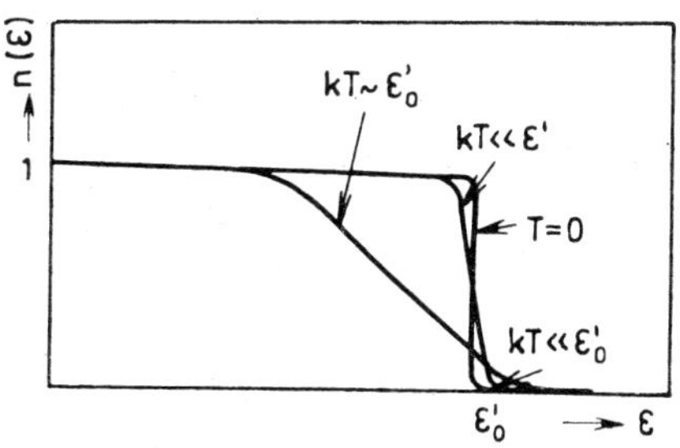

Fig. I.3. The Fermi–Dirac distribution function $W(\varepsilon)$, for different T values (schematic)

As is well known from thermodynamics, the molar heat capacity measured at constant volume is the derivative of the internal energy with respect to T, i.e.:

$$C_V = \left(\frac{\partial U}{\partial T}\right)_V \qquad \text{(I.48)}$$

For metals, the electronic contribution to the molar heat is obtained by multiplying Eq. (I.47) by the Avogadro number, and differentiating with respect to T:

$$C_{V,e} = N\frac{\partial \bar{\varepsilon}}{\partial T} = k\frac{\pi^2}{2}\frac{RT}{\varepsilon_0'} \qquad \text{(I.49)}$$

where the relationship $R = Nk$ was used. Even when the simple relationship of Eq. (I.24) is inapplicable to the calculation of the level density $g(\varepsilon)$, the expression for the electronic contribution to the molar heat can still be simply derived, and the result is [16]

$$C_{V,e} \approx \frac{\pi^2 k}{3}\frac{RT\, g(\varepsilon)}{N} \qquad \text{(I.50)}$$

(it must be noted that $N \neq N_A$; N is the number of atoms in the metal). Eq. (I.49) can be derived from Eq. (I.50) in the special case $g(\varepsilon) = C\sqrt{\varepsilon}$.

At normal temperatures the electronic part of the molar heat at constant volume in metals is usually small compared to the contribution from the lattice vibrations. At low temperatures, however, the term due to lattice vibrations, given by

$$C_{V,\mathrm{vibr}} = k\frac{12\,\pi^4}{5}\left(\frac{T}{\theta_D}\right)^3 \qquad \text{(I.51)}$$

(where the constant θ_D is the Debye characteristic temperature), is of roughly the same magnitude as $C_{V,e}$. At low temperatures, as can be seen from a comparison of Eqs (I.50) and (I.51), $C_{V,\mathrm{vibr}}$ decreases at a greater rate with decreasing T than does $C_{V,e}$. When $\varepsilon_0' \approx 1$ eV and $\theta_D \approx 100$ K, $C_{V,\mathrm{vibr}} \approx C_{V,e}$ at around 1 K.

The measured specific heat functions for different metals can generally be written in the form

$$C_V = \alpha T + 3Rf_D\left(\frac{\theta_D}{T}\right) \tag{I.52}$$

where the Debye function f_D is defined by a fairly complicated integral [17]. Putting the first term of Eq. (I.52) equal to Eq. (I.50):

$$\varepsilon_0' = \frac{\pi^2 k}{2\alpha} R \tag{I.53}$$

In the case of metals containing z valence electrons per atom ($z > 1$), the expressions (I.50) and (I.53) for $C_{V,e}$ and ε_0' should be multiplied by z. If the theoretical value for ε_0' calculated via the substitution $m_e^* = m_e$ according to Eq. (I.43) is divided by the corresponding experimental value determined from Eq. (I.53), the ratio of the effective and true electronic masses can be obtained:

$$\frac{\varepsilon_0' \text{ (theor.)}}{\varepsilon_0' \text{ (exptl.)}} = \frac{\dfrac{h^2}{2\, m_e}\left(\dfrac{3\, n_0}{8\pi}\right)^{2/3}}{\dfrac{h^2}{2\, m_e^*}\left(\dfrac{3\, n_0}{8\pi}\right)^{2/3}} = \frac{m_e^*}{m_e} \tag{I.54}$$

The above value is usually between 1 and 3 for simple metals not having an incomplete d-shell. Thus, m_e^*/m_e is 1.47 for copper, and 1.61 for aluminium.

The electronic contribution to the molar heat for transition metals possessing incomplete d-subshells (Fe, Co, Ni) is roughly ten times that of simple metals. This can be interpreted on the basis of Eq. (I.50), by considering that for these metals the five d-bands arising from the five possible atomic d-states merge into each other, and therefore the level density will be very high both inside and at the upper limit of the band (according to calculations the density is about 10 times as high as that of the s-bands). In transition metals this d-band is not completely filled, and therefore the level density for the uppermost filled level, which determines $C_{V,e}$ according to Eq. (I.50), arises partly from the next s-band, but mostly from the d-band (see Fig. I.4a), i.e.

$$g(\varepsilon') = g_s(\varepsilon') + g_d(\varepsilon') \tag{I.55}$$

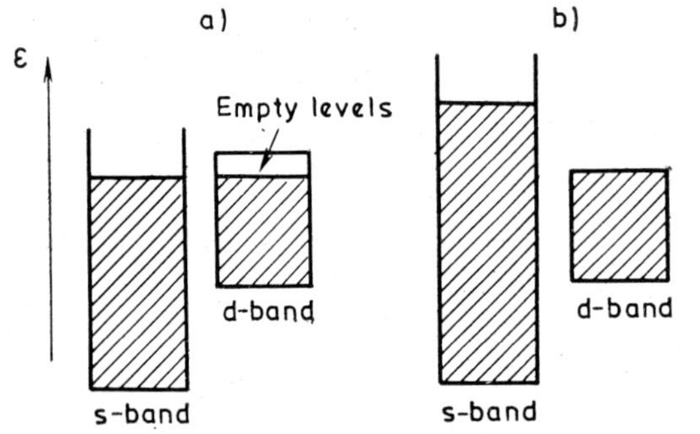

Fig. I.4. (a) d- and s-bands of transition metals; (b) d- and s-bands of noble metals (shaded areas are occupied)

where $g_s(\varepsilon')$ is the level density arising from the s-band, and $g_d(\varepsilon')$ is the level density due to the d-bands. Since the $g(\varepsilon')$ value so obtained is about ten times higher than the corresponding value for simple metals, the tenfold increase of the electronic contribution to the molar heat is understandable.

On the other hand, when the 3d-band is completely filled with 10 electrons, as for example, in copper, and the eleventh valence electron occupies the next 4s-band, the highest filled levels arise only from the 4s-band (see Fig. I. 4b), i.e.

$$g(\varepsilon') = g_s(\varepsilon')$$

and therefore $g(\varepsilon')$, and consequently $C_{V,e}$, will be of the usual magnitude.

Due to the high electronic contribution to the molar heat of transition metals the coefficient α of the first term in Eq. (I.52) is large. For this reason the experimental value of ε_0', which can be calculated from Eq. (I.53), will be small, and since it occurs in the denominator in Eq. (I.54), a large value is obtained for the effective mass. Figuratively speaking, it can be said that in energy bands of high level density the electrons behave as heavy particles, since their effective mass increases. Thus, for nickel the value of m_e^* calculated from the measured value of the electronic contribution to the molar heat, according to Eq. (I.52)—(I.54), is $m_e^* = 28\, m_e$ [18].

SECTION 2.2. THE ENERGY BAND MODEL OF METALS

In the one-dimensional case the expression for the Bloch crystal orbitals given by Eq. (I.38) (see also Section I.1.5) is simplified to

$$\psi(x) = e^{ikx} u(x) \tag{I.56}$$

If the lattice periodicity of the function $u(x)$ is taken into account, i.e. if $u(x + a') = u(x)$ (where a' is the lattice constant), it follows from Eq. (I.56) that

$$\psi(x + a') = e^{ika'}\, \psi(x) \tag{I.57}$$

The first, and perhaps the simplest, periodic potential function used for the solution of the one-dimensional Schrödinger equation which considers Eq. (I.57) is the Kronig-Penney potential function [19]. This is depicted in Fig. I.5. As can be seen from the figure, the potential peaks of height V_0 and breadth b are located at a distance a, and the value of the potential in the gaps is zero. By substituting the potential into the one-dimensional Schrödinger equation

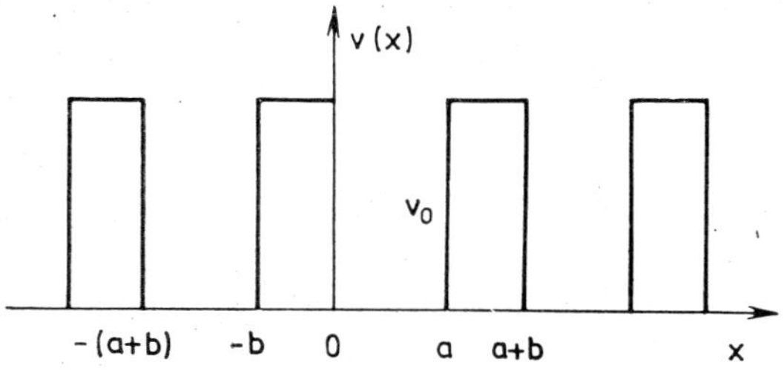

Fig. I.5. One-dimensional Kronig–Penney potential function

$$\frac{-\hbar^2}{2\,m_e}\,\frac{d^2\,\psi(x)}{dx^2} + V\,\psi(x) = E\,\psi(x) \tag{I.58}$$

it can easily be ascertained that in the zero potential ranges, as, for example, in the case where $0 \leq x \leq a$, the general solution of Eq. (I.58) is given by

$$\psi_1 = A\,\cos(\beta x) + B\,\sin(\beta x);$$

$$\beta = \left(\frac{2m_e E}{\hbar^2}\right)^{1/2} \tag{I.59}$$

(where A and B are constants). Similarly, it can be shown that in the V_0 potential ranges, as, for example, in the range $a \leq x \leq (a + b)$, the solution of Eq. (I.58) is

$$\psi_2 = C\,\cosh(\gamma x) + D\,\sinh(\gamma x);\ \gamma = \left[\frac{2m_e(V_0 - E)}{\hbar^2}\right]^{1/2} \tag{I.60}$$

where C and D are constants.

Since the wave function and its first derivative with respect to x must also be continuous at the point $x = a$, the relationships

$$\begin{aligned} \psi_1(a) &= \psi_2(a) \\ \psi_1'(a) &= \psi_2'(a) \end{aligned} \tag{I.61}$$

should hold, where the symbol $'$ indicates differentiation with respect to x. Using the above expressions, the constants C and D in Eq. (I.60) can be expressed as functions of A and B. Considering Eq. (I.57) (with $a' = a + b$), a homogeneous linear system of equations consisting of two equations can then be derived for the determination of A and B. It is well known that the condition for obtaining a non-trivial solution for a system of homogeneous linear equations is that the determinant formed from the coefficients of the unknowns should be zero. If the determinant obtained from the system of equations containing A and B as unknowns is expanded and put equal to zero, one obtains

$$\cos k(a + b) = \cos(\beta a)\cosh(\gamma b) + \frac{\gamma^2 - \beta^2}{2\beta}\sin(\beta a)\sinh(\gamma b) \tag{I.62}$$

and this result can be used for the determination of the possible values of E.

In order to simplify the determination of the possible values of E, let us assume that $V_0 \to \infty$ and simultaneously $b \to 0$, but in such a manner that the product $V_0 b$ remains a certain finite number (for example, the value of $V_0 b$ prior to performing the limiting operation i.e. the area of a rectangle of the Kronig-Penney potential function). Introducing the quantity

$$P = \lim_{\substack{V_0 \to \infty \\ b \to 0}} (m_e a b V_0/\hbar^2) \tag{I.63}$$

and performing the limiting operation, Eq. (I.62) can finally be transformed into the simple form [19]

$$\cos(ka) = \cos(\beta a) + \frac{P}{\beta a}\sin(\beta a) \tag{I.64}$$

Eq. (I.64) is satisfied for real values of k if the value of the right hand side lies between -1 and $+1$. On plotting the right hand side of Eq. (I.64) against $\beta a = (2m_e E/\hbar^2)^{1/2}$ for a given value of P (see Fig. I.6), it can be seen that in the case of a given value of a Eq. (I.64) can be satisfied only for certain values of β and E (the allowed ranges of βa are indicated by heavy lines in the diagram).

Physically, this means that the energy of an electron moving in the field of the Kronig-Penney periodic potential may have only certain values. The energy spectrum, shown in Fig. I.6, consists of alternately allowed and forbidden regions.

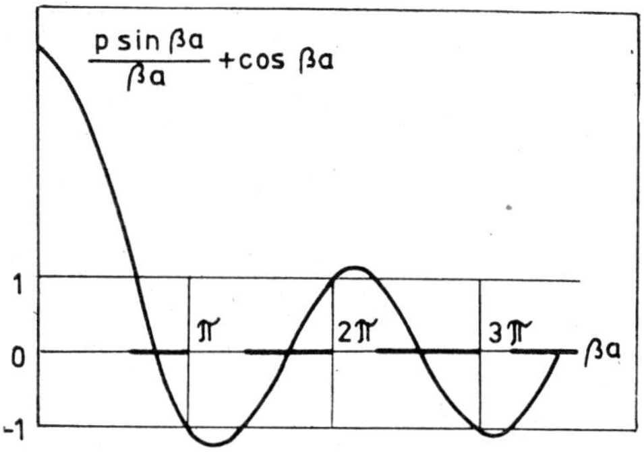

Fig. I.6. Energy spectrum for the one-dimensional Kronig–Penney potential [19]

In a qualitative sense a similar result is obtained in the one-dimensional case if the simple Kronig-Penney potential function is replaced by another potential function in Eq. (I.58), the one-dimensional Schrödinger equation. Thus, for instance, if $V(x)$ of the form

$$V(x) = V_0(1 - \cos 2\pi x) \tag{I.65}$$

is substituted into Eq. (I.58), the Mathieu differential equation, which has an exact solution, is obtained. By a more detailed analysis of the problem, Slater [20] has shown that the energy values for which the Schrödinger equation under the given limiting conditions can be solved form bands again, and the band width increases strongly with the decrease of the lattice constant. By expanding the potential function in Eq. (I.65) around the point $x = 0$, Slater also proved that the results obtained from the solution of the Mathieu equation are valid for the real one-dimensional lattice, too.

As we have seen in Section I.1.5, in the three-dimensional case the wave functions of an electron moving in a periodic potential field can be ordered according to the wavenumber vector, $\mathbf{k}$. The potential function and the function $u_{\mathbf{k}}(\mathbf{r})$ having the periodicity of the lattice [see Eq. (I.37) in Section I.1.5] can be expanded into a Fourier series with respect to the lattice vectors $\mathbf{K}_j$ of the reciprocal lattice:

$$V(\mathbf{r}) = \sum_{\mathbf{K}_j} v(\mathbf{K}_j) e^{i\mathbf{K}_j\mathbf{r}} \tag{I.66}$$

and

$$u_{\mathbf{k}}(\mathbf{r}) = \sum_{\mathbf{K}_j} c(\mathbf{K}_j)\, e^{i\mathbf{K}_j\mathbf{r}} \tag{I.67}$$

where $v(\mathbf{K}_j)$ and $c(\mathbf{K}_j)$ are the appropriate Fourier coefficients:

$$v(\mathbf{K}_j) = \frac{1}{\Omega} \int V(\mathbf{r})\, \mathrm{e}^{-\mathrm{i}\mathbf{K}_j\mathbf{r}}\, \mathrm{d}\tau \tag{I.68}$$

and

$$c(\mathbf{K}_j) = \frac{1}{\Omega} \int u_{\mathbf{k}}(\mathbf{r})\, \mathrm{e}^{-i\mathbf{K}_j\mathbf{r}}\, \mathrm{d}\tau \tag{I.69}$$

In these expressions Ω is the volume of the unit cell, and the integration is carried out over the unit cell. By using the expansions given by Eqs (I.66) and (I.67) it can be proved by applying perturbation theory (for the somewhat involved calculation see Ref. [21]) that, when starting in an arbitrary direction in the $\mathbf{k}$-space, the function $E_{\mathbf{k}}(\mathbf{k})$ has a discontinuity at the boundary surface of the first Brillouin zone. According to the details of the calculation, the width of the range of the forbidden $E_{\mathbf{k}}$ values is $2v(\mathbf{K}_j)$, that is, just double the Fourier coefficient for the vector $\mathbf{K}_j$ characterizing the corresponding boundary plane of the Brillouin zone (see Section I.1.2).

However, even if a forbidden band occurs in a curve $E_{\mathbf{k}}(\mathbf{k})$ for a given direction in $\mathbf{k}$-space, this does not necessarily mean that the forbidden energy values would not be allowed in other directions of $\mathbf{k}$-space. In the total energy value spectrum of the crystal (i.e. belonging to all directions of the $\mathbf{k}$-space), therefore, often no forbidden bands occur at all, and the energy bands corresponding to certain free atomic levels (see Section I.1.4) may even overlap each other*. However, if the range of forbidden $E_{\mathbf{k}}$ values in a given direction of the $\mathbf{k}$-space is broad on the boundary surface of the Brillouin zone, a forbidden band can generally also be observed in the whole energy spectrum.

SECTION 2.3. LCAO APPROXIMATION OF BLOCH CRYSTAL ORBITALS

The Bloch crystal orbitals defined in Section I.1.5 can be most simply approximated by a linear combination of the atomic orbitals of the atoms forming the crystal. If the wave function of the valence electron of the atom situated in the lattice point characterized by the lattice vector $\mathbf{R}_j$ is denoted by $\chi_n(\mathbf{r} - \mathbf{R}_j)$**, then according to this approximation the Bloch crystal orbital $\psi_{\mathbf{k}}$ can be given in the form

$$\psi_{\mathbf{k}}(\mathbf{r}) = \sum_j C_j(\mathbf{k})\, \chi_n(\mathbf{r} - \mathbf{R}_j) \tag{I.70}$$

* This is particularly true for crystals built up from atoms containing several levels of the same energy, or levels lying close to each other, e.g. the transition metals.

** The subscript n in $\chi_n(\mathbf{r} - \mathbf{R}_j)$ refers to the kind of wave function taken into account in the construction of the Bloch crystal orbitals.

where the summation extends to all the atoms forming the crystal. If the coefficients $C_j(\mathbf{k})$ are given by

$$C_j(\mathbf{k}) = e^{i\mathbf{k}\mathbf{R}_j} \tag{I.71}$$

the Bloch orbital $\psi_k(\mathbf{r})$ can be given in the usual form

$$\psi_{\mathbf{k}}(\mathbf{r}) = e^{i\mathbf{k}\mathbf{r}}\, u_{\mathbf{k}}(\mathbf{r}) = e^{i\mathbf{k}\mathbf{r}} \sum_j e^{-i\mathbf{k}(\mathbf{r}-\mathbf{R}_j)}\, \chi_n(\mathbf{r} - \mathbf{R}_j) \tag{I.72}$$

where the sum on the right hand side of the equation (the Bloch sum) shows the periodicity of the lattice.

The LCAO approximation of crystal orbitals is primarily successful in the case when the overlap between the atomic wave functions of neighbouring lattice points is small, i.e. if the equation

$$\int \chi_n^*(\mathbf{r} - \mathbf{R}_i)\, \chi_n(\mathbf{r} - \mathbf{R}_j)\, dV = \delta_{ij} \tag{I.73}$$

is approximately valid, where δ_{ij} is the Kronecker symbol.

Assuming the validity of Eq. (I.73) for solids containing only one atom in the unit cell, the form of $\psi_{\mathbf{k}}(\mathbf{r})$ will be

$$\psi_{\mathbf{k}}(\mathbf{r}) = N^{-1/2} \sum_j e^{i\mathbf{k}\mathbf{R}_j}\, \chi_n(\mathbf{r} - \mathbf{R}_j) \tag{I.74}$$

where N is the number of atoms forming the crystal. In the construction of Eq. (I.74) it should also be assumed that the difference $E_m^0 - E_n^0$ between the energy level E_n^0 of the atomic orbital $\chi_n(\mathbf{r} - \mathbf{R}_j)$ satisfying the Schrödinger equation

$$-\frac{\hbar^2}{2m_e} \Delta\chi_n(\mathbf{r} - \mathbf{R}_j) + W(\mathbf{r} - \mathbf{R}_j)\, \chi_n(\mathbf{r} - \mathbf{R}_j) = E_n^0\, \chi_n(\mathbf{r} - \mathbf{R}_j) \tag{I.75}$$

and the closest energy level E_m^0, is much higher than the width of the energy band arising from the level E_n^0. If this condition is not fulfilled, $\psi_{\mathbf{k}}$ can be given, not by Eq. (I.74), but only by a linear combination of several Bloch sums:

$$\psi_{\mathbf{k}} = \sum_j e^{i\mathbf{k}\mathbf{R}_j} [A_1\, \chi_1(\mathbf{r} - \mathbf{R}_j) + A_2\, \chi_2(\mathbf{r} - \mathbf{R}_j) + \ldots] \tag{I.76}$$

where $A_1, A_2, \ldots$ are constants. In this case, the resulting number of energy bands will be the same as the number of Bloch sums in Eq. (I.76), which is, in turn, equal to the number of atomic levels of the same, or very slightly differing energies.

To return to the case of the more simple Eq. (I.74) crystal orbitals, it can be proved that the relation

$$\int \psi_{\mathbf{k}}^*(\mathbf{r})\, \psi_{\mathbf{k}}(\mathbf{r}) dV = N^{-1} \sum_j e^{i(\mathbf{k}-\mathbf{k}')\mathbf{R}_j} = \delta_{\mathbf{k},\mathbf{k}'} \tag{I.77}$$

holds. In order to prove the first equality, Eq. (I.74) should be substituted into Eq. (I.77), and Eq. (I.73) should be taken into account. The second equality can be proved relatively simply if summation is replaced by integration. On substituting Eq. (I.74) into the one-electron Schrödinger equation:

$$H\psi_{\mathbf{k}} = -\frac{\hbar^2}{2m_e}\Delta\psi_{\mathbf{k}} + V(\mathbf{r})\,\psi_{\mathbf{k}} = E_{\mathbf{k}}\psi_{\mathbf{k}} \tag{I.78}$$

multiplying from the left by $\chi_n^*(\mathbf{r} - \mathbf{R}_i)$, and integrating over the entire space we obtain

$$N^{-1/2}\sum_j e^{i\mathbf{k}\mathbf{R}_j}\int \chi_n^*(\mathbf{r}-\mathbf{R}_i)\left\{\frac{-\hbar^2}{2m_e}\Delta\chi_n(\mathbf{r}-\mathbf{R}_j)+[V(\mathbf{r})-E_{\mathbf{k}}]\chi_n(\mathbf{r}-\mathbf{R}_j)\right\}dV=0 \tag{I.79}$$

On substituting $-\frac{\hbar^2}{2\,m_e}\Delta\chi_n\,(\mathbf{r} - \mathbf{R}_j)$ from Eq. (I.75) into Eq. (I.79):

$$\sum_j e^{i\mathbf{k}\mathbf{R}_j}\int\chi_n^*(\mathbf{r} - \mathbf{R}_i)\,[\varepsilon_{\mathbf{k}} - W'(\mathbf{r} - \mathbf{R}_j)]\,\chi_n\,(\mathbf{r} - \mathbf{R}_j)\,dV = 0 \tag{I.80}$$

where

$$\varepsilon_{\mathbf{k}} = E_{\mathbf{k}} - E_n^0 \tag{I.81}$$

is the deviation from the atomic energy level, and

$$W'(\mathbf{r} - \mathbf{R}_j) = V(\mathbf{r}) - W(\mathbf{r} - \mathbf{R}_j) \tag{I.82}$$

is the deviation of the crystal potential from the atomic potential. By taking Eq. (I.73) into account, Eq. (I.80) can be further simplified into the form

$$\varepsilon_{\mathbf{k}} = -\,\alpha_n - \sum_{\mathbf{R}_{ij}\neq 0} e^{i\mathbf{k}\mathbf{R}_{ij}}\,\beta_n(\mathbf{R}_{ij}) \tag{I.83}$$

where $\mathbf{R}_{ij} = \mathbf{R}_j - \mathbf{R}_i$ and

$$\alpha_n = -\int\chi_n^2(\mathbf{r} - \mathbf{R}_i)\,W'(\mathbf{r} - \mathbf{R}_i)\,dV \tag{I.84}$$

$$\beta_n\,(\mathbf{R}_{ij}) = -\int\chi_n^*(\mathbf{r} - \mathbf{R}_i)\,W'(\mathbf{r} - \mathbf{R}_j)\,\chi_n(\mathbf{r} - \mathbf{R}_j)\,dV \tag{I.85}$$

The values of the atomic wave functions $\chi_n(\mathbf{r} - \mathbf{R}_j)$ decrease very rapidly, in general exponentially, with increasing distance $\mathbf{r} - \mathbf{R}_j$. It follows from this, on the basis of Eq. (I.85), that the terms $\beta_n(\mathbf{R}_{ij})$ also decrease rapidly with increasing internuclear distances $\mathbf{R}_{ij}$. In many cases, therefore, a good approximation is obtained even if only the immediate neighbour interaction terms are retained in Eq. (I.83) (in this case $\mathbf{R}_i$ and $\mathbf{R}_j$ point to directly neighbouring lattice points). For example, in the case of the simple cubic lattice, if χ_n is an s-function of spherical symmetry, and only the interactions with the 6 immediate neighbours are taken into account, it follows from Eq. (I.83), by simple trigonometry, that

$$E_{\mathbf{k}} = E_n^0 - \alpha_n - 2\beta_n(a')\,(\cos k_x a' + \cos k_y a' + \cos k_z a') \tag{I.86}$$

where a' is the lattice constant. The limits of the band given by Eq. (I.86) are at the vectors $\mathbf{k} = \mathbf{0}$ and $\mathbf{k} = \frac{\pi}{a'}\,(1, 1, 1)$, while the band width is

$12\beta_n(a')$. By expanding the cosine functions in Eq. (I.86) around $\mathbf{k} = \mathbf{0}$, and taking into account only two terms from the expansion:

$$E_{\mathbf{k}} = E_n^0 - \alpha_n - 2\beta_n(a') \left(1 - \frac{k_x^2 a'^2}{2} + 1 - \frac{k_y^2 a'^2}{2} + 1 - \frac{k_z^2 a'^2}{2}\right) =$$

$$= E_n^0 - \alpha_n - 6\beta_n(a') + \beta_n(a') k^2 a'^2 = E_0 + \beta_n(a')\, k^2 a'^2 \qquad \text{(I.87)}$$

where E_0 denotes the sum of terms independent of k. If this expression is compared to the energy expression* $-\frac{\hbar^2}{2m_e} k^2$ of the free electrons described by the wave function $e^{i\mathbf{kr}}$, it can be seen that in the approximation $\mathbf{k} = \mathbf{0}$ (at the bottom of the band) the electrons may be regarded as being free electrons of effective mass

$$m_e^* = \frac{-\hbar^2}{2a^2\beta_n(a')}$$

The corresponding expressions for the possible energy levels of bands formed from atomic s-states in the case of body- and face-centred cubic and closely-packed hexagonal lattices can also be derived from Eq. (I.83) using elementary trigonometry if only nearest-neighbour interactions are taken into account [22].

The situation is slightly more complicated if the atomic wave function χ_n is not spherically symmetric and there are several atomic orbitals of the same energy. Thus, in the case of p-functions, the three atomic orbitals, p_x, p_y and p_z, have the same energy. In this case, instead of Eq. (I.74) a function $\psi_{\mathbf{k}}$ consisting of 3 sums, and having the form of Eq. (I.76) must be used. In a similar manner to the derivation yielding Eq. (I.83), the expression

$$\begin{aligned} A_1 \Big(\varepsilon_{\mathbf{k}} + \alpha_{11} + \sum_{\mathbf{R}_{ij} \neq 0} e^{i\mathbf{k}\mathbf{R}_{ij}} \beta_{11}(\mathbf{R}_{ij})\Big) + & \\ + A_2\Big(\alpha_{12} + \sum_{\mathbf{R}_{ij} \neq 0} e^{i\mathbf{k}\mathbf{R}_{ij}} \beta_{12}(\mathbf{R}_{ij})\Big) + & \qquad \text{(I.88)} \\ + A_3\Big(\alpha_{13} + \sum_{\mathbf{R}_{ij} \neq 0} e^{i\mathbf{k}\mathbf{R}_{ij}} \beta_{13}(\mathbf{R}_{ij})\Big) = 0 & \end{aligned}$$

and two other similar expressions are now obtained. In the latter two expressions the first subscript of the β's and α's is 2 or 3, respectively ($\varepsilon_{\mathbf{k}}$ always appears in the diagonal term). In these equations again

$$\mathbf{R}_{ij} = \mathbf{R}_j - \mathbf{R}_i$$

* In the case of free electrons $W = 0$, i.e. $H\psi = -\frac{\hbar^2}{2m_e} \Delta\psi_{\mathbf{k}} = E_{\mathbf{k}}\psi_{\mathbf{k}}$. By back-substitution it can be seen that if $E_{\mathbf{k}} = \frac{\hbar^2}{2m_e} k^2$, the solution of the equation is $\psi_{\mathbf{k}} = e^{i\mathbf{kr}}$.

and

$$\alpha_{nm} = - \int \chi_n^*(\mathbf{r} - \mathbf{R}_i)\, W'(\mathbf{r} - \mathbf{R}_i)\, \chi_m(\mathbf{r} - \mathbf{R}_i)\, \mathrm{d}V \qquad (I.89)$$

$$\beta_{nm}(\mathbf{R}_{ij}) = - \int \chi_n^*(\mathbf{r} - \mathbf{R}_i)\, W'(\mathbf{r} - \mathbf{R}_j)\, \chi_m(\mathbf{r} - \mathbf{R}_j)\, \mathrm{d}V \qquad (I.90)$$

with n, $m = 1, 2, 3$. The condition for a non-trivial solution of the unknowns A_1, A_2 and A_3 of the system of these three linear equations is that the determinant of the coefficients should be zero (the elements of the coefficient matrix are given by the expressions in square brackets). This condition yields three roots for each value of **k**. The roots for the different **k** values give, in this way, three energy bands.

In the case of a simple cubic lattice the non-diagonal elements α_{nm} ($n \neq m$) between the p-functions can easily be shown to be zero. If only the nearest-neighbour interactions are taken into account, it can be proved that the non-diagonal terms β_{nm} also vanish. In Eq. (I.98) and the two other similar equations, therefore, only the diagonal elements remain, and accordingly:

$$E_{\mathbf{k}} = E_1^{(0)} - \alpha_{11} - \beta_{11}^{(1)}(a) \cos k_x a - 2\, \beta_{11}^{(2)}(a)(\cos k_y a + \cos k_z a) \qquad (I.91)$$

In this expression $\beta_{11}^{(1)}$ and $\beta_{11}^{(2)}$ are $\beta_{11}(a)$ values belonging to the σ-type and π-type overlaps, respectively, of the p-functions. Consequently, for the case of simple cubic lattices, three p-bands containing the same energy levels (triply degenerate bands) are obtained.

In the application of the LCAO crystal orbital method a fairly difficult problem arises in the choice of the integrals α_{nm} and β_{nm}. [Since the potentials $W'(\mathbf{r} - \mathbf{R}_j)$ are not exactly known, these integrals can generally not be calculated explicitly.] This problem is most frequently solved, after the proposal of Slater and Koster [23], by regarding α_{nm} and β_{nm} as parameters. They are evaluated by determining the value of $E_{\mathbf{k}}$ by applying other, more accurate approximations for certain high-symmetry points of the **k**-space [e.g. in the case of a face-centred cubic lattice for points $\mathbf{k} = \mathbf{0}$ and $\mathbf{k} = \frac{\pi}{a}(0, 0, 1)$], and by substituting the $E_{\mathbf{k}}$ values obtained into Eq. (I.83). $E_{\mathbf{k}}$ values can then be determined for arbitrary points in the **k**-space. Calculations of this type have been carried out for several metals, which include copper and nickel.

SECTION 2.4. OTHER APPROXIMATIONS FOR THE CALCULATION OF THE ENERGY BANDS OF METALS

Besides the LCAO approximation the cell method of Wigner and Seitz is most frequently used for the construction of Bloch crystal orbitals. According to this method all the atoms forming the crystal lattice are conceived as being surrounded by a polyhedron (the so-called Wigner-Seitz cell). The cells are constructed so that the lines joining an atom with its nearest neighbours are orthogonally bisected by planes, and of the many polyhedra formed by these planes the one having the smallest volume is taken. It is obvious that in the case of metals containing a single atom in the unit cell, the Wigner-Seitz cell is the unit cell itself, whereas if the unit cell contains more than one atom, the number of Wigner-Seitz cells will be equal to the number of atoms in the unit cell. The volume of the cell can therefore be given as

$$V_{W-S} = \frac{1}{t}\, \mathbf{a}_1(\mathbf{a}_2 \times \mathbf{a}_3) \tag{I.92}$$

where t is the number of atoms in the unit cell. The individual atoms are always situated in the centre of the cell, and the symmetry of the cells is identical to that of the crystal.

Taking into account that the Bloch crystal orbitals

$$\psi_{\mathbf{k}}(\mathbf{r}) = \mathrm{e}^{i\mathbf{kr}}\, u_{\mathbf{k}}(\mathbf{r}) \tag{I.93}$$

have the property

$$\psi_{\mathbf{k}}(\mathbf{r} + \mathbf{R}_j) = \mathrm{e}^{i\mathbf{kR}_j}\, \psi_{\mathbf{k}}(\mathbf{r}) \tag{I.94}$$

which is due to the lattice periodicity of the function $u(\mathbf{r})$; it follows for solids containing only one atom in the unit cell that it is sufficient to solve the Schrödinger equation for only a single Wigner-Seitz cell. The potential function of the crystal within the cell can be proved to be spherically symmetric to a reasonably good approximation, and therefore the solution of the one-electron Schrödinger equation can be sought by expanding the function $\psi_{\mathbf{k}}(\mathbf{r})$ within the cell in terms of spherical harmonics.

In the classical calculations of Wigner and Seitz on alkali metals, assuming a function of spherical symmetry $V(\mathbf{r})$, the function ψ_0 for $\mathbf{k} = \mathbf{0}$ in the cell was first considered in the simple form

$$\psi_0 = u_0(\mathbf{r}) = f_s(\mathbf{r}) \tag{I.95}$$

(i.e. from the power series in terms of the spherical harmonics only the first term was retained), and the functions $\psi_{\mathbf{k}}$ for $\mathbf{k} \neq \mathbf{0}$ and corresponding energy values $E_{\mathbf{k}}$ were calculated by perturbation theory. On substituting the spherically symmetric function $f_s(\mathbf{r})$ into the Schrödinger equation

$$-\frac{\hbar^2}{2m_e}\Delta\psi_0 + V(r)\,\psi_0 = E_0\psi_0 \tag{I.96}$$

the expression

$$-\frac{\hbar^2}{2m_e}\left(\frac{\mathrm{d}^2 f_s(r)}{\mathrm{d}r^2} + \frac{2}{r}\,\frac{\mathrm{d}f_s(r)}{\mathrm{d}r}\right) + V(r) f_s(r) = E_0 f_s(r) \tag{I.97}$$

was obtained*, which can be solved exactly by taking into account the boundary condition

$$\left(\frac{\mathrm{d}f_s}{\mathrm{d}r}\right)_{r=r_s} = 0 \tag{I.98}$$

* The Laplace operator $\frac{\partial^2}{\partial x^2} + \frac{\partial^2}{\partial y^2} + \frac{\partial^2}{\partial z^2}$ is simplified to $\frac{\mathrm{d}^2}{\mathrm{d}r^2} + \frac{2}{r}\frac{\mathrm{d}}{\mathrm{d}r}$ in the case of spherical symmetry.

which follows from Eq. (I.94)*. r_s in Eq. (I.98) is the radius of the sphere whose volume is identical to that of the W—S cell in question, and according to which the cell can be substituted to a good approximation in the case of spherically symmetric potential. Eq. (I.97) is solved by assuming different trial values for E_0, and integrating the equation numerically for a given function $V(r)$. The procedure is repeated until a value of E_0 is found for which Eq. (I.98) is satisfied for a given value of r_s. The value of E_0 thus obtained and the corresponding $f_s(r)$ function give a physically correct solution of the problem.

When E_0 and $f_s(r)$ are known, the functions $\psi_{\mathbf{k}}$ for $\mathbf{k} \neq \mathbf{0}$ and the energy levels $E_{\mathbf{k}}$, the set of which specifies the energy band, are calculated by perturbation theory. According to the results of the slightly sophisticated calculations, which are therefore not reproduced here,

$$\psi_{\mathbf{k}}(\mathbf{r}) = \left[f_s(\mathbf{r})\,(1 - \mathrm{i}\mathbf{k}\mathbf{r}) + \mathrm{i}\mathbf{k}\mathbf{r}\,\frac{P(r)}{r^2}\right] \mathrm{e}^{\mathrm{i}\mathbf{k}\mathbf{r}} \tag{I.99}$$

where the function $P(r)$ is the solution of the exactly solvable differential equation

$$\frac{\mathrm{d}^2P}{\mathrm{d}r^2} - \frac{2}{r}\frac{\mathrm{d}P}{\mathrm{d}r} + \frac{2m_e}{\hbar^2}\,[E_0 - V(r)]\,P = 0 \tag{I.100}$$

under the boundary condition

$$\left(\frac{P(r)}{r^2}\right)_{r=r_s} = f_s(r)_{r=r_s} \tag{I.101}$$

[which can be derived again from Eq. (I.94)]. The energy levels $E_{\mathbf{k}}$ are given by

$$E_{\mathbf{k}} = E_0 + \frac{\alpha\,\hbar^2 k^2}{2m_e} \tag{I.102}$$

where

$$\alpha = \frac{4\pi}{3}\, r_s^3[f_s(r_s)]^2 \left[\frac{r_s}{P(r_s)}\left(\frac{\mathrm{d}P}{\mathrm{d}r}\right)_{r=r_s} - 1\right] \tag{I.103}$$

It can be seen from the above result that the valence electrons of the alkali metals can again be formally regarded as being free electrons of effective mass $m_e^* = \dfrac{m_e}{\alpha}$.

If the W—S cell cannot be substituted by a sphere, and the deviations of the Schrödinger equations for $\mathbf{k} = \mathbf{0}$ and $\mathbf{k} \neq \mathbf{0}$, respectively, (i.e. the perturbation terms) are so large that the perturbation theory cannot be applied, the Schrödinger equation should be solved independently for the

* Eq. (I.94) transforms into $f_s(\mathbf{r} + \mathbf{R}_j) = f_s(\mathbf{r})$ in the case of $\mathbf{k} = \mathbf{0}$, upon the substitution $\psi_0 = f_s$; this can be satisfied only if f_s has an extremum at the boundary surface of the sphere equivalent to the W — S cell, i.e. if $\left(\dfrac{\mathrm{d}f_s}{\mathrm{d}r}\right)_{r=r_s} = 0$.

individual values of **k**, by taking into account the appropriate boundary conditions valid at the boundary surfaces between the cells which are derivable from Eq. (I.94). For this purpose, every function $\psi_{\mathbf{k}}$ should be expanded in terms of its spherical harmonics. The convergence of the expansion can be promoted by taking into account the symmetry of the crystal. These slightly more complicated considerations, however, cannot be discussed in detail here.

When the unit cell contains several W—S cells, the Bloch condition of Eq. (I.94) should be modified. For the sake of simplicity let us consider the case when the unit cell contains two W—S cells. Let these be denoted by p and q. In cell p, instead of the function $u_{\mathbf{k}}(\mathbf{r})$ let us introduce function $u^{(p)}(\mathbf{r} - \mathbf{S}_p)$, which is everywhere identical to the original function

$$u_{\mathbf{k}}(\mathbf{r}) = u_{\mathbf{k}}^{(p)}(\mathbf{r} - \mathbf{S}_p) \text{ (in W—S cell } p) \tag{I.104}$$

where $\mathbf{S}_p$ is the position vector of the nucleus in cell p.

Similarly,

$$u_{\mathbf{k}}(\mathbf{r}) = u_{\mathbf{k}}^{(q)}(\mathbf{r} - \mathbf{S}_q) \text{ (in W—S cell } q) \tag{I.105}$$

Since $u_{\mathbf{k}}(\mathbf{r})$ is a continuous function, it follows that at the common boundary surface of the two cells

$$u_{\mathbf{k}}^{(p)}(\mathbf{r}_b - \mathbf{S}_p) = u_{\mathbf{k}}^{(q)}(\mathbf{r}_b - \mathbf{S}_q) \tag{I.106}$$

should hold, where the vector $\mathbf{r}_b$ is the position vector of a point on the common boundary surface. On substituting Eqs (I.104) and (I.105), respectively, into Eq. (I.93) one obtains

$$\psi_{\mathbf{k}}(p) = e^{i\mathbf{k}(\mathbf{r}-\mathbf{S}_p)}\, u_{\mathbf{k}}^{(p)}(\mathbf{r} - \mathbf{S}_p) \tag{I.107}$$

$$\psi_{\mathbf{k}}(q) = e^{i\mathbf{k}(\mathbf{r}-\mathbf{S}_q)}\, u_{\mathbf{k}}^{(q)}(\mathbf{r} - \mathbf{S}_q) \tag{I.108}$$

Applying Eq. (I.107) to the boundary surface, and taking into account Eq. (I.106), the boundary condition

$$\psi_{\mathbf{k}}^{(p)}(\mathbf{r}_b) = e^{i\mathbf{k}(\mathbf{S}_q-\mathbf{S}_p)}\, \psi_{\mathbf{k}}^{(q)}(\mathbf{r}_b) \tag{I.109}$$

is obtained. By means of this method a number of diatomic solids such as LiF, LiH, PbS and Si have been considered on the basis of the Wigner-Seitz cell method.

It should be added that it has been shown by Kuhn and Van Vleck that the potential $V(\mathbf{r})$ can also be constructed from experimental values of the first s- and p-levels of the corresponding free atom. This modification of the cell method, when the potential $V(\mathbf{r})$ is calculated from the experimental data instead of the electron distribution in the cell, is called the "quantum defect method".

Crystal orbitals are often constructed by the method of "orthogonalized plane waves". The method is based on the idea that in the region between the individual atoms forming the lattice crystal orbitals $\psi_{\mathbf{k}}$ can be well approximated by a Fourier series of plane waves (terms having the form $e^{i(\mathbf{k}+\mathbf{K}_j)\mathbf{r}}$), whereas in the neighbourhood of atoms they should exhibit atomic wave function properties. Based on a proposal of Herring, therefore, a suitable LCAO crystal orbital is added to each plane wave, to speed up the convergence of the expansion. However, in order to ensure that the crystal orbital $\psi_{\mathbf{k}}$ approximated by this method should be the wave function for an energy level of one of the energy bands of the valence electrons, and not the wave function of the electrons of the atomic core, the sum of the plane wave and the LCAO crystal orbital is formed in such a way that the expression obtained is orthogonal* to the crystal orbitals of the electrons of the atomic core (orthogonalized plane wave). Since this and other methods elaborated for the calculation of the energy bands of metals are more complicated from a mathematical point of view than the approximations discussed so far, their further discussion exceeds the scope of this book.

* Two functions are orthogonal if the integral of their product extended over the entire space is zero, i.e. ψ_i and ψ_j are orthogonal if $\int \psi_i \psi_j \mathrm{d}V = 0$.

CHAPTER 3. BAND THEORY OF SEMICONDUCTORS

F. Solymosi

For an understanding of the band theory of semiconductors, let us consider the band structure of copper(I) oxide. We shall start with a discussion of the band structure of copper metal (Fig. I.7).

When copper atoms approach each other, the uppermost energy levels (4p) first interact and split to yield bands, and then, successively, the lower levels, too; when the distance between the individual nuclei reaches the value a_0 corresponding to crystalline copper (the half face diagonal of the close-packed cubic lattice), the bands overlap as shown in the diagram. There is no appreciable interaction between the electrons in the lowest levels. The valence electron of copper metal occupies the 4s-level, and up to the 3d-band in the crystal all the bands are filled. The 4s-band is filled only to the level that is below the upper end of the overlapping 3d-band.

The energy band structure of a lattice consisting of copper and oxygen atoms can be seen in Fig. I.8. In this case, instead of 6, only 4 electrons per atom occupy the uppermost 2p-band. In the crystal lattice the valence electron of the copper atom is transferred from the 4s-orbital to the 2p-orbital of the oxygen atom, thereby forming the band structure given in the central of the diagram; Since two copper atoms are linked to one oxygen atom in the lattice of copper(I) oxide, of the bands of the copper(I) oxide

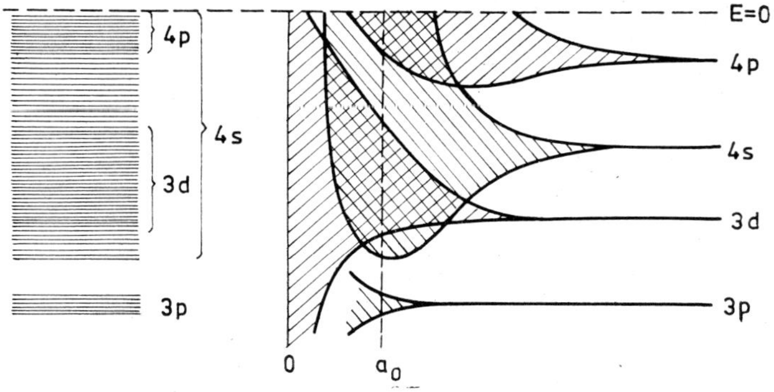

Fig. I.7. Energy bands of copper metal

the 2p-band of the oxygen is completely filled, and the 4s-band of the copper is vacant. The band structure of copper in copper(I) oxide is different from that in copper metal given in the previous diagram in the sense that the 3d and 4s-orbitals do not overlap due to the greater distance between the

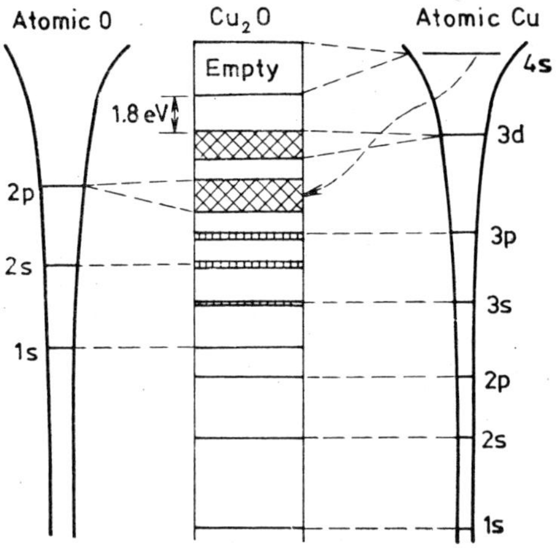

Fig. I.8. Energy bands of copper(I) oxide

copper atoms. Since the energy bands of copper(I) oxide do not overlap (see Fig. I. 8), at 0 K the system has a completely filled valence band and an empty conduction band.

SECTION 3.1. THE MECHANISM OF ELECTRICAL CONDUCTION IN SEMICONDUCTORS

The mechanism of electrical conduction in semiconductors can be most easily understood if the differences between the band structures of metals, semiconductors and insulators are studied. There are partially filled bands in metals only. Contrary to the classical view, of the electrons filling the band it is only those of highest energy that take part in electrical conduction. At *absolute zero* temperature the band structures of semiconductors and insulators consist of completely filled and entirely vacant bands. The difference between the two kinds of materials is that the gap between the filled and vacant bands, the forbidden band, is much wider in insulators than in semiconductors. Since the completely filled and vacant bands make no contribution to the conductivity, the conductivity of insulators is practically nil.

At absolute zero all the electrons of semiconductors containing no impurities are paired, and occupy the valence band. Above this temperature, however, the atoms of the crystal vibrate, and the energy of vibration can easily be transferred to the electrons. If the energy thus gained is high enough, an electron is raised through the forbidden band to the next vacant band, the conduction band. Upon application of an electrical field both the

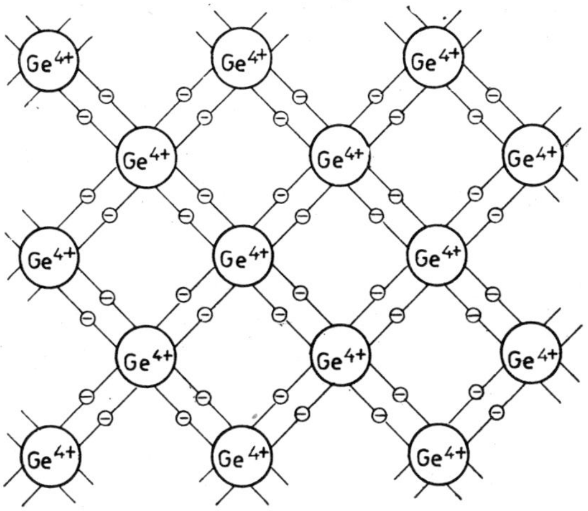

Fig. I.9. The crystal lattice of germanium

electron and the residual positive hole move, giving rise to a certain conductivity. With further increase in temperature more and more electrons jump from the valence band to the conduction band; the conductivity of the semiconductor increases. This can be well illustrated by taking the example of germanium semiconductors (Fig. I.9). In the regular structured germanium crystal every germanium atom is surrounded by four other germanium atoms in a tetrahedral arrangement. Due to the sp^3 hybridization four electron pairs are formed around the germanium atom, that constitute the completely filled valence band of germanium. On an increase in energy some of the electrons bonding the germanium atoms may escape and become free electrons, leaving positive holes in their place. In the band model this is equivalent to transferring electrons from the valence band into the conduction band. The number of electrons involved, n, depends on the width of the forbidden band, E_g, and the temperature, T:

$$n \sim \exp(-E_g/2\,kT) \qquad \text{(I.110)}$$

where k is the Boltzmann constant. This conduction mode is the *intrinsic conduction* of the crystal (for a more elaborate discussion of the intrinsic conduction of semiconductors see Chapter I.4).

SECTION 3.2. IMPURITY SEMICONDUCTORS

Conductivity due to impurities in the crystal occurs even more frequently than intrinsic conduction. For example, if a germanium atom in the germanium crystal is replaced by arsenic, of the 5 valence electrons of arsenic only four electrons take part in the bond formed with the neighbouring germanium atoms. The fifth electron is not included in the valence electrons of the crystal, but is located around the arsenic core. This situation can be seen in Fig. I.10, where the fifth valence electron occupies a localized level in the potential field of the arsenic atom, below the lower limit of the conduction band of germanium. Upon a small energy gain the electron may leave this energy level, and jump into the conduction band. Impurities of this kind, which yield electrons, are called *defects of donor character*. The process can be characterized by the equation

$$D^0 \rightleftarrows D^+ + \ominus$$

where D^0 is the non-dissociated donor, and D^+ is the dissociated defect site. The fifth valence electron of arsenic thereby becomes a conducting electron, and increases the conductivity of germanium to a considerable extent. *The conduction provided by the excess electron* is called *n-type conduction*, and semiconductors containing donor defect sites are called *n-type semiconductors*.

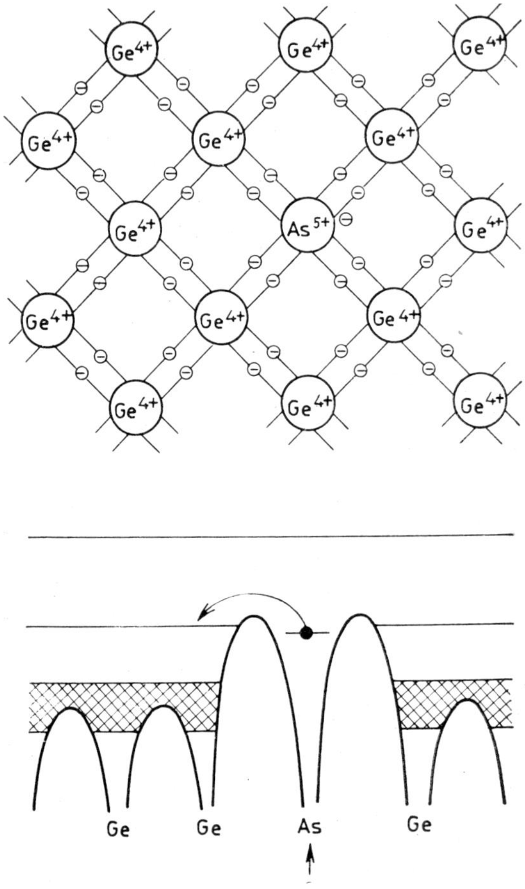

Fig. I.10. The crystal lattice and energy band model of germanium containing an arsenic atom

When an indium atom, containing fewer valence electrons, is built into the germanium crystal, one electron will be missing from the bonds oriented towards the neighbouring germanium atoms. Upon gaining a relatively small amount of thermal energy, however, an electron can be shifted from one of the neighbouring bonds to this electron defect site. The vacancy thereby formed can be filled by the electron of another germanium atom, and so on. In terms of the energy band system this can be interpreted as follows (see Fig. I.11). There is a vacant energy level in the potential field of the indium atom next to the upper limit of the valence band. On thermal excitation an electron is raised to this energy level, leaving a vacancy, i.e. a positive hole, in its place. *The conduction provided by this process* is called *positive hole- or defect-conduction*. Defect sites capable of taking up the electron of the valence band are called *acceptors*. The process can be characterized by the equation

$$A^0 \rightleftarrows A^- + \oplus$$

Semiconductors containing acceptor-type impurities are called *defect- or p-type semiconductors.*

n- and p-type conduction may appear in the germanium crystal not only if the crystal contains impurities, but also if there are irregularities in the crystal lattice. Such an irregularity may result, for instance, if a germanium atom is inserted between the lattice points of an otherwise perfect lattice, or if an atom is missing from the lattice points.

A germanium atom inserted between the lattice points results in a donor-type defect site. Consequently, provided that no other impurities are present, this germanium crystal can be regarded as an n-type conductor. Taking into account the donor-reaction, one may write

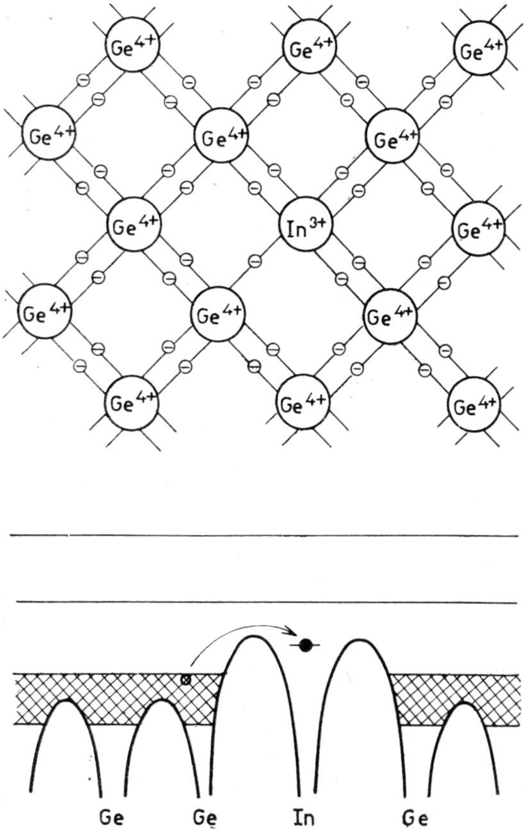

Fig. I.11. The crystal lattice and energy band model of germanium containing an indium atom

$$\text{Ge}\,\bigcirc^0 \rightleftarrows \text{Ge}\,\bigcirc^+ + \ominus$$

The germanium atom between the lattice points of a germanium crystal gives up its electron upon a relatively small, 0.0314 eV, energy gain. The dielectric constant of the germanium lattice, as a dielectric medium, is 16.1, so the ionization energy of germanium is lower by a factor of $1/\varepsilon_{\text{Ge}}^2 = 1/259$ than in vacuum (8.13 eV).

Missing lattice points in the germanium crystal result in acceptor-type defect sites, and due to their presence defect-type conduction is favoured.

In semiconductors of different types the movement of electrons and positive holes must be conceived in a dynamic sense. The thermal vibration of the crystal lattice results in a constant circulation of electrons and positive holes. The energy levels are filled and emptied, and the motion and exchange of electrons is continuous. Since even a low amount of energy is sufficient for the ionization of donor and acceptor levels, the conduction character of impurity semiconductors at low temperature is always determined by the nature of the impurities. With increasing temperature more and more donor or acceptor levels become ionized, and consequently the conductivity increases. At a certain temperature, however, a state is reached where the donor levels are exhausted and the conductivity no longer increases. Depending on the nature of the semiconductor, the transfer of electrons from the valence band becomes preferred, i.e. the occurrence of intrinsic conduction. Since the valence band of a semiconductor contains an almost inexhaustible number of electrons, the number of transferred electrons and the positive holes left in their place far exceeds the number of charge carriers arising from the impurities. Accordingly, at higher temperatures the intrinsic conduction of the crystal predominates.

Despite this, the conductivity of a semiconductor cannot be increased indefinitely, since the conductivity depends not only on the number of charge carriers, but also on the mobility of the charged particles.

SECTION 3.3. APPLICATION OF FERMI STATISTICS TO SEMICONDUCTORS

This section deals with the behaviour of electrons and positive holes in semiconductors, the treatment being based on Fermi-Dirac statistics. This mathematical model will lead us to the concept of the Fermi level, which plays an extremely important role not only in the physical application of semiconductors but also in their catalytic behaviour, as will be shown in Section III.2.2.

It is generally known that a system left on its own tends to a state corresponding to an energy minimum; similarly, the electrons of a solid-state material tend to occupy the lowest available energy levels. According to the Pauli principle every level may be occupied by two electrons. The lowest levels of the band structure are practically filled. In the higher energy regions, however, levels can be found which contain only one electron, or no electrons at all, while the uppermost levels are practically vacant. Thus, there is a gradual distribution. At low temperatures (around 0 K) the transition between vacant and filled levels is sharp, but with increasing temperature it becomes more and more diffuse due to the transfer of electrons. For the low energy levels the probability of these being completely filled is unity, whereas near the upper edge of the band structure this probability drops to zero. The course of the probability function is given by Fermi statistics:

$$P(E) = \{1 + \exp[(E - E_F)/kT]\}^{-1} \quad \text{(I.111)}$$

$$P(U) = \{1 + \exp[(U - U_F)/kT]\}^{-1} \quad \text{(I.112)}$$

where E is the energy, U is the potential in volts corresponding to the energy of the level whose probability of occupation is being considered, k is the Boltzmann constant, and e is the charge of the electron. E_F, the Fermi level or Fermi energy, denotes the energy of the level where the probability of occupation is 1/2. At absolute zero the levels up to this energy are fully occupied, while at higher temperatures this is the mean energy of the diffuse upper part of the electron distribution.

In the case of semiconductors, when the conduction band is separated from the valence band by a forbidden band, the Fermi level usually lies within the forbidden band. In this case, therefore, the Fermi level cannot be defined as the level whose occupation probability is 1/2 (levels in the forbidden band may not be occupied). Hence, in the treatment of semiconductors one should start from the fact that for intrinsic semiconductors the number of electrons in the conduction band is equal to the number of positive holes in the valence band ($n = p$). These numbers can be evaluated by means of the Fermi-Dirac distribution function $P(E)$ given by Eq. (I.111), and the level density functions $N_2(E)$ for the conduction band and $N_1(E)$ for the valence band via the expression

$$n = \int_{E_{c,l}}^{E_{c,u}} P(E)\, N_2(E)\, \mathrm{d}E = \int_{E_{v,l}}^{E_{v,u}} [1 - P(E)]\, N_1(E)\, \mathrm{d}E = p \qquad \text{(I.113)}$$

where $E_{c,l}$ and $E_{c,u}$ denote the lower and upper limits of the conduction band, and $E_{v,l}$ and $E_{v,u}$ the lower and upper limits of the valence band, respectively.

On substituting expression (I.111), and the easily derivable expression for the level density [24]

$$N_a(E) = \frac{4\pi}{h^3}(2\, m_q^*)^{3/2}\, (E - E_{p,l})^{1/2} \qquad \text{(I.114)}$$

$$\begin{pmatrix} \text{if } a = 2 \text{ then } q = \mathrm{e} \text{ and } p = \mathrm{c} \\ \text{if } a = 1 \text{ then } q = \mathrm{h} \text{ and } p = \mathrm{v} \end{pmatrix}$$

into Eq. (I.113), after integration:

$$n = 2\left(\frac{2\,\pi m_\mathrm{e}^*\, kT}{h^2}\right)^{3/2} \mathrm{e}^{-(E_{c,l} - E_F)/kT} = 2\left(\frac{2\,\pi m_\mathrm{h}^*\, kT}{h^2}\right)^{3/2} \mathrm{e}^{-(E_F - E_{v,u})/kT} = p \qquad \text{(I.115)}$$

(m_e^* is the effective mass of the electrons in the conduction band, and m_h^* that of the positive holes in the valence band), from which the Fermi level E_F can be determined. As a result:

$$E_F = \frac{1}{2}\left(E_{c,l} + E_{v,u} + kT \ln \frac{N_v}{N_c}\right) \qquad \text{(I.116)}$$

where

$$N_v = 2\left(\frac{2\, m_\mathrm{h}^*\, kT}{h^2}\right)^{3/2}$$

and

$$N_c = 2\left(\frac{2\, m_\mathrm{e}^*\, kT}{h^2}\right)^{3/2} \qquad \text{(I.117)}$$

It can be seen from Eqs (I. 116) and (I. 113) that if $m_h^* = m_e^*$, and therefore $N_v = N_c$, their Fermi level is in the middle of the forbidden band between the conduction and valence bands. However, if $N_v \neq N_c$, or $p \neq n$ (not intrinsic semiconductors), the Fermi level shifts from the middle of the forbidden band.

From thermodynamic considerations not discussed here in detail [25] it can further be shown that the energy of the Fermi level is identical to the chemical potential of the crystal. On this basis it can also be proved that the potential difference, readable on a voltmeter, of two metals joined to the two poles of an electric battery is equal to the Fermi potential difference corresponding to the difference of the Fermi levels in the two metals (due to the electric source). The potential corresponding to the Fermi level is therefore called the electrochemical potential. Fermi-type probability functions calculated for two different temperatures are given in Fig. I.12. It can be seen that at a given temperature the energy range between the filled and empty energy levels is of the order of only $2kT - 3kT$.

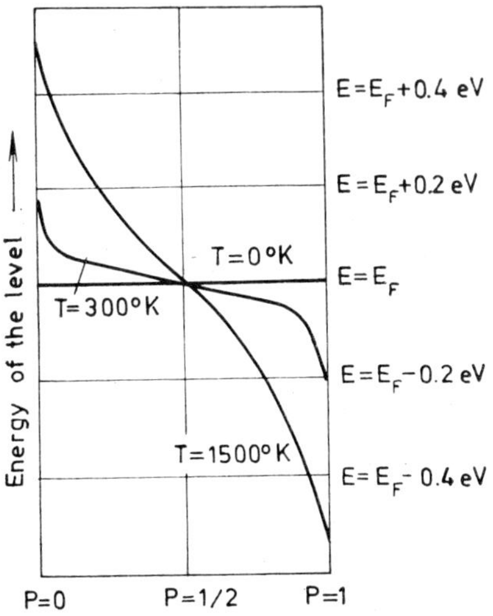

Fig. I.12. The hape of the Fermi probability function at different temperatures

The electron and positive hole concentrations of semiconductors, and their variation with temperature and with the concentration of impurities, can be evaluated by means of the Fermi probability function. The following sections deal with the investigation of the electron distribution and the temperature dependence of the electron concentration in different types of semiconductors, the treatment being based on Fermi statistics.

SECTION 3.4. INTRINSIC SEMICONDUCTORS

The left hand part of Fig. I.13 shows the band structure of an intrinsic semiconductor, and the right hand part of the diagram the Fermi distribution function. Let us denote the energy gap between the upper limit of the valence band and the lower limit of the conduction band by E_g, and the level density vs. energy function by $N_1(E)$ in the valence band, and by $N_2(E)$ in the conduction band. $N_1(E)$ and $N_2(E)$ give the number of levels per unit energy interval when going vertically in Fig. I.13.

Some important consequences can be drawn from these diagrams.

In the absence of donor- or acceptor-type defect sites, electrons may enter the conduction band only from the valence band, involving the formation of positive holes.

At any temperature above zero there should be electrons in the conduction band and the same number of positive holes in the valence band. With increasing temperature the numbers of electrons and positive holes increase, but their ratio still remains unity.

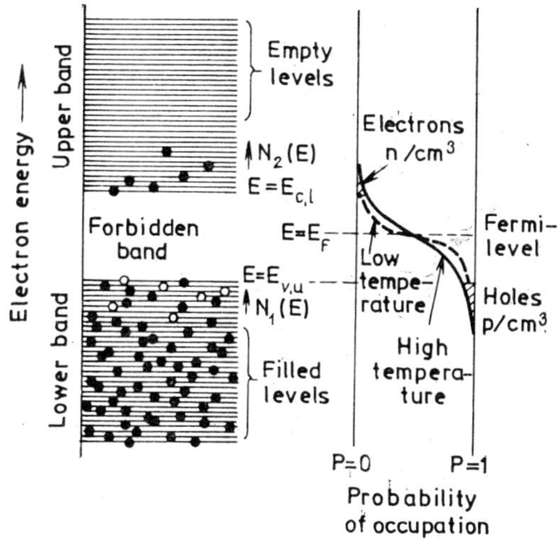

Fig. I.13. The energy band structure of an intrinsic semiconductor, and the Fermi probability function

The concentrations of electrons and positive holes as a function of temperature can be determined more simply than in the preceding section by the following approximation. Since the electrons in the conduction band occupy primarily the lowest energy levels of this band, the level set of the conduction band may be replaced by a suitable number of so-called effective equivalent energy levels. Thus, the energy levels in the conduction band are represented by N_c equivalent levels of energy $E_g + E_{v,u}$, all occupied by two electrons. The number of these levels is given by

$$\frac{N_c}{2} = \left(\frac{2\pi m_e^* kT}{h^2}\right)^{3/2} \text{cm}^{-3} \tag{I.118}$$

where m_e^* is the effective mass of the electron, and the subscript c refers to the conduction band (cf. Section 1.3.3). In an analogous manner, the energy levels of the valence band filled with positive holes can be replaced by N_v equivalent energy levels of energy $E_{v,u}$, where

$$\frac{N_v}{2} = \left(\frac{2\pi m_h^* kT}{h^2}\right)^{3/2} \text{cm}^{-3} \tag{I.119}$$

The subscript v refers to the valence band, and m_h^* is the effective mass of the positive hole. Assuming that $m_e^* = m_h^* = m_e$, i.e. the rest mass of a free electron:

$$N_c = N_v = 2\left(\frac{2\pi m_e kT}{h^2}\right)^{3/2} \text{cm}^{-3} \tag{I.120}$$

If the product of temperature-independent terms is denoted by M, the final results will be

$$N_c = N_v = 2\,MT^{3/2}\ \text{cm}^{-3} \tag{I.121}$$

Based on this result the number of electrons in the conduction band can be determined by multiplying N_c by the occupation probability of the individual levels:

$$n = N_c\{1 + \exp[(E_{c,1} - E_F)/kT]\}^{-1} \text{ cm}^{-3} \qquad (I.122)$$

To obtain the number of positive holes in the valence band, the number of electrons possible in the band is first determined, and from this value the number of electrons really present is subtracted:

$$p = N_v(1 - \{1 + \exp[(E_{v,u} - E_F)/kT]\}^{-1}) \text{ cm}^{-3} \qquad (I.123)$$

Since these considerations apply to intrinsic semiconductors, the concentrations of electrons and positive holes should be the same. As a final result:

$$n = p = \frac{2\,MT^{3/2}}{1 + e^{E_g/2kT}} \text{ cm}^{-3} \qquad (I.124)$$

In order to calculate the conductivity of an intrinsic semiconductor the expressions obtained for n and p should be substituted into

$$\sigma = (n\mu_n + p\mu_p)\,e \qquad (I.125)$$

(see Chapter 1.4), where μ_n and μ_p are the mobilities of the electron and positive hole, respectively, and e is the charge of the electron.

SECTION 3.5. p-TYPE SEMICONDUCTORS

The band system of p-type semiconductors differs from that of intrinsic semiconductors in three respects: the conduction band is practically vacant (below the temperature range of intrinsic conduction); a set of acceptor levels can be found in the forbidden band; and the valence band is approximately filled. Let us denote the energy of localized acceptor levels by E_a, and the number of acceptor levels per cm³ of semiconductor by N_a (Fig. I.14). If the amount of impurity material is small,

$$N_a \ll N_v \qquad (I.126)$$

should hold. In the determination of the temperature-dependence of the Fermi level the following facts must be taken into consideration. Positive holes are formed if the electrons of the valence band jump to the acceptor levels due to energy gain. These processes need much less energy than the raising of an electron to the conduction band. For just this reason this process starts even at low temperatures. Thus, below the temperature range of intrinsic conduction there will be more holes in the valence band than electrons in the conduction band. Accordingly, the Fermi level will be somewhere in the lower part of the forbidden band. This situation is reflected in the following equation:

$$p = N_v(1 - \{1 + \exp[(E_{v,u} - E_F)/kT]\}^{-1}) = N_a\{1 + \exp[(E_a - E_F)/kT]\}^{-1} + N_c\{1 + \exp[(E_{c,1} - E_F)/kT]\}^{-1} \qquad (I.127)$$

where it has been taken into account that due to the overall electroneutrality the number of holes in the lower band should be equal to the sum of electrons in the upper band and the acceptor levels. (A localized acceptor level may be occupied by only a single electron.) The solution of this equation for E_F yields an extremely complicated expression, which can be simplified,

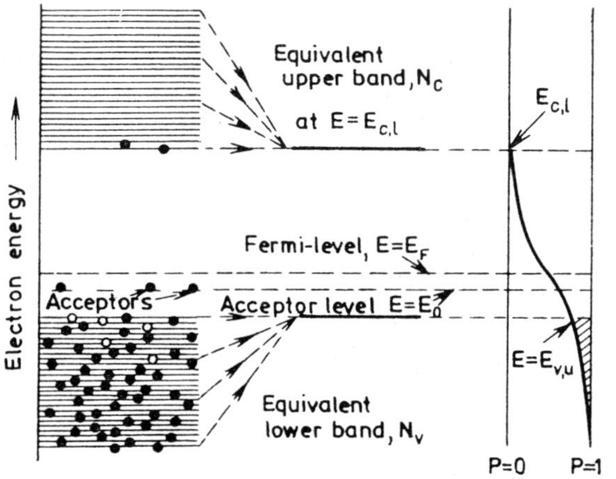

Fig. I.14. The energy band structure of a p-type semiconductor and the Fermi probability function

however, if the solution is investigated separately for the different temperature ranges.

1. At temperatures much lower than the temperature of intrinsic conduction, when the number of electrons in the conduction band can be neglected in comparison with the number of positive holes, it follows that

$$E_F \approx kT \ln \left\{ \frac{N_v}{N_a^2} - \frac{1}{2} + \frac{1}{2} \left[\left(1 - \frac{N_v}{N_a} \right)^2 + \frac{4 N_v}{N_a} e^{E_a/kT} \right]^{1/2} \right\} \tag{I.128}$$

if the final term of the previous equation is omitted. When the temperature is sufficiently low, the exponential term is much larger than the other terms, and the expression for the Fermi level becomes

$$E_F \approx \frac{kT}{2} \ln \frac{N_v}{N_a} + \frac{E_a}{2} \approx \frac{kT}{2} \ln \frac{2\, MT^{3/2}}{N_a} + \frac{E_a}{2} \tag{I.129}$$

It can be seen from the equation that in the neighbourhood of absolute zero temperature the Fermi level is at $E_a/2$, i.e. at the mid-point of the energy gap between the acceptor level and the valence band. The concentration of holes may be obtained from Eqs (I.127) and (I.129):

$$p \approx \sqrt{N_a} \sqrt{2\, MT^{3/2}}\, e^{-E_a/2kT} \text{ cm}^{-3} \tag{I.130}$$

In this very low temperature range the conductivity depends only on the concentration of impurity material.

2. With increase of temperature the Fermi level becomes higher, and at a given T it coincides with the acceptor level. In this case $\mathrm{p} = N_a/2$, i.e. half of the acceptor levels are occupied by electrons.

Taking this fact and the equality $E_F = E_a$ into account, on the basis of Eq. (I.127) the temperature at which these conditions are fulfilled is obtained as

$$T = \frac{E_a}{k \ln\left(\dfrac{4\, MT^{3/2}}{N_a} + 1\right)} \text{ (K)} \tag{I.131}$$

3. With further increase of temperature the acceptor levels become more and more occupied. If the temperature is not yet high enough to raise electrons from the valence band into the conduction band, the Fermi level can be given as

$$E_F \approx kT \ln \frac{2\, MT^{3/2}}{N_a} \tag{I.132}$$

and the defect electron concentration as

$$\mathrm{p} \approx \frac{2\, MT^{3/2}\, N_a}{2\, MT^{3/2} + N_a \exp(E_a/kT)} \text{ cm}^{-3} \tag{I.133}$$

At still higher temperatures the acceptor levels become fully occupied; in this temperature range the temperature influences only the mobility of the positive holes.

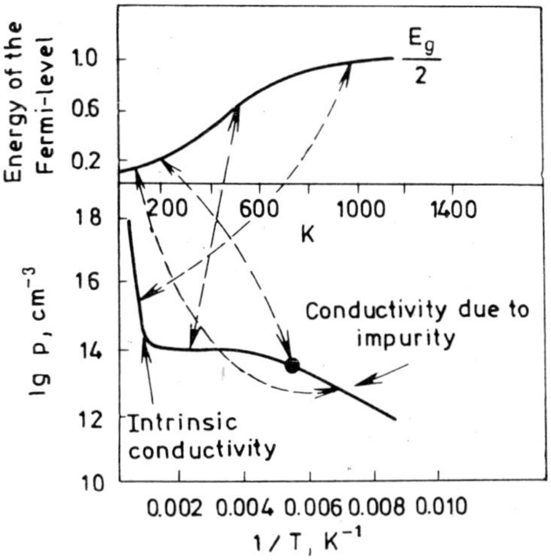

Fig. I.15. The variation of the defect electron concentration and of the position of the Fermi level as functions of temperature ($E_g = 2$ eV, $E_a = 0.2$ eV, $N_a = 10^{14}$ cm^{-3})

4. If the temperature is sufficiently high to raise the Fermi level to approximately the middle of the forbidden band, electron transfer may take place from the valence band to the conduction band. Now, from Eq. (I.127) one can obtain for the Fermi level

$$E_F = kT \ln \frac{4\,MT^{3/2}}{N_a + (N_a^2 + 16\,M^2T^3\,e^{-E_{c,1}/kT})^{1/2}} \tag{I.134}$$

since the first term of the right hand side of Eq. (I.127) simplifies to N_a.

At sufficiently high temperatures the concentration of positive holes may be given in the same way as for intrinsic semiconductors:

$$p = N_a + n = 2\,MT^{3/2}\,e^{-E_g/2kT}\ \text{cm}^{-3} \quad (E_F = E_g/2) \tag{I.135}$$

This means that at high temperatures p-type semiconductors become intrinsic conductors, i.e. the number of electrons promoted from the valence band into the conduction band exceeds the number of electrons populating the acceptor level by several orders of magnitude. The variation of the Fermi level, calculated according to the above equations, and that of the concentration of positive holes with the reciprocal of the absolute temperature are given in Fig. I.15. The corresponding sections of the Fermi level and positive hole concentration functions are joined by broken lines.

SECTION 3.6. n-TYPE SEMICONDUCTORS

The simplifying assumptions made in the preceding section are applied again, i.e. the lower and upper bands are substituted by N_c or N_v equivalent levels, respectively, with the common energy values $E = E_{v,1}$ and $E = E_{v,c}$, respectively, and it is also assumed that the donor concentration

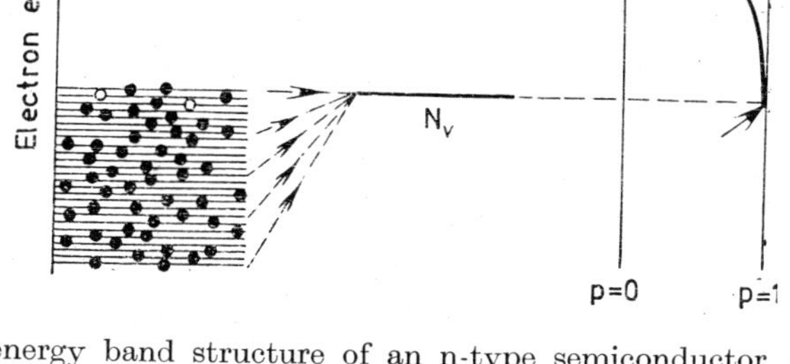

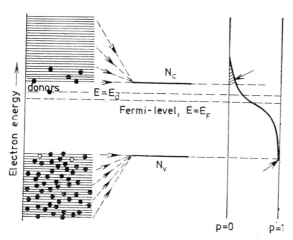

Fig. I.16. The energy band structure of an n-type semiconductor, and the Fermi probability function

is much lower than N_c. The number of electrons in the conduction band is obviously given by the sum of the dissociated donor levels and the number of positive holes in the valence band:

$$\begin{aligned} n &= N_c\{1 + \exp[(E_{c,1} - E_F)/kT]\}^{-1} = \\ &= N_d(1 - \{1 + \exp[(E_d - E_F)/kT]\}^{-1} + \\ &+ N_v(1 - \{1 + \exp[(E_{v,u} - E_F)/kT]\}^{-1}) \end{aligned} \tag{I.136}$$

This distribution is represented by the band system shown in Fig. I.16. Again, the effect of temperature on the position of the Fermi level can be most conveniently studied separately in the different temperature ranges.

1. In the neighbourhood of absolute zero temperature the Fermi level is halfway between the valence band and the donor levels. For E_F and n:

$$E_F = \frac{kT}{2} \ln \frac{2\,MT^{3/2}}{N_d} + \frac{E_{c,1} + E_d}{2} \tag{I.137}$$

and

$$n = \sqrt{N_d}\sqrt{2\,MT^{3/2}} \exp[-(E_{c,1} + E_d)/2\,kT] \text{ cm}^{-3} \tag{I.138}$$

2. With increase of temperature the Fermi level shifts toward the middle of the forbidden band. The donor level is reached at a temperature

$$T = \frac{E_{c,1} - E_d}{k \ln\left(\frac{4\,MT^{3/2}}{N_d} - 1\right)} \tag{I.139}$$

The number of electrons is, of course, $\frac{N_d}{2}$. At a still higher temperature the transition range is also reached in this case when almost all the donor levels dissociate, and a further temperature rise produces practically no increase in the number of conducting electrons. In this temperature range

$$E_F = E_{c,1} - kT \ln \frac{2\,MT^{3/2}}{N_d} \tag{I.140}$$

for the Fermi level, and

$$n = N_d \text{ cm}^{-3} \tag{I.141}$$

for the electron concentration. As the temperature is increased the transfer of valence band electrons becomes more and more predominant, and the system finally reaches the temperature range of intrinsic conduction.

CHAPTER 4. ELECTRICAL AND MAGNETIC PROPERTIES OF METALS AND SEMICONDUCTORS

J. Ladik

SECTION 4.1. THE THEORY OF THE ELECTRICAL CONDUCTIVITIES OF METALS AND SEMICONDUCTORS

According to classical electrodynamics, the current density in a given medium, j_x, caused by a field strength E_x in the x-direction can be given by

$$j_x = \sigma E_x \tag{I.142}$$

where σ is the specific conductivity of the medium. If the current is due to electrons or positive holes, respectively, the current due to one electron or hole is

$$i_x = \pm\, e v_x \tag{I.143}$$

where v_x is the x-component of the velocity of the particle, able to move freely in the direction of the field. The density of the macroscopic current passing the plane perpendicular to the x-direction at a point x, y, z can be calculated by the integral

$$j_x(x, y, z) = \pm \int e\, v_x f_n(x, y, z, v_x, v_y, v_z)\, \mathrm{d}v_x\, \mathrm{d}v_y\, \mathrm{d}v_z \tag{I.144}$$

where the integration should embrace all possible values of v_x, v_y and v_z. The distribution function f_n gives the number of particles whose coordinates lie between x and $x + \mathrm{d}x$, y and $y + \mathrm{d}y$, and z and $z + \mathrm{d}z$, respectively, in the steady-state condition, and, at the same time, whose velocity components lie between v_x and $v_x + \mathrm{d}v_x$, v_y and $v_y + \mathrm{d}v_y$ and v_z and $v_z + \mathrm{d}v_z$, respectively*. The distribution function f_n can be obtained from the solution of the Boltzmann equation of state. The derivation of this equation, and a discussion of methods for its solution, however, are beyond the scope of this treatment. (For a discussion of the problem in more detail see for example Ref. [26].)

By assuming a field along the x-axis and a constant temperature, the function f_n can be determined from the Boltzmann equation on the basis of the free electron model. On substituting the function thus obtained into Eq. (I.144), and pursuing a slightly tedious calculation which takes Eq. (I.142) into account the specific conductivity may be shown to be

$$\sigma = \frac{j_x}{E_x} = \frac{e^2\, n_0 l(\varepsilon_0')}{m_e\, v(\varepsilon_0')} \tag{I.145}$$

* The expression is valid only if either electrons only or positive holes only take part in the conduction. In the general case, when both kinds of particles take part in the conduction, j_x must be given as the difference of the expressions for the two kinds of particles.

$l(\varepsilon_0')$ and $v(\varepsilon_0')$ are the values of the mean free path and velocity, respectively, for $\varepsilon = \varepsilon_0'$, where

$$\varepsilon_0' = \frac{h^2}{2m_e^*}\left(\frac{3n_0}{8\pi}\right)^{2/3} \tag{I.43}$$

and $n_0 = \frac{N}{V}$ is the number of electrons in unit volume. Eq. (I.145) gives a good estimate of the conductivity only for metals, since it was derived on the basis of the free electron model.

Eq. (I.145) for the conductivity of metals yields a value whose order of magnitude agrees with the conductivity of common metals measured at room temperature if l is chosen as 10—100 times the internuclear distance. The increase in conductivity with decreasing temperature can be interpreted only by assuming that $l \to \infty$ if $T \to 0$. This, however, contradicts the picture suggested by the simple free electron model, since in the derivation yielding Eq. (I.145) l was regarded as constant.

When the quantum mechanical band model is applied to metals, attention must be paid to two important points. On the one hand, according to the band model only the electrons occupying the partially filled bands of the metal may take part in the conduction, since electrons occupying completely filled bands cannot be accelerated by the electric field. Due to the acceleration the energy of the electrons increases, but if there are no vacant levels in the band that may be occupied by the electrons of increased energy, the acceleration and hence the conduction cannot take place. This is why solids having partially filled energy bands are conductors, whereas solids having only completely filled bands are insulators (see Fig. I.17).

On the other hand, if the electrons moving in metals are described by Bloch-type plane waves of the form

$$\psi_{\mathbf{k}}(\mathbf{r}) = e^{i\mathbf{kr}}\, u_{\mathbf{k}}(\mathbf{r}) \tag{I.38}$$

as in the case of a perfect lattice (containing no defect sites or impurities), it can easily be proved that the speed of the electrons in a given state $\varepsilon(\mathbf{k}$-) will be constant if lattice vibrations are not taken into account. Consequently, the values of the mean free path, l, and σ become infinite. According to quantum mechanics, the mean free path and conductivity may be finite only if the scattering of electrons by the quanta of lattice vibrations (phonons) is taken into account, or if the perturbation (scattering) effects of impurities and defect sites on the electron waves are taken into consideration.

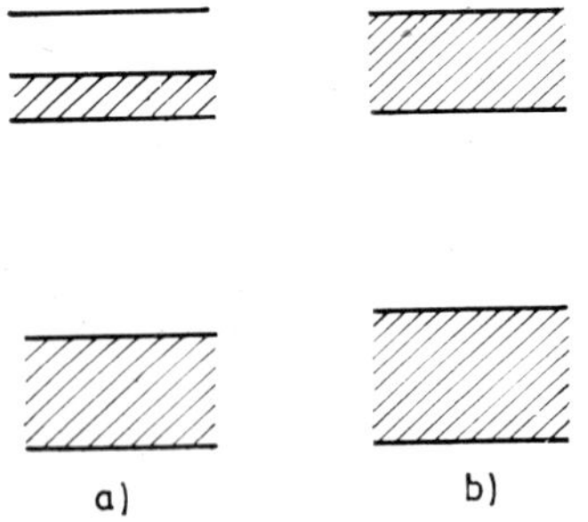

Fig. I.17. a) conductor with a partially filled band, b) insulator with completely filled bands

The variation of the potential acting on the electrons due to the lattice vibrations is fairly well approximated by an expression given by Bardeen. On the basis of this it has been shown that if an electron is scattered by one of the normal vibrations of the lattice, the quantum number of the normal vibration in question may change only by ± 1. As a result of this selection rule it is possible to determine the probability, on scattering, of the transition of an electron from a given initialstate to different final states allowed by the selection rules. According to the results of the fairly sophisticated derivations (see e.g. Ref. [27]), Eq. (I.140) for the conductivity can be formally retained if $l(\varepsilon_0') = l(\mathbf{k}_0)$ ($\mathbf{k}_0$ is the **k**-value belonging to the uppermost filled level, see Fig. I.18) is replaced by

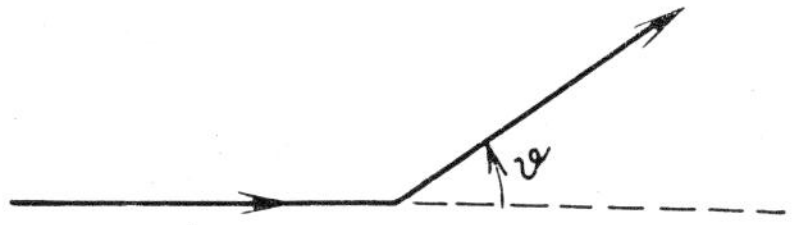

Fig. I.18. The definition of the angle theta

$$l(\mathbf{k}_0) = \frac{9c^2Nm}{32\pi^3kT}\left(\frac{\mathrm{d}\varepsilon}{\mathrm{d}\mathbf{k}}\right)^2_{\mathbf{k}=\mathbf{k}_0}\frac{2^{5/2}}{k_0^2}C^{-2}. \qquad \text{(I.146)}$$

In this expression k is the Boltzmann constant, N is the number of uniform atoms of mass m forming the crystal, and the constant C^2 is a complex function of the angle between the directions of the electron before and after scattering (see Fig. I.18; for the exact expression of C^2, which contains integrals with respect to ϑ, see e.g. Ref. [27]). By assuming the simple relationship

$$\varepsilon(\mathbf{k}) = \frac{h^2k^2}{2m_e} = \frac{1}{2}m_e v(\mathbf{k})^2 \qquad \text{(I.147)}$$

between the energy and wavenumber of the electron, expressions for $\left(\frac{\mathrm{d}\varepsilon}{\mathrm{d}\mathbf{k}}\right)_{\mathbf{k}=\mathbf{k}_0}$ and $v(\varepsilon_0') = v(\mathbf{k}_0)$ can be given in a simple way.

On substituting these expressions into Eqs (I.146) and (I.145) [28]:

$$\sigma = \frac{2^{8/3}\,e^2}{\pi}\,\frac{h^3}{m_e^2}\,k_0^5\,\frac{m}{C^2}\,\frac{1}{kT} \qquad \text{(I.148)}$$

For simple metals (not transition metals) this expression gives fair agreement with the experimental results, and the temperature-dependence also agrees with experiment.

The conductivities of transition metals can be calculated in a similar manner. In this case it is taken into account that both s-p-, and d-electrons take part in the conduction. The derivation [involving the assumption that due to scattering by the lattice vibrations s-p- and d-electrons may jump not only to the vacant levels of their own band but also to the vacant levels of other bands (e.g. a scattered s-p-electron may jump to a vacant s-p-level, but also to a vacant d-level)] yields an expression for σ that is again able to describe the experimental results fairly well. By assuming that electron waves are scattered by impurity atoms and lattice imperfections, expressions have been derived in a similar manner for the resistance of metals due to impurities and lattice imperfections. These calculations, however, cannot be reproduced here.

As shown in Chapter I.3, the conductivity of a semiconductor may arise from the fact that due to thermal excitation electrons may be promoted from the completely filled valence band into the vacant conduction band, through the not too wide forbidden band (intrinsic conduction). On the other hand, due to donor- or acceptor-type impurities, respectively, electrons may enter the conduction band, or positive holes may be formed in the valence band (impurity conduction).

In the case of intrinsic conduction, since both the electrons and the positive holes take part in the conduction, the specific conductivity is

$$\sigma = e(n\mu_e + p\mu_h) \tag{I.149}$$

where n and p are the densities of the mobile electrons or positive holes, respectively, at a given temperature, while μ_e and μ_h are the corresponding mobilities. For the calculation of the densities of the mobile particles and the mobilities, the energy band structure of the semiconductor must be known. Since the resistance in the case of semiconductors is also due to the scattering of charge carriers by the different lattice vibrations, it can be determined by calculating the interaction of the electrons or positive holes with phonons.

Simple approximate expressions for the mobilities determined from the above effect have been given by Shockley [29] (the deformation potential approximation). Due to the nature of the basic assumptions made in the derivations these expressions can be used only for semiconductors, and even then only if the valence and conduction bands are not too narrow (their width must be at least 0.05 eV). According to the results of Shockley:

$$\mu_e = \frac{2^{3/2}\,\pi^{1/2}}{3}\,\frac{C_{l,l}\,\hbar^4 e}{\varepsilon_{le}^2\, m_e^{*5/2}\,(kT)^{3/2}} \tag{I.150}$$

and

$$\mu_h = \frac{2^{3/2}\,\pi^{1/2}}{3}\,\frac{C_{l,l}\,\hbar^4 e}{\varepsilon_{lh}^2\, m_h^{*5/2}\,(kT)^{3/2}} \tag{I.151}$$

where $C_{l,l}$ is the elastic constant for the vibration causing the dilatation (or contraction) of the lattice, k is the Boltzmann constant, and

$$m_e^* = \frac{\hbar^2\pi^2}{2a^2\,\Delta E_c} \quad \text{and} \quad m_h^* = \frac{\hbar^2\pi^2}{2a^2\,\Delta E_v} \tag{I.152}$$

are the corresponding effective masses (a is the lattice constant, and ΔE_c and ΔE_v are the widths of the conduction and valence bands, respectively). Finally, the definitions of the deformation potentials ε_{le} and ε_{lh} are

$$\varepsilon_{le} = \frac{\delta E_{c,l}}{\Delta} \quad \text{and} \quad \varepsilon_{lh} = \frac{\delta E_{v,u}}{\Delta} \tag{I.153}$$

where $\delta E_{c,l}$ and $\delta E_{v,u}$ are the shifts of the lower limit of the conduction band and the upper limit of the valence band, respectively, in the case of the dilatation Δ of the lattice. It can be shown that the contributions of other modes of lattice vibration to the conductivity are negligible [30].

According to Eq. (I. 115) the densities of mobile electrons and holes are

$$n = \mathrm{e}^{-(E_{c,l}-E_{\mathrm{F}})/kT}\, 2(2\,\pi m_e^*\, kT/h^2)^{3/2} \tag{I.154}$$

and

$$p = \mathrm{e}^{-(E_{\mathrm{F}}-E_{v,u})/kT} 2(2\pi m_h^*\, kT/h^2)^{3/2} \tag{I.155}$$

respectively, where the Fermi level E_F for intrinsic semiconductors can be well approximated by the expression

$$E_{\mathrm{F}} = \frac{E_{c,l} + E_{v,u}}{2} \tag{I.156}$$

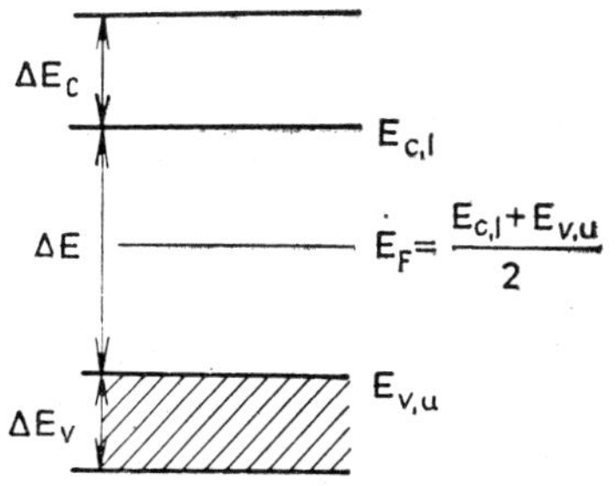

Fig. I.19. The position of the Fermi level in intrinsic semiconductors

(see Fig. I.19). On substituting Eq. (I.156) into Eqs (I.154) and (I.155), and these equations together with Eqs (I.150) and (I.151) into Eq. (I.149) it follows for the specific conductivity that

$$\sigma = \frac{2}{3\pi^2} e^2 \hbar C_{l,l} \left(\frac{1}{m_h^*\, \varepsilon_{l,h}^2} + \frac{1}{m_e^*\, \varepsilon_{l,e}^2}\right) \mathrm{e}^{-E_g/2kT} = \sigma_0\, \mathrm{e}^{-E_g/2kT} \tag{I.157}$$

where E_g is the width of the forbidden band between the valence and conduction bands.

In the case of donor- or acceptor-type impurities the conductivity is determined by the contribution of electrons or holes, respectively, alone. Eqs (I.150) and (I.151) can still be used, but the simple expression (I.156) for E_{F} cannot be substituted into Eqs (I.154) and (I.155), since in the determination of the Fermi level the locations of the energy levels of the impurities must also be taken into account. Accordingly, for example in the case of acceptor-type impurities:

$$\sigma = \frac{2}{3\,\pi^2}\, e^2 \hbar C_{l,l} \frac{1}{m_h^*\, \varepsilon_{l,h}^2}\, \mathrm{e}^{-(E_{\mathrm{F}}'-E_{v,u})/kT} \tag{I.158}$$

where E_{F}', is the energy corresponding to the Fermi level in the lattice with impurities.

SECTION 4.2. HALL-EFFECT IN METALS AND SEMICONDUCTORS

It is known from the elements of electrodynamics that if a magnetic field of strength **H** acts on a particle of velocity v and charge e, the force exerted by the magnetic field on the moving particle can be given as

$$\mathbf{F} = \frac{e}{c}\,[\mathbf{v} \times \mathbf{H}] \tag{I.159}$$

5*

(the Lorentz force law). If the definition of a vector product is taken into account, it can be seen from Eq. (I.159) that the force, **F**, is always perpendicular to the plane determined by vectors **v** and **H**. If a macroscopic conductor is subjected to an electric field along the x-axis (E_x) causing a current I_x in the x-direction, and, simultaneously, to a magnetic field of strength H_z along the z-axis, an electric field of strength

$$E_y = -\varrho I_x H_z \tag{I.160}$$

will also be generated along the y-axis, due to the Lorentz force. This phenomenon is called the Hall-effect and the constant in Eq. (I.160) the Hall constant.

The values of the Hall constant for metals can be evaluated from the Boltzmann equation of state given in the preceding Section by assuming an electric field along the x-axis, a magnetic field perpendicular to the electric field, along the z-axis, and a constant temperature. This rather complicated calculation can be greatly simplified if the distribution function, f_n, obtained from the solution of the equation is assumed to have the form

$$f_n = f_n^0 + v_x \chi_1(v) + v_y \chi_2(v) \tag{I.161}$$

$$v = \sqrt{v_x^2 + v_y^2 + v_z^2}$$

and the function f_n^0 corresponding to the field-free state is expressed by means of the Fermi-Dirac distribution function

$$f(\varepsilon) = \{\exp[(\varepsilon - \varepsilon')/kT] + 1\}^{-1} \tag{I.162}$$

(see Ref. [33]) [it can be shown that $f_n^0(v_x, v_y, v_z) = \dfrac{2\, m_e^3}{h^3} f(\varepsilon)$]. By applying this approximation, which corresponds to the free electron model, to the Hall constant, after a somewhat lengthy calculation (the details of which cannot be discussed here) one obtains [31]

$$\varrho = \frac{el(\varepsilon_0')}{m_e\, v(\varepsilon_0')\, c\sigma(H_z)} \tag{I.163}$$

In this equation $\sigma(H_z)$ is the specific electrical conductivity in the case of a magnetic field of strength H_z, which can be proved [31] to be given by

$$\sigma(H_z) = \sigma(0)\left[1 - \frac{BH_z^2}{1 + CH_z^2}\right] \tag{I.164}$$

where

$$B = \frac{\pi^3}{3}\,\frac{ekTl(\varepsilon_0')}{m_e^2 v^3(\varepsilon_0')} \quad \text{and} \quad C = \sigma(0)^2\varrho^2$$

[for the definition of $l(\varepsilon_0')$ see the text in the preceding Section following Eq. (I.145), as well as Eq. (I.43) in the same section]. In the case of a weak magnetic field, $\sigma(H_z) \approx \sigma(0)$. On substituting expression (I.145) for the latter quantity into Eq. (I.163) the extremely simple result

$$\varrho = \frac{1}{n_0 ec} \quad \left(n_0 = \frac{N}{V}\right) \tag{I.165}$$

can be obtained.

The values of the Hall constant calculated for several monovalent metals on the basis of Eqs (I.163) and (I.164) agree, as regards order of magnitude, with the corresponding experimental values, whereas in the case of divalent and transition metals not even the order of magnitude is correct. By means of calculations based on the band model, however, (which cannot be given here due to their complexity) a better agreement has been obtained between the calculated and experimental values for these cases too.

If Eq. (I.145) for the specific conductivity is multiplied by expression (I.165) for the Hall constant, and Eq. (I.146) of the preceding Section is taken into account, we obtain

$$\sigma\varrho = \frac{\sigma}{n_0 ec} = \frac{\mu}{c} \tag{I.166}$$

where μ is the mobility of the charge carriers. Consequently, by measuring simultaneously the conductivity and the Hall-effect, the mobility in metals can also be determined.

From similar considerations the values of the Hall constant can be calculated for semiconductors, too. By assuming that there are n_b bound states of energy E_v in unit volume of a semiconductor, and, separated from these by a forbidden band of width E_g, there are free electron states completely vacant at $T = 0$ K, appropriate distribution functions can also be derived for the case of semiconductors [see Eqs (I.111) and (I.112) in Chapter I.3.]. Using this latter function for the solution of the Boltzmann equation of state, we obtain after a fairly lengthy calculation for the Hall constant:

$$\varrho = \frac{e\pi l_0}{\sqrt{4\pi c m_e kT}} \frac{1}{\sigma(H_z)} \tag{I.167}$$

or, making the assumption $\sigma(H_z) = \sigma(0)$:

$$\varrho = \frac{3\pi}{8} \frac{1}{n_f ec} \tag{I.168}$$

l_0 in Eq. (I.167) is the mean free path, assumed to be independent of the energy, while n_f in Eq. (I.168) is the number of free electrons in unit volume, which can be proved [32] to be

$$n_f = n_b^{1/2} \left[\frac{\pi^{1/2}}{2} (kT)^{3/2} \tilde{C}\right]^{1/2} e^{-E_g/2kT} \tag{I.169}$$

where

$$\tilde{C} = \frac{4\pi(2m_e^*)^{3/2}}{h^3}$$

If the calculation is based on Eq. (I.168), agreement as to the order of magnitude can be obtained between the measured and calculated values of ϱ. Also, in the case of semiconductors more accurate values were obtained for the Hall constant by means of the band model than by the free electron model. Discussion of these calculations, however, is beyond the scope of this Section.

SECTION 4.3. A BRIEF REVIEW OF THE THEORY OF THE MAGNETIC BEHAVIOUR OF SOLIDS

If a magnetic field of strength H* acts on a solid, a magnetic moment of

$$J = \chi H \tag{I.170}$$

is induced in unit volume of the solid. If the susceptibility, χ, defined by Eq. (I.170) is negative, the material is diamagnetic, and if positive, the solid is paramagnetic or ferromagnetic.

It is known from classical electrodynamics that the movement of a charge travelling in a closed orbit is modified by the magnetic field in such a way that the field related to the magnetic moment of the moving charge opposes the external field. According to the laws of quantum mechanics this orientation of the magnetic moment ensures the most stable state for the system. The orbital motion of electrons therefore provides a diamagnetic contribution to the susceptibility in all materials. Consequently, if the magnetic moments arising from the electron spins happen to compensate each other, the system in question is diamagnetic. If the system has a resultant spin, however, the corresponding magnetic moment points along the field, and the system is paramagnetic. Finally, in certain cases (mainly Fe, Co, and Ni, having incomplete d-subshells) the magnetic moments of the individual atoms, even in the absence of an external magnetic field are aligned in parallel in smaller or larger parts of the crystal (the Weiss zones), and the large resultant magnetic moments arising in this way tend to follow the direction of the field to an extent depending on the strength of the external magnetic field. This phenomenon is called ferromagnetism.

For the theoretical calculation of the magnetic susceptibility of any system it must be known how this depends on the energy of the system. In order to determine this relationship let us assume that the strength of the external magnetic field, H, changes by a value ΔH**. In this case the change of the energy in unit volume of the system can be given in the form

$$\frac{\Delta E}{V} = -J\Delta H = -\chi H \Delta H \tag{I.171}$$

* For the sake of simplicity the z-axis of the coordinate system points in the direction of the field, and only the absolute value of $H_z = H$ need be taken into account.
** ΔH denotes the z-component of the change ΔH of the magnetic field strength, i.e. ΔH_z.

where V is the volume of the system. On replacing the difference quotient in Eq. (I.171) by the corresponding partial derivative:

$$\chi = -\frac{1}{VH}\frac{\partial E}{\partial H} \tag{I.172}$$

If the temperature of the system is 0 K, $E = E_0$, where E_0 is the energy of the ground state of the system. This leads to the simple relation

$$\chi = -\frac{1}{VH}\frac{\partial E_0}{\partial H} \tag{I.173}$$

If $T \neq 0$, the average value of $\dfrac{\partial E}{\partial H}$ will be

$$\overline{\frac{\partial E}{\partial H}} = \frac{\sum_i \frac{\partial E_i}{\partial H} \exp(-E_i/kT)}{\sum_i \exp(-E_i/kT)} = -kT\frac{\partial \ln \Omega}{\partial H} \tag{I.174}$$

where the partition function, Ω, is defined as

$$\Omega = \sum_i \exp(-E_i/kT) \tag{I.175}$$

and the different values of E_i are the possible energy states of the system. On substituting the right hand side of Eq. (I.174) into Eq. (I.172) one obtains

$$\chi = \frac{kT}{VH}\frac{\partial \ln \Omega}{\partial H} \tag{I.176}$$

As an application, let us first consider a system consisting of N atoms between which no interactions occur, and where the square of the resultant angular moment of the atoms is $\hbar^2 I(I+1)$ ($I = 0, 1/2, 1, 3/2, \ldots$). If the square of the resultant spin moment of the atoms is $\hbar^2 S(S+1)$ ($S = 0, 1/2, 1, 3/2, \ldots$), and that of the resultant orbital angular moment is $\hbar^2 L(L+1)$ ($L = 0, 1, 2, \ldots$), it can be shown by quantum mechanical considerations [33] that the part of the energy of the system depending on the magnetic field will be

$$E_{\text{magn}} = H_z \mu_B\, g(I, S, L) I_z + \frac{H_z^2 e^2}{12 m_e c^2} n \overline{r^2} \tag{I.177}$$

where H_z is the strength of the magnetic field (again, the z-axis is parallel to the field), $\mu_B = \dfrac{e\hbar}{2 m_e c}$ is the Bohr magneton, g is the Landé factor defined as

$$g(I, S, L) = 1 + \frac{I(I+1) + S(S+1) - L(L+1)}{2I(I+1)} \tag{I.178}$$

I_z is the quantum number for the z-component of the total angular moment ($I_z = 0, 1/2, 1, 3/2, \ldots$), n is the total number of electrons in the atom, and $\overline{r^2}$ is the average, over all the electrons, of the squared orbital radii of the electrons (more precisely, that of the quantum mechanical expectation value of r^2).

If $I \neq 0$, the second term of Eq. (I.177) can be shown to be negligible in comparison with the first term. In the case of atoms having an odd number of electrons or open inner shells, therefore, it is a good approximation to consider only the first term of Eq. (I.177). If it is assumed, for the sake of simplicity, that kT is much smaller than the splitting of the $2I + 1$ states for a particular value of I due to the magnetic field, then in the calculation of Ω it is sufficient to take into account only these $2I + 1$ states. Hence:

$$\Omega = \left[\sum_{I_z=-I}^{+I} \exp\left[- H_z\, g(I, S, L)\, I_z/kT\right]\right]^N \tag{I.179}$$

By performing the summation of Eq. (I.179) and substituting the result into Eq. (I.174), after a calculation which cannot be given here in detail:

$$\chi = \frac{N}{V} \frac{\mu_B^2 g^2 I(I + 1)}{3kT} \tag{I.180}$$

This expression was derived by Langevin [34] at the beginning of this century on the basis of classical mechanics, with the difference that in place of $\mu_B^2 g^2 I(I + 1)$ the square of the permanent magnetic moment of the atom was used.

In the case of atoms having closed shells $I = I_z = 0$; thus, the first term of Eq. (I.177) makes no contribution, but the second term still does not vanish. The susceptibility corresponding to this term can be derived as

$$\chi = -\frac{Nn}{V} \frac{e^2}{6m_e c^2} \overline{r^2} \tag{I.181}$$

Hence, quantum mechanical calculations, too, lead to the result that atoms of closed shells are diamagnetic. Eq. (I.181) can be applied advantageously for the calculation of the susceptibilities of ionic crystals consisting of ions with closed electron shells. In this case it should be taken into account that there are several types of ions in the crystal. With this extension Eq. (I.181) becomes

$$\chi = -\frac{e^2}{6m_e c^2} \sum_\alpha \frac{N_\alpha n_\alpha}{V} \overline{r_\alpha^2} \tag{I.182}$$

where n_α and $\overline{r_\alpha^2}$ are the number of electrons and mean square orbital radii of the αth kind of ion, respectively, and N_α is their number in the crystal. Finally, it should be mentioned that Eq. (I.182) has also proved to be very useful for the calculation of the diamagnetic contribution arising from the closed electron shells of the building units of any solid.

It is a harder task to calculate the magnetic susceptibility arising from the more or less delocalized valence electrons of solids. If the valence electrons are regarded as free electrons and the Schrödinger equation of the form valid in the case of a magnetic field is solved, the expression

$$E_n = \frac{h^2 k_z^2}{2m_e} + H_z \, \mu_B (2n + 1) \tag{I.183}$$

can be derived for the energy levels. By means of this expression and the Fermi-Dirac quantum statistics the partition function can be constructed which, when substituted into Eq. (I.174) gives

$$\chi = -\frac{4 m_e \, \mu_B^2}{3h^2} \left(\frac{3N}{\pi V} \right)^{1/3} \tag{I.184}$$

for the susceptibility. (For the details of the calculation see Ref. [35].)

The Landau derivation qualitatively discussed above has been extended by Peierls [36] to the case of quasi-bound electrons. The extension is based on the assumption that the interaction between the atoms forming the solid is weak, and can thus be treated as a perturbation. According to the results of the calculation the diamagnetic susceptibility arising from the valence electrons can be given as the sum of three terms:

$$\chi = \chi_1 + \chi_2 + \chi_3 \tag{I.185}$$

where χ_1 is given by Eq. (I.181),

$$\chi_2 \approx \frac{m_e}{m_e^*} \chi_1 \tag{I.186}$$

(m_e^* is the effective mass for that part of the energy band occupied by electrons), and

$$\chi_3 = -\frac{\alpha^2}{6\pi^2} \int_F \left(\frac{\partial^2 \varepsilon}{\partial k_x^2} \frac{\partial^2 \varepsilon}{\partial k_y^2} - \frac{\partial^2 \varepsilon}{\partial k_x \, \partial k_y} \right) \left| \frac{1}{\mathrm{grad}_k \, \varepsilon} \right| \mathrm{d}F \tag{I.187}$$

The integration should be extended to the surface of the filled part of the wavenumber space, and

$$\alpha = \frac{\varepsilon_0'}{kT} \tag{I.188}$$

where ε_0' is the value of the function $\varepsilon(\mathbf{k})$ at the uppermost occupied level. By assuming the simple relationship

$$\varepsilon(\mathbf{k}) = \frac{h^2}{2m_e} [\alpha_1 (k_x^2 + k_y^2) + \alpha_3 k_z^2] \tag{I.189}$$

one obtains

$$\chi_3 = -0.122 \sqrt{\varepsilon_0'} \sqrt{\frac{\alpha_1^2}{\alpha_3}} \, 10^{-6} \text{ CGS units} \tag{I.190}$$

where $\alpha_1 = m_e/m_e^*$ along the x- and y-directions, and $\alpha_3 = m_e/m_e^*$ along the z-direction, provided that the latter coincides with the main crystallographic axis. The application of Eq. (I.190) led to good agreement with the experimental values for bismuth if $\alpha_1 \approx 40$ and $\alpha_3 \approx 1$ (for further details see Ref. [37]).

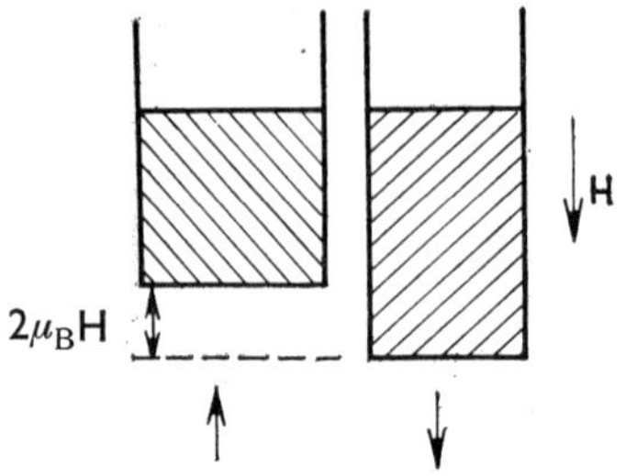

Fig. I.20. The shifts of the electron levels of different spins in a magnetic field

The simplest form of the theory of paramagnetism can be attributed to Pauli [38]. He assumed that under the effect of an external magnetic field of strength H the energy levels of electrons with magnetic moments (arising from spin) parallel to the field are shifted downwards by $\mu_B H$ due to the interaction with the field (see Fig. I.20), whereas the energy levels of electrons of opposite spin are shifted upwards by the same value. There will, therefore, be more electrons in the system with magnetic moments pointing along the field, than electrons with the opposite moment. The number of electrons leaving the band with a moment opposite to the field, Δn, will just equal to the number of states situated at the upper end of the band in a range of width $\mu_B H$, i.e.:

$$\Delta n = \mu_B H g_s(\varepsilon') \tag{I.191}$$

where $g_s(\varepsilon')$ is the level density at the upper end of the filled energy band. The difference between the number of electrons occupying the two different bands is therefore

$$2\Delta n = 2\mu_B H g_s(\varepsilon') \tag{I.192}$$

from which the magnetic moment in unit volume is

$$J = \frac{2\mu_B^2 H g_s(\varepsilon')}{V} \tag{I.193}$$

and the corresponding susceptibility

$$\chi = \frac{2\mu_B^2 g_s(\varepsilon')}{V} \tag{I.194}$$

For free electrons:

$$g_s(\varepsilon') = \frac{g(\varepsilon')}{2} \approx \frac{3}{4} N \frac{1}{\varepsilon_0'} \tag{I.195}$$

from which the paramagnetic susceptibility is

$$\chi = +\frac{3}{2} n_0 \mu_B^2 \frac{1}{\varepsilon_0'} \quad \text{with} \quad n_0 = \frac{N}{V} \tag{I.196}$$

The susceptibility values calculated from Eq. (I.196) agree, in order of magnitude, with the experimental values found for alkali metals and alkaline earth metals, though their numerical values are generally higher than the experimental data. This indicates that diamagnetic corrections must not be neglected.

In order to construct a more exact model it is necessary to consider that,

(1) the function $g_s(\varepsilon')$ is not identical to the function valid for free electrons,
(2) in the calculation of the total energy of solids the electron exchange and correlation energy terms, which influence susceptibility, have to be taken into account by quantum chemical methods, and
(3) the diamagnetic contribution to the susceptibility is taken into consideration.

With these modifications a better agreement is achieved between the calculated and experimental values of susceptibility. (For details see Ref. [39].) In a way that cannot be discussed here in detail the theory is also able to reproduce the temperature-dependence of the paramagnetic susceptibility, which is of the form

$$\chi = \frac{C}{T + \Delta} \tag{I.197}$$

where C is the Curie constant, and Δ the empirical Weiss constant.

The simplest quantum mechanical theory of ferromagnetism was developed as early as 1928 by Heisenberg [40]. Analogous to the work of Heitler and London concerning the H_2 molecule he found that the spins of the electrons of a system containing two atoms having one electron each will be parallel (triplet state) if the exchange integral

$$J_{AB} = \int \psi_A^*(1)\, \psi_B^*(2) V_{AB}\, \psi_A(2)\psi_B(1) \mathrm{d}V_1 \mathrm{d}V_2 \tag{I.198}$$

in the total energy expression

$$E = E_c - J_{AB} \tag{I.199}$$

for the triplet state of the system is positive. In the exchange integral

$$V_{AB} = \frac{e^2}{R_{AB}} + \frac{e^2}{r_{12}} - \frac{e^2}{r_{A1}} - \frac{e^2}{r_{B2}} \tag{I.200}$$

is the interaction potential between the two atoms (for the notation of distances see Fig. I.21). In Eq. (I.198) ψ_A and ψ_B are the wave functions of the electrons of atoms A and B, respectively.

The term E_c in Eq. (I.199) is given by

$$E_c = 2\varepsilon + C_{AB} \tag{I.201}$$

with

$$C_{AB} = \int |\psi_A(1)|^2\, V_{AB}\, |\psi_B(2)|^2\, \mathrm{d}V_1 \mathrm{d}V_2$$

where ε is the energy of the free atom.

According to the work of Bethe [41] the value of the integral J_{AB} is generally positive if R_{AB} is large compared to the effective radius of the

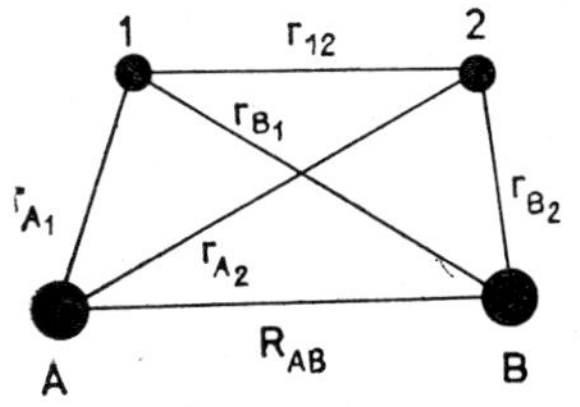

Fig. I.21. The notation of distances in a two atom — two electron system

atom, and the actual values of the wave functions are small in the neighbourhood of the nuclei. It can be shown that these conditions are best met by the interactions of d- and f-electrons occupying open shells. This is the case with the members of the iron group and the rare earth metals. Expression (I.199) applicable for the case of two atoms can be generalized for the case of polyatomic systems, within the framework of the Heitler-London approximation, and in this case:

$$E' = E'_{\mathrm{c}} - \sum_{i<j} J_{ij} \tag{I.202}$$

where

$$E'_c = \sum_{i=1} \varepsilon_i + \sum_{i<j=1} C_{ij}$$

and the summations are performed over all $i \neq j$ electron pairs. The approximation is quite valid if only the nearest-neighbour interactions are taken into account. In the case of crystals consisting of identical atoms only one kind of exchange integral appears in Eq. (I.202) (the exchange integral between the immediate neighbours), and it is the sign of this integral that determines whether the crystal is ferromagnetic or not.

If J is positive, the individual electron spins, even in the absence of an external field, are in a parallel orientation in smaller or larger parts of the crystal (the Weiss zones), causing high resultant magnetic moments. Due to the effect of an external field these resultant moments tend to follow the direction of the field to an extent depending on the strength of the magnetic field. Rise of temperature disturbs this orientation. Hence, it is understandable that with increasing temperature the resultant magnetic moments of ferromagnetic materials decrease, and at a critical temperature (the Curie point) the ferromagnetism disappears, and the crystal exhibits normal paramagnetic behaviour. All these facts can be derived from Heisenberg's theory.

Since the first quantum mechanical theory of ferromagnetic behaviour developed by Heisenberg, more accurate models have been constructed by Bloch, Slater and other research workers, and in most cases these are able to reproduce the experimental values quantitatively. These rather sophisticated models, however, cannot be discussed here (see e.g. Ref. [42]).

Solids exhibiting para- and ferromagnetic behaviour may, in general, give rise to electron spin resonance spectra. The theory of these spectra follows in Chapter VI.7.

CHAPTER 5. QUANTUM MECHANICAL STUDY OF THE SURFACE ELECTRON STATES OF SOLIDS

J. Ladik

SECTION 5.1. TAMM-TYPE AND SHOCKLEY-TYPE SURFACE STATES

As a first step in the theoretical treatment of the complicated processes which play a part in catalysis, the surface electron states of the solid on which the catalytic reaction takes place must be studied by quantum mechanical methods. Methods for the calculation of the electronic structure of the complete system containing catalyst and chemisorbed reaction partners, i.e. the more exact electron theory of catalysis, may be developed only after these studies.

There have already been attempts at the quantum mechanical investigation of the surface electron states of solids in the simplest cases. This Section presents a brief account of these calculations.

The possibility of the formation of surface states was first pointed out by Tamm [43], who used a one-dimensional model for the calculations. The potential function used is shown in Fig. I.22. The diagram, in fact, shows the Kronig-Penney periodic potential (see Fig. I.5) for positive values of x [19, 44]. (In the diagram a is the lattice constant, and b and E_0 are the width and height of the potential barrier.) In Tamm's calculations the potential function within the crystal was approximated by this function, and for the crystal surface at $x = 0$ a constant repulsion potential, W, was assumed. The possible energy states of the electron moving in the field of this potential were calculated by solving the Schrödinger equation. It turned out from the calculations that whereas in the case of an infinite crystal the possible energy states form bands (see Section I.2.2.), in the case of the semi-infinite crystal, represented by Fig. I.22, if the energy of the electron E is lower than W, a new possible energy level, the surface level,

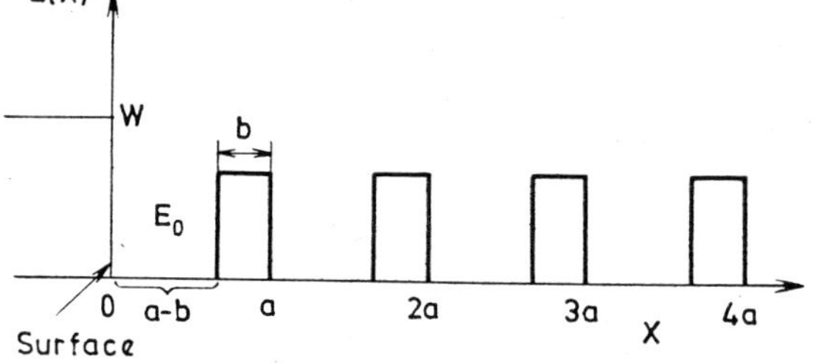

Fig. I.22. The potential function used by Tamm [44]

appears between every two neighbouring bands (see Fig. I.23). As far as the wave functions for these surface states are concerned, it has been found that their values decrease exponentially on both sides of the surface with increasing distance from the surface; in other words the electrons occupying the surface levels are practically localized to the surface.

The solution of the Schrödinger equation with the potential function given in Fig. I.22 shows furthermore that the value of E may be arbitrary

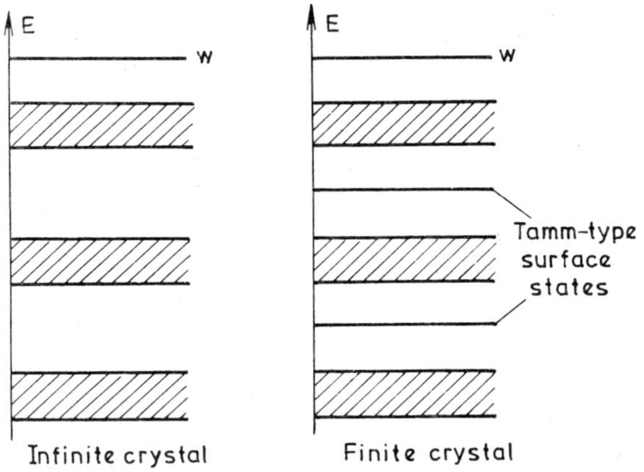

Fig. I.23. Tamm-type surface states

if $E > W$, and the wave functions for these states are periodic outside the lattice, whereas they decrease exponentially with the distance from the surface inside the lattice. These wave functions give the probability of the occurrence of electrons striking the surface and reflected from it. It was later shown by Fowler [45] that in the case of a finite one-dimensional crystal (the potential function given in Fig. I.22 pertains to a crystal infinite in *one* direction) two surface states of the same energy (doubly degenerate states) occur in each of the forbidden bands. One of the states of a degenerate pair is localized on one end on the chain, the other on the other end.

According to the more elaborate investigations of Shockley [46, 47], also carried out on a one-dimensional model, there are two cases where surface states may occur in finite crystals. Either the form of the potential is different on the surface from that inside the crystal (this is the situation with Tamm-type surface states), or the volume bands (belonging to the infinite crystal) arising from the different atomic levels overlap (Shockley-type surface states). This latter case, of course, may occur only if the atoms forming the crystal are close enough to each other, and so the bands will be wide (see Fig. I.24).* Furthermore, according to the results of Shockley

* Consequently, the formation of Shockley-type surface states is a function of the geometry of the lattice. Thus, surface states of this type generally occur in metals but not in insulators.

the surface states due to the overlapping bands are again doubly degenerate, one of the states arising from the perturbation of the lower volume band and the other from that of the upper band.

The above results for the one-dimensional case can easily be generalized to the three-dimensional case [48]. According to the results the number of

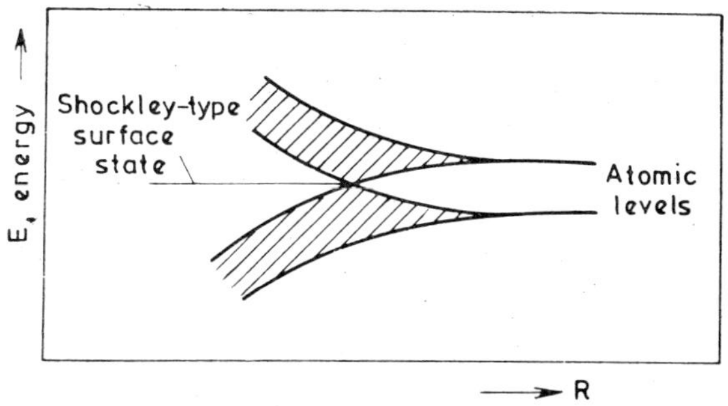

Fig. I.24. The formation of Shockley-type surface states [47]

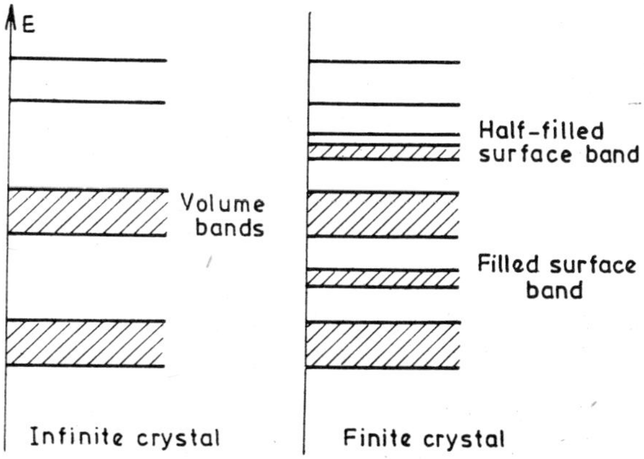

Fig. I.25. Surface energy bands

surface states in each of the forbidden bands is equal to the number of atoms on the surface of the crystal. If the surface area is large enough, these levels lie very close to each other, and also merge into bands (surface bands)*. The energy levels of the surface band lying between the highest filled volume band (containing two electrons with antiparallel spins in each of its levels) and the lowest empty volume band, will generally be half-occupied since the number of electrons entering the surface band from

* Since the number of surface atoms in a real crystal is about one million times lower than the total number of atoms forming the crystal, the formation of surface bands exerts practically no influence on the shapes and level density distributions of the volume bands.

the volume bands should correspond to the number of levels arising from the filled volume bands (see Fig. I.25). In the ideal case this would lead to surface conduction. However, since the surface of even an ideal crystal is never completely smooth, and always contains adsorbed impurities, this surface conduction does not occur in practice.

In the past 15 years a number of research workers have studied the problem of surface states. The periodic potential functions used in these investigations approximate the real situation better than the Kronig-Penney potential. Sine the Schrödinger equation can generally not be solved exactly, in the case of these more complicated potential functions approximations must be used. The most frequently applied approximations are the following: the simple Hückel method (see Section I.2.3), the SCF (Self-Consistent Field), LCAO (Linear Combination of Atomic Orbitals), MO (Molecular Orbital) method, the free electron model (cf. Section I.2.1), and the OPW (Orthogonalized Plane Waves, cf. Section I.2.4) method.

SECTION 5.2. MODEL CALCULATIONS ON SURFACE STATES

Since the relatively very simple Hückel-type LCAO MO method gives fairly good results whenever the same properties of molecules of similar type are compared, this method has been rather extensively applied by several authors, for instance by Goodwin [49], Artmann [50], Grimley [51], Koutecký [52], Davison et al. [53] and Levine et al. [54] to study the surface states of different models and simple solids. In the case of an infinite chain consisting of identical atoms, only two parameters, α_i^0 (the Coulomb integral for the ith atom), and β_{ij}^0 (the resonance integral between neighbouring i and j atoms, see Section I.2.3), play a role in this approximation provided that each atom contributes only a single state to the delocalized crystal orbital.

As far as the formation of Tamm-type surface states is concerned, three basic cases can be distinguished [52]:

(1) it is assumed that the Coulomb integral of the terminal atom, and the resonance integral between the terminal atom and the next one are the same as the corresponding integrals in the infinite chain (in other words, there is no potential change at the surface);

(2) it is assumed that the integral α_i of the surface atom differs from the corresponding integral (α_i^0) of the atoms inside the chain; and

(3) the integral β_{ij} between the terminal atom and its nearest neighbour differs from the value β_{ij}^0 assigned to atom pairs inside the chain. Accordingly in cases (2) and (3) it is assumed that the potential function at the surface is not the same as inside the chain.

According to the results of more elaborate investigations, the surface state never occurs in case (1), whereas for the other two cases criteria can be derived concerning the parameters α_i and β_{ij} that determine whether, or not, Tamm-type surface states occur [52]. It follows from the above considerations that states localized to the surface atoms may occur if the electron affinity of these atoms is significantly different from that of the bulk atoms, or the strengths of the bonds between the surface atoms and their nearest neighbours are significantly different from those of the corresponding bonds inside the crystal.* This bond strength difference may be due to the variation of the interatomic distance, or to the fact that the

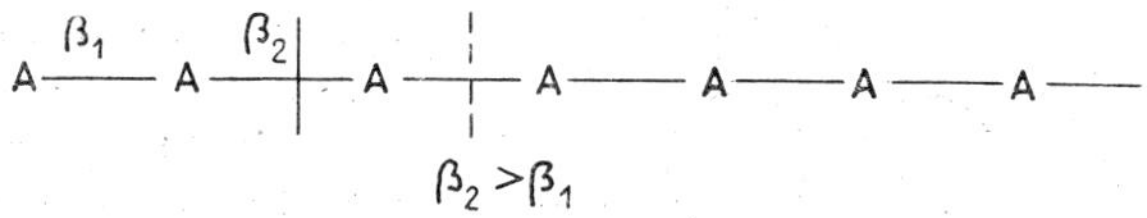

Fig. I.26. A chain containing identical atoms bound by alternate strong and weak bonds [49]

chemical structure of the surface atoms is different from that of the bulk atoms.

In connection with the Tamm-type surface states formed under the above conditions, it should be added that according to the results of detailed calculations the change of the potential function at the surface does not necessarily cause the formation of surface charges [55].

The conditions of the formation of Shockley-type surface states were investigated by Koutecký [56], again by means of the Hückel-type LCAO MO method. It was shown that in the case of a chain containing identical atoms bound alternately by strong and weak bonds (see Fig. I.26) surface states occur only if the surface cuts the strong bond, no surface state being formed if the weak bond is intersected by the surface. On the basis of this result the definition of the unit cell of the crystal may be extended by the condition that its boundary surfaces should, where possible, intersect only weak bonds. In other words, surface states can occur only in the case when the boundary surfaces of the crystal intersect unit cells, taken in the above sense.

Furthermore, it can be shown [52] that a chain containing identical atoms bound alternately by strong and weak bonds is equivalent, as regards surface states, to a cubic system containing identical atoms bound by bonds of uniform strength, in which one 2s and one 2p-function belong to all the atoms (Goodwin-Artmann model [49, 50], see Fig. I.27). According to the results of Hückel-type calculations on this model, surface states occur if the volume energy bands arising from the 2s and 2p-atomic states overlap each

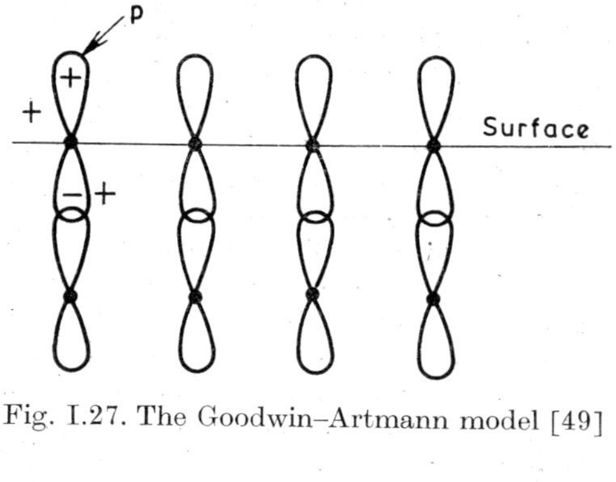

Fig. I.27. The Goodwin–Artmann model [49]

* In the latter case two surface states of different energy are formed in the case of a "semi-infinite" (finite on one side) chain, one corresponds to the bonding state and the other to the antibonding state of a diatomic molecule.

other. These states, as well as the surface states occurring in the case of a chain containing alternately strong and weak bonds, are therefore Shockley-type surface states.

A further interesting result of the calculations carried out on the Goodwin-Artmann model is that if the lattice atoms are so close to each other that the s and p-bands overlap and hence surface states occur, the hybridization of the atomic s and p-states also takes place. It can be seen from the signs of the s and p-functions shown in Fig. I.27 that the electron density arising from one of the two atomic wave functions formed by hybridization is higher in the part of the individual unit cells closer to the surface than in their other part. In other words, one of the hybridized wave functions of the surface atoms creates free valences directed from the surface towards the external space (see Fig. I.28).

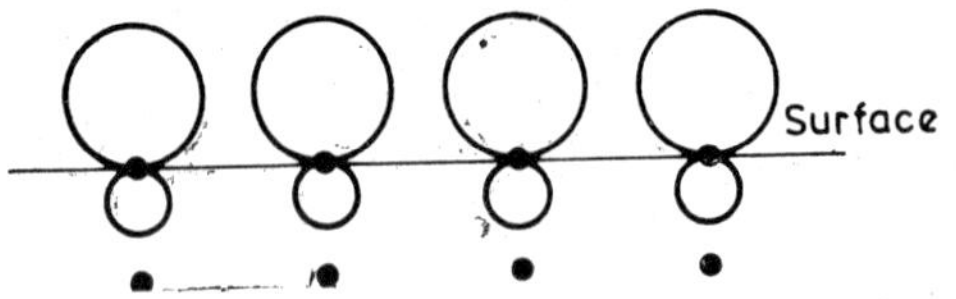

Fig. I.28. sp hybrid wave functions of surface atoms, oriented from the surface toward the vacuum

If all the surface atoms contribute two electrons for the occupation of the two hybrid states arising from the atomic s and p-wave functions, all the states will only be singly occupied, and the wave functions shown in Fig. I.28 will create free valences. This radical nature of the surface is obviously extremely important as regards the catalytic behaviour of the surface.

SECTION 5.3. THE INVESTIGATION OF THE SURFACE STATES OF DIAMOND, ZINC BLENDE, GRAPHITE, NAPHTHALENE AND ANTHRACENE

Shockley-type surface states play an important role in the surface properties of diamond and graphite. Very interesting investigations were carried out on these systems by Koutecký and Tomasek, and by Koutecký, using the simple LCAO MO method [57, 58]. The surface states of a three-dimensional diamond lattice were calculated for two different cases. In one of these it was assumed that the lattice is bordered by a (111) plane (this plane intersects only one of the bonds formed by the sp^3 hybrids of the carbon atoms). The results indicate that in the states belonging to the surface band formed between the completely filled valence band and the completely empty conduction band (as there are 2 carbon atoms in the unit cells of diamond, 2 volume energy bands can be obtained from the calculation), the electrons are practically localized to those sp^3 hybrid states of the surface atoms which point from the surface toward the vacuum. Since each of the surface atoms contributes only one electron to the surface state, the surface band is only half occupied. Consequently, the diamond surface bordered by a (111) plane behaves as a polyradical having non-bonded valences (electrons of unpaired spins).

A different result was obtained with the assumption that the diamond lattice is bordered by a (100) plane. (This plane intersects two C—C bonds in the diamond lattice.) Shockley-type surface states are formed again, and the electrons occupying these states are again practically localized in hybrid states pointing from the surface atoms toward the external space.

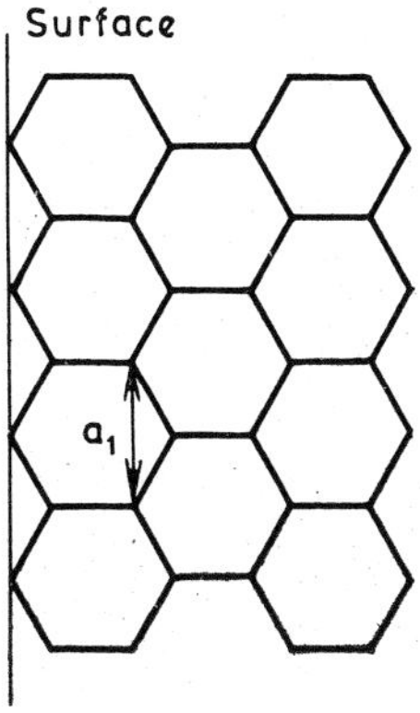

Fig. I.29. The boundary surface parallel to the elementary translation vector, $\mathbf{a}_1$, in the two-dimensional graphite model [54]

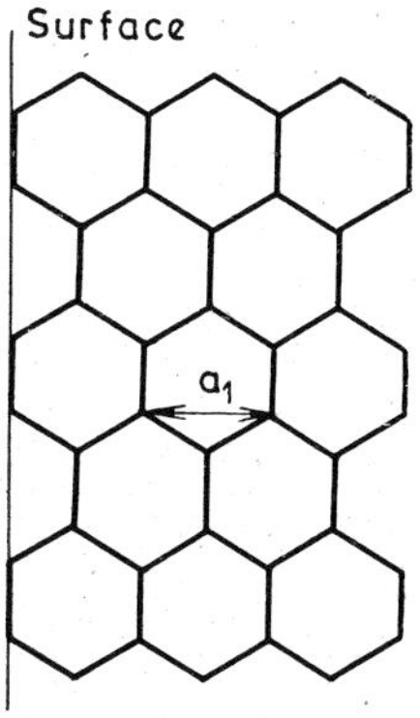

Fig. I.30. The boundary surface perpendicular to the elementary translation vector, $\mathbf{a}_1$, in the two-dimensional graphite model [54]

However, in this case two bonds of each surface atom are intersected by the boundary surface, and so these outward-oriented states are occupied by a pair of electrons with antiparallel spins (at least in the ground state) and not by an unpaired electron. This boundary surface of the diamond lattice in the ground state may thus form only donor-acceptor type bonds with the molecules of the gas phase.

The general form of the resolvent method was developed by Freeman [59] in the framework of the Hückel-type LCAO MO approximation for semi-infinite crystals containing an arbitrary number of atoms. This theory was applied to the (100) plane of zinc blende by Levine and Freeman [60]. According to the results an acceptor-like surface band exists below the conduction band by a few tenths of eV.

The two-dimensional model of the graphite lattice was investigated by the simple LCAO MO method [57]. It was found that if the boundary surface is assumed to be parallel with the elementary translation vector $\mathbf{a}_1$ (Fig. I.29), Shockley-type surface states occur, resulting in a polyradical structure oriented outwards, whereas if the boundary surface is assumed to be perpendicular to the vector $\mathbf{a}_1$ (Fig. I.30), no surface state is formed at all.

The surface states of anthracene and naphthalene were recently calculated by Stern and Green [61] using a slightly modified version of the simple LCAO MO approximation. In naphthalene only one surface band was found for the electrons and one for the holes, and the bands were located in different crystallographic planes. No separated surface bands, only isolated surface states, were found for anthracene. These results are very important in the interpretation of the electron and hole injection effects found experimentally (see, for instance, Ref. [62] and the references therein).

It should be noted that some attempts have recently been made [63, 64] to treat the localized states relativistically. It is easy to understand from the rather complicated structure of the Dirac equation that these attempts must still be on the level of oversimplified model calculations.

The results briefly reviewed here are, of course, only the first steps toward an understanding of the surface electron states of solids. Development is expected in many respects. It would be worthwhile to extend the investigations from the model systems and covalent atomic lattices to metals and semiconductors. This extension, however, requires the further improvement of the methods of calculation and the elaboration of more accurate approximations. Since the interaction between the electrons on the surface and those in the bulk of the crystals is strong, it is very probable that acceptable results can be obtained only by the self-consistent field (SCF) procedure. The SCF LCAO equations thus obtained appear to be solved most conveniently by matrix formalism. The application of this technique also ensures that an arbitrary number of orbitals in the unit cell can be taken into account. It should be mentioned that in the application of matrix methods for the problem of localized states for polymers some progress was made by Biczó, Kertész and Suhai [65]. In addition, investigations are expected on simple models where the deformation of the potential on the surface and the overlap of the volume energy bands are simultaneously taken into account in the calculation (mixing of Tamm- and Shockley-type surface states).

Only after a suitable further development in the theory can the quantum mechanical problem of the solid — adsorbed (chemisorbed) molecule system be expected to be approximately solved, thereby opening the way towards the quantum mechanical theory of heterogeneous catalysis.

An excellent picture of recent development of surface state theory can be obtained from the papers of Davison and Levine [66], Koutecký [67], Koutecký and Davison [68], and Levine and Mark [69].

CHAPTER 6. THE DEFECT STRUCTURES AND ELECTRICAL BEHAVIOUR OF SEMICONDUCTOR OXIDES

F. Solymosi

SECTION 6.1. DEFECT TYPES IN SOLIDS

The discovery of the defect structures of solids can be attributed primarily to Frenkel, Schottky and Wagner. On the basis of their investigations it is now generally known that solids having ideally formed crystal lattices, in which all the lattice points are occupied by the appropriate crystal constituents, are extremely rare. In many instances the laws of constant and multiple proportions can be regarded only as limiting cases. The majority of inorganic substances are inclined to exhibit some disorder (vacancies, or atoms or ions inserted between the lattice points) even when satisfying the requirements of stoichiometry; alternatively, not even the stoichiometry is retained, i.e. one or another of the components of the solid is in excess. These imperfections are generally referred to as *point defects*. Another extensive group of imperfections is formed by *dislocations* or *line defects*. In the following discussion the main characteristics of dislocations are first given.

The theory of dislocations regards the crystal initially as a regular unit from the viewpoint of geometrical structure, and deals with the deviations from this ideal arrangement.

The concept of dislocations was first introduced into chemistry in 1949, when Burton, Cabrera and Frank assumed that dislocations play a role in crystal growth. Optical and electron microscopic studies soon confirmed the correctness of this assumption, and with the improvement of experimental techniques the research field dealing with the properties and roles of dislocations has rapidly developed. Although dislocations have only an indirect role in the electrical behaviour of solids, numerous studies have indicated that these defects, as unbalanced points of the lattice, affect not only the growth and formation of crystals, but also the reactivity and at the same time the catalytic behaviour of solids.

There are two types of line defects, the *edge dislocation* and the *screw dislocation*. The first type was defined by Taylor, the second by Burgers. The geometries of the above dislocations are shown in Fig. I.31. The crystal denoted by a) in the figure contains positive and negative edge dislocations (extra half planes). Crystal b) exhibits screw dislocation, and in c) can be seen the atomic arrangement characteristic for the screw dislocation. Full circles denote atoms in the plane of the paper, and empty circles those in the adjacent plane below. Dislocations can be characterized by their Burgers vector, **b**. A line dislocation can be regarded as the boundary between the relatively displaced two parts of the crystal; the direction and magnitude of the displacement are determined by **b**. The dislocation line is of the screw-type if **b** is parallel to the line, and of the edge-type if **b** is perpendicular to it.

In the case of any other direction of **b** the dislocation is of a mixed-type. It has been shown by Frank [70] that if the value of **b** is sufficiently large (10 Å), an open crater is formed around the screw dislocation on the surface of the crystal extending to a certain depth. This arrangement promotes diffusion into the volume of the crystal. If **b** is a lattice vector, the dislocation is called perfect, and otherwise imperfect.

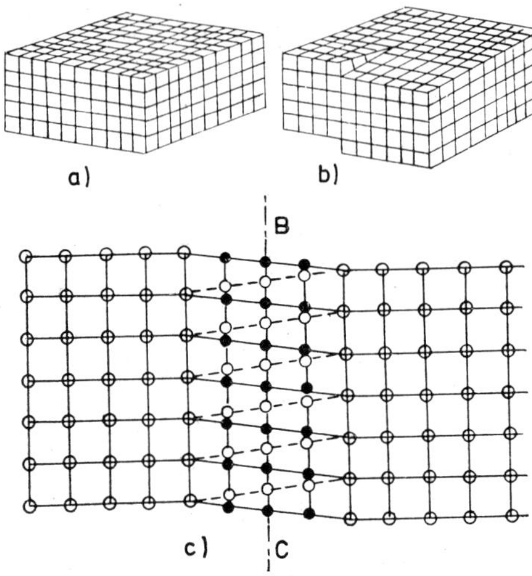

Fig. I.31. Edge and screw dislocations of the crystal

The energy of the dislocation is proportional to b^2. This energy is considerable, indicating that the crystal is not in thermal equilibrium with the dislocation. The formation of dislocations is a more-or-less unavoidable consequence of crystal growth.

Dislocations generate a stress field around themselves, resulting in interactions. Dislocations of equal sign repel, and those of opposite sign attract each other. As a result of the interactions two- and three-dimensional dislocation nets may be formed. The two simplest types are the tilt and twist boundary lines. The first type is due to a series of parallel edge dislocations (Fig. I.32). The second type is a grid of two series of non-parallel screw dislocations.

In metals treated by gradual cooling the density of dislocations, i.e. the number of lines crossing 1 cm^2 of the surface, is of the order of 10^6. This

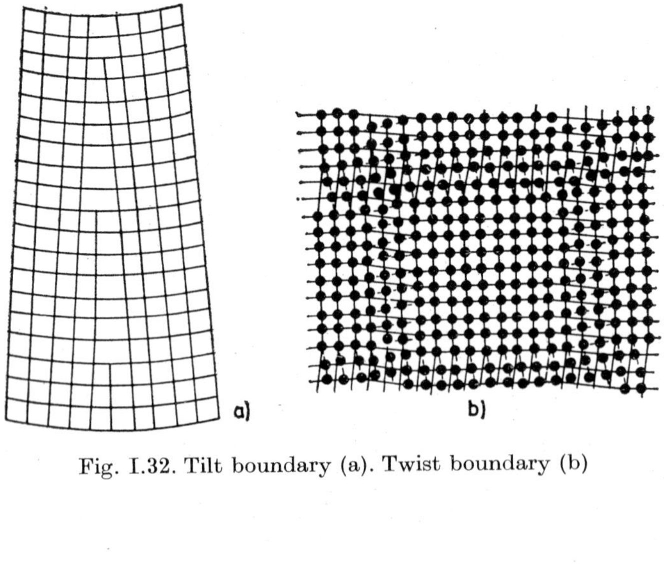

Fig. I.32. Tilt boundary (a). Twist boundary (b)

figure may increase to 10^{10} in a deformed specimen. Pure germanium and silicon single crystals contain 10^2—10^3 dislocations per cm^2, although in certain cases the crystal can be grown without the formation of dislocations. Due to the surrounding stress field, point and line defects may interact with each other.

According to the observations of Cottrell [71] foreign atoms in the crystal move towards the dislocations and form an atmosphere around them, similarly to the Debye–Hückel clouds in the solutions of electrolytes. At the point where the binding energy excess is U the concentration of impurities around the edge dislocations is

$$c' = c_0 \exp (U/kT) \qquad (I.203)$$

where c_0 is the average concentration in the crystal well away from the dislocation. This indicates that as far as purity is concerned, crystals may be very heterogeneous.

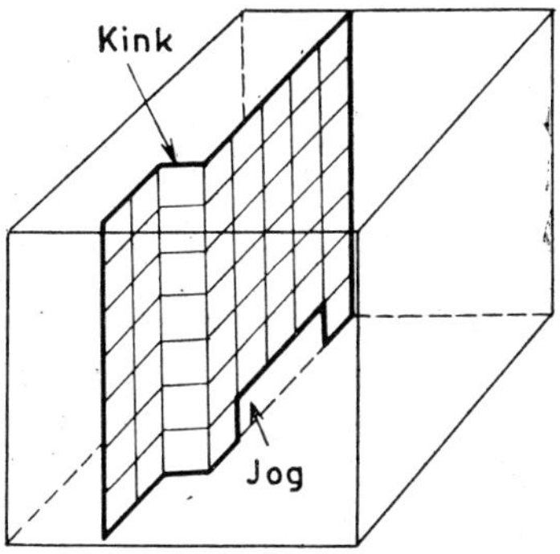

Fig. I.33. Edge dislocation containing jogs and kinks

Dislocations in the crystal are mobile. There are two basically different types of motion, *glide* and *climb*. Glide is a conservative fast motion (its velocity in the crystal is a fraction of the velocity of sound), which takes place if pressure is applied to a crystal, the dislocations generally multiplying. Moving dislocations intersect others, which leads to the formation of vacancies or interstitials, particularly if a screw dislocation happens to cut another screw dislocation. Climb, i.e. motion by diffusion, may, in principle, be attributed to dislocations of strong edge character, but since screw dislocations generally also contain edge-parts due to the thermal motion, the phenomenon can be treated as general. This motion can be most conveniently illustrated with the example of a pure edge dislocation (Fig. I.33). When an atom is added or subtracted, the line acquires a jog. By such a repeated process the extra half-plane can shorten (positive climb) or lengthen (negative climb).

SECTION 6.2. THE DEFECT STRUCTURES OF CRYSTALS OF STOICHIOMETRIC COMPOSITION

It is convenient to treat the formation conditions of point defects separately for stoichiometric and non-stoichiometric ionic crystals, but first the notation of defect sites should be reviewed.

Point defects are nowadays generally denoted by means of the Schottky symbols. For the sake of completeness, the notations introduced by Wagner, and by Hauffe and Schottky, which can be found mainly in earlier papers, are also given in Table I.2.

TABLE I.2

The notations of defect sites

Type	Notation of Wagner	Notation of Hauffe and Schottky	Notation of Schottky
Cation vacancy	(K′)	K□′	∣ K ∣′
Anion vacancy	(A˙)	A□˙	∣ A ∣˙
Interstitial cation	$K_{(\dot{z})}$	K○˙	K˙
Interstitial anion	$A^{-}_{(z)}$	A○′	A′
Bivalent cation in the lattice point of a monovalent cation	$K^{2+}_{1(g)}$	K_1●˙(K_2)	K_1 ∣ K_2 ∣˙
Monovalent cation in the lattice point of a bivalent cation	$K^{1-}_{2(g)}$	K_2●′(K_1)	K_2 ∣ K_1 ∣′
Uncharged defect site (as a superscript)	—	x	x
Defect site of a single positive charge (as a superscript)	—	˙	˙
Defect site of a single negative charge (as a superscript)	—	′	′
Free electron	$e^{-}_{(z)}$	⊖	e′
Defect electron (positive hole)	˙(e^{-})	⊕	∣ e ∣˙

According to Frenkel and Schottky there are four types of stoichiometric crystals:

(1) Interstitial cations and vacancies in the cation lattice. This limiting type is shown in Fig. I.34a. Some of the cations leave the lattice sites and move into interstitial positions with the formation of ∣ K ∣′ and K˙, whereas the anion lattice remains unchanged. By the principle of electroneutrality

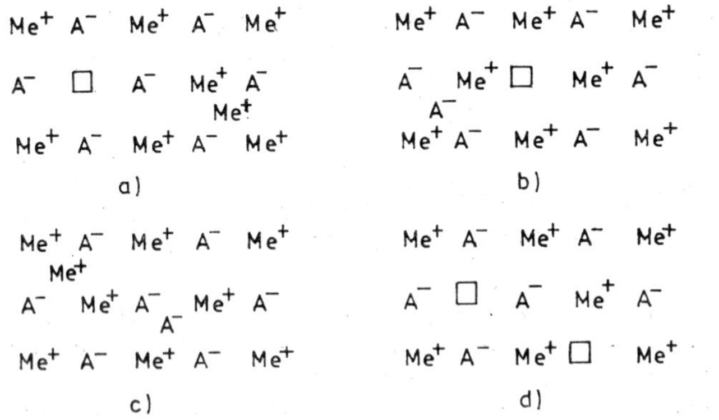

Fig. I.34. Defect structure types of stoichiometric compounds: (a) interstitial cations and vacancies in the cation lattice; (b) interstitial anions and vacancies in the anion lattice; (c) interstitial cations and anions; (d) vacancies in the cation and anion lattice

the number of interstitial cations is the same as the number of vacant cation lattice points:

$$x_{K^\bullet} = x_{|K|'} \tag{I.204}$$

This defect structure is shown, for example, by silver bromide (*Frenkel*-defects).

(2) Interstitial anions and vacancies in the anion lattice. In this case, as can be seen in Fig. I.34b, the cation lattice remains unchanged. A condition similar to the above must also be fulfilled here, namely, the number of interstitial anions must be the same as the number of vacant anion lattice points:

$$x_{A'} = x_{|A|^\bullet} \tag{I.205}$$

X-ray investigations on strontium fluoride (SrF_2) have proved the existence of this limiting case (*Frenkel*-defects).

(3) The third limiting case, the *anti-Schottky* type (Fig. I.34c), is very rare. In this type there is no vacant lattice point, but both anions and cations may be situated interstitially. Their amounts should be the same in accordance with the requirements of electroneutrality:

$$x_{K^\bullet} = x_{A'} \tag{I.206}$$

Studies carried out so far suggest that it is very probable that the structures of magnesium-lead (Mg_2Pb) and magnesium-tin (Mg_2Sn) alloys correspond to this type (*Schottky*-defects).

(4) This limiting case, the *Schottky*-type, is shown in Fig. I.34d. Here only vacant lattice points occur, both in the anion and in the cation lattice. According to the energy considerations of Schottky and Jost, the alkali halides may exhibit fine structures of this type. The condition

$$x_{K'} = x_{|A|^\bullet} \tag{I.207}$$

must, of course, be fulfilled (*Schottky*-defects).

Although examples are known for all the limiting cases, experience indicates that in the cation lattice generally vacancies and interstitial defects may occur equally, whereas in the anion lattice primarily the formation of vacancies must be taken into account.

The number of Schottky-defects (n') in equilibrium with the total crystal at a given temperature is

$$n' = N' \exp(-U_S/kT) \tag{I.208}$$

where U_S is the energy of formation of a vacancy, and N' is the number of possible positions. Schottky-defects are mobile. The mobility, as well as the energy of formation of the vacancies, may differ widely depending on the nature of the atoms and ions in the crystal. In ionic crystals, as we have seen, anion vacancies represent a positive, and cation vacancies a negative, charge. Under the influence of an external electric field ionic conduction takes place; defects of opposite charge attract each other, forming neutral ion pairs, which are much more mobile than the individual isolated defects, but do not contribute to the ionic conduction.

In equilibrium, the number of Frenkel-defects is given by

$$n' = (NN')^{1/2} \exp(-U_F/2kT) \tag{I.209}$$

where N and N' are the numbers of lattice sites and possible interstitial positions, respectively, and U_F is the energy of formation of the arrangement corresponding to the Frenkel-defect.

Whether Frenkel- or Schottky-type defects predominate in the solid depends on the relative values of U_S and U_F, which are ultimately determined by geometrical factors. It can be stated generally that Frenkel-type defect sites occur in crystals where the ion sizes differ considerably and due to the strong polarization effects the van der Waals energy and the dielectric constant are high. On the other hand, Schottky-type defects occur if the van der Waals energy and the dielectric constant in the crystal are not too high and the ion sizes are not remarkably different, either. Frenkel- and Schottky-type defects can be identified on the basis of extremely accurate density measurements, since the densities of solids are influenced (decreased) only by the formation of Schottky-type defect sites.

SECTION 6.3. THE DEFECT STRUCTURES OF CRYSTALS OF NON-STOICHIOMETRIC COMPOSITION

Most ionic crystals, and, in particular, the oxides and sulphides which are important in catalysis, have non-stoichiometric compositions, and often contain anion or cation excesses in amounts barely detectable by analytical methods. In these crystals generally only one ion type is displaced. In comparison with the types discussed so far it is a more essential difference that in this case electrons or positive holes (defect electrons) may also occur, which compensate the charge of excess anions or cations. This also means that the non-stoichiometric crystals, due to the high mobilities of the electrons and defect electrons, are electron conductors, whereas in crystals of stoichiometric composition ionic conduction predominates. In non-stoichiometric crystals the following defect types may occur:

(1) metal excess due to the anion vacancy;
(2) metal excess due to interstitially situated metal;
(3) anion excess due to interstitially situated anions;
(4) anion excess due to cation vacancy.

A metal excess in ionic crystals is generally caused by processes in which, at a high temperature for instance, the crystal loses a given quantity of its negative component (oxygen, sulphur or iodine) with the formation of anion vacancies or interstitial incorporation. In practice, both defect types occur simultaneously to a varying extent.

The first case is shown in Fig. I.35a, where the apparent metal excess arising from anion deficiency is represented in the form of anion vacancies and quasi-free electrons. This case is realized in the structures of potassium

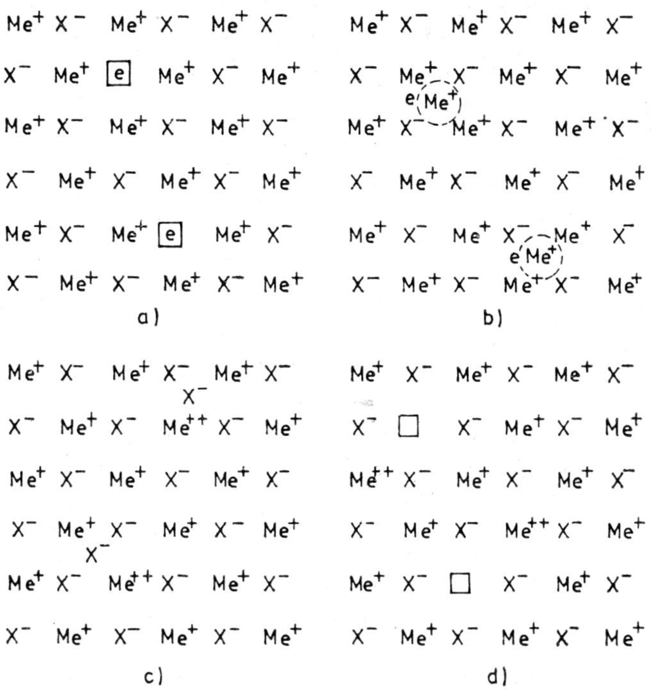

Fig. I.35. Defect structure types of non-stoichiometric compounds: (a) a compound containing metal excess due to anion deficiency; (b) a compound containing metal excess due to interstitial cations; (c) a compound having interstitial cations; (d) a compound having cation deficiency

chloride, sodium chloride, potassium bromide, titanium dioxide, thorium dioxide, lead sulphide, etc.

The second type is realized, among others, in zinc oxide and cadmium oxide. Their structures are shown in Fig. I.35b.

It depends almost exclusively on the experimental temperature as to whether the metal is incorporated into the interstitial place in the form of a neutral atom or as an ion and an electron.

A defect structure of the third group is shown in Fig. I.35c. The charge of interstitial anions is compensated by the higher valence state of a proportion of the metal ions. Uranium dioxide has a structure corresponding to this type.

The defect site model arising from cation vacancies in the lattice points is shown in Fig. I.35d. The negative charge may be compensated in two ways: by the larger charge of some of the cations, which is the more frequent case; or through a process in which the anion in the lattice point loses an electron, its charge decreasing; in the case of a monovalent anion, it becomes neutral. Of the oxides, nickel(II) oxide, cobalt(II) oxide and copper(I) oxide have defect structures corresponding to the first possibility.

In the following paragraphs the defect structure, electrical behaviour, and the correlation between these properties for the most important semiconductors, which can also be regarded as catalysts, will be discussed in somewhat more detail. Knowledge of these properties is nowadays an indispensable requirement for an understanding of catalytic behaviour.

SECTION 6.4. THE DEFECT STRUCTURES AND CONDUCTION PROPERTIES OF n-TYPE OXIDES

PARAGRAPH 6.4.1. ZINC OXIDE

The most characteristic representative of n-type conductors is zinc oxide. Conduction through the excess electrons may take place in the following four ways:

$$\mathrm{ZnO} \rightleftarrows \mathrm{Zn}^{\cdot} + e' + 1/2\,\mathrm{O}_2^{(g)} \tag{I.210a}$$

$$\mathrm{ZnO} \rightleftarrows \mathrm{Zn}^{\cdot\cdot} + 2e' + 1/2\,\mathrm{O}_2^{(g)} \tag{I.210b}$$

$$\mathrm{ZnO}\ \text{(unperturbed lattice)} \rightleftarrows |\mathrm{O}|^{\cdot} + e' + 1/2\,\mathrm{O}_2^{(g)} \tag{I.211a}$$

$$\mathrm{ZnO}\ \text{(unperturbed lattice)} \rightleftarrows |\mathrm{O}|^{\cdot\cdot} + 2e' + 1/2\,\mathrm{O}_2^{(g)} \tag{I.211b}$$

The choice between the two main possibilities is at present very difficult, since both mechanisms are supported by experimental data. Diffusion and X-ray analysis studies, however, increasingly tend to confirm the formation of interstitial zinc ions. The corresponding defect structure is shown in Fig. I.36. Applying the mass-action law to Eq. (210a):

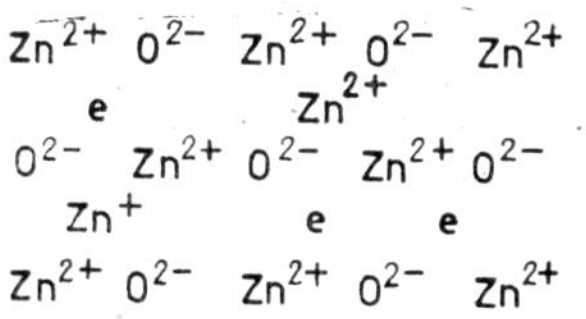

Fig. I.36. The defect structure of pure zinc oxide

$$K' = \frac{x_{\mathrm{Zn}^{\cdot}} \cdot x_{e'}\, p_{\mathrm{O}_2}^{1/2}}{x_{\mathrm{ZnO}}} \tag{I.212}$$

(x denotes mole fraction), from which it follows that

$$p_{\mathrm{O}_2}^{-1/2}\, K'\, x_{\mathrm{ZnO}} = K\, p_{\mathrm{O}_2}^{-1/2} = x_{\mathrm{Zn}^{\cdot}} \cdot x_{e'} \tag{I.213}$$

Since the concentration of conduction electrons is the same as the concentration of interstitial zinc ions, and since the electrical conductivity is proportional to x_e, the dependence of the conductivity on the pressure of oxygen can be given as

$$\sigma = \text{const.}\ p_{\mathrm{O}_2}^{-1/4} \tag{I.214}$$

if the validity of Eq. (210a) is assumed. On the other hand, if dissociation equilibrium, Eq. (210b), is taken into account, the dependence of the conductivity on the pressure of oxygen will be

$$\sigma = \text{const.}\ p_{\mathrm{O}_2}^{-1/6} \tag{I. 215a}$$

or, in the general case

$$\sigma = \text{const.}\ p_{\mathrm{O}_2}^{-1/n} \tag{I.215b}$$

The experimental results of Baumbach and Wagner [72], which have been confirmed by several authors, are in good agreement with the above considerations. As can be seen from Fig. I.37, the conductivity of zinc oxide decreases with the partial pressure of oxygen, and n in the exponent has a value between 4 and 5.

The defect structure and the electrical behaviour of zinc oxide are extremely sensitive to the conditions of preparation. The variation of the

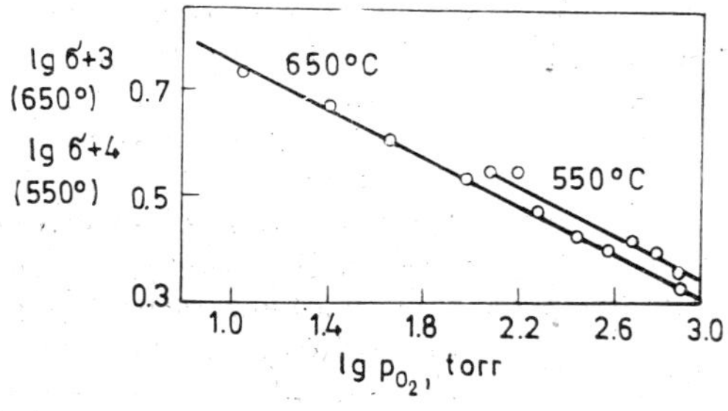

Fig. I.37. The variation of the conductivity of zinc oxide with the equilibrium partial pressure of oxygen [72]

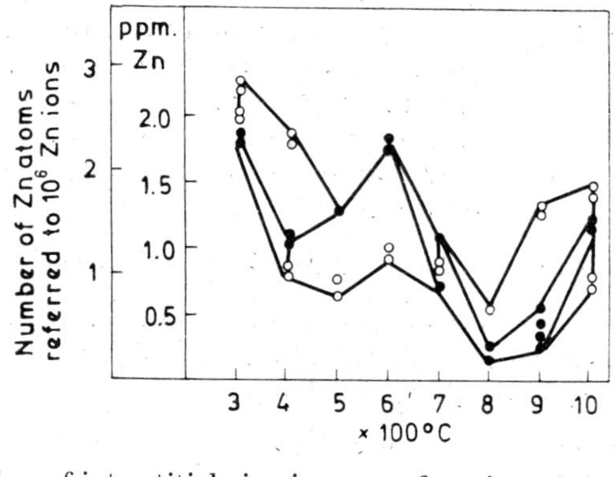

Fig. I.38. The number of interstitial zinc ions as a function of the temperature of pretreatment of zinc oxide [73] (● denotes the most probable values)

zinc excess in zinc oxide as a function of the preparation temperature [73] is given in Fig. I.38. Compositions deviating most from the stoichiometric one were found in samples treated at 300 °C and 600 °C, respectively. Above 600 °C the zinc excess rapidly decreases, and after a minimum at 800 °C again increases.

Different pretreatment processes also influence the lattice constant of zinc oxide, which, in the majority of cases, indicates a change in the number of interstitial zinc ions [74]. Pretreatment with hydrogen, e.g. at 600 °C, considerably increases the lattice constant. The lattice constant also varies sensitively with the pretreatment of the zinc oxide; the lattice constant of powdered zinc oxide reaches a maximum at 750 °C, whereas that of zinc oxide pressed into discs has a maximum at 900 °C. These experimental facts are particularly important in the interpretation of the catalytic behaviour of zinc oxide, and, as will be shown in Paragraph III.2.2.2, the catalytic efficacy of zinc oxide is in close correlation with its zinc excess.

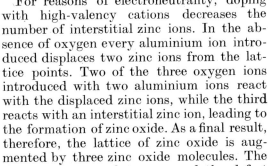

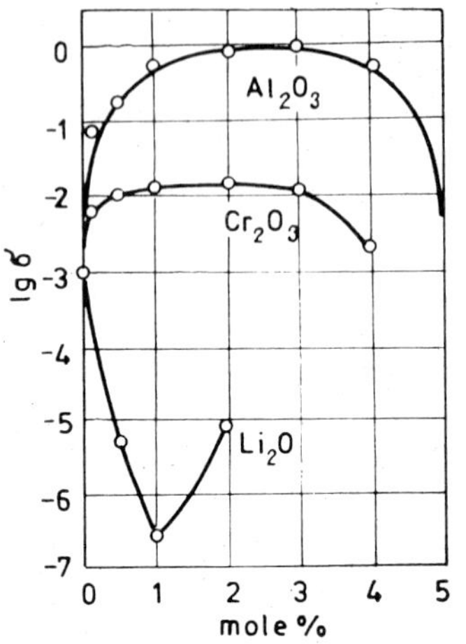

Fig. I.39. The effects of foreign ions on the conductivity of zinc oxide [75]

The effects of foreign ions on the defect structure and conductivity of zinc oxide have been studied by Hauffe [75], who found that the conductivity increases upon the addition of aluminium oxide or chromium oxide (Fig. I.39).

For reasons of electroneutrality, doping with high-valency cations decreases the number of interstitial zinc ions. In the absence of oxygen every aluminium ion introduced displaces two zinc ions from the lattice points. Two of the three oxygen ions introduced with two aluminium ions react with the displaced zinc ions, while the third reacts with an interstitial zinc ion, leading to the formation of zinc oxide. As a final result, therefore, the lattice of zinc oxide is augmented by three zinc oxide molecules. The extent of this reaction depends on the pressure of oxygen, and the defect structure of the zinc oxide. The above incorporation of aluminium oxide can be described by the equation

$$Al_2O_3 + Zn^{\cdot\cdot} \rightleftarrows 2\,Al\,|\,Zn\,|^{\cdot} + 3\,ZnO \qquad (I.216)$$

According to this, the electron concentration does not increase. However, if only one of the oxygen ions of the aluminium oxide reacts with an interstitial zinc ion of a single positive charge:

$$Al_2O_3 + Zn^{\cdot} \rightleftarrows 2\,Al\,|\,Zn\,|^{\cdot} + e' + 3\,ZnO \qquad (I.217)$$

the electron concentration will increase. If, upon doping with aluminium oxide, the oxygen ion does not react with an interstitial zinc ion, the following reaction will take place:

$$Al_2O_3 \rightleftarrows 2\,Al\,|\,Zn\,|^{\cdot} + 2\,e' + 2\,ZnO + 1/2\,O_2^{(g)} \qquad (I.218)$$

According to this reaction the aluminium ions form a substitution defect site of single positive charge in the zinc oxide lattice, with the simultaneous formation of two zinc oxide molecules, two electrons, and the evolution of an oxygen atom.

In the above reaction the equilibrium is strongly shifted towards the right hand side, even at a high oxygen pressure. A similar change is caused in the lattice of zinc oxide by introducing chromium oxide or iron oxide.

Upon doping with aluminium oxide the defect structure of the zinc oxide lattice given in Fig. I.36 changes into the structure shown in Fig. I.40.

When low-valency ions, e.g. lithium or sodium, are introduced, the bivalent zinc ions in the lattice points are substituted by monovalent lithium

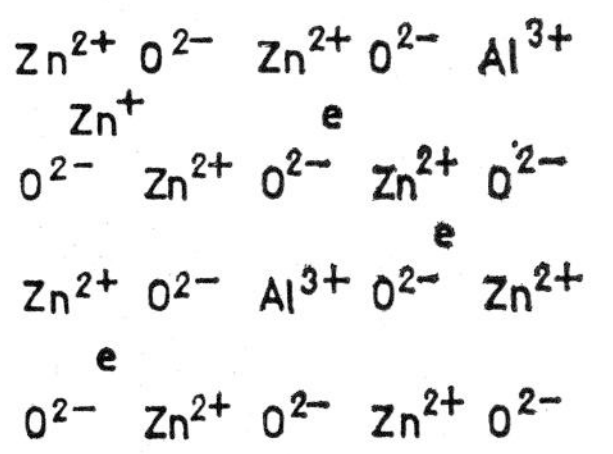

Fig. I.40. The defect structure of zinc oxide doped with aluminium ions

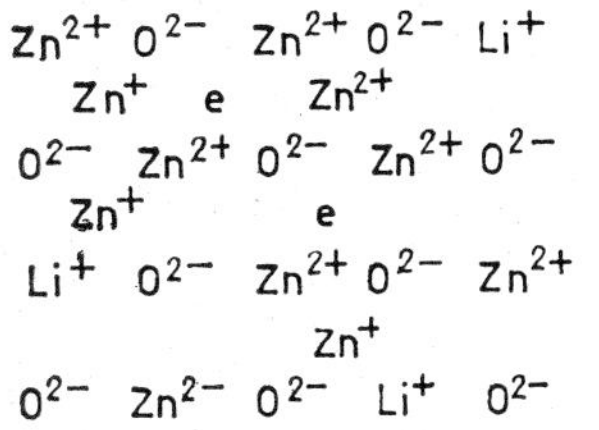

Fig. I.41. The defect structure of zinc oxide doped with lithium ions

ions. Conductivity measurements show that, in this case, the conductivity of the zinc oxide decreases. On the basis of the equation

$$x_{Zn^{\cdot}}\, x_{e'} = K \qquad (p_{O_2} = \text{const.}) \qquad (I.219)$$

this means that the number of interstitial zinc ions increases. The defect structure formed upon doping with lithium oxide is shown in Fig. I.41, while the doping can be represented by the following overall equation:

$$Li_2O + 2e' + 1/2\, O_2^{(g)} \rightleftarrows 2\, Li \mid Zn \mid' + 2\, ZnO \qquad (I.220)$$

However, the most recent investigations indicate that caution must be exercised with regard to the introduction of high-valency ions. The introduction of lithium and gallium ions was investigated by Bremer et al. [76] by determining the zinc excess in zinc oxide [77, 78]. Contrary to expectation the number of interstitial zinc ions changed only upon doping with lithium oxide (0.07 mole per cent of lithium oxide was incorporated into the zinc oxide lattice at 850 °C), and did not change upon doping with gallium oxide. This experiment led to the conclusion that gallium oxide, in contrast to lithium oxide, does not enter the lattice of zinc oxide, but reacts with it to yield the spinel $ZnGa_2O_4$, the formation of which could be detected by X-ray studies.

PARAGRAPH 6.4.2. TITANIUM DIOXIDE

Hall effect measurements indicate that both anatase and rutile modifications of titanium dioxide belong to the group of n-type semiconductors. Here, too, there are two possibilities for the interpretation of the electron conduction, namely the assumption either of interstitial titanium atoms, or of oxygen vacancy. The variation of conductivity with oxygen pressure has been determined, and it has been found that at low oxygen pressures (below 30 torr) and elevated temperatures the relationship can be given by

$$\sigma = \text{const. } p_{O_2}^{-1/2} \qquad (I.221)$$

and at low temperatures by

$$\sigma = \text{const. } p_{O_2}^{-1/3} \tag{I.222}$$

At higher oxygen pressure it has been found that

$$\sigma = \text{const. } p_{O_2}^{-1/4.5} \tag{I.223}$$

From Eq. (I.221) Earle derived the defect site reaction:

$$TiO_2 \rightarrow Ti^{\cdot} + e' + O_2^{(g)} \tag{I.224}$$

According to this reaction the presence of a singly charged interstitial titanium ion ($Ti^{\cdot\cdot\cdot\cdot} + 3\,e'$) ought to be assumed.

The X-ray and density measurements of Ehrlich [79], however, indicated the formation of oxygen vacancy. Accordingly, the equation for the formation of this defect site should be written in the following form:

$$TiO_2 \text{ (unperturbed lattice)} = |O|^{\cdot} + e' + 1/2\,O_2 \tag{I.225}$$

$$TiO_2 \text{ (unperturbed lattice)} = |O|^{\cdot\cdot} + 2\,e' + 1/2\,O_2 \tag{I.226}$$

If these equations hold, the conductivity vs. oxygen pressure functions should have the form

$$\sigma = \text{const. } p_{O_2}^{-1/4} \tag{I.227}$$

and

$$\sigma = \text{const. } p_{O_2}^{-1/6} \tag{I.228}$$

respectively. Hauffe et al. [80] have found that between 800° and 1000 °C, in the oxygen pressure range of $10^{-3} - 10^{-2}$ torr, the slopes of the log σ vs. log p_{O_2} curves are between $-1/4.5$ and $-1/5$.

Despite elaborate studies on titanium dioxide no unequivocal conclusions can be drawn as to the type of its defect structure. Numerous experimental facts support the assumption of oxygen vacancies [81—84], others, however, the presence of interstitial titanium ions [85—87].

Kofstad [88] considers that titanium dioxide contains both bivalent oxygen vacancies and tri- or tetravalent interstitial titanium ions. It has been concluded from the thermodynamic characteristics of titanium dioxide that at a low temperature and high oxygen pressure it is mainly oxygen vacancies that occur, whereas at low pressure and high temperature the formation of interstitial titanium ions predominates.

Doping with ions of different valency also influences the conductivity in the case of titanium dioxide. The effects of tantalum(V) oxide, niobium(V) oxide, tungsten oxide, nickel oxide, aluminium oxide and chromium(III) oxide have so far been investigated. The oxides of high-valency ions increase the conductivity of titanium dioxide. Their effects can be characterized by the following equations:

$$Ta_2O_5 = 2\,Ta\,|\,Ti\,|^{\cdot} + 2\,e' + 2\,TiO_2 + 1/2\,O_2 \tag{I.229}$$

$$WO_3 = W\,|\,Ti\,|^{\cdot\cdot} + 2\,e' + TiO_2 + 1/2\,O_2 \tag{I.230}$$

and

$$|\,O\,|^{\cdot} + WO_3 \rightleftarrows W\,|\,Ti\,|^{\cdot\cdot} + e' + TiO_2 \tag{I.231}$$

Consequently, if the number of electrons increases, the number of oxygen vacancies will decrease.

In accordance with the electron conduction character of titanium dioxide the introduction of lower-valency ions ought to decrease the conductivity. An observable reduction of conductivity takes place on doping with nickel oxide, but on doping with gallium oxide the conductivity remains practically unchanged. Completely surprising behaviour has been observed, however, in the titanium dioxide — chromium oxide system. According to the theory of defect structures, upon doping with chromium oxide the following reactions should take place:

$$1/2\ O_2 + 2\ e' + Cr_2O_3 = $$
$$= 2\ Cr|\ Ti\ |' + 2\ TiO_2 \qquad (I.232)$$

$$Cr_2O_3 \rightleftarrows 2\ Cr|\ Ti\ |' + 2\ TiO_2 + |\ O\ |^{\cdot\cdot} \qquad (I.233)$$

leading to a decrease of the electron concentration of the titanium dioxide. The measurements indicate, however, that the introduction of chromium oxide increases the conductivity of titanium dioxide to a marked extent (Fig. I. 42).

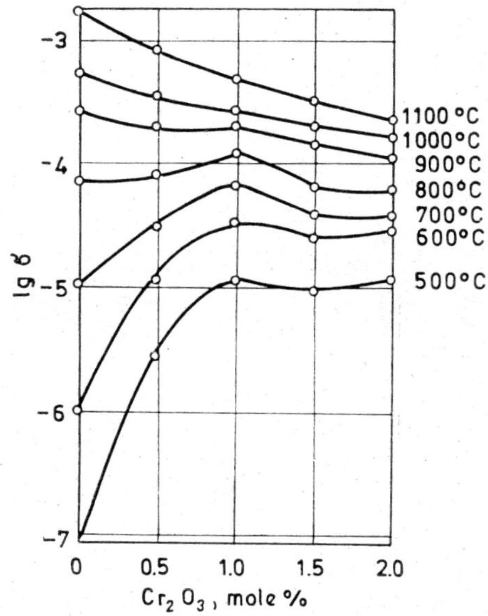

Fig. I.42. The conductivity, at different temperatures, of titanium dioxide doped with chromium oxide [80]

In order to interpret these results it has been assumed by Hauffe et al. [80] that the trivalent chromium ion is oxidized to a higher valence state in air at atmospheric pressure, and is incorporated into the titanium dioxide lattice in the tetra- or pentavalent state. The chromium ion then increases the electron concentration and conductivity of titanium dioxide as a higher valence state ion:

$$1/2\ O_2 + Cr_2O_3 = 2\ Cr|\ Ti\ |^{\cdot} + 2\ e' + 2\ TiO_2 \qquad (I.234)$$

A difficulty arises, however, in that at low temperatures the conductivity of titanium dioxide containing chromium oxide does not change initially with increasing oxygen pressure, but increases considerably in a later phase. On the other hand, at higher temperatures the conductivity first decreases to a minimum, and then increases again (Fig. I.43). This behaviour contradicts the *n*-type conduction character of the mixed oxide, since it is characteristic of *n*-type conducting oxides that their conductivities decrease with increasing oxygen pressure. This can be observed, for instance, in the case of pure titanium dioxide. The results obtained for chromium oxide indicate that the mechanism of conduction of titanium dioxide doped with chromium oxide is entirely different from that of pure titanium dioxide [89].

In our opinion, the interpretation given by Hauffe is correct only to the extent that the introduction of chromium oxide does indeed result in the formation of chromium ions of a higher valence state (this is supported by the measured active oxygen content of the titanium dioxide — chromium oxide system). The increase in conductivity, however, cannot be attributed to the reaction of Eq. (I.234), i.e. to the increase of the electron concentration, but to the change in the character of the surface conduction, i.e. to the occurrence of defect conduction. In the case when titanium dioxide containing chromium oxide is treated at 400 °C with hydrogen or formic acid vapour, the conductivity decreases considerably. After the establishment of a steady state the curve given in Fig. I.44 is obtained for the variation of the conductivity of titanium dioxide with the chromium oxide content. Consequently, in a reducing atmosphere the introduction of chromium oxide decreases the conductivity of titanium dioxide, just as expected from the n-type conduction character of the latter. Very similar behaviour is found when chromium oxide is introduced in vacuum. On the above basis it can be assumed that at first chromium oxide is incorporated into the titanium dioxide lattice according to the equation

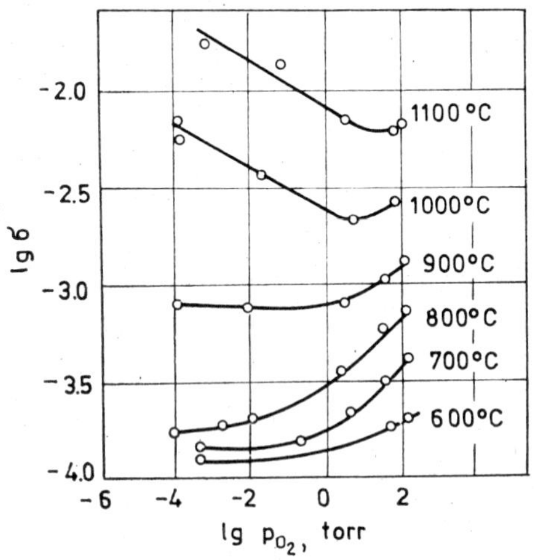

Fig. I.43. The conductivity of titanium dioxide doped with chromium oxide as a function of the partial pressure of oxygen [80]

$$2\,e' + 1/2\,O_2 + Cr_2O_3 \rightleftarrows 2\,Cr|\,Ti\,|' + 2\,TiO_2 \qquad (I.235)$$

The oxidizability of the chromium ion, however, increases very strongly when it is introduced into titanium dioxide, and at a higher temperature, in the presence of oxygen the chromium(III) ion (or rather $Cr|\,Ti\,|'$) is oxidized to a tetra- or pentavalent ion. The overall equation of the oxidation and the introduction of Cr_2O_3 is then

$$Cr_2O_3 + 1/2\,O_2 \rightleftarrows 2\,Cr|\,Ti\,|^{\times} + 2\,TiO_2 \qquad (I.236)$$

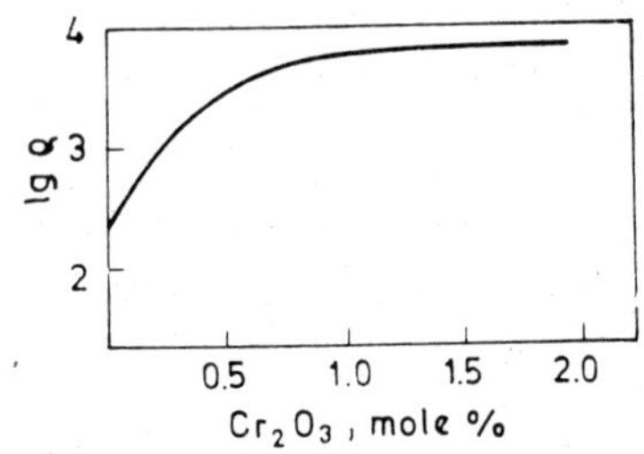

Fig. I.44. The electrical resistance of titanium dioxide doped with chromium oxide, at 400 °C, after pretreatment with formic acid [89]

or upon further chemisorption of oxygen:

$$1/2\,O_2 + Cr|\,Ti\,|^{x} \rightleftarrows 1/2\,O^{-}_{2(chem)} + Cr|\,Ti\,|^{\cdot} \tag{I.237}$$

(where $1/2\,O^{-}_{2(chem)}$ denotes a chemisorbed oxygen atom). Substitution defect sites $Cr|\,Ti\,|^{x}$ and particularly $Cr|\,Ti\,|^{\cdot}$ may dissociate very easily:

$$Cr\,|\,Ti\,|^{x} \rightleftarrows Cr\,|\,Ti\,|' + |\,e\,|^{\cdot} \tag{I.238}$$

$$Cr\,|\,Ti\,|^{\cdot} \rightleftarrows Cr\,|\,Ti\,|^{x} + |\,e\,|^{\cdot} \tag{I.239}$$

the reactions yielding defect electrons. With this scheme it is understandable why the conductivity of titanium dioxide doped with chromium oxide increases with the increasing partial pressure of oxygen. The effect of hydrogen or formic acid shown in Fig. I.45 can be attributed to the fact that the defect electron concentration of the surface decreases, partly due to chemisorption of a donor character and partly to the decreased amount of active oxygen on the surface, whereby the conductivity of the mixed oxide also decreases. When the active oxygen content of the surface (i.e. chromium ions of a higher valence state) has been exhausted in the reaction the n-type conduction character of the system reappears, and in this phase the chemisorption of hydrogen and formic acid according to the reaction

$$1/2\,H_2 = H^{+}_{(chem)} + e' \tag{I.240}$$

increase the electron concentration of the titanium dioxide. At a higher temperature the surface is reduced to a varying extent, in addition to the chemisorption which yields electrons. This reduction further decreases the oxygen deficiency of the titanium dioxide, thereby increasing its conductivity.

The above picture of the mechanism of chromium oxide incorporation is supported by measurements carried out on pure titanium dioxide and titanium dioxide doped with tungsten oxide. In these cases the electrical resistance always decreases on the action of hydrogen or formic acid and increases on the action of oxygen, indicating the n-type conduction character of the oxide mixture. The extremely favourable catalytic properties of titanium dioxide doped with chromium oxide (primarily in donor-type reactions) supply further evidence of the above mechanism, since these properties can only be interpreted, in a manner similar to above, by assuming surface defect conduction (see Chapter VI.5).

The surface p-type conduction of titanium dioxide doped with chromium oxide and the above interpretation of the mechanism of this conduction are supported by the recent experiments and considerations of Hauffe [90, 91].

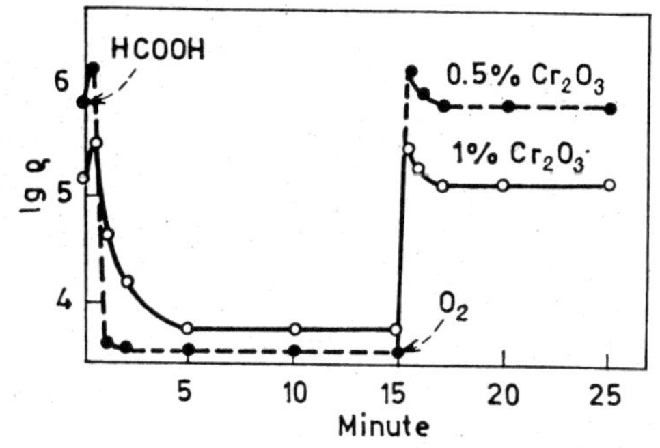

Fig. I.45. The variation of the electrical resistance of titanium dioxide doped with chromium oxide, under the effect of oxygen or formic acid [89]

The electrical conductivity of tin dioxide, which is also an n-type conductor, decreases upon doping with chromium oxide, despite the fact that the introduced chromium is also oxidized to a higher valence state [92] (see Fig. I.46).

The reason for the different behaviour of tin dioxide and titanium dioxide is probably the fact that the electrical conductivity, n-type conduction,

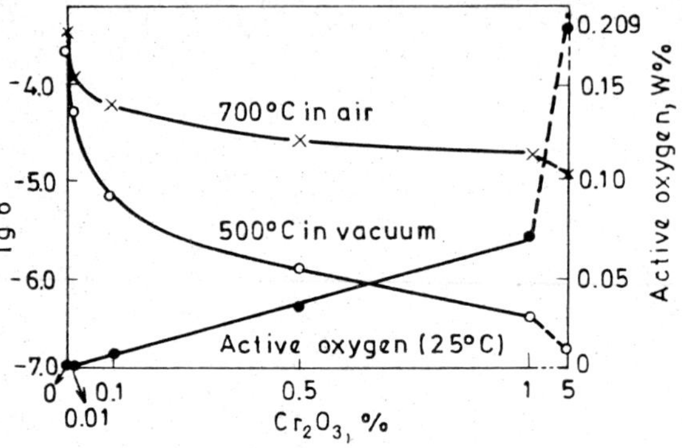

Fig. I.46. The electrical conductivity and active oxygen content of a $SnO_2 + Cr_2O_3$ mixture. Oxides prepared at 900 °C [92]

of tin dioxide is several orders higher than that of titanium dioxide, and the number of electron holes formed on the oxidation of the chromium(III) ions is insufficient to compensate the n-type conduction of the tin dioxide.

PARAGRAPH 6.4.3. CADMIUM OXIDE

Cadmium oxide also belongs to the group of *n*-type conducting oxides, since its conductivity decreases with increasing oxygen pressure. This conduction character is supported by the Hall effect and the behaviour of the thermoelectric potential. Cadmium oxide differs from other semiconductor oxides in its relatively high conductivity ($1.5 \times 10\ \Omega^{-1}\ \text{cm}^{-1}$ at 20 °C), and in the fact that the conductivity decreases with increasing temperature. It can probably be attributed to the high conductivity that higher-valency ions do not increase the conductivity of cadmium oxide, doping with bismuth oxide even decreases the conductivity slightly, probably due to the reduction of the mobility of the electrons.

The defect structure of cadmium oxide and the mechanisms of doping with foreign ions have been studied by X-ray investigations.[93]. Similarly to the case of zinc oxide, the cadmium excess is located interstitially, since the lattice constant increases with increasing cadmium excess. Different heat-treatment procedures do not influence the lattice constant of cadmium oxide. Different processes for the preparation have similarly slight effects. As can be seen in Fig. I.47, doping with indium oxide increases the lattice constant after an initial decrease, whereas the incorporation of silver ions always decreases the lattice constant. These changes have been interpreted by assuming the presence of interstitial cadmium ions. The introduction of indium oxide creates an equivalent number of cation vacancies. Due to the high concentration of interstitial cadmium ions this defect structure cannot be regarded as being stable, and driven by the tendency to thermodynamic equilibrium between the interstitial defect sites and vacancies, the interstitial cadmium ions occupy the vacant cation sites. This process obviously decreases the lattice constant. When the supply of interstitial cadmium ions is exhausted the incorporation of further indium ions proceeds according to the controlled valence process, i.e. a number of bivalent cadmium ions equivalent to the trivalent indium ions become monovalent. The ionic size and the lattice constant increase with de-

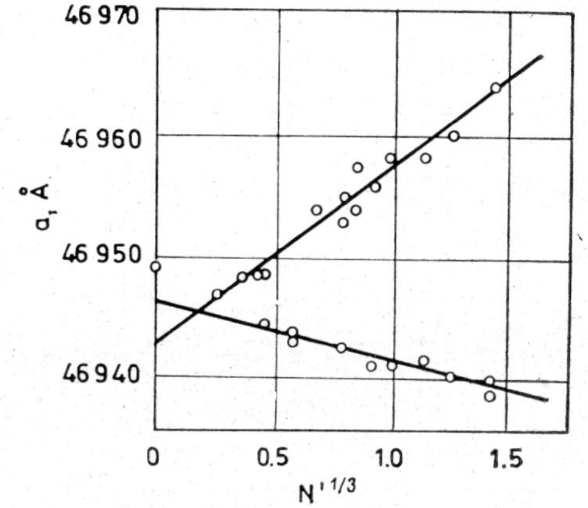

Fig. I.47. The variation of the lattice constant of cadmium oxide as a function of the amount of indium oxide or silver oxide incorporated [93] (N' is the atom per cent of foreign ions)

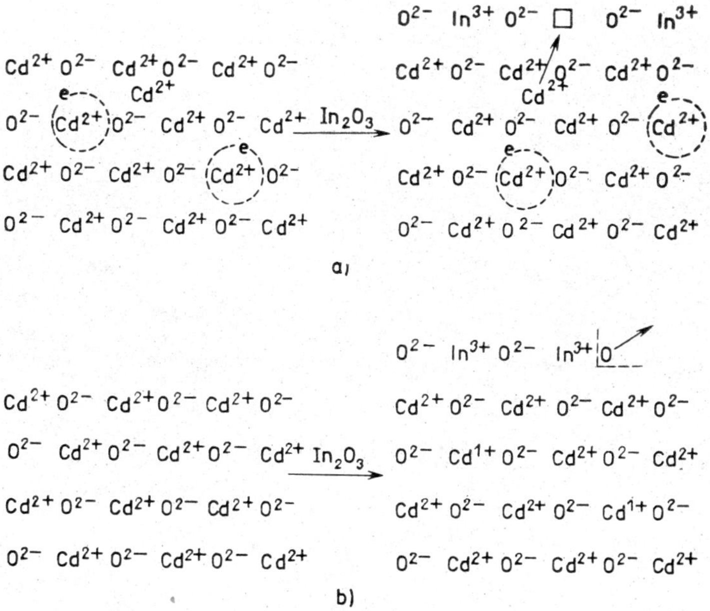

Fig. I.48. The incorporation of indium oxide into the lattice of cadmium oxide [93]

creasing ionic charge. The above processes of incorporation of indium oxide are illustrated in Fig. I.48.

The introduction of lower-valency ions decreases the lattice constant, and this, too, was interpreted on the basis of the controlled valence process. From the differences between the lattice constant of pure cadmium oxide and those of doped samples extrapolated to the zero value (i.e. for a perfect lattice) the cadmium excess of cadmium oxide has been determined to be 0.04%.

PARAGRAPH 6.4.4. ALUMINIUM OXIDE

At low temperatures aluminium oxide is practically an insulator. At higher temperatures, mainly after heat-treatment in vacuum, however, its conductivity increases significantly. The measurements of Hartmann [94] are

TABLE I.3

The conductivities and Hall coefficients of pure and doped aluminium oxide

Composition	Conductivity at 900 °C Ω^{-1} cm^{-1}	Hall coefficient at 400 °C
Al_2O_3 + 5 mole% TiO_2	7.1×10^{-6}	$+(4 \pm 0.4) \times 10^{-3}$
Al_2O_3 + 2 mole% GeO_2	5.3×10^{-6}	—
Al_2O_3 (pure)	4.2×10^{-6}	$+(5.5 \pm 0.5) \times 10^{-3}$
Al_2O_3 + 2 mole% BeO	3.6×10^{-6}	—
Al_2O_3 + 5 mole% NiO	2.5×10^{-6}	$+(8 \pm 0.7) \times 10^{-3}$

given in Fig. I.49, where curves a) and b) show the variation of the conductivity with temperature. The samples were treated in vacuo at 450 °C for different periods. The effect of vacuum heat-treatment on the conductivity can be interpreted by the dissociation of the oxide, i.e. by assuming the formation of oxygen vacancies and electrons:

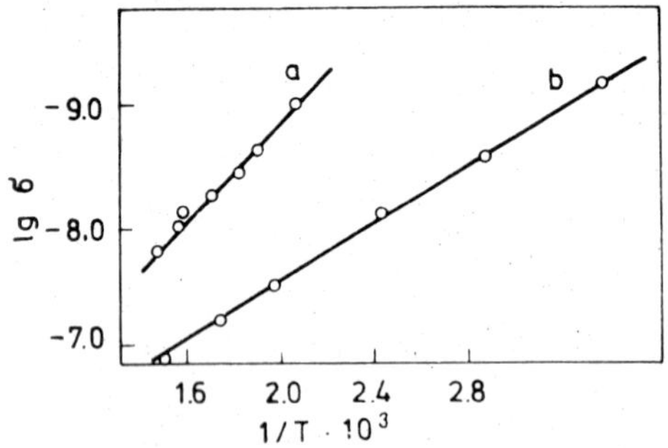

Fig. I.49. The conductivity of aluminium oxide as a function of temperature in vacuo [94]; a) after pretreatment at 450 °C for 80 hr; b) after pretreatment at 450 °C for 120 hr

$$Al_2O_3 \text{ (unperturbed lattice)} = |O|^{\cdot\cdot} + 2e' + 1/2\, O_2 \quad \text{(I.241)}$$

If the oxide pretreated in vacuo at 450 °C for several hours is brought into contact with oxygen, the conductivity decreases again. The conductivity of aluminium oxide also changes according to an n-type conduction character on the introduction of foreign ions. The conductivity of aluminium oxide doped with foreign ions, and the values of the Hall coefficients are listed in Table I. 3.

Despite the fact that the differences in the conductivities of different samples are fairly small, it will be shown later in the discussion of investigations into the effects of the supports (Section III.2.3) that these small changes of conductivity lead to considerable variations in the support properties of aluminium oxide.

SECTION 6.5. THE DEFECT STRUCTURES AND CONDUCTION PROPERTIES OF p-TYPE OXIDES

PARAGRAPH 6.5.1. NICKEL OXIDE

The most characteristic representative of *p*-type oxides is nickel oxide. Its conductivity increases with increasing partial pressures of oxygen, in proportion to the 1/4.5th power of the pressure. Taking into account the possible transformation processes the introduction of oxygen can be characterized by the following equations:

$$O_2 \rightleftarrows 2\,NiO + 2\,|\,Ni\,|'' + 4\,|\,e\,|^{\cdot} \qquad (I.242)$$

or, using a different notation

$$O_2^{(g)} \rightleftarrows NiO + 2\,|\,Ni\,|'' + 4\,Ni\,|\,Ni\,|^{\cdot} \qquad (I.243)$$

where Ni | Ni |· denotes a trivalent nickel ion situated in the lattice point of a bivalent nickel ion. From the mass-action law

$$x^2_{|Ni|''}\, x^4_{|e|^{\cdot}} = K p_{O_2} \qquad (I.244)$$

and taking into account

$$x_{|Ni|''} = 1/2 x_{|e|^{\cdot}} \qquad (I.245)$$

the following equation is obtained for the oxygen-dependence of the defect electron concentration and conductivity

$$x_{|e|^{\cdot}} = 1/4\, K p_{O_2}^{1/6} \qquad (I.246)$$

The general equation for the oxygen-dependence of the conductivity of semiconductors is

$$\sigma \sim p_{O_2}^{1/n} \qquad (I.247)$$

The defect structure of nickel oxide is shown in Fig. I.52a. The oxygen excess is seen to be manifested in the fact that a certain number of metal ions are missing from the lattice points. To compensate the oxygen excess (nickel vacancies), some of the nickel ions transform into the trivalent state.

The composition of nickel oxide, particularly that of the surface layer, depends very sensitively on the mode of preparation and the pretreatment temperature. Since the electrical behaviour of semiconductor oxides depend on their defect structures, it is understandable that the electrical behaviour of nickel oxide is also very sensitive to its previous treatment.

The composition of nickel oxide obtained by the decomposition of nickel carbonate prepared by different processes was studied by Derén et al. [95]. The results of these experiments are given in Table. I.4.

TABLE I.4

The active oxygen content and the activation energy of conductivity of nickel oxide prepared at different temperatures

Number of experiment	Temperature of decomposition of $Ni(CO_3)_2$ °C	Excess O_2 %	Surface size $m^2\ g^{-1}$	Surface concentration of excess O_2		Activation energy of conductivity eV
				atom/$m^2 \times 10^{-18}$	the monomolecular layer, %	
1	2	3	4	5	6	7
I.	500	0.24	7.7	2.5	23	0.73
	600	0.14	3.0	3.7	33	0.68
	700	0.09	2.6	2.7	25	0.79
	800	0.04	1.8	1.9	17	0.64
	900	0.02	1.4	1.0	9	0.76
	1000	0.02	0.5			0.68
II.	500	0.43	24.0	1.4	13	0.66
	600	0.25	8.1	2.4	22	0.77
	700	0.14	5.5	2.0	18	0.71
	800	0.07	2.4	1.8	16	0.74
	900	0.04	1.9	1.5	14	0.73
	1000	0.02	0.9			0.74
III.	500	0.47	22.3	1.7	15	0.72
	600	0.27	7.1	3.1	28	0.66
	700	0.09	3.5	2.1	19	0.66
	800	0.06	2.0	2.4	22	0.68
	900	0.04	1.8	1.7	15	0.64
	1000	0.02	1.1	1.3	12	0.71

Nickel oxide prepared from nickel carbonate at a low temperature, 500 °C, contains a considerable amount of excess oxygen; the content of this decreases with increasing temperature of preparation, and at about 1000 °C the nickel oxide attains the stoichiometric composition. The deviation from stoichiometry also appears in the colour of the nickel oxide. Nickel oxide prepared at 500 °C is greyish-black, which turns into dark green and yellowish-green at 800 °C and 1100 °C, respectively. It can be seen from the data of the Table that with an increasing temperature of preparation not only the oxygen excess, but also the surface area of the nickel oxide, decreases, the concentration of the excess oxygen changing in parallel with the surface area of the nickel oxide. The changes are generally much sharper at lower than at higher temperatures. The parallel behaviour becomes particularly apparent if the oxygen excess is related to unit surface area (Table I.4, column 6). The activation energy of the conductivity of nickel oxide is practically independent of its previous treatment (Table I.4, column 7). The electrical conductivity data, on the other hand, vary widely in this respect: the higher the preparation temperature, the lower the conductivity of the sample.

If it is assumed after Dry and Stone [96] that the number of oxygen atoms corresponding to a monomolecular layer is equal to the number of surface Ni^{2+} ions, one can determine the fraction of the monolayer that corresponds to the excess amount of oxygen given in column 5 of Table I.4. This percentage value is to be found in column 6 of the Table. The order of magnitude of the coverage is in good agreement with the value obtained by the investigation of the chemisorption of nickel oxide. These data indicate that the excess oxygen does not enter the lattice of the nickel oxide, but is partly bound by chemisorption to the surface of the crystal, and is partly built into the surface layer. This assumption is supported by data obtained when nickel oxide samples prepared at different temperatures are treated at 400 °C under a pressure of 10^{-3} torr. After the pretreatment all the samples become yellowish-green, and the composition corresponds to the formula NiO. Since the temperature is far below the Tamman temperature of nickel oxide (1000 °C), the appreciable diffusion of oxygen from the nickel oxide lattice can be excluded.

On the basis of the above discussion it can be stated that in the decomposition of nickel carbonate nickel oxide of nearly stoichiometric composition is formed, which in turn reacts rapidly with the oxygen of the gas phase. It should be emphasized, however, that these considerations pertain only to nickel oxide prepared from nickel carbonate. Nickel oxide obtained from nickel nitrate at 400–500 °C contains excess oxygen in much higher amounts than the samples discussed above, due to the oxidative properties of the nitrate anion, and hence the oxygen excess is incorporated into the nickel oxide lattice in a nearly homogeneous distribution. If the pretreatment temperature is raised above about 1000 °C, however, nickel oxide of stoichiometric composition can be obtained in this case too.

The defect structure and conductivity of nickel oxide are also influenced by the introduction of ions of different valence states. Of the lower-valence ions lithium increases the conductivity of nickel oxide by several orders of magnitude, but potassium ions, on the other hand, have no effect at all (Fig. I.50). This can be attributed to the fact that the size of the lithium ion is almost the same as that of Ni^{2+}, whereas the potassium ion has a quite different size, so that it enters the lattice to only a limited extent, or not at all. Lithium ions in the lattice of nickel oxide occupy the positions of bivalent nickel ions, and create a substitution de-

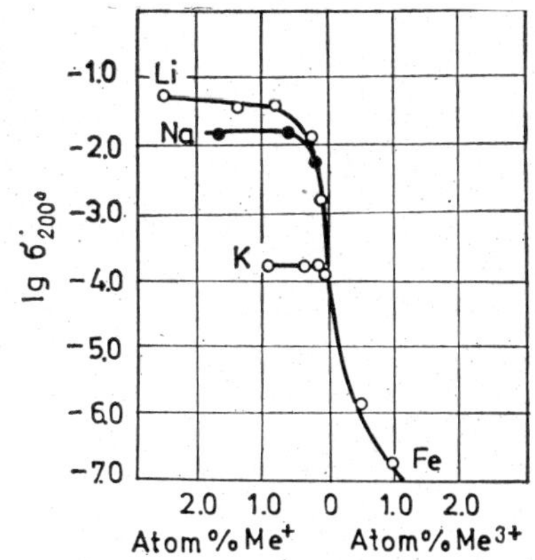

Fig. I.50. The effects of foreign ions on the conductivity of nickel oxide [95]

fect site of single negative charge. Doping with lithium oxide can be described by the following equation:

$$Li_2O + 1/2\, O_2^{(g)} = 2\, Li\, |\, Ni\, |' + 2\, |\, e\, |^{\cdot} + 2\, NiO \qquad (I.248)$$

It can be seen from this that simultaneously with the introduction of lithium oxide it is also necessary to take into account the introduction of oxygen and the formation of a trivalent nickel ion. It follows from the mass-action law represented by Eq. (I.244) that at a constant oxygen pressure the number of nickel vacancies should decrease with increasing defect electron concentration. Utilizing relation (I.244) between the concentrations nickel vacancies and defect electrons, and also the condition of electroneutrality

$$x_{|e|^{\cdot}} = 2\, x_{|Ni|''} + x_{Li|Ni|'} \qquad (I.249)$$

the ratio of the conductivities of the pure phase and the phase doped with lithium oxide can be expressed as

$$\frac{\sigma}{\sigma_0} = \frac{x_{Li|Ni|'}}{x^{\circ}_{|e|^{\cdot}}} \qquad (I.250)$$

(σ_0 is the conductivity of pure nickel oxide), provided that the condition

$$x_{Li|Ni|'} \gg x^{\circ}_{|e|^{\cdot}} \qquad (I.251)$$

is satisfied. Hence, if the ratio of the conductivities is plotted against 1/2 Li_2O, a straight line is obtained, from the slope of which the defect electron concentration of pure nickel oxide can easily be determined. Such a plot for the experimental results of Verwey [97] is shown in Fig. I.51, from which the defect electron density of nickel oxide is 5×10^{-6}. Obviously, this calculation of the defect electron concentrations of oxides is permissible only in cases when the foreign ion is completely incorporated into the lattice of the host oxide. According to the X-ray investigations of Fensham [98]

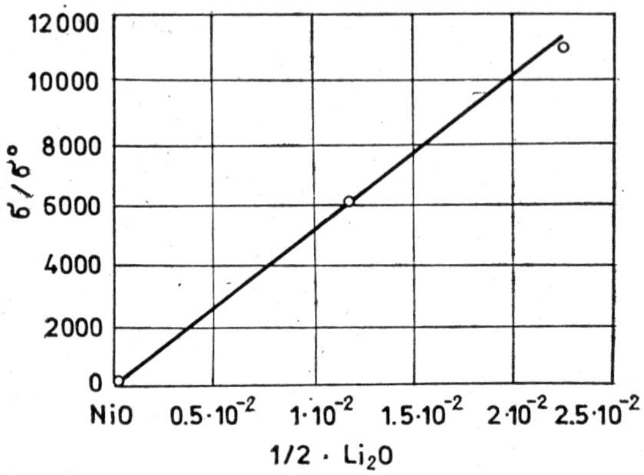

Fig. I.51. The ratio of the conductivities of nickel oxide doped with lithium oxide and of pure nickel oxide as a function of the lithium oxide content of nickel oxide [97]

for the nickel oxide-lithium oxide system this condition is met at 1200 °C (this is also the temperature applied by Verwey in his experiments).

As expected, higher-valency ions such as chromium(III), aluminium(III) and indium(III) decrease the conductivity of nickel oxide. This may be described by the following general equation (Me denotes trivalent metal ion):

$$Me_2O_3 + 2 \mid e \mid^{\cdot} \rightleftarrows 2\, Me \mid Ni \mid^{\cdot} + 2\, NiO + 1/2\, O_2 \qquad (I.252)$$

According to Eq. (I.244) the introduction of higher-valency ions increases the number of nickel vacancies through the reduction of the defect electron concentration. This was observed in the investigations of the oxidation of nickel-chromium alloys, and of the reactivity of nickel oxide doped with chromium oxide in different solid-phase reactions. The defect structures of pure and doped nickel oxides assumed on the basis of these investigations are shown in Fig. I. 52.

In connection with the mechanism of the incorporation of higher-valency ions, further details have been discovered by recent investigations [99—101]. The measurements of Meier and Rapp [101] show that the conductivity of nickel oxide doped with chromium oxide and prepared at 850 °C is proportional to the 1/4th power of the partial pressure of oxygen, whereas

Ni^{2+} O $\quad Ni^{2+} O^{2-}$ $Ni^{2+} O^{2-}$ Ni^{3+} $\qquad$ $Ni^{2+} O^{2-}$ $Ni^{2+} O^{2-}$ $Ni^{3+} O^{2-}$ Ni^{3+}

O^{2-} $Ni^{3+} O^{2-}$ $Ni^{2+} O^{2-}$ $Ni^{2+} O^{2-}$ $\qquad$ O^{2-} $Ni^{3+} O^{2-}$ $Ni^{2+} O^{2-}$ $Ni^{2+} O^{2-}$

$Ni^{3+} O^{2-}$ $Ni^{2+} O^{2-}$ □ O^{2-} Ni^{2+} $\qquad$ $Ni^{3+} O^{2-}$ Li^+ O^{2-} □ O^{2-} Li^+

O^{2-} $Ni^{2+} O^{2-}$ $Ni^{2+} O^{2-}$ $Ni^{2+} O^{2-}$ $\qquad$ O^{2-} $Ni^{2+} O^{2-}$ $Ni^{3+} O^{2-}$ $Ni^{2+} O^{2-}$

$Ni^{2+} O^{2-}$ □ O^{2-} $Ni^{2+} O^{2-}$ Ni^{3+} $\qquad$ Li^+ O^{2-} □ O^{2-} $Ni^{2+} O^{2-}$ Ni^{3+}

O $\quad Ni^{2+} O^{2-}$ $Ni^{2+} O^{2-}$ $Ni^{2+} O^{2-}$ $\qquad$ O^{2-} $Ni^{2+} O^{2-}$ $Ni^{2+} O^{2-}$ $Ni^{2+} O^{2-}$

a) $\qquad\qquad$ b)

Ni^{2+} O^{2-} $Ni^{2+} O^{2-}$ $Ni^{2+} O^{2-}$ Ni^{3+}

O^{2-} $Ni^{2+} O^{2-}$ $Ni^{2+} O^{2-}$ $Ni^{2+} O^{2-}$

$Ni^{3+} O^{2-}$ $Ni^{2+} O^{2-}$ □ O^{2-} Cr^{3+}

O^{2-} $Ni^{2+} O^{2-}$ $Ni^{2+} O^{2-}$ $Ni^{2+} O^{2-}$

$Cr^{3+} O^{2-}$ □ O^{2-} $Ni^{2+} O^{2-}$ Ni^{2+}

O^{2-} $Ni^{2+} O^{2-}$ $Ni^{2+} O^{2-}$ $Ni^{2+} O^{2-}$

c)

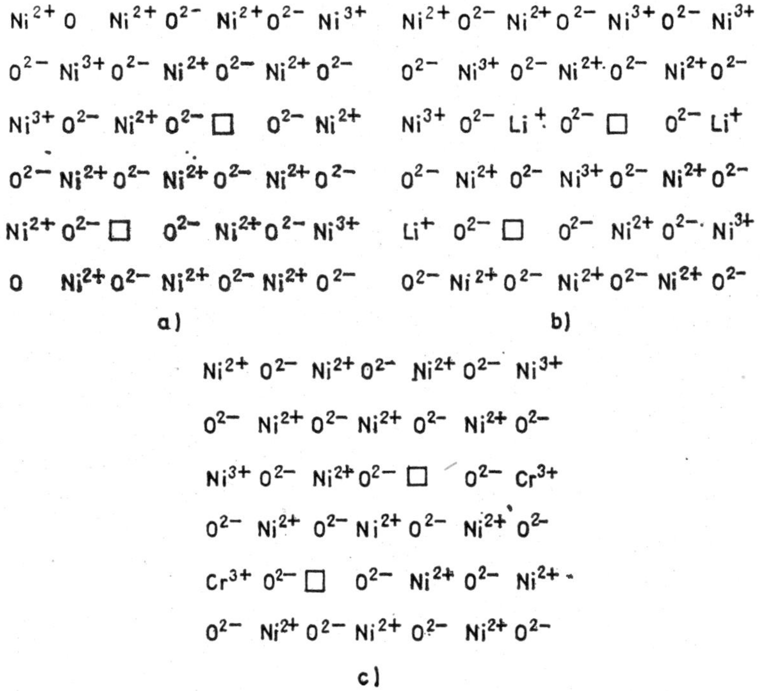

Fig. I.52. The defect structure of nickel oxide; a) pure nickel oxide; b) nickel oxide doped with lithium oxide; c) nickel oxide doped with chromium oxide

that of pure nickel oxide prepared under similar conditions is proportional to its 1/6th power. The conclusion was drawn from this that chromium ions in the nickel oxide lattice are compensated by completely ionized nickel vacancies. It is interesting that the solubility of chromium oxide in nickel oxide is independent of the partial pressure of oxygen. The activation energy of the conductivity is 14.5 kcal mole^{-1} for pure nickel oxide and 20 kcal mole^{-1} for nickel oxide doped with chromium oxide.

Defect structures and conduction properties similar to those of nickel oxide have also been found for cobalt(II) oxide and manganese(II) oxide. The properties of these oxides are therefore not discussed in detail here.

PARAGRAPH 6.5.2. COPPER(I) OXIDE

The conductivity of copper(I) oxide at high temperature is proportional to the partial pressure of oxygen. Hence, copper(I) oxide too can be classified as a defect conductor. The value of n in the electrical conductivity vs. partial pressure of oxygen function [see Eq. (I.247)] is 7. The probable defect structure of copper(I) oxide is shown in Fig. I.53. The oxygen excess (copper defect) is compensated by an appropriate number of bivalent copper ions. The excess of oxygen in copper(I) oxide samples pretreated in different ways was determined by Wagner et al. [102]; it was found that at 1000 °C the excess oxygen content of copper(I) oxide is proportional to the 1/5th power of the oxygen pressure. The oxygen excess may enter the copper(I) oxide lattice in two steps. In the first the oxygen molecule is chemisorbed on the copper(I) oxide surface:

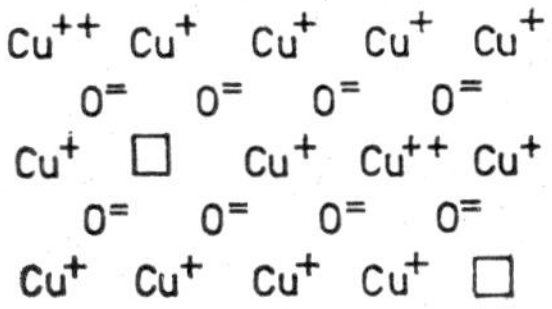

Fig. I.53. The defect structure of copper(I) oxide

$$O_2 = 2\,O^-_{(\text{chem})} + 2\,|\,e\,|^{\bullet} \tag{I.253}$$

In the second step 4 Cu^+ ions of the copper(I) oxide phase move to the surface from the interior of the oxide, with the simultaneous shift of two further electrons, to form a copper(I) oxide lattice with the oxygen atoms:

$$O^-_{(\text{chem})} = Cu_2O + 2\,|\,Cu\,|' + |\,e\,|^{\bullet} \tag{I.254}$$

Every oxygen molecule absorbed creates 4 electron defects and 4 Cu^+ vacancies in the interior of the copper(I) oxide crystal.

On addition of the above two equations:

$$O_2^{(g)} = 2\,Cu_2O + 4\,|\,Cu\,|' + 4\,|\,e\,|^{\bullet} \tag{I.255}$$

If the mass-action law is applied, the dependence of the conductivity on the partial pressure of oxygen is

$$\sigma = \text{const.}\ p_{O_2}^{1/8} \tag{I.256}$$

which is in fair agreement with the experimentally found $p_{O_2}^{1/7}$. More recent investigations on copper(I) oxide indicate that at a lower oxygen pressure intrinsic conduction must also play a role.

On the basis of the experimental results of Mayer and Friedrich [103] chromium(III) oxide has been classified as a defect conductor. This is supported by the experiments of Anderson et al. [104], which indicate a slight increase in the conductivity with increasing oxygen pressure. On the other hand, the conductivity was found to decrease when the oxygen atmosphere was replaced by a $H_2 - H_2O$ or $CO - CO_2$ system. Apart from the effect of the partial pressure of oxygen on the conductivity, doping experiments with foreign ions also prove the defect conductor character of chromium(III) oxide very convincingly. On analogy with other defect conductor oxides, defect conduction may be attributed to the following defect site reaction:

$$3/4\ O_2 = |\ Cr\ |''' + 3\ |\ e\ |^{\cdot} + 1/2\ Cr_2O_3 \qquad (I.257)$$

The dependence of the defect electron concentration and conductivity on the oxygen pressure can hence be calculated in the conventional manner. The value found for n [see Eq. (I.247)] is 5.3. The experiments of Hauffe and Block [105], however, led to an extremely high value for n (30!), which indicates that although the dependence of the conductivity on the oxygen pressure suggests the defect conductor character of chromium(III) oxide, this defect conduction can by no means be attributed to the above defect site reaction. Therefore it was assumed that in chromium(III) oxide two equilibria exist simultaneously. The first, which can be described by the above equation, yields only a small number of defect electrons. The

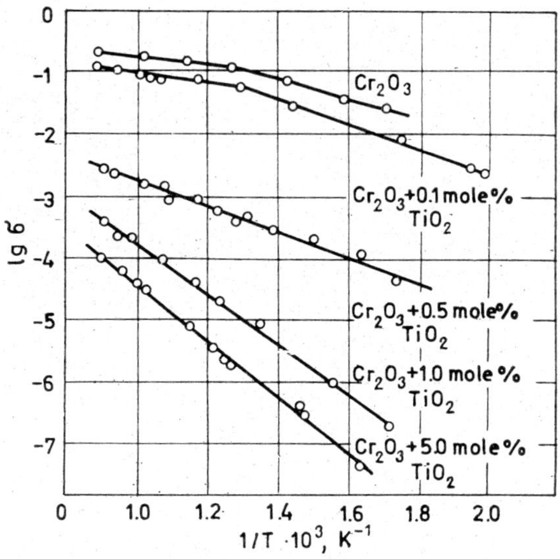

Fig. I.54. The conductivity of chromium oxide doped with titanium dioxide as a function of the reciprocal temperature [105]

defect conduction of chromium(III) oxide is due, in fact, to a defect site reaction independent of the partial pressure of oxygen. This reaction was described by the following equations:

$$Cr_2O_3 \text{ (unperturbed lattice)} = |\,Cr\,|''' + Cr^{\cdot\cdot} + |\,e\,|^{\cdot} \quad (I.258)$$

$$Cr_2O_3 \text{ (unperturbed lattice)} = |\,Cr\,|''' + Cr^{\cdot} + 2\,|\,e\,|^{\cdot} \quad (I.259)$$

Accordingly, both interstitial chromium ions and chromium vacancies should be taken into account in the interpretation.

If the mass-action law is applied to Eq. (I.258) we obtain:

$$x_{|Cr|'''}\, x_{Cr^{\cdot\cdot}}\, x_{|e|^{\cdot}} = K_2 \quad (I.260)$$

Furthermore, for pure chromium oxide it is valid:

$$x^{o}_{|Cr|'''} = x^{o}_{Cr^{\cdot\cdot}} = x^{o}_{|e|^{\cdot}} = \sqrt[3]{K_2} \quad (I.261)$$

The conductivity of chromium(III) oxide decreases with the incorporation of higher-valency ions (Fig. I.54), whereas it increases with the incorporation of nickel ions with their lower valency. On the basis of the defect structure discussed in the foregoing the effect of doping with foreign ions can be given by the following equations:

$$|\,e\,|^{\cdot} + TiO_2 = Ti\,|\,Cr\,|^{\cdot} + 1/2\;Cr_2O_3 + 1/4\;O_2 \quad (I.262)$$

$$3\,|\,e\,|^{\cdot} + WO_3 = W\,|\,Cr\,|^{\cdot\cdot\cdot} + 1/2\;Cr_2O_3 + 3/2\;O_2 \quad (I.263)$$

$$Cr^{\cdot\cdot} + |\,e\,|^{\cdot} + WO_3 = W\,|\,Cr\,|^{\cdot\cdot\cdot} + Cr_2O_3 \quad (I.264)$$

and

$$1/4\;O_2 + NiO = Ni\,|\,Cr\,|' + |\,e\,|^{\cdot} + 1/2\;Cr_2O_3 \quad (I.265)$$

It is worth noting that the conductivity of chromium (III) oxide remains practically unchanged upon doping with zinc oxide, which can be interpreted by the following defect site reaction:

$$2\;ZnO = 2\;Zn\,|\,Cr\,|' + Cr^{\cdot\cdot} + 1/2\;Cr_2O_3 + 1/4\;O_2 \quad (I.266)$$

According to this, Eq. (I.261) no longer holds; the number of interstitial chromium ions increases and the number of chromium vacancies decreases with increasing zinc oxide content.

The investigations of Fischer et al. [106] have revealed further details of the electrical behaviour of chromium(III) oxide. Thermoelectric potential and conductivity measurements between 700 and 1750 °C indicate that chromium can be regarded as a defect conductor only up to 1250 °C. Above this temperature it behaves as a typical intrinsic conductor oxide. In this temperature range both the electrical resistance and the thermoelectric potential are entirely independent of the partial pressure of oxygen. The introduction of lower-valency ions, copper(I) and magnesium(II), increases the defect electron concentration in both temperature ranges. At 1200°C the electrical resistance and the thermoelectric potential of chromium(III) oxide doped with 1% of copper(I) oxide decrease with the partial pressure of oxygen, and the experimentally determined value of n [see Eq. (I.215b)] is in agreement with the value calculated from the mass-action law and the principle of electroneutrality.

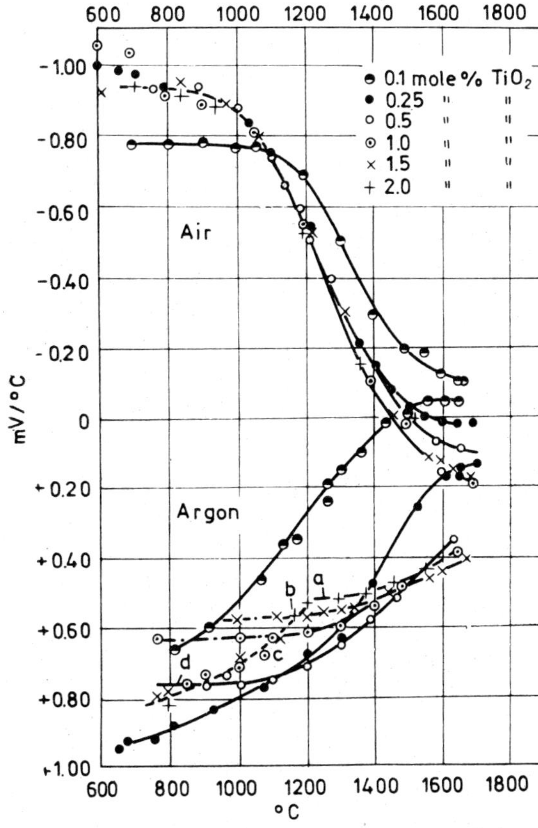

Fig. I.55. The thermoelectric potential of chromium oxide containing titanium oxide, as a function of temperature [106]

The effect of titanium dioxide on the electrical behaviour of chromium oxide is considerably more complex. Irrespective of the amount of titanium dioxide incorporated, chromium oxide behaves as a defect conductor between 600 and 1400 °C in air. In an argon atmosphere (partial pressure 10^{-2} torr), however, as shown in Fig. I.55, the value of the thermoelectric potential is positive in all cases (the heated electrode is positive), which supports the assumption of an n-type conducting character. This observation indicates that in an oxygen-free atmosphere the incorporation of titanium dioxide proceeds according to the equation

$$TiO_2 = Ti \mid Cr \mid^{\cdot} + e' + 1/2\ Cr_2O_3 + 1/2\ O_2 \qquad (I.267)$$

[This mechanism, of course, occurs only when the titanium dioxide has consumed all the defect electrons of the chromium oxide according to Eq. (I.262).] The electrical resistance of chromium oxide is influenced by titanium dioxide in agreement with the above. In an argon atmosphere the resistance of chromium oxide decreases, whereas in air it slightly increases with increasing titanium dioxide content.

SECTION 6.6. DEFECT STRUCTURES AND CONDUCTION PROPERTIES OF INTRINSIC CONDUCTOR AND AMPHOTERIC OXIDES

Although the majority of semiconductor oxides are n- or p-type conductors, there are certain oxides which, on the basis of their electrical behaviour, cannot be classified rigorously as either n- or p-type conductors. Their most characteristic feature is that the conduction properties depend very sensitively on their pretreatment, the conditions of the measurements, etc. The differences are due to the fact that in the cases discussed so far the deviations from stoichiometry were fairly pronounced, and could, at most, only be increased or decreased by external effects (doping with foreign ions, use of a different gas atmosphere, etc.).

The group of oxides which will now be discussed, however, have nearly stoichiometric compositions; depending on the external conditions, different defect structures may occur, and this of course, influences their conduction character, too. Consequently, different pretreatment methods may result not only in quantitative but also in qualitative changes. By means of the properties one can distinguish between intrinsic conductors and amphoteric semiconductors. Intrinsic conduction may take place if, although the crystal contains no excess metal or non-metal component, one of the constituents of the crystal can relatively easily lose an electron, and so equal numbers of electrons and defect electrons take part in the conduction. Materials which, depending on the experimental conditions, may contain either metal or non-metal components in excess are regarded as amphoteric semiconductors.

PARAGRAPH 6.6.1. COPPER OXIDE

Although the occurrence of electron conduction is generally closely related with the non-stoichiometric composition of the crystal, copper(II) oxide, despite its stoichiometric composition, can be regarded as an electron conductor in a strict sense. From a knowledge of the transition number of copper ions it can be assumed that copper ions take practically no part in the conduction. The same holds for oxygen, due to its large diameter and deformability. Between 800 and 1000 °C, the conductivity of copper(II) oxide scarcely depends on the partial pressure of oxygen, and thus copper oxide has been classified as an intrinsic conductor by Wagner. Intrinsic conduction takes place in copper(II) oxide according to the equations

$$2\,Cu^{2+} = Cu^{+} + Cu^{3+} \tag{I.268}$$

$$\text{zero} = e' + |\,e\,|^{\cdot} \tag{I.269}$$

i.e. one of the copper(II) ions becomes trivalent and the other monovalent. The trivalent ion represents the defect electron, the monovalent one the electron.

The intrinsic conductor behaviour of copper oxide is supported by the results of doping experiments. The incorporation of lower-valency ions leads to the formation of a certain number of trivalent copper ions, while the incorporation of higher-valency ions involves the formation of monovalent copper ions. Consequently, in copper oxide doped with lower-valency or higher-valency ions, defect conduction or electron conduction, respectively, predominates.

The conduction character of copper oxide, can thus be influenced by doping. This is expressed by the following equations:

$$1/2\,O_2 + Li_2O \rightleftarrows 2\,Li\,|\,Cu\,|' + 2\,|\,e\,|^{\cdot} + 2\,CuO \tag{I.270}$$

$$Cr_2O_3 \rightleftarrows 2\,Cr\,|\,Cu\,|^{\cdot} + 2\,e' + 2\,CuO \tag{I.271}$$

The related results of Hauffe and Grünewald [107] are given in Fig. I.56.

The conductivity does not increase to the same extent upon doping with lower or higher-valency ions. According to Hauffe and Grünewald this must be attributed either to the different mobilities of electron defects or

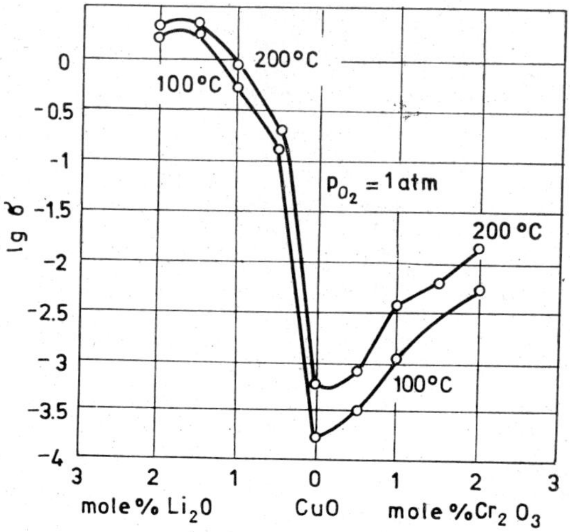

Fig. I.56. The effects of foreign ions on the conductivity of copper(II) oxide [107]

to the effects of foreign ions on the mobilities. Although these investigations support the intrinsic conductor character of copper oxide, recent considerations and results indicate that one has to be cautious concerning the type of the conduction [108]. The fact that the conductivity of copper oxide is independent of the partial pressure of oxygen at high temperatures is not a certain criterion of intrinsic conduction. It might be supposed that the defect electron excess formed through the chemisorption and introduction of oxygen

$$1/2\ O_2 \rightleftarrows O^-_{(chem)} + |\ e\ |^{\cdot} \qquad (I.272)$$

is negligible in comparison with the high defect electron concentration of the oxide. This might also explain why the chemisorption of oxygen does not change the value of the conductivity. This assumption is supported, for example, by the experimental results obtained for the nickel oxide — lithium oxide system. The conductivity of nickel oxide doped with 1% lithium oxide is completely independent of the partial pressure of oxygen, though the mixed oxide cannot be regarded as an intrinsic conductor; indeed, it is a very good defect conductor. The same assumption is supported by conductivity measurements on copper oxide at a lower temperature (500 °C). These experiments led to the result that the conductivity of copper oxide in vacuo increases considerably with increasing oxygen pressure. In this case, due to the low temperature, the defect electron concentration of copper oxide is significantly lower. Consequently, the defect electron excess arising from the chemisorption of oxygen according to Eq. (I.272) is already manifested in the value of the conductivity. The recent doping experiments on copper oxide also contradict the assumption of intrinsic conduction [108]. The doping of copper oxide samples prepared in different ways with higher-valency ions (iron, aluminium, chromium, etc.) does not increase the conductivity of the copper oxide in any of the cases; rather, it slightly decreases both in vacuo and in air. This can be interpreted on the basis of the following dopping equation:

$$2\ |\ e\ |^{\cdot} + Fe_2O_3 = 2\ Fe\ |\ Cu\ |^{\cdot} + 2\ CuO + 1/2\ O_2 \qquad (I.273)$$

When the conductivities of the copper oxide + 1% aluminium oxide systems are plotted against the oxygen pressure, positive n [see in Eq. (I.247)] values are obtained, although according to the assumptions of Hauffe and Grünewald negative n values would be expected. The defect conduction character of copper oxide is also proved by recent thermoelectric potential measurements.

On the basis of the above it may be concluded that under the majority of experimental conditions copper oxide contains a slight oxygen excess, and behaves as a defect conductor. Intrinsic conduction can be observed only under special conditions. The exact nature of these conditions is not yet known.

The variation of the conductivity of calcium oxide with the oxygen pressure can be seen in Fig. I.57. The conductivity decreases with decreasing partial pressure, and attains a minimum at about 10^{-2} torr [109]. Above this value the conductivity increases sharply with further decreasing pressure.

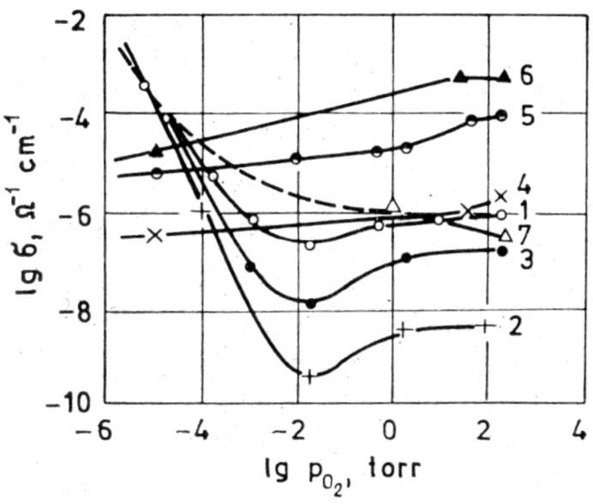

Fig. I.57. The conductivities of pure and doped calcium oxide as functions of the partial pressure of oxygen [109];

1: CaO
2: CaO + 0.1 mole% Y_2O_3
3: CaO + 2.0 mole% Y_2O_3
4: CaO + 0.5 mole% Li_2O
5: CaO + 1.0 mole% Li_2O
6: CaO + 2.0 mole% Li_2O
7: CaO + 1.0 mole% La_2O_3

This behaviour indicates that, depending on the partial pressure of oxygen, not only the number of charge carriers but also the character of the conduction changes (amphoteric semiconductor). At higher oxygen pressure calcium oxide behaves as a defect conductor, and the increasing effect of oxygen on the conductivity can be explained by the increasing defect electron concentration due to the chemisorption of oxygen:

$$1/2\ O_2^{(g)} = O^-_{(chem)} + |\,e\,|^{\cdot} \qquad (I.274)$$

or at sufficiently high temperature to the introduction of oxygen:

$$1/2\ O_2^{(g)} = CaO + |\,Ca\,|'' + 2\,|\,e\,|^{\cdot} \qquad (I.275)$$

At an oxygen pressure lower than 10^{-2} torr the rapid increase of the conductivity with decreasing pressure indicates an n-type conductor character, which can be attributed to the defect site reaction

$$CaO\ (\text{unperturbed lattice}) = |\,O\,|^{\cdot\cdot} + 2\,e' + 1/2\ O_2^{(g)} \qquad (I.276)$$

In the range around 10^{-2} torr oxygen pressure, at the conductivity minimum, a quasi-intrinsic conduction appears.

The electron conduction of calcium oxide can easily be interpreted by the metal excess due to the formation of oxygen ion vacancies. On the other hand, the interpretation of defect conduction is a much harder task, since in the cases discussed so far defect conduction takes place only if there are different valence states of the cation. Accordingly, the appearance of defect conduction can be interpreted only by assuming the formation of monovalent oxygen ions [109]. Further evidence of the amphoteric behaviour of calcium oxide is that the conductivity changes if ions of a different valence state are introduced. The variations of the conductivity with the partial pressure of oxygen for calcium oxide doped with lithium oxide, yttrium oxide or lanthanum oxide are also given in Fig. I.57. It can be observed that in the range of defect conduction the higher-valency ions decrease the conductivity of the calcium oxide; this can be described by the following reactions:

$$2\ |\ e\ |^{\cdot} + Y_2O_3 \rightleftarrows 2\ Y\ |\ Ca\ |^{\cdot} + 2\ CaO + 1/2\ O_2 \qquad \text{(I.277)}$$

$$2\ |\ e\ |^{\cdot} + La_2O_3 \rightleftarrows 2\ La\ |\ Ca\ |^{\cdot} + 2\ CaO + 1/2\ O_2 \qquad \text{(I.278)}$$

On the basis of the behaviour of n-type conductor oxides it would be expected that in the range of n-type conduction higher-valency ions would increase the electron concentration:

$$La_2O_3 \rightleftarrows 2\ La\ |\ Ca\ |^{\cdot} + 2\ e' + 2\ CaO + 1/2\ O_2 \qquad \text{(I.279)}$$

and, consequently, also the conductivity of calcium oxide. However, this is the situation only in the case of lanthanum oxide, whereas the conductivity of calcium oxide doped with yttrium oxide does not exceed the conductivity of pure calcium oxide even at a very low oxygen pressure. This behaviour was attributed to the low solubility of yttrium oxide. Thus, the electron excess formed in a manner similar to that described by the above equation is substantially lower than the number of electrons arising from reaction (I.274) on calcium oxide in vacuum. According to expectation lithium oxide increases the defect conductivity of calcium oxide:

$$1/2\ O_2 + Li_2O \rightleftarrows 2\ Li\ |\ Ca\ |' + 2\ |\ e\ |^{\cdot} + 2\ CaO \qquad \text{(I.280)}$$

The variation of the conductivity of a calcium oxide-lithium oxide mixture with the oxygen pressure, however, indicates that the type of conduction does not change with decreasing oxygen pressure, and calcium oxide doped with lithium oxide is predominantly a defect conductor even in vacuum.

PARAGRAPH 6.6.3. BISMUTH OXIDE

The electrical behaviour of bismuth oxide is in many respects similar to that of calcium oxide, since the character of conduction varies with the partial pressure of oxygen [110, 111]. Below 650 °C, in an oxygen atmosphere, it behaves as a defect conductor, but at a higher temperature, and an oxygen pressure of 10^{-3} torr, bismuth oxide is an n-type conductor. Between 500 and 710 °C the conductivity varies with the partial pressure of oxygen according to $p_{O_2}^{1/4.5}$, from which the occurrence of the defect site reaction:

$$3/2\ O_2 = Bi_2O_3 + 2\ |\ Bi\ |'' + 6\ |\ e\ |^{\cdot} \qquad (I.281)$$

has been concluded.

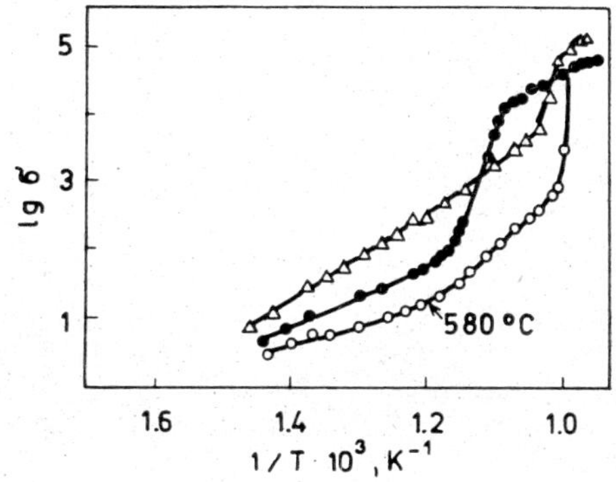

Fig. I.58. The logarithm of the electrical conductivity of bismuth oxide as a function of the reciprocal temperature [116]

○ in the presence of 0.2 torr oxygen, on heating
● in the presence of 0.2 torr oxygen, on cooling
△ in the presence of 157 torr oxygen, on heating

In the temperature range 300—580 °C the oxygen pressure-dependence of the conductivity was found to be

$$\sigma = k\, p_{O_2}^{1/4} \qquad (I.282)$$

The electrical conductivity of bismuth oxide plotted against $1/T$ can be seen in Fig. I.58 [116].

Under an oxygen pressure of 0.2 torr at 580 °C, and 157 torr at 650 °C, break-points can be observed in the $\log\sigma$ vs. $1/T$ curves. It is apparent from thermoelectric potential measurements that the changes in the slope of the curves are due to a p-n-type change.

PARAGRAPH 6.6.4. IRON(III) OXIDE

Opinions concerning the conduction type of iron(III) oxide are rather contradictory. Since its conductivity is independent of the oxygen pressure, iron(III) oxide was classified by Wagner as an intrinsic conductor, whereas it has been classified by other research workers as an electron excess conductor. The real situation is that the composition of iron(III) oxide, too, depends very sensitively on the conditions of pretreatment, and this in turn influences the conduction properties. In order to demonstrate the effect of pretreatment, the results of Morin [113] are cited. In these studies the deviation from stoichiometry was determined by density measurements on iron(III) oxide. The density of iron(III) oxide treated in air (specimen 5 in Fig. I.59) was regarded as a basis, since in this case the deviation from stoichiometry was so low ($Fe_{2.0005}O_3$) that it could not be detected by density measurements. The first four samples given in Table I.5 were treated

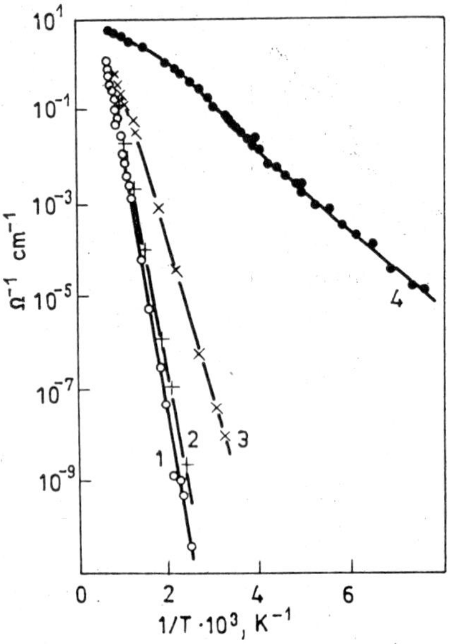

Fig. I.59. The conductivities of pure iron(III) oxide and of iron(III) oxide doped with titanium dioxide as functions of the reciprocal temperature. Samples were heat-treated in an oxygen atmosphere. The compositions of the samples are given in Table 5 [113]

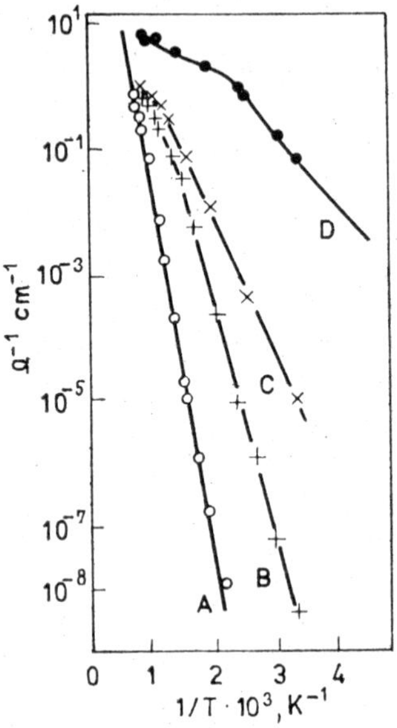

Fig. I.60. The conductivities of pure iron(III) oxide and of iron(III) oxide doped with titanium dioxide as functions of the reciprocal temperature. Samples were heat-treated in a mild reducing atmosphere (74 N_2 + 26 O_2). The compositions of the samples are given in Table 5 [113]

TABLE I.5

The compositions of pure and doped iron(III) oxide

Number of sample	Amount of titanium atom%	Density	Number of Fe atoms cm^{-3}	Pretreatment
1	0.00	5.179	3.977×10^{22}	
2	0.05	—	3.975×10^{22}	
3	0.20	—	3.969×10^{22}	ignited in O_2
4	1.00	—	3.937×10^{22}	
5	—	5.190	4.000×10^{22}	
A	0.00	5.211	4.023×10^{22}	
B	0.05	—	4.021×10^{22}	ignited in air
C	0.20	—	4.015×10^{22}	
D	1.00	—	3.983×10^{22}	

in oxygen at atmospheric pressure, whereas samples A to D were treated in air at 1100 °C. If a perfect oxygen lattice is assumed in iron(III) oxide, it can be calculated that the oxidized samples have an iron deficiency of about 2.3×10^{20} atom/cm^3, whereas the samples treated in air contain about the same iron excess. Correspondingly, iron(III) oxide may exhibit p- and n-type conduction properties, depending on the pretreatment. This fact has also been proved by thermoelectric potential measurements. As can be seen in Figs I.59 and I.60, however, the conductivities of reduced and oxidized iron(III) oxide samples are not essentially different. The conductivities of samples doped with titanium dioxide are significantly higher at low temperatures than that of pure iron(III) oxide, and depend on the quantity of titanium dioxide incorporated. The conductivities of samples prepared in a slightly reducing atmosphere are always higher than those of samples treated in an oxidizing atmosphere. There is no significant difference between the conductivities of samples 4 and D, indicating that the number of defects created by different pretreatment methods can be neglected in comparison with the number of defects formed upon doping with 1% of titanium.

The effects of lower-valency ions on the electrical behaviour of ferric oxide have been studied by Cormack, Gardner and Moss [112]. Thermoelectric potential and conductivity measurements showed that on the incorporation of 0.01 atom% magnesium into weakly *n*-type ferric oxide the conduction character changes to p-type. It was also found that if the pretreatment is carried out at 1300 °C in air, with up to 0.2% magnesium concentration, every magnesium ion gives rise to one defect electron. When a higher magnesium concentration is applied, the formation of a magnesium-ferrite spinel is preferred.

It is generally known that the catalytic behaviour and reactivity of ferric oxide change very sensitively with the temperature of pretreatment. Samples treated at 650—700 °C show an exceptionally high reactivity. The electron diffraction studies of Finch and Sinha [114] showed that this is the temperature range where the $\beta \rightarrow \gamma$ transition of iron(III) oxide takes place. It appeared probable that the high reactivity of ferric oxide is a consequence of the defects formed during the transformation. In order to prove this assumption the electrical parameters of ferric oxide samples treated at different temperatures were determined [115]. Some of the results obtained are given in Fig. I.61.

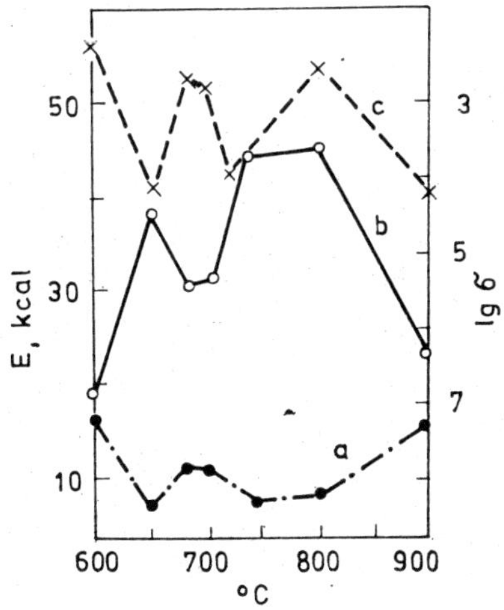

Fig. I.61. Electrical behaviour of ferric oxide as a function of the conditions of heat-treatment [115] a) the logarithm of the specific conductivity of ferric oxide at 300 °C, in air, b) the activation energy of the conductivity of ferric oxide, in air, c) the logarithm of the specific conductivity of ferric oxide at 500 °C, in vacuum

The electrical conductivity of ferric oxide as a function of the temperature of pretreatment has a maximum between 685 and 700 °C if the treatment is carried out either in air or in vacuum. In agreement with the variation of the conductivity the activation energy of the conductivity has a minimum at 685—700 °C. The conductivities of ferric oxide samples treated at different temperatures also depend on the partial pressure of oxygen in different ways. In the oxygen pressure range between 10^{-4} and 10^{-2} torr the conductivities of ferric oxide samples are practically independent of the partial pressure of oxygen. Between 10^{-2} and 1 torr the conductivity decreases significantly for all samples. The highest reduction, 2.5—3 orders of magnitude, can be found for samples treated at 600—650 °C, and the lowest reduction, 0.5—0.8 orders of magnitude, for samples treated at temperatures above 720 °C. At pressures above 1 torr the conductivity of ferric oxide increases slightly with increasing oxygen pressure.

The density of a ferric oxide sample also changes with the temperature of pretreatment: it increases up to 650 °C, rapidly decreases between 567 and 720 °C, and increases again above 720 °C.

It has been concluded from the above results that the increased conductivity of ferric oxide samples treated at 675—700 °C can be attributed to the defect site reaction:

$$3/4\ O_2 = |\ Fe\ |''' + 3\ |\ e\ |^{\cdot} + 1/2\ Fe_2O_3 \qquad (I.283)$$

whereas the high reactivity of ferric oxide can be attributed to the increasing number of iron defect sites, and, thus, to the higher mobility of the ferric ions.

REFERENCES TO PART I

I. CITED REFERENCES

1. See e.g. Wannier, G. H.: Elements of Solid State Theory. Cambridge University Press, Cambridge, England, 1960, p. 12
2. See e.g. Kittel, C.: Introduction to Solid State Physics. John Wiley and Sons Inc., New York; Chapman and Hall Ltd., London, 1953, p. 9
3. See e.g. Wannier, G. H.: loc. cit. (1), p. 14
4. See e.g. Kittel, C.: loc. cit. (2), p. 12
5. Reitz, J. R.: Solid State Physics. Eds. Seitz, F. and Turnbull, D. Academic Press, New York, 1955. I. p. 1
6. Erdey-Gruz, T. and Schay G.: Elméleti fizikai kémia. Tankönyvkiadó, Budapest, 1964. I., p. 411
7. Coulson, C. A.: Valence. Clarendon Press, Oxford, 1953, p. 271
8. Seitz, F.: Modern Theory of Solids. McGraw-Hill, New York–London, 1940, p. 142
9. Coulson, C. A.: loc. cit. (7), p. 279
10. Ladik, J. and Biczó, G.: Acta Phys. Hung. **47,** 263 (1966)
11. Pauling, L.: Proc. Roy. Soc. **196,** 343 (1949)
12. Coulson, C. A.: loc. cit. (7) p. 289
13. Pauling, L.: J. Am. Chem. Soc. **69,** 542 (1947)
14. See e.g. Reitz, J. R.: loc. cit. (5), p. 190
15. Seitz, F.: loc. cit. (8), p. 146
16. Seitz, F.: loc. cit. (8), p. 151
17. Seitz, F.: loc. cit. (8), p. 109
18. Seitz, F.: loc. cit. (8), p. 159
19. Kronig, R. and Penney, W. G.: Proc. Roy. Soc. **A 130,** 499 (1931)
20. Slater, J. C.: Phys. Rev. **87,** 807 (1952)
21. Reitz, J. D.: loc. cit. (5), p. 29
22. Reitz, J. D.: loc. cit. (5), p. 49
23. Slater, J. C. and Koster, G. F.: Phys. Rev. **94,** 1498 (1954)
24. Shockley, W.: Electrons and Holes in Semiconductors. Van Nostrand, Princeton, 1955, p. 221
25. Shockley, W.: loc. cit. (24), p. 463
26. Seitz, F.: loc. cit. (8), p. 168
27. Seitz, F.: loc. cit. (8), p. 520; Bardeen, J.: Phys. Rev. **52,** 688 (1937)
28. Seitz, F.: loc. cit. (8), p. 529
29. Shockley, W.: loc. cit. (24), p. 278
30. Shockley, W.: loc. cit. (24), p. 264
31. Seitz, F.: loc. cit. (8), p. 181
32. Seitz, F.: loc. cit. (8), p. 186
33. Seitz, F.: loc. cit. (8), p. 576
34. Langevin, P.: J. phys. **4,** 678 (1905)
35. Seitz, F.: loc. cit. (8), p. 583
36. Peierls, R.: Z. Physik **81,** 186 (1933)
37. Seitz, F.: loc. cit. (8), p. 595
38. Pauli, W.: Z. Physik **41,** 81 (1927)
39. Seitz, F.: loc. cit. (8), p. 599
40. Heisenberg, W.: Z. Physik **49,** 619 (1928)

41. See e.g. Seitz, F.: loc. cit. (8), p. 613
42. Kittel, C. and Galt, J. K.: Solid State Physics. Eds. Seitz, F. and Turnbull, D. Academic Press, New York, 1957, Vol. III. p. 421
43. Tamm, I.: Physik. Z. Sowj. **1,** 733 (1932)
44. Seitz, F.: loc. cit. (8), p. 321
45. Fowler, R. H.: Proc. Roy. Soc. **141,** 56 (1953)
46. Shockley, W.: Phys. Rev. **56,** 317 (1939)
47. Seitz, F.: loc. cit. (8), p. 323
48. Seitz, F.: loc. cit. (8), p. 324
49. Goodwin, E. T.: Proc. Camb. Phil. Soc. **35,** 205, 221, 232 (1939)
50. Artmann, K.: Z. Physik **131,** 244 (1952)
51. Grimley, T. B.: Adv. Catalysis **12,** 1 (1960)
52. Koutecký, J.: Physica status solidi **1,** 554 (1964)
53. Davison, S. G. and Koutecký J.: Proc. Phys. Soc. (London) **89,** 237 (1966)
54. Levine, J. D. and Davison, S. G.: Phys. Rev. **174,** 911 (1968)
55. Koutecký, J.: Phys. Rev. **108,** 13 (1957)
56. Koutecký, J.: Czech. J. Phys. **B11,** 565 (1961)
57. Koutecký, J. and Tomasek, M.: Phys. Rev. **120,** 1212 (1960)
58. Koutecký, J.: Czech. J. Phys. **B12,** 184 (1962)
59. Freeman, S.: Phys. Rev. **B2,** 3272 (1970)
60. Levine, J. D. and Freeman, S.: Phys. Rev. **B2,** 3255 (1970)
61. Stern, P. S. and Green, M. E.: J. Chem. Phys. **58,** 2507 (1973)
62. Eley, D. D. and Newman, O. M. G.: Trans. Faraday Soc. **66,** 110 (1970)
63. Davison, S. G. and Steślicka, M.: Int. J. Quantum Chem. **4,** 445 (1971)
64. Subramanian, R. and Bhagwat, K. V.: J. Phys. C.: Solid State Phys. **5,** 798 (1972)
65. Biczó, G. Kertész, M. and Suhai, S.: Z. Chem. **15,** 203 (1975)
66. Davison, S. G. and Levine, J. D.: Solid State Physics. Ed. Seitz, Turnbull, Ehrenreich; Academic Press, New York, 1970, Vol. 25, p. 1
67. Koutecký, J.: Advances in Chemical Physics. Ed. Prigogine, I. Wiley–Interscience, New York, 1965 **9,** 85
68. Koutecký, J. and Davison, S. G.: Int. J. Quantum Chem. **2,** 291 (1968)
69. Levine, J. D. and Mark, P.: Phys. Rev. **182,** 926 (1969)
70. Frank, F. C.: Acta Cryst. **4,** 497 (1951)
71. Cottrell, A. H.: Plastic Flow in Crystals. Clarendon Press, Oxford, 1953
72. Baumbach, H. H. and Wagner, C.: Z. phys. Chem. **B22,** 199 (1933)
73. Arnold, K.: Erweitertes Kolloquium über Fragen der heterogenen Katalyse. Berlin, 1962
74. Cimino, A., Mazzone, G. and Porta, P.: Z. phys. Chem. **41,** 154 (1964)
75. Hauffe, K. and Block, J.: Z. phys. Chem. **190,** 438 (1950); Hauffe, K. and Vierk, A. L.: Z. phys. Chem. **196,** 160 (1950)
76. Bremer, H., Vogt, F. and Wendlant, K. P.: Z. Chem. **7,** 441 (1967)
77. Allsopp, H. J.: J. Soc. Anal. Chem. **82,** 474 (1957)
78. Derén, J. and Kowalska, A.: Chem. Analit. (Warsaw) **7,** 563 (1962)
79. Ehrlich, P.: Z. Elektrochem. **45,** 362 (1939)
80. Hauffe, K., Grünewald, H. and Tränckler–Gresse, R.: Z. Elektrochem. **56,** 937 (1952)
81. Greener, E. H., Barone, F. J. and Hirthe, W. M.: J. Am. Ceram. Soc. **48,** 623 (1965)
82. Förland, K. S.: Acta Chem. Scand. **18,** 1267 (1964)
83. Kofstad, P.: J. Phys. Chem. Solids **23,** 1579 (1962)
84. Haul, R., Just, D. and Dümbgen, G.: Proc. 4th Int. Symp. on the Reactivity of Solids, Amsterdam, 1960. Elsevier, Amsterdam, 1961. p. 65
85. Tannhauser, D. S.: Solid State Commun. **1,** 223 (1963)
86. Blumenthal, R. N., Coburn, J., Baukus, J. and Hirthe, W. M.: J. Phys. Chem. Solids **27,** 643 (1966); J. Electrochem. Soc. **114,** 172 (1967)
87. Hurlen, T.: Acta Chem. Sand. **13,** 365 (1959)
88. Kofstad, P.: J. Less-Common Metals **13,** 635 (1967)
89. Szabó, Z. G. and Solymosi, F.: Acta Chim. Hung. **25,** 145 (1960)
90. Hauffe, K.: Coloquio sobre Química Física de procesosen superficies sólidas, p. 277. Librería Científica Medinaceli, Madrid, 1965

91. Hauffe, K.: Reaktionen in und an festen Stoffen, Springer-Verlag, Berlin, 1966 p. 212
92. Solymosi, F. and Bánsági, T.: Space Engineering (Ed. G. A. Partel), D. Reidel Publishing Company, Doldrecht, 1970, p. 145
93. Cimino, A. and Marezio, M.: J. Phys. Chem. Solids **17,** 57 (1960)
94. Hartmann, W.: Z. Physik **102,** 709 (1936)
95. Bielanski, A., Derén, J., Haber, J. and Sloczynski, J.: Trans. Faraday Soc. **16,** 165 (1962); Derén, J., Haber, J. and Sloczynski, J.: Bull. Acad. Sci. Poland, Sér. sci. chim. **8,** 391 (1960)
96. Dry, M. E. and Stone, F. S.: Disc. Faraday Soc. **28,** 192 (1959)
97. Verwey, E. J. W., Haayman, P. W., Romeyn, F. C. and Ousterhout, G. W.: Philips Research Reports **5,** 173 (1950)
98. Fensham, J.: J. Am. Chem. Soc. **76,** 969 (1954)
99. Schlosser, E. G.: Z. Elektrochem., Ber. Bunsenges. physik. Chem. **65,** 453 (1961)
100. Schwab, G. M. and Schmid, H.: J. Appl. Physics **33,** 426 (1962)
101. Meier, G. H. and Rapp, R. A.: Z. phys. Chem. N. F. **74,** 168 (1971)
102. Baumbach, H. H., Dünwald, H. and Wagner, C.: Z. phys. Chem. **B 22,** 226 (1933)
103. Mayer, W.: Z. Elektrochem. angew. physik. Chem. **50,** 274 (1944); Friedrich, E.: Z. Physik. **34,** 637 (1926)
104. Bevan, D. M., Schelton, J. B. and Anderson, J. S.: J. Chem. Soc. 1729 (1948)
105. Hauffe, K. and Block, J.: Z. phys. Chem. **118,** 232 (1951)
106. Lorenz, G. and Fischer, W. A.: Z. phys. Chem. **18,** 265 (1958); Fischer, W. A. and Lorenz, G.: Arch. Eisenhüttenwes. **28,** 497 (1957); Fischer, W. A. and Dietrich, H.: Z. phys. Chem. **41,** 205 (1964); Dietrich, H. and Fischer, W. A.: Z. phys. Chem. **41,** 287 (1964)
107. Hauffe, K. and Grünewald, H.: Z. phys. Chem. **198,** 248 (1951)
108. Batta, I., Solymosi, F. and Szabó, Z. G.: J. Catalysis **1,** 103 (1962)
109. Hauffe, K. and Tränckler, G.: Z. Physik **136,** 1093 (1953)
110. Hauffe, K. and Peters, H.: Z. phys. Chem. **201,** 121 (1952)
111. Mansfield, R.: Proc. Phys. Soc. **B62,** 476 (1949)
112. Cormack, D., Gardner, R. F. G. and Moss, R. L.: J. Catalysis, **17,** 219 (1970)
113. Morin, F. J.: Phys. Rev. **83,** 1005 (1951)
114. Finch, G. I. and Sinha, K. P.: Proc. Roy. Soc. **A 241,** 1 (1957)
115. Solymosi, F., Jáky, K. and Szabó, Z. G.: Z. anorg. allg. Chem. **368,** 211 (1969)
116. Rao, C. N. R., Subba Rao, G. V. and Ramdas, S.: J. Phys. Chem. **73,** 672 (1969)

II. HANDBOOKS

1. Van Bueren, H. G.: Imperfections in Crystals. North-Holland, Amsterdam, 1961
2. Dekker, A. J.: Solid State Physics. MacMillan, London, 1958
3. Gray, T. J.: The Defect Solid State. Interscience Publ. Inc. New York, 1957
4. Kittel, C.: Introduction to Solid State Physics. John Wiley, New York, Chapman and Hall Ltd. London, 1953
5. Reitz, J. R.: Solid State Physics. Eds. Seitz, F. and Turnbull, D. Academic Press, New York, 1955
6. Seitz, F.: Modern Theory of Solids. McGraw-Hill, New York–London, 1940
7. Shive, J. N.: The Properties, Physics and Design of Semiconductor Devices. Van Nostrand, New York, 1960
8. Shockley, W.: Electrons and Holes in Semiconductors. Van Nostrand, Princeton, 1955
9. Spenke, E.: Elektronische Halbleiter. Springer Verlag, Berlin, 1955
10. Wannier, G. H.: Elements of Solid State Theory. Cambridge University Press, Cambridge, England,1960

III. ADDITIONAL REFERENCES

to Chapter 6. The defect structures and electrical behaviour of semiconductor oxides

1. Koch, F. and Cohen, J. B.: Acta Cryst., **B25**, 275 (1969).
2. Bursill, L. A., Hyde, B. G., Terasaki, S. and Watanabe, D.: Phil. Mag., **20**, 347 (1969)
3. Thornber, M. R. and Bevan, D. J. M.: J. Solid State Chem., **1**, 536, 545 (1970)
4. Allpress, J. G. and Gado, P.: Crystal Lattice Defects, **1**, 331 (1970)
5. Tilley, J. D.: Materials Res. Bull., **5**, 813 (1970)
6. Allpress, R. G., Tilley, R. J. and Sienko, M. J.: J. Solid State Chem. **3**, 440 (1971)
7. Berak, J. and Sienko, M. J.: J. Solid State Chem. **2**, 109 (1970)
8. Bursill, L. A. and Hyde, B. G.: Proc. Roy. Soc., **A320**, 147 (1970)
9. Mulay, L. N. and Danley, W. J.: J. Appl. Phys., **41**, 877 (1970)
10. Chu, E. G. C. W.: Phys. Rev. (B), **1**, 4700 (1970)
11. Martin, W., Gruehn, R. and Schäfer, H.: J. Solid State Chem. **1**, 425 (1970)
12. Eror, N. G. and Smyth, D. M.: Chemistry of Extended Defects in Non-metallic Solids. Ed. Eyring, L. and Keeffe, M. O.: North Holland, Amsterdam, 1970, p. 62
13. Magnéli, A.: Chemistry of Extended Defects in Non-metallic Solids. Ed. Eyring, L. and Keeffe, M. O.: North Holland, Amsterdam, 1970, p. 148
14. Sienko, M. J. and Berak, J. M.: Chemistry of Extended Defects in Non-metallic Solids. Ed. Eyring, L. and Keeffe, M. O.: North Holland, Amsterdam, 1970, p. 541
15. Nakahira, M. and Saeki, E. M.: Proc. 7th. Symp. on Reactivity on Solids. Ed. Anderson, J. S., Roberts, M. W. and Stone, F. S.: Chapman and Hall, London, 1972, p. 97
16. Süptitz, P. and Teltow, J.: Festkörperchemie. Ed. Boldyrev, V. and Meyer, K. VEB Deutscher Verlag für Grundstoffindustrie, Leipzig, 1973, p. 19
17. Anderson, J. S.: Surface and Defect Properties of Solids. Ed. Roberts, M. W. and Thomas, J. M. The Chemical Society, 1972, p. 1

PART II

ADSORPTION ON SOLID-GAS INTERFACES

P. Fejes

INTRODUCTION

The most important steps determining the rates of heterogeneous catalytic reactions are listed in almost all textbooks as follows:

1. diffusion of the reactants to the surface of the catalyst;
2. sorption of the starting materials;
3. the actual catalytic reaction in the sorbed layer;
4. desorption of the reaction products;
5. removal of the desorbed products from the vicinity of the surface.

With properly dimensioned reactors and appropriately chosen catalysts, the diffusion of the reactants to the catalyst surface and the removal of the products from its vicinity proceed at sufficiently high rates to permit neglecting steps 1 and 5. (Porous adsorbents with high specific surface areas are exceptions to this rule. This problem will be dealt with in Section IV.1.4.) Hence, elucidation of the mechanism of a catalytic reaction in effect involves the investigation of steps 2, 3 and 4.

As regards sorption phenomena the laws of chemisorption will be discussed in greater detail in Part III. Physical adsorption, bringing about weak bonds with the surface, plays a rather subordinate role in heterogeneous catalysis. Nevertheless, it was deemed expedient, for didactical reasons, to illustrate the most important fundamental concepts with examples chosen from the field of physical adsorption. Experience gained here will provide a useful basis for the understanding and application of catalyst investigation methods based on up-do-date adsorption principles and techniques.

1. DEFINITION

Gases and vapours will always "condense" on the surface of solids. This concentration increase on the gas-solid interface is called adsorption. The fact of adsorption will be easily understood if it is borne in mind that normal to the surface the field of the surface lattice elements is not balanced and is thus capable of impeding the movement of molecules of foreign substances for various periods of time, and of retaining them on the surface. The phenomenon is called monomolecular or plurimolecular adsorption, depending upon whether the adsorbed phase is one or (perhaps only in places) several molecular layers thick.

Though the nature of surface fields and of adsorption interaction may vary within extremely wide limits and, in all probability, continuously for different materials, it is advisable to distinguish two types of adsorption.

One type is characterized by non-specific interactions between the surface lattice elements of the solid (called the adsorbent) and the molecules of the other, homogeneous (in this discussion almost always gas, called the adsorbate) phase; the forces of attraction and repulsion are similar to those which, in non-ideal gases, cause deviation from the ideal gas law, liquefaction of gases, etc. This type of adsorption, with reference to the nature of the acting forces, is called physical or van der Waals adsorption. As regards its nature and mechanism, the phenomenon of physical adsorption resembles the condensation of a vapour on the surface of its own liquid phase.

In contrast to physical adsorption, we speak of chemisorption* when the gas molecules interact with the surface so strongly that some of the bonds of the adsorbate (and adsorbent) are weakened or even disappear and, instead, new bonds are formed between the adsorbent and the adsorbate. This may involve the displacement of the formed molecular fragments from the surface and — provided the steric and energetic conditions are appropriate — their penetration into the lattice structure of the solid. In the latter case we are dealing with absorption phenomena which, of course, are not always necessarily preceded by chemisorption. (The comparatively large bromine molecules, for instance, without the splitting of the bond between the two atoms, have enough room to penetrate between relatively distant layers of the graphite lattice.)

2. PHYSICAL ADSORPTION AND CHEMISORPTION

There are certain differences (though perhaps not always clearly distinct) in the nature of physical adsorption and chemisorption which can be used for their discrimination.

One of these criteria is the measure of the heat of adsorption.

Adsorption is a spontaneous process proceeding with the decrease of the Gibbs' free enthalpy, in other words, the free enthalpy of the adsorption system (adsorbent + adsorbed gas) and of the gas phase in thermodynamic equilibrium with it is always lower than the overall free enthalpy of the initial system consisting of the separate, not covered adsorbent and the gas phase containing the *total quantity* of the adsorbate. Since binding of the molecules to the surface is equivalent to the loss of some of their degrees of freedom in the gas state, the adsorption process is accompanied by a drop in entropy too.** $\Delta G = \Delta H - T\Delta S$ and thus, of necessity, the enthalpy of the system must also decrease. Therefore, adsorption, i.e. chemisorption is always an exothermic process, and the enthalpy change of the system is the heat of adsorption.

It might be expected from the nature of the forces involved in physical adsorption that the heat of adsorption would hardly differ from the heat of condensation for gases. This is in fact so, but nevertheless there are ex-

*By similar word-formation physical adsorption can be called physisorption.

** For more exact definitions see Chapter 3 on the thermodynamics of adsorption.

ceptional cases in which absolute values over 10 kcal mole^{-1} have been measured.

In contrast, the heat of chemisorption is in most cases of almost the same order of magnitude as the heat effects of chemical reactions. (The chemisorption of oxygen on metals is accompanied by a heat of chemisorption of around -100 kcal mole^{-1} [1].)

Though this distinction may be generally valid, in certain cases (e.g. in the case of hydrogen and certain metals) the heat of chemisorption barely exceeds -3 kcal mole^{-1} [2, 3].

Accordingly, high heats of adsorption (with absolute values over 10 kcal mole^{-1}) are always indicative of chemisorption, whereas in the event of low heats of adsorption there may be some uncertainty in deciding the type of sorption.

The temperature range in which the sorption process takes place may also characterize one or another type of adsorption. It follows from the similarity between physical adsorption and the condensation of gases that physical adsorption is likely to occur mainly near or below the boiling point of the adsorbate. This condition implies that if the equilibrium vapour pressure of the liquid adsorbate at the temperature of measurement is p_0, the prevailing equilibrium pressure is p, and the relative pressure $p/p_0 < 0.01$, then the coverage of the surface in the case of physical adsorption is only a fraction of the monomolecular coverage.

This condition is not sufficiently strict: porous adsorbents with fine capillary structures are capable of binding considerable quantities of gases by means of physical adsorption at relative pressures as low as $p/p_0 = 10^{-8}$ [4], while certain chemisorption types characterized by comparatively low heats of adsorption appear only at fairly high relative pressures where physical adsorption too is already quite considerable.

Chemisorption, like chemical reactions, requires activation energy, and thus proceeds at a measurable rate only at temperatures exceeding a minimum value. Physisorption, in contrast, needs no activation energy and therefore takes place at any temperature (if it takes place at all) at the rate at which the transport processes involved are capable of ensuring the supply of material. This shows that unequivocal results cannot be expected from the measurement of sorption rate: with porous adsorbents the processes known collectively as "pore diffusion transport" may be very slow in physical adsorption too, while (like the many chemical processes which do not require an appreciable activation energy) chemisorption may take place instantaneously at quite low temperatures on surfaces of a "strongly unsaturated nature" [1].

It has already been mentioned that the forces responsible for the physisorption of gases are non-specific, and thus under appropriate experimental conditions physical adsorption may take place on any surface. In the case of chemisorption, on the other hand, much depends on the purity of the surface (whether the surface contains earlier-adsorbed foreign substances) and, due to the specific character of adsorbent-adsorbate interactions, on the fact that not even a clean surface will necessarily interact with any adsorbate.

It has often been suggested that chemisorbed atoms or molecules are localized on the surface, while those physically adsorbed possess a certain freedom of movement (they are delocalized). It will be shown later that localized and delocalized adsorption may occur in both cases.

In certain cases, despite often contradictory views, it is easy to decide whether we are dealing with physisorption or chemisorption. In other cases the problem is not so simple, showing that no dividing line can be drawn between the two types of sorption.

3. ADSORPTION AND ABSORPTION

When two different phases come into contact, both adsorption and absorption may occur. Absorption is characterized by the penetration of one (usually gas) phase into the interior of the other homogeneous phase; in other words, the extension of the "interface" is not limited exclusively to the "surface" of the second phase. If the sorbent is a liquid, the process is that of simple dissolution, but with solids the situation is somewhat more complicated.

Absorption phenomena involving solids can be roughly divided into three groups. The first type is pore diffusion in adsorbents (activated carbon, clay minerals, etc.) having an extensive internal surface interlaced by a fine pore system. This type of absorption is usually accompanied by (physical) adsorption on the pore walls. It is not a characteristic type, though it fulfills the above criteria of absorption.

Nor is the diffusion transport observed along the boundaries of the crystallites in adsorbents having crystalline structures (primarily metals) a characteristic type of absorption, since it is followed by physisorption or chemisorption.

In the third, and really characteristic, absorption process, the sorbate penetrates the crystal structure of the adsorbent (usally some metal), and hence a true solution is formed. In fact, only this last phenomenon is usually called absorption in the strictest sense of the word.

Occurrence of the various types of absorption is closely linked to the structure of the adsorbent. For metals the ratio of the quantities of substances bound by adsorption and absorption (provided this latter takes place at all) is determined by the ratio surface/volume. With coarse powders of relatively low specific surface area absorption is more important, whereas with adsorbents having a high specific surface area, e.g. vacuum-evaporated films, adsorption predominates.

Like any other diffusion process, absorption takes place at a finite rate (at not too high a temperature) and can therefore easily be mistaken for an equally slow phenomenon, activated adsorption (chemisorption) on the free surface. In certain cases, however, it is possible to distinguish between the two processes.

(1) The formation of solid solutions, i.e. true absorption, is usually endothermic, while adsorption is almost always an exothermic process.

(2) The solution of Fick's diffusion equation for more simple cases (e.g. a model referring exclusively to "internal" diffusion or the moving bound-

ary problem characteristic of the oxidation of metals) gives the quantity of adsorbed gas at low coverages as a linear function of the square root of time, and this function contains the diffusion constant characteristic of the rate of the diffusion process.

Interestingly enough, this approximate law is occasionally also valid for higher coverages [5]. The diffusion constant can be determined by comparing experimental results with the theoretical equation, and the temperature-dependence of the diffusion constant may provide information on the nature of the process. Diffusion processes — with the exception of surface migration or dissolution phenomena in the crystal lattice, which require certain activation energies — are usually not activated,* while chemisorption processes are activated ones, their rates depending exponentially on temperature.

(3) For physical adsorption, in the state of adsorptive saturation, the product of the number of sorbed molecules and the area of one molecule must agree as regards order with the adsorbent surface**. Since gas quantities far in excess of monomolecular coverage are required for the formation of true or pseudo-compounds, solid solutions, etc. a marked difference between this product and the surface of the adsorbent is always indicative of absorption (e.g. the formation of palladium, zirconium, titanium hydrides, etc.).

4. EQUILIBRIUM CONDITIONS

For an adsorbent of given mass m_a the specifically adsorbed quantity χ is a function of only the equilibrium pressure p and the temperature T:

$$\chi = f_1(p, T) \tag{II.1}$$

Because of the mathematical equivalence of the three variables, any one of them can be fixed at a given value and the dependence of the other two on each other can be investigated. In most cases the adsorbed quantity is measured as a function of the equilibrium pressure at several fixed temperatures. The adsorption isotherms thus obtained can be described by means of the general isotherm equation:

$$\chi = f_1(p);\ T = \text{constant} \tag{II.2}$$

If the adsorbed quantity is determined as a function of temperature at given constant equilibrium pressures, adsorption isobars of the form:

$$\chi = f_1(T);\ p = \text{constant} \tag{II.3}$$

are obtained.

* More accurately: the apparent activation energy agrees with that calculated from the temperature-dependence of the diffusion constant.

** This problem will be treated more rigorously in Chapter VI.5 in relation to the methods of measuring specific surfaces.

From any of the former two sets of curves it is possible to plot the adsorption isosteres, which cannot be directly measured, but are characterized by the following equation:

$$p = f_2(T); \ \chi = \text{constant} \tag{II.4}$$

In contrast to chemisorption, the determination of equilibrium data for physisorption involves no particular theoretical or experimental difficulties.

By equilibrium we mean that the rates of the two opposite processes, adsorption and desorption, are the same. In many chemisorption processes, only at relatively high temperatures does any significant desorption take place, and the desorbed substances are usually not identical with the initial adsorbate, but various decomposition products of the latter. It is quite possible that decomposition already begins during the chemisorption and not under the influence of subsequent heating, but the strongly bound molecular fragments are able to leave the surface only at higher temperatures. (The decomposition products of hydrocarbons, for instance, quite often form a "tar-like" coating on the adsorbent surface [6]. CO chemisorbed on metal oxides can often be recovered only in the form of CO_2, indicating surface compound formation in the adsorption layer [7].) The presence of decomposition products in the gas phase makes the determination of the chemisorbed quantity by means of the adsorption technique quite impossible. At the same time, there cannot be any possibility of the system reaching thermodynamic equilibrium either. It is unnecessary to emphasize here that thermodynamic principles cannot be applied without reservation to pseudo-equilibrium data obtained under unsuitable experimental conditions.

According to what has been said so far, the danger of arriving at incorrect conclusions by the inadvertent application (on the basis of experimental results) of thermodynamic correlations exists, particularly for inhibited chemisorption processes requiring activation energies.

Summing up, equilibrium data (in addition to physisorption) can in general be obtained merely for "weak" chemisorption, since only in this case is there the possibility of equilibration at not too high experimental temperatures.

Experimental adsorption isotherms can be used, among others, for the determination of the heat of adsorption. From two or more isotherms determined at slightly different temperatures the isosteric heat of adsorption, Q_{st}, pertaining to the adsorbed quantity χ can be calculated by means of a formula analogous to the Clausius-Clapeyron equation:

$$\left(\frac{\mathrm{d} \ln p}{\mathrm{d}T}\right)_\chi = -\frac{Q_{st}}{RT^2} \tag{II.5}$$

The various definitions of the heat of adsorption will be discussed in Section 4.1; all that need be said here is that, as regards its nature, Q_{st} is the differential heat of adsorption for some given coverage. This heat effect would be observed experimentally if the adsorbent were brought into contact with small (theoretically infinitesimally small) "gas doses" and the heat effect accompanying the adsorption were referred to 1 mole of adsorbed gas. Admission of finite gas doses leads to the observation of the overall, integral effect, and thus the heat of adsorption measured under these conditions is called the "integral heat of adsorption"; this is some average value of the differential heat of adsorption depending on the initial and final coverages.

For the theoretical calculation of the heat of adsorption it is necessary to know the potential energy of the molecules bound on the adsorbent surface, or, in other words, the interaction energy of the adsorbent-adsorbate system. The latter reflects the measure and nature of the adsorption forces between the molecules of the adsorbent and of the adsorbate, and as such (in addition to the value of the heat of adsorption) has a decisive influence on the other thermodynamic parameters of the adsorbed layer, too. Theoretical calculation of the adsorption isotherm can finally be traced back to seeking the configuration (i.e. a given "arrangement" of the adsorbate molecules on the adsorbent surface) associated with the minimum potential energy (more precisely: Gibbs' free enthalpy) of the system.

The interaction energy of the adsorbent surface changes from one point to another. If these are taken to be points lying in an x, y, plane*, the interaction energy can be described mathematically by means of a function of the general form $\psi(x, y, z_0)$ where z_0 is fixed. This function is that of a surface, the so-called potential energy surface. The simpler case when ψ is independent of x and y, that is, when the surface is mathematically homogeneous, will be discussed in some detail in Section 1.1.

5. LOCALIZED AND DELOCALIZED ADSORPTION

Whether the adsorbed molecules are bound to one place (localized) or, like a two-dimensional gas, move freely on the surface, depends on the parameters of the potential energy surface characterizing the adsorbent and the adsorbate. If the height of the potential barrier separating the corresponding potential energy minima for the adsorption sites (i.e. certain lattice elements) is greater than the average thermal energy for the corresponding degree of freedom, then after adsorption the molecule is unable to get out of the "potential well" and consequently the adsorption is localized.

If the energy barrier is low, or negligible compared to the average thermal energy of the adsorbed molecules, we are no longer dealing with the adsorption sites which were previously identified with the minima of the potential energy surface. In this case nothing prevents the free displacement of the adsorbed molecules along the surface, and hence this type is called delocalized adsorption.

* It should be noted that in molecular dimensions no adsorbent "surface" can be considered a plane. In this sense the plane concerned is only a mathematical abstraction to help future discussion.

The terms "localized" and "delocalized" have been used in a relative sense, compared to the potential energy surface and the actual thermal energy. This is also manifested in the occasional occurrence of certain displacements in the localized layers too; however, the mobility here is limited to discrete "jumps" now requiring an activation energy, and which take place when the molecules have acquired a thermal energy greater than the height of the potential barrier. In this sense, the delocalized layers are always mobile, while, depending on the circumstances, the localized layers are either mobile or immobile.

In order to avoid any possible misunderstanding, in the following we shall only speak about localized and delocalized layers, respectively, with the reservation that the possibility of displacement is not excluded in the case of localized adsorption either.

CHAPTER 1. TYPES OF ADSORBENT-ADSORBATE INTERACTION

Following the well-proven methods of other branches of physics and chemistry, in some simpler cases the adsorption forces will be considered to be independent of each other. The London dispersion forces, the polarization forces of an electrostatic nature due to the interaction of atoms or molecules with ions or dipoles, the Coulomb forces of attraction and repulsion between ions and dipoles, the exchange forces resulting in covalent bonds, etc. will be treated separately. Below, a more detailed consideration will be given only to some of the more important of these forces.

Provided the potential functions corresponding to these forces were known, it would be theoretically possible to determine the potential energy change (adsorption energy) accompanying adsorption.

The first difficulty encountered here is our unsatisfactory knowledge concerning the structure of the adsorbing surface and of the adsorption bond distances.

The distance between an adsorbed atom and the surface is determined by the equilibrium between the forces of attraction and repulsion. Its value can be approximately determined from some arrangement of the atoms or molecules of the adsorbate similar to that in physical adsorption (e.g. from the van der Waals radii). The adsorption bond distance is the sum of the atomic radius thus obtained and the radius of the surface atoms.

Secondly, accurate calculation of the adsorption energy is impeded by the fact that the repulsive forces arising from the mutual penetration of the electron clouds of different atoms are not sufficiently well known.

In the calculation of the lattice energies of ionic crystals, e.g. alkali metal halides, alkaline earth metal oxides, etc., Born and Mayer accounted for the potential of repulsive forces by means of the following equation [8]:

$$\psi_r = R' \exp(-r/\varrho) \tag{II.6}$$

where R' and ϱ are constants, and r is the distance between the atoms.

In other cases, as for example in the investigation of the formation of alkali halide ion-pairs in the vapour state, the same authors achieved good agreement with the following empirical correlation:

$$\psi_r = R/r^{12} \tag{II.7}$$

where R was a constant [9]. It seems that for greater ionic distances (e.g. the lattice distance) Eq. (II.6) is the more appropriate, while for smaller distances, such as the distance between the ions of ion-pairs, Eq. (II.7) furnishes the more satisfactory results.

It will be shown in the following Section that the potential of non-polar van der Waals attractive forces can be described by the equation

$$\psi_{at}^{(d)} = -C_1/r^6 \tag{II.8}$$

where C is a constant. Thus, considering the overall effect of the repulsive forces, the potential function will take the form

$$\psi^{(d)} = \psi_{at}^{(d)} + \psi_r = -\frac{C_1}{r^6} + \frac{B}{r^{12}} \tag{II.9}$$

From the relation $\frac{d\psi^{(d)}}{dr} = 0$, the equilibrium atomic distance r_0 is obtained as

$$r_0 = \sqrt[6]{\frac{2R}{C_1}}$$

Consequently, the potential energy decrease on the interaction of two (one surface and one absorbed) atoms is:

$$\psi_{r=r_0}^{(d)} = -\frac{1}{2}\frac{C_1}{r_0^6}$$

Figure II.1 shows the shapes of the corresponding potential curves. It appears clearly from the figure that without accounting for the repulsive forces the potential energy decrease would be twice that given above. In rough estimations the repulsion potential is either completely neglected, or the adsorption energy is calculated from the attraction potential alone and a fixed percentage (e.g. 40%) is deducted from the result [10].

Despite these difficulties, very accurate adsorption energies have been calculated for certain more simple adsorbates and adsorbents, essentially by means of a refinement of Eq. (II.9) [11].

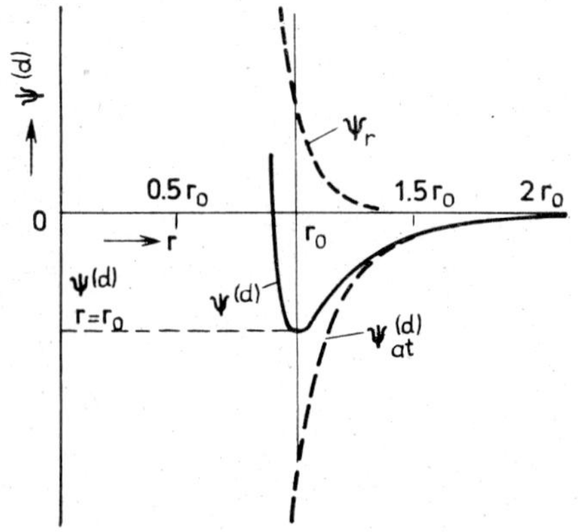

Fig. II.1. Potential curve produced by the interaction of two atoms

SECTION 1.1. NON-POLAR VAN DER WAALS ATTRACTIVE FORCES*

According to London, the type of van der Waals forces known as "dispersion forces" result from the interaction of constantly changing inducing and induced dipoles. The potential of the attractive forces is inversely proportional to the sixth power of the distance:

$$\psi_{at}^{(d)} = -\frac{C_1}{r^6} \tag{II.10}$$

The equation also accounting for the induced dipole-quadrupole and quadrupole-quadrupole interactions, and the repulsion interaction of Eq. (II. 7) is [11]:

$$\psi^{(d)} = -\frac{C_1}{r^6} - \frac{C_2}{r^8} - \frac{C_3}{r^{10}} + \frac{R}{r^{12}} \tag{II.11}$$

where C_1, C_2 and C_3 are the coefficients of the induced dipole-dipole, dipole-quadrupole and quadrupole-quadrupole interactions, respectively, and R is the repulsion constant.

By means of quantum-mechanical calculations the following relations are obtained for the constants in Eq. (II.11) [13, 14]:

$$C_1 = 6\, mc^2(\alpha_i/\chi_i + \alpha_j/\chi_j)^{-1} \tag{II.12}$$

$$C_2 = \frac{45\, h^2}{32\, \pi^2 m}\, \alpha_i\, \alpha_j \left\{ \frac{1}{2\left[\dfrac{\alpha_j}{\chi_j}\Big/\dfrac{\alpha_i}{\chi_i} + 1\right]} + \frac{1}{2\left[\dfrac{\alpha_i}{\chi_i}\Big/\dfrac{\alpha_j}{\chi_j} + 1\right]} \right\} \tag{II.13}$$

$$C_3 = \frac{105\, h^4}{256\, \pi^4 m^3 c}\, \alpha_i\, \alpha_j \left\{ \frac{\alpha_i/\chi_i}{3\left[\dfrac{\alpha_j}{\chi_j}\Big/\dfrac{\alpha_i}{\chi_i} + 1\right]} + \frac{3}{4\left[\dfrac{\chi_i}{\alpha_i} + \dfrac{\chi_j}{\alpha_j}\right]} + \frac{\alpha_j/\chi_j}{3\left[\dfrac{\alpha_i}{\chi_i}\Big/\dfrac{\alpha_j}{\chi_j} + 1\right]} \right\} \tag{II.14}$$

where the α terms are the average polarizabilities of the atoms (the indices i and j refer to the adsorbate and adsorbent, respectively)

the χ terms are the average diamagnetic susceptibilities

m is the mass of the adsorbate atom

c is the velocity of light and

h is Planck's constant.

The theoretical value of the repulsion constant R is not known. It will be seen later that it can be determined with the help of the potential function $\psi^{(d)}(z)$ describing the interaction between the adsorbed atom and the (entire) adsorbent, since for the equilibrium bond distance z_0 $[d\psi^{(d)}(z)/dz]_{z=z_0} = 0$, so that R can be expressed.

* The word "non-polar" indicates that interaction takes place between molecules without permanent dipole moments (for details see Ref. [12]).

Equation (II.9) describes the interaction between only two atoms. In the calculation of the adsorption energy the interaction of the adsorbed atom with every adsorbent atom has to be calculated, and the results summed. (This is possible, since to a first approximation the dispersion forces are additive.)

With particular regard to Eq. (II.11) this means the determination of the following sum:

$$\psi^{(d)} = -C_1 \sum_j r_{i,j}^{-6} - C_2 \sum_j r_{i,j}^{-8} - C_3 \sum_j r_{i,j}^{-10} + R \sum_j r_{i,j}^{-12} \qquad \text{(II.15)}$$

where the index i stands for the gas molecule, and
j is a running index for the atoms of the adsorbent.

If we are dealing with the adsorption not of atoms, but of molecules, the addition has to be extended to all atoms of the molecule, but in this case there may be deviations from additivity.

The intricate calculations were simplified by Polányi and London who did not perform the summation, but determined the volume integral [15].

If it is assumed that the atoms of the adsorbent form a continuum, then by using Eq. (II.9) the calculated energy of interaction of a single adsorbed atom with an adsorbent assumed to be of infinite extension in the negative half-space is:

$$\psi^{(d)}(z) = \frac{RN\pi}{45 z^9} - \frac{C_1 N \pi}{6z^3} \qquad \text{(II.16)}$$

where N is the number of atoms of the adsorbent in unit volume.

With the condition that at $z = z_0$ $d\psi/dz = 0$, it follows that

$$R = \frac{5}{2} C_1 z_0^6 \qquad \text{(II.17)}$$

and thus

$$\psi^{(d)}(z) = \frac{C_1 \pi N}{6 z_0^3} \left[\frac{1}{3} \left(\frac{z_0}{z}\right)^9 - \left(\frac{z_0}{z}\right)^3 \right] \qquad \text{(II.18)}$$

The maximum decrease in the potential energy caused by the dispersion forces or, in other words, the depth of the "potential well", is

$$\psi^{(d)}(z_0) = \frac{C_1 \pi N}{9 z_0^3} \qquad \text{(II.19)}$$

Analytical expressions similar to Eq. (II.18) have the advantage that they permit the determination of the frequency of oscillations normal to the surface of the adsorbed molecules.

In the case of low-amplitude oscillations the system behaves like a harmonic oscillator. The frequency of oscillation can be obtained from the equation:

$$\nu = \frac{1}{2\pi} \sqrt{\frac{f}{m}} \qquad \text{(II.20)}$$

where m is the mass of the adsorbed molecule and f is the force constant defined by the equation:

$$f = \left[\frac{d^2\psi^{(d)}}{dz^2}\right]_{z=z_0} \tag{II.21}$$

Thus, from Eq. (II.21) we obtain the force constant

$$f = \frac{3\, C_1\, \pi N}{z_0^5} \tag{II.22}$$

that is, the frequency of the oscillation normal to the surface

$$\nu = \frac{3\sqrt{3}}{2\,\pi\, z_0}\sqrt{\frac{\psi^{(d)}(z_0)}{m}} \tag{II.23}$$

The value of $\psi^{(d)}(z_0)$ at $z = z_0$ [Eq. (II.19)] gives the adsorption energy for one atom at 0 K. Apart from the fact that because of the neglect of the dipole-quadrupole and quadrupole-quadrupole interactions the potential function according to Eq. (II.18) will not give a satisfactory approximation, for more accurate calculations the thermal energies of the molecules in the gas state and of the adsorbed molecules must also be taken into account, as at temperatures other than absolute zero not only will the potential energy decrease, but there will also be a considerable change in the thermal energies of the different degrees of freedom [11].

As a result of adsorption the translational movement of the gas atom in the z-direction is replaced by an oscillation the frequency of which, at a potential energy decrease [according to Eq. (II.23)] of some kcal mole^{-1} (and thus corresponding to physical adsorption), is of the order of 10^{12}s^{-1}.

This method of calculation has since been improved; of the modified methods the reader's attention is called to those of Avgul [14] and Crowell [16].

These calculations, and particularly the improved ones (whose details are not given here), provide sufficient basis for the description of the movement of the adsorbed atom in the z-direction, but say nothing about the mode of displacement along the surface in the x, y plane. It is quite obvious that if the adsorbent is not treated as a continuum the function $\psi^{(d)}(z)$ must also depend on the choice of x and y. This may be expressed formally by considering the potential function to be a function of the coordinates x and y too. If z in $\psi(x, y, z)$ is arbitrarily* fixed at a certain value z_0, in the case of a homogeneous and plane surface the function $\psi(x, y, z_0)$ thus formed exhibits a periodicity in the variables x and y, corresponding to the atomic structure of the surface. For real surfaces, which are never ideally plane, even after the most careful machining, this periodicity is understandably absent, since at certain points of the surface the adsorbed atom or molecule may interact not with a single, but simultaneously with several,

* In the case of heterogeneous adsorbing surfaces, (since there is no single equilibrium bond distance z_0) the plane $z = z_0$ is arbitrarily chosen. This, however, does not affect the validity of our arguments.

surface atoms, so that at these points the potential function has "deeper" minima than those characteristic of the homogeneous surface. Adsorption "activities" higher than those of the plane faces are displayed particularly by the adsorbent atoms situated on the internal surfaces of submicroscopic capillaries (where for geometrical reasons the contacting surfaces are also greater) and by those on the peaks and edges having unbalanced force fields (where interaction with a single atom may result in a greater potential energy decrease than that occurring as a result of simultaneous interaction with several "normal" surface atoms) [17]. Since the equilibrium of adsorption systems is characterized by an energy minimum (or, more correctly, by a minimum of the free enthalpy), the energy minimum of the initially uncovered and later (under the effect of consecutive gas additions) partially covered surface will be achieved if the molecules to be bound are able to occupy adsorption sites corresponding to the deepest possible potential well. (A precondition of this is the ability of the sorbed molecules to change their positions, since on arriving from the gas phase they will not be bound in this most favourable configuration.) This explains the higher differential heats of adsorption of the first gas doses at these "active" sites or surface sections than the heats of adsorption of later doses, the differential heat of adsorption in general exhibiting a tendency to decrease with increasing coverage. Phenomena related to capillary structures will not be discussed here, since these would unnecessarily complicate further discussion, while the relevant theories cannot yet be considered final. The problem of heterogeneous surfaces too will be dealt with only in connection with one or two special problems.

SECTION 1.2. ADSORPTION OF IONS ON METAL SURFACES

If an ion is adsorbed on the surface of a metal, the electric charge of the ion polarizes the metal. This interaction can be described as the generation of a charge of opposite sign in the metal at a depth from the surface which corresponds to the distance between the inducing charge and the surface (electric image). The energy of interaction (attraction) is:

$$\psi_{\mathrm{at}}^{(f)}(z) = -\frac{n_i e^2}{4z} \qquad \text{(II.24)}$$

where n_i is the number of elementary charges of the ion.

To a first approximation the potential energy of the ion adsorbed on the surface is obtained as the sum of Eq. (II.24) and the first term on the right side of Eq. (II.16), i.e. the repulsion potential. If this sum is determined numerically for the adsorption of Cs^+ on W surfaces, for example, the experimentally determined value of 54.5 kcal mole^{-1} for the heat of adsorption leads to a value of 1.74 Å for the equilibrium bond distance z_0 [18], indicating that the "surface" of the metal coincides with the plane joining the centres of the surface tungsten atoms.

SECTION 1.3. POLARIZATION OF MOLECULES ADSORBED ON CONDUCTOR SURFACES

By the measurement of contact potentials (see Section 7.2) Mignolet demonstrated the considerable dipole moment acquired by non-polar molecules physically adsorbed on metal surfaces [19]. For instance, the induced dipole moment of xenon adsorbed on a nickel surface was found to be 0.42 *D*.

A similar conclusion was drawn when the properties of gases adsorbed on activated carbon were studied by other methods. It should be mentioned here that the differences in behaviour of gases physically adsorbed on activated carbon from the requirements of the two-dimensional van der Waals equation of state (see Paragraph 2.1.2) can presumably be attributed to the same effect, the mutual repulsion of the surface dipoles; thus, calculations of the two-dimensional van der Waals constants, neglecting these forces, generally yield results not in agreement with those determined experimentally [20]. The polarization of adsorbed molecules is probably closely connected with the double layer formed on the metal surface.

The induced dipole moment can be determined from the difference $\Delta a'$ between the measured and calculated values of the two-dimensional van der Waals constant a' using the equation:

$$\Delta a' = -\frac{\pi \mu^2}{2D} \qquad \text{(II.25)}$$

where μ is the induced dipole moment and
$2D$ is the diameter of the molecule [20].

For physically adsorbed gases this effect on the surface of conductors represents an adsorption energy of about 4—6 kcal mole^{-1}, which, if there is no other type of interaction, comprises the bulk of the total adsorption energy.

SECTION 1.4. IONIZATION OF METAL ATOMS ADSORBED ON CONDUCTOR SURFACES

Chemisorption of alkali metal and alkaline earth metal atoms on the surfaces of other metals leads to the formation of ions on the adsorbent surface. Ionic chemisorption takes place if the work function of the metal adsorbent, $e\Phi$ is greater than the ionization energy of the adsorbate, eV_i:

$$eV_i < e\Phi \qquad \text{(II.26)}$$

The potential curve of the sodium atom-tungsten surface system is shown schematically in Fig. II.2 (De Boer, [21]). The minimum at B corresponds to the adsorption of Na atoms, and that at E to the adsorption of Na^+ ions. The difference between levels D and A is equal to the differencebetween the ionization energy of sodium atoms and the work function of tungsten, that is to $(eV_i - e\Phi)$.

The potential energy change accompanying the chemisorption of the Na atom is described by the curve *ACEF*, and its desorption by the curve *ECA*. In other words, the Na^+ ion is desorbed in the form of the Na atom. In the case of caesium the *A* level is higher than the *D* level, and desorption accordingly takes place in the ionic form.

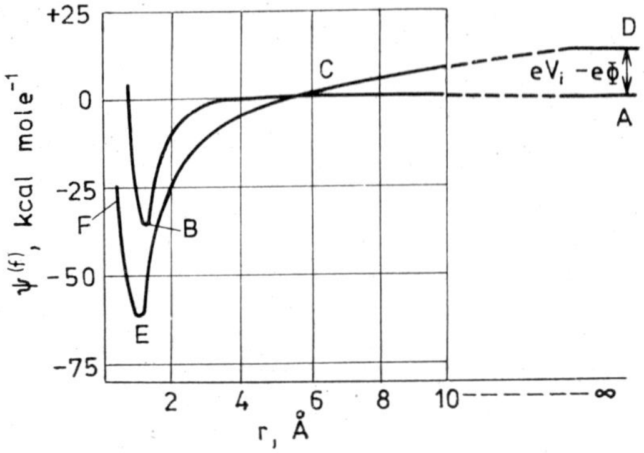

Fig. II.2. Potential curve of the Na atom — W surface

The ionic nature of chemisorption is confirmed by the value of the resulting (apparent) surface dipole moment, which is far larger than expected from theoretical considerations for covalent or coordinate bonds.

Table II.1 shows the surface dipole moments of certain metals adsorbed on a tungsten surface for a coverage extrapolated to zero. The Table also contains the calculated dipole moments (product of the electric charge and the distance *d* between the positive and negative "image" charges).

TABLE II.1

Calculated and experimentally determined surface dipole moments

System	μ(experimental)	d(Å)	μ(calculated)
Na on W	11.3	1.83	8.78
K on W	11.5	2.22	10.90
Cs on W	8.1	2.62	12.50
Th on W	2.0	1.78	8.70

The Table confirms the good agreement between the calculated and experimentally determined values for Na, K and Cs layers, which is further proof of the ionic nature of the bond [22—25].

The few interaction types discussed above are mainly characteristic of physical adsorption, although ionic adsorption leads already into the field of chemisorption. The nature of the bonds formed in the course of chemisorption on the surface of metals and semiconductors, and the structures and properties of chemisorbed layers will be treated in Chapter 8.

CHAPTER 2. ADSORPTION ISOTHERMS

The quantity of substance chemisorbed would be one of the most important pieces of information in heterogeneous catalysis provided an instrument were available whereby the quantity adsorbed could be measured separately for each reaction partner under reaction conditions and a method were known whereby the data thus obtained could be directly correlated with the rate of the catalytic reaction. However tempting this possibility, its realization is hampered by innumerable theoretical and practical difficulties. It may be worthwhile to point out the most important of these, since without being acquainted with them the reader might not perceive clearly the true objective of this Chapter.

Flow reactors (integral or differential type) or static reactors (with or without recirculation) are mainly used for the study of heterogeneous catalytic reactions. In the case of steady state operation, flow reactors fail to furnish any direct information whatsoever regarding the quantities of substances sorbed, as sorption processes occur in the transient period of operation preceding the steady state. The steady state is characterized by coverages which, though changing along the length coordinate of the reactor, are constant in time at a given point. Theoretically it might be possible to measure the *total* quantity of material sorbed at the beginning of the reaction in static reactors, provided this quantity is not too small and the pressure drop caused by sorption can be distinguished from the change due to the reaction. In fact only the order of magnitude of these quantities can be estimated in this way, even in the best case. Thus, we have to resort to indirect methods (mainly to various spectroscopic methods); however, because of difficulties in calibration these are more suitable for recording changes in coverage than for the determination of the absolute quantities.

Let us nevertheless assume that the absolute quantities of the sorbed materials are known. This information, too, would lead us nowhere, since some of the material might be physically sorbed, while the chemisorbed fraction might be present in various geometrical arrangements on the different crystal faces and accordingly be bound by different bonds, only some of which might participate in the catalytic process. At the same time, it is common knowledge that the order of the catalytic reaction, as well as the part orders for the different components, depend sensitively upon the partial pressures, indicating that the role of the coverage is of primary and decisive importance.

As far as the determination of the quantities of material sorbed during catalytic reactions is concerned, the difficulties can be partly avoided by the separate study of the chemisorption of the different components at temperatures at which no appreciable autodegradation takes place. To speak about avoiding autodegradation, however, may be extremely misleading, since chemisorption leads *a priori* to a certain decomposition of the molecules. As a compromise we might consider the investigation of the sorption properties of gases for which the decomposition process is not extensive or, even if accompanied by the splitting of the bond, the original bonding state of the molecule can be re-established. This condition is primarily fulfilled for the simple diatomic gases (H_2, N_2, O_2). But here again it must be borne in mind that on the active sites of the various crystal surfaces several layers with different properties can exist simultaneously, and since certain elementary steps of the chemisorption under the chosen experimental conditions are not necessarily reversible, the determined quantity of sorbed material vs. partial pressure correlations, will not necessarily lead to isotherms, even though, in certain cases, they can be described by an isotherm equation. One of the important consequences of this is that, as already mentioned in the Introduction, these pseudo-isotherms cannot be used for the calculation of thermodynamic quantities [e.g. isosteric heats of adsorption (see later on p. 184)].

Adsorption equilibrium can be described quantitatively with the help of isotherms, isobars and isosteres. In this Chapter we shall take a closer look at the most descriptive means of graphical presentation, the adsorption isotherm.

Any isotherm can be classified as one of the five types shown in Fig. II.3. Of these, types II—V characterize plurimolecular, and type I monomolecular adsorption. Chemisorption isotherms always belong to type I.

In principle there are two possibilities for the derivation of theoretical isotherm equations.

In the kinetic method ad- and desorption rates are expressed with the relevant partial pressures and coverages, the two are then made equal in accordance with the requirement of equilibrium and the resulting algebraic equation is solved for the coverage. Chapter 6 will show that because of the uncertainties in the functions $f(\theta)$ and $\bar{f}(\theta)$ this task can be solved for the very simplest cases only.

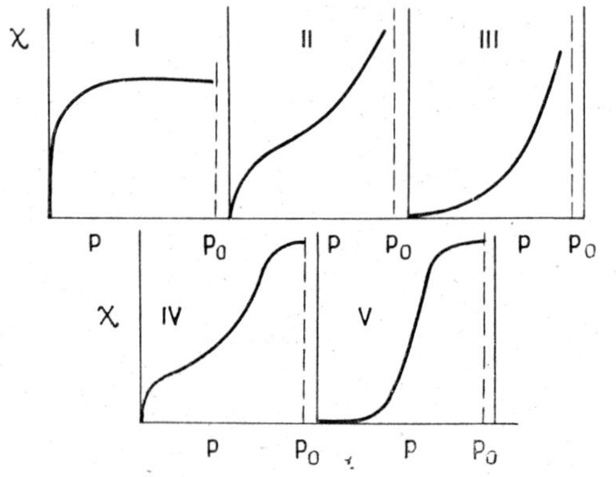

Fig. II.3. Types of adsorption isotherms according to Brunauer, Deming and Teller; p_0 = equilibrium vapour pressure of the liquid adsorbate

The second possibility is the description by thermodynamics. If the methods of statistical thermodynamics are followed, the chemical potentials of the gas and adsorbed phase must be expressed by the partition functions. The algebraic equation obtained by equating the two expressions can be solved for the coverage. The same procedure can be applied to the classical thermodynamic treatment. In some cases the Gibbs adsorption equation [see Eq. (II.87)] can be made good use of, as this enables the calculation of the isotherm, starting from the equation of state of the adsorbed layer.

Whichever solution is chosen, the derivation of isotherm equations must start from a specific adsorption model representing an *a priori* commitment as far as the mobility or localization and structure of the layer, the types of possible interactions (adsorbent — adsorbate, adsorbate — adsorbate), the coverage-dependence of the interactions, etc. are concerned. It is unfortunate that there is no unequivocal relationship between the chosen model and the isotherm derived from it, that is to say, one and the same isotherm can usually be correlated to more than one model.

The absolutely minimum requirement as regards the agreement between experimental and theoretical isotherms is that the course of the isosteric heats of adsorption calculated from these isotherms should satisfy the requirements of the model.

Since isotherms characteristic of physical adsorption play an important role in up-to-date catalyst studies, these cannot be neglected. The limited size of this book, however, permits no more than a brief presentation of the simplest isotherm types for homogeneous adsorbing surfaces. The reader may find information on details not given here in textbooks on adsorption (see e.g. ref. [10]).

SECTION 2.1. MONOMOLECULAR PHYSISORPTION

PARAGRAPH 2.1.1. DEGREES OF FREEDOM OF ADSORBED ATOMS AND MOLECULES

In Chapter 1 we discussed the adsorption forces acting between separate, non-interacting molecules and the adsorbent, and presented an interpretation of the concept of adsorption energy and its mathematical formulation for the most important cases.

We next inspect the degrees of freedom of movement of an atom adsorbed on the surface or, in other words, the change in the number of degrees of freedom compared to that in the gas phase, when, as a result of adsorption, the atom can move only along the surface in two dimensions.

In the gas phase each atom possesses three translational degrees of freedom. At low temperatures, in the case of localized adsorbed atoms, these are replaced by three vibrational degrees of freedom. The frequency $\nu = \nu_z$ has already been mentioned. The frequencies ν_x, ν_y in the plane of the adsorbent surface are determined by the shape of the potential function $\psi^{(d)}(x, y, z_0)$. The frequencies ν_x and ν_y are of the order of $10^{11}\ \mathrm{s}^{-1}$ [26]. At sufficiently high temperatures the average thermal energy may exceed the height of the energy barrier in the vicinity of the minimum of $\psi^{(d)}(x,y,z_0)$, so that in accordance with what was said earlier, the adsorption may become delocalized (mobile). Delocalized layers consisting of atoms possess two translational and one vibrational degree of freedom. Accounting for the degrees of freedom of adsorbed polyatomic molecules is a more complicated problem. It has been mentioned in the discussion of adsorbent — adsorbate interactions that in certain, more simple, cases interactions are additive with respect to the atoms making up the molecule. In the case of diatomic molecules the three translational degrees of freedom of the centre of mass in the gas phase are modified similarly to the degrees of freedom of a single atom and, in addition, any changes due to adsorption in the two rotational and one (internal) vibrational degree of freedom must also be taken into consideration. Internal vibration is hardly influenced at all by the fact of adsorption, because the adsorption forces can be neglected compared with the binding forces of the molecule. Rotation, however, may become impeded and in extreme cases the two rotational degrees of freedom will be limited to rotation around an axis perpendicular to the surface and in a plane intersecting the axis of the molecule, and to an additional vibration normal to the surface.

PARAGRAPH 2.1.2. THE ADSORPTION ISOTHERM FOR DELOCALIZED ADSORPTION

Experience has shown that at very low pressures ($p \to 0$) and on homogeneous adsorbing surfaces the number of adsorbed molecules, N_a, is proportional to the equilibrium gas pressure (Henry's law):

$$N_a = c\,p; \; T = \text{constant} \tag{II.27}$$

where c is a constant characteristic of the adsorbent and the adsorbate at a given temperature.

By substituting Eq. (II.27) into the Gibbs adsorption law (see Section 3.2):

$$\pi_f = kT \int_0^p \frac{N_a(p)}{A_s} \, \mathrm{d} \ln p \tag{II.28}$$

we obtain for the equation of state of the two-dimensional adsorbed phase:

$$\pi_f A_s = N_a kT \tag{II.29}$$

where A_s is the surface area of the adsorbent and
π_f is the two-dimensional spreading pressure.

Reversing the procedure: Equation of state (II.29) leads by use of the Gibbs adsorption equation to the linear adsorption isotherm expressed by Eq. (II.27). Contrary to what is claimed in several handbooks, the equation of state (II.29) can be used only for the description of delocalized adsorption, since in the case of localized adsorption the spreading pressure loses physical meaning.

It should be mentioned, for the sake of completeness, that within the limits of its validity Eq. (II.27) also gives a correct description of localized adsorption, and this is probably the reason for several authors claiming the applicability of equation of state (II.29) for this case too.

At higher equilibrium pressures and associated higher surface concentrations $\Gamma = N_a/A_s$, the interaction between the adsorbed molecules must also be taken into consideration.

Let us first restrict our arguments to the case of monomolecular adsorption and for the sake of simplicity assume a homogeneous adsorbent surface. Let us further assume in accordance with Cassel [27] and Hill [28] that the adsorbed molecules fulfill the requirements of the following two-dimensional van der Waals equation of state:

$$\left(\pi_f + \frac{a' N_a^2}{A_s^2}\right)(A_s - N_a a_0) = N_a kT \tag{II.30}$$

where the constants a' and a_0 are related to the constants a and b* of the three-dimensional van der Waals equation:

$$\left(p + \frac{a N_g^2}{v^2}\right)(v - N_g b) = N_g kT \tag{II.31}$$

$$a' = \frac{\pi a}{4 b}\left(\frac{3 b}{2 \pi}\right)^{2/3} \tag{II.32}$$

$$a_0 = \frac{\pi}{2}\left(\frac{3 b}{2 \pi}\right)^{2/3} \tag{II.33}$$

The constant a_0 is characteristic of the area of the adsorbate molecule, and as such, plays an important role in the determination of specific surface areas by means of the adsorption method (see Part VI, Paragraph 5.3.2).

According to the law of corresponding states, the constant b in the three-dimensional van der Waals equation is related to the critical temperature T_c and the critical pressure p_c of the adsorbate:

$$b = \frac{kT_c}{8 p_c}$$

and thus a_0 can be expressed by means of these data too**:

$$a_0 = 6.354 \left(\frac{T_c}{p_c}\right)^{2/3} \quad (\text{in Å}^2) \tag{II.34}$$

* It is clear from the form of the equations that a' and a_0, and also a and b are not molar quantities, but refer to a single molecule.

** It is noted as a matter of interest that the molecular areas calculated from Eq. (II.34) agree roughly with those calculated from the density of the adsorbate in the liquid state by means of the following equation:

$$a_0 = 3.464 \left(\frac{M}{4\sqrt{2} N_a d}\right)^{2/3} \tag{II.35}$$

where M is the molar mass of the adsorbate and
d is the density of the adsorbate in the liquid state.

Similarly to the van der Waals constants the two- and three-dimensional critical parameters can also be correlated [27, 28]. Thus, for instance, the following simple correlation is valid between the two-dimensional critical temperature T'_c and the three-dimensional one, T_c:

$$T'_c = \frac{1}{2} T_c \tag{II.36}$$

T'_c can also be expressed from the two-dimensional van der Waals constants:

$$T'_c = \frac{8\,a'}{27\,ka_0} \tag{II.37}$$

etc.

The adsorption isotherm corresponding to the equation of state (II.30) can essentially be determined by the former method [29]. For this let us write Eq. (II.28) in the following differential form:

$$\frac{N_a}{A_s} = \frac{p}{kT}\,\frac{d\pi_f}{dp} = \frac{p}{kT}\,\frac{d\pi_f}{dN_a}\,\frac{dN_a}{dp} \tag{II.38}$$

According to Eq. (II.30) the differential quotient $d\pi_f/dN_a$ is:

$$\frac{d\pi_f}{dN_a} = \frac{A_s kT}{(A_s - a_0 N_a)^2} - \frac{2a' N_a}{A_s^2} \tag{II.39}$$

After substitution of Eq. (II. 39) into Eq. (II.38), rearrangement and integration we obtain:

$$\ln p = \ln \frac{N_a}{(A_s - a_0 N_a)} + \frac{a_0\,N_a}{(A_s - a_0 N_a)} - \frac{2\,a' N_a}{kT\,A_s} + \text{const} \tag{II.40}$$

where the constant ($\equiv \ln k_b$) characterizes the adsorbent—adsorbate interaction, as no such data were included in the initial equations.

Taking into account that

$$a_m \equiv \frac{A_s}{N_{a,\infty}} \approx a_0$$

where a_m is the area occupied by one molecule in the monomolecular layer, i.e. the space requirement, and $N_{a,\infty}$ is the number of adsorbed molecules required for monomolecular coverage, and further if θ is the ratio of the number of molecules N_a and $N_{a,\infty}$ (the coverage), by means of the former substitution from Eq. (II.40) we obtain the following isotherm equation:

$$p = k_t \frac{\theta}{1-\theta} \exp\left(\frac{\theta}{1-\theta} - \frac{2\,a'\theta}{a_0 kT}\right) \tag{II.41}$$

Neglecting the exponential factor on the right hand side of Eq. (II.41), we arrive at the well-known Langmuir isotherm equation.*

Certain investigators have recently attempted to replace the uncritical use of the various isotherm equations by the development of experimentally supported theories for the simplest adsorption models, and to adapt these to more complicated cases. The theoretical difficulties arising from the heterogeneity of the surface are circumvented by using adsorbents having homogeneous surfaces, and experimental temperatures are chosen at which the adsorbed layers are delocalized, and hence the fine structure of the surface cannot assert itself. The only remaining serious theoretical difficulty is related to the equation of state describing the behaviour of the adsorbed layer. The relevant experiments of several authors have proved that even under the most extreme experimental conditions the two-dimensional van der Waals equation of state is surprisingly well applicable, and when used with circumspection gives not merely a semi-quantitative, but sometimes even a quantitative description of the adsorbed layer. In certain cases the results obtained for homogeneous adsorbents can be adapted without difficulty to fairly simple heterogeneous surfaces.

In this paragraph we wish to describe physical adsorption on homogeneous surfaces and, within this, the most characteristic properties of the delocalized layers, and thus it will be of use to consider the isotherm equation (II.41) in somewhat more detail.

In order to have a clear idea of the meaning of the constants in Eq. (II.41) let us divide both sides of this equation by the vapour pressure of the liquid adsorbate, p_0, and introduce the symbols

$$\frac{k_0}{p_0} = k_2 \text{ and } \frac{2a'}{a_0 kT} = k_1$$

Hence, instead of Eq. (II.41) we may write:

$$x = k_2 \frac{\theta}{1-\theta} e^{\frac{\theta}{1-\theta}} e^{-k_1\theta} \tag{II.42}$$

where $x = p/p_0$, the relative pressure defined earlier.

Except for the temperature, k_1 depends only on parameters characteristic of the adsorbate. The limiting cases of $k_1 \approx 1$ and $k_1 \approx 10$ correspond to weak and strong intermolecular interaction respectively.**

* There is no strict theoretical justification for this omission. Although the condition $a' = 0$ is in agreement with the requirements of the Langmuir adsorption isotherm (see Paragraph 2.2.1), the simplification $\exp[\theta/(1-\theta)] \approx 1$ can be true only for coverages $\theta \ll 1$, and this would unnecessarily limit the applicability of the Langmuir isotherm. It would be in contradiction with the experimental finding that, provided the isotherm equation correctly describes the experimental results, its applicability is not limited to regions of low coverage.

** For instance, with $a' = 5\ 10^{-32}$ erg cm^2 and $a_0 = 25\ 10^{-16}$ cm^2 at room temperature ($kT = 4\ 10^{-14}$ erg): $k_1 = 1$; while with $a' = 50\ 10^{-32}$ erg cm^2: $k_1 = 10$.

In contrast to k_1, k_2 is a quantity characteristic of the adsorbent-adsorbate interaction.*

From the isotherm equation (II.41), q_{st}, the heat of adsorption of one molecule [see also Eq. (II.5)] is given by:

$$q_{st} = -kT^2 \left(\frac{\partial \ln p}{\partial T}\right)_\theta = -\left(kT^2 \frac{d \ln k_b}{dT} + \frac{2a'\theta}{a_0}\right) \qquad (II.43)$$

The first term on the right hand side is the isosteric heat of adsorption for $\theta = 0$, and the second term is the contribution of the adsorbate–adsorbate interactions.

Figures II.4 and II.5 show some isotherms calculated by means of Eq. (II.42). In the first case the value of k_2 was taken as 5 ($k_2 = 5$ corresponds to a weak adsorbent—adsorbate interaction), while, in the other case, k_2 was taken as 0.1, and k_1 was varied.

According to Fig. II.4 two-dimensional condensation takes place at $k_1 > 6.5$, resulting in a break ("jump") on the isotherm. Under appropriate experimental conditions this type of condensation of van der Waals gases is similar to the normal three-dimensional condensation; at pressures corresponding to the jump two phases, the two-dimensional gas and the condensed phase, exist at the same time.

On the curves for $k_1 = 1$ and 5 in Fig II.5 there is no condensation, even theoretically; the adsorbed molecules are in a state similar to that of a compressed gas. At $k_1 = 7$ and 10 surface condensation takes place, but the corresponding relative pressure is so small that the jump does not appear in the scale of the diagram. Thus, the visible sections of these curves correspond to the two-dimensional condensed state of the adsorbed molecules.

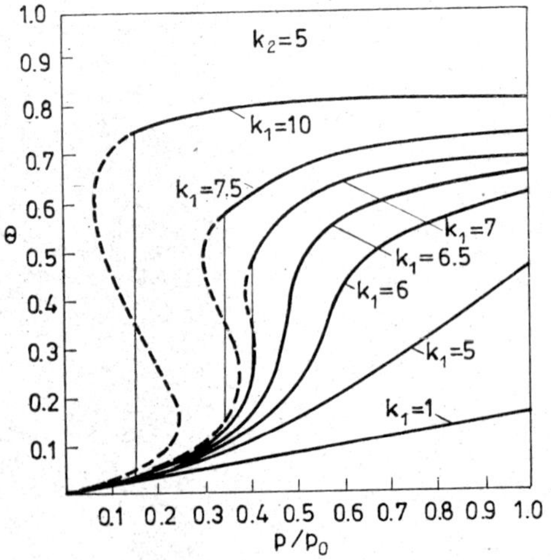

Fig. II.4. Adsorption isotherms calculated from Eq. (II.42); $k_2 = 5$; k_1 variable

In certain cases (e.g. when $k_1 = 0$ and $k_2 = 10^{-3}$, or $k_1 = 6$ and $k_2 = 10^{-1}$), isotherms of apparently the same shape are obtained for various k_1 and k_2 values, but at very low pressures there is a marked difference between the two isotherms. For a clear demonstration of these differences the determination of the equilibrium parameters must be extended to very low pressures.

* Equation (II.42) gives a guide as to the absolute value of k_2. If $\theta \to 0$, $k_2 = x/\theta$; thus, for $\theta = 0.01$ at a relative pressure of $x = 10^{-4}$, k_2 must be equal to 10^{-2}. In practice k_2 is usually smaller than 0.1.

Two-dimensional surface condensation takes place only if the temperature is below the two-dimensional critical temperature. Comparison of Eq. (II.37) for T_c' with a formula given earlier, for k_1 shows that the con-

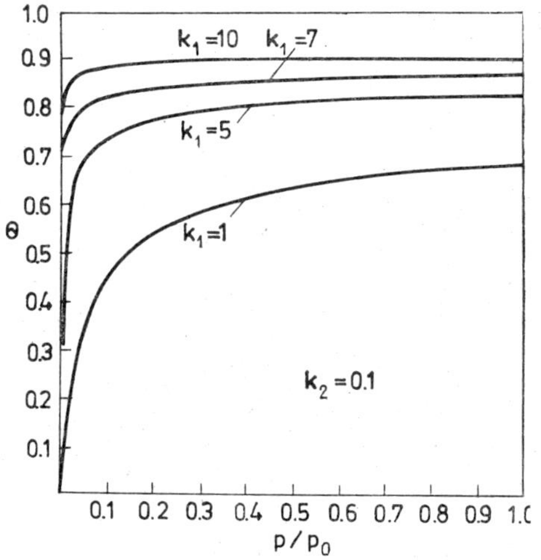

Fig. II.5. Adsorption isotherms calculated from Eq. (II.42); $k_2 = 0.1$; k_1 variable

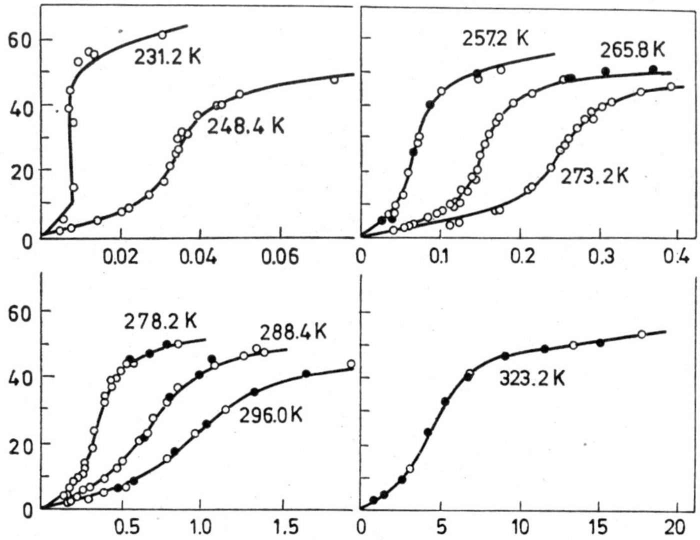

Fig. II.6. Adsorption isotherms of CCl_4 on graphite, in temperature ranges below and above the two-dimensional critical temperature. Adsorbed quantity in μmole g^{-1}, equilibrium pressure in torr. Continuous curve calculated from Eq. (II.41); circles represent experimentally obtained points, ○ adsorption, ● desorption dition for condensation is $k_1 > 27/4$. When $k_1 < 27/4$ the adsorption layer will be in the supracritical state.

Example. The above will be illustrated with data of Machin et al. [270], who determined the adsorption isotherms of CCl_4, $CHCl_3$, and $CFCl_3$ in the temperature range 195–323 K. The adsorbent was graphitized carbon black, type P–33, with a practically homogeneous surface. Some of the experimentally determined isotherms for CCl_4 are shown in Fig II.6. The full lines were calculated directly from Eq. (II.41)*. The two-dimensional phase transition can actually be observed on the isotherm plotted below the two-dimensional critical temperature (see Table II.2).

TABLE II.2

Some adsorption parameters of CCl_4, $CHCl_3$ and $CFCl_3$ adsorbed on graphite

	CCl_4	$CHCl_3$	$CFCl_3$
m' μmole g^{-1}	78 ±2	81 ±1	82 ±1
$2a'/a_0$, kcal mole^{-1}	3.2±0.1	2.8±0.1	2.70±0.04
$q_{st}(\theta \approx 1)$, kcal mole^{-1}	9.25	8.07	7.91
T_c',K	239	209	201
1/2 T_c,K	282	270	237

If ln k_b is plotted vs. $1/T$ all isotherms are straight lines. The heats of adsorption calculated from Eq. (II.43), and some other characteristic data of the isotherms are given in Table II.2. It is surprising that the critical temperatures calculated from Eqs. (II.36) and (II.37) are not identical. Machin claims that the difference in the T_c' values obtained by the different calculations is probably a result of the polarization of the adsorbate molecules.

It has been shown in Section 1.3 that the difference between the measured and calculated values of the van der Waals constants is [Eq. (II.25)]:

$$\Delta a' = a' - a'_{\text{calc}} = -\frac{\pi\mu^2}{2D}$$

and is characteristic of the value of the induced dipole moment. Taking into consideration that the relation $b = \frac{16\pi}{3} D^3$ holds between constant

* The isotherm equation (II.41) contains four unknown parameters: χ_m (hidden in θ, as $\theta = \chi/\chi_m$), k_b, a' and a_0. Since Eq. (II.41) contains only their quotient, the last two cannot be determined separately. Machin obtained the constants from the analytically unsolvable equation by an essentially nomographic method. Today the very great number of experimental data determining the nine experimental curves could be processed by optimization, using an electronic computer working on the principle of non-linear least squares [30]: the parameters could thus be obtained very quickly and with very high accuracy.

b in the three-dimensional van der Waals equation and the van der Waals radius D, the calculated value of a' according to Eq. (II.32) is

$$a'_{\mathrm{calc}} = \frac{3\,a}{16\,D}$$

It follows from the relation $\mu = E\bar{\alpha}$ that there is an unequivocal relationship between the induced dipole moment μ and the surface field strength E of the adsorbent ($\bar{\alpha}$ being the mean value of the polarizability of the molecule), and hence if a' is experimentally determined E can be calculated.

From the data for CCl_4 a surface field strength of 1 10^5 electrostatic unit cm^{-2} was obtained for graphite.

In the above Sections, discussion of adsorption processes on porous adsorbents with high specific surfaces was deliberately avoided, as the relevant theories are still not complete and the knowledge in this field that can be considered final is accessible in the various textbooks.

Localized adsorbed layers will be discussed in relation to chemisorption phenomena, since they are more characteristic of chemisorption than of physical adsorption.

SECTION 2.2. MONOMOLECULAR CHEMISORPTION

Consideration will now be given to those isotherm equations which can be used mainly for the description of the chemisorption equilibrium. However, this certainly does not mean that the validity of any of the isotherms indicates the presence of chemisorption. If the adsorbed layer is localized, the same isotherms are also suitable for the description of the physisorption equilibrium. The Langmuir isotherms give a correct description of adsorption equilibria on homogeneous surfaces, while the Freundlich and Tempkin isotherms do the same for adsorption on surfaces of a heterogeneous type.

PARAGRAPH 2.2.1. THE LANGMUIR ISOTHERM

This isotherm equation was originally applied to the description of the sorption of gases physically adsorbed on mica. The conditions from which the isotherms was derived were:

(1) the adsorbed layer is localized; adsorption takes place only if a molecule collides with an empty adsorption site;

(2) each molecule occupies only a single adsorption site;

(3) the adsorption energy is the same on each adsorption site; there are no secondary adsorbate—adsorbate interactions (the heat of adsorption is independent of the coverage).

Today these seem quite special, but the rates of sorption processes in heterogeneous catalysis can still be described by the isotherm just as Langmuir derived it at the beginning of the century, with practically no modification. The underlying concepts, therefore, have proved of more lasting value than the chosen model.

To derive the isotherms of the Langmuir type let us begin with the expressions for ad- and desorption rates on homogeneous adsorbing surfaces (see Paragraph 6.2.3). Putting J_a [Eq. (II.127)] and J_d [Eq. (II.129)] equal, and taking into account that

$$E_d^{\ddagger} - E_a^{\ddagger} = -Q \tag{II.44}$$

after appropriate transformations we obtain:

$$p = \frac{K}{\sigma}\sqrt{2\pi m kT}\,\frac{\bar{f}(\theta)}{f(\theta)}\exp(Q/RT) \tag{II.45}$$

If Fowler's third condition for the above adsorption model [31], is accepted as correct, Eq. (II.45) depends on θ only through $f(\theta)$ and $\bar{f}(\theta)$, and if the terms independent of θ are incorporated into the constant b, then instead of Eq. (II.45) we may write:

$$p = \frac{1}{b}\frac{\bar{f}(\theta)}{f(\theta)}$$

Depending on how $f(\theta)$ and $\bar{f}(\theta)$ are substituted, different isotherm equations are obtained.

If the adsorbed molecule occupies only a single adsorption site, independently of the localized or delocalized nature of the formed layer, then according to point (1) of Paragraph 6.3.1:

$$f(\theta) = 1 - \theta \quad \text{and} \quad \bar{f}(\theta) = \theta$$

and hence:

$$p = \frac{\theta}{b(1-\theta)}$$

or after rearrangement:

$$\theta = \frac{bp}{1+bp} \tag{II.46}$$

In the event of dissociative adsorption, if the layer is localized and statistically random:

$$f(\theta) = \frac{z}{z-\theta}(1-\theta)^2 \quad \text{and}$$

$$\bar{f}(\theta) = \frac{(z-1)^2}{z(z-\theta)}\theta^2, \quad \text{and thus}$$

$$\theta = \frac{\sqrt{b'p}}{1+\sqrt{b'p}} \tag{II.47}$$

where $b' = \left(\frac{z}{z-1}\right)^2 b$.

For layers having the same properties as before, but delocalized, a repetition of the derivation leads to:

$$\theta = \frac{\sqrt{bp}}{1+\sqrt{bp}} \tag{II.48}$$

Isotherm equations can be deduced for adsorbates consisting of bulky molecules and consequently covering more than one adsorption site but as will be shown in Paragraph 6.3.1 these have little practical importance.

If the potential adsorption sites are ordered, but only a single such ordered layer can develop on the surface, and then only at high coverage [in other words, the adsorption sites are occupied in a statistically random manner (see also Paragraph 6.2.1)], then in the case of non-dissociative adsorption the relevant expressions for $f(\theta)$ and $\bar{f}(\theta)$ are presumably still valid but for a change in the definition of the "monomolecular layer"(see Section 2.4). When the layers are formed as a result of dissociative adsorption, with different coverages possibly possessing interlaced structures, the situation is too complicated to survey, not to mention that in these cases the condition of "surface"homogeneity can *a priori* not be fulfilled.

Among others, the inhomogeneity of the adsorbing surface might be responsible for the fact that theoretically it is never possible to describe adsorption on polycrystalline (or, in general, on inhomogeneous) surfaces by the Langmuir isotherm equation. One or two examples exist in the literature where this appears to be successful, but these should be regarded as being exceptions. On the other hand, it should be pointed out that in practical experience (mainly in measurements on oxide-type adsorbents and activated carbons) there is almost always some coverage above which the isotherm can be described by the Langmuir equation.

Although this problem will be returned to in the following Sections, it might be worthwhile to add a few remarks to the conditions on which the derivation of the original Langmuir isotherm is based. The localization of point (1) is in fact not a *sine qua non* of the derivation. Recent investigations have proved, for both non-dissociative and dissociative adsorption, that layers with definite structures very frequently occur. This can only occur as a result of the excited molecule or atom changing its place on the surface. There is also a possibility of the site of dissociation not being the same as that of the bond formation, leading to a different interpretation of the concept of the localized layer, since not only the adsorbent—adsorbate interactions play an important role in the development of the surface structure, but also the (mainly repulsive) interactions between the adsorbed atoms.

The second condition is also in need of thorough revision (see Paragraph 8.1.2), though this possibility cannot *ab ovo* be excluded in the case of certain systems. As already mentioned, the fulfillment of condition (3) is the least likely.

The best way to find out whether the simple isotherm [Eq. (II.46)] is indeed applicable is to consider it in the form in which it will give a straight line when plotted in a rectangular coordinate system:

$$\frac{p}{\theta} = \frac{1}{b} + p \tag{II.49}$$

Thus, if p/θ calculated from the experimental data is plotted as a function of p, a straight line of unit slope should be obtained with an intercept of $1/b$.

Since
$$\frac{1}{b} = \frac{K\sqrt{2\pi mkT}}{\sigma} \exp(Q/RT)$$

the intercepts thus obtained from isotherms for different temperatures must depend approximately exponentially upon the reciprocal temperature.

This method of representation presupposes a knowledge of θ, but in practice this value is usually not directly available. The procedure then is to use the equation defining θ:

$$\theta = \chi/\chi_{\mathrm{m}}$$

where χ is the specific sorbed quantity pertaining to the equilibrium pressure p [in $\mathrm{cm^3}$ (S.T.P.) $\mathrm{g^{-1}}$]

and χ_{m} is the limiting value of the above for $p \to \infty$, that is for the case of complete monomolecular coverage [in $\mathrm{cm^3}$ (S.T.P.) $\mathrm{g^{-1}}$].

Substituting θ by χ/χ_{m} we obtain:

$$\frac{p}{\chi} = \frac{1}{b\,\chi_{\mathrm{m}}} + \frac{p}{\chi_{\mathrm{m}}}$$

b and χ_{m} can then be determined from the slope and intercept of this straight line.

The relation [Eq. (II.47)] derived for dissociative adsorption can be rearranged in a similar way:

$$\frac{\sqrt{p}}{\chi} = \frac{1}{\sqrt{b'}\,\chi_{\mathrm{m}}} + \frac{\sqrt{p}}{\chi_{\mathrm{m}}} \qquad \text{(II.50)}$$

In accordance with our earlier arguments it is found in practice that plotting the experimental data in this way fails to lead to straight lines in most cases. This is due to the fact that for heterogeneous adsorbent surfaces the minimum free enthalpy of the adsorption system is achieved when the gas molecules always occupy the most active available adsorption sites (that is those characterized by the highest absolute values of the heat of adsorption). It follows from the definition of b that increasing values of the heat of adsorption are associated with decreasingly smaller intercepts; this results in the isotherm curve, bending towards the abscissa with decreasing pressure.

Though adsorption systems in which the process itself and the adsorption isotherm strictly fulfil the above conditions occur very rarely in practice, it follows from the nature of chemisorption isotherms (type I) that with appropriately chosen values of the parameters b and χ_{m} the experimentally obtained isotherms can quite often be described over fairly wide pressure ranges. For the parameters b and χ_{m} to be physically realistic it is first of all necessary that the heat of adsorption calculated from the adsorption isosteres should not depend upon the coverage [see conditions (3)]. χ_{m} and b can be calculated from two, preferably closely situated isotherm points:

$$\chi_{\mathrm{m}} = \frac{p_2 - p_1}{p_2/\chi_2 - p_1/\chi_1} \quad \text{and} \quad b = \frac{1}{\chi_{\mathrm{m}}(p_i/\chi_i - p_i/\chi_{\mathrm{m}})} \qquad (i = 1{,}2)$$

It is quite obvious that, provided the Langmuir equation is valid and irregular fluctuations due to experimental errors are discounted, the series of χ_m and b values thus obtained cannot depend upon the average adsorbed quantity $\chi = (\chi_1 + \chi_2)/2$. There must also be agreement within the limits of error between the values of χ_m obtained from isotherms for different temperatures. Finally, it is required that the isotherm equation be applicable to the entire coverage range between 0 and 1.

Example. From a study of the chemisorption of H_2, Ward obtained the equilibrium data shown in Table II.3, for two samples (A and B) [271]. The heat of adsorption appeared to be independent of coverage. Figure II.7 suggests that the function p/χ vs. p is a straight line for both samples. Closer scrutiny of the figure, however, clearly reveals that the experimental data in fact lie on a curve. Deviation from linearity at both low and high pressures is quite obvious. The curves $\sqrt{p}/\chi$ are not linear and, in accordance with what was said above [see Eq.(II.50)], this indicates that the chemisorption is not dissociative. This finding, however, is in contrast with our ideas about the chemisorption of hydrogen on metals.

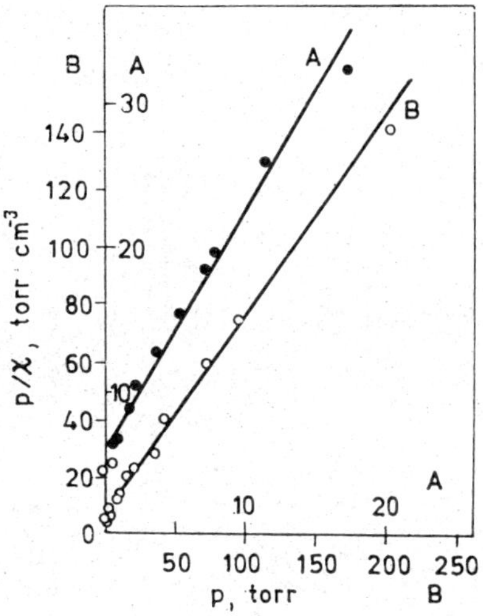

Fig. II.7. Adsorption of hydrogen on copper powder. Isotherms plotted as p/χ vs. p

TABLE II.3

Chemisorption of hydrogen on copper powder at 25 °C

Sample A		Sample B	
Pressure torr	Adsorbed volume cm^3	Pressure torr	Adsorbed volume cm^3
0.19	0.042	1.05	0.239
0.66	0.138	1.70	0.464
0.97	0.163	2.95	0.564
1.01	0.161	3.25	0.559
1.10	0.171	5.40	0.659
1.90	0.221	8.90	0.761
2.65	0.256	10.65	0.800
4.05	0.321	17.65	0.941
5.55	0.371	21.50	0.995
7.50	0.411	36.2	1.121
8.15	0.421	45.1	1.160
11.95	0.471	74.5	1.281
17.55	0.550	95.8	1.300
		204.8	1.471

In the case of adsorbents with heterogeneous adsorbing surfaces, particularly in the range of low equilibrium pressures, the Langmuir isotherm fails to furnish a correct description of the experimental data. Freundlich has successfully applied the following empirical isotherm equation to these cases:

$$\chi = cp^{1/n} \qquad n > 1 \tag{II.51}$$

where the constants c and n are temperature-dependent and c also depends, among other factors, on the dimensions in which the quantity adsorbed is expressed.

It is quite easy to see that at medium pressures and using appropriately chosen constants the Langmuir and Freundlich isotherms will overlap. However, in certain cases the range of validity of the Freundlich isotherm is the greater, and hence the question arises as to what theoretical assumptions must be made with regard to the adsorbent and the mechanism of adsorption in order to justify the Freundlich isotherm.

For a description of adsorption on heterogeneous surfaces Fowler suggested the equation [32]

$$\theta = \int \theta(Q) g(Q) dQ \tag{II.52}$$

where

$$g(Q) \equiv N_s^{-1} \frac{dN(Q)}{dQ}$$

is the probability density function of the adsorption sites possessing an adsorption energy between Q and $Q + dQ$ (see Chapter 4), and N_s and $N(Q)$ are the total number of adsorption sites and the number of sites with adsorption energies between Q and $Q + dQ$, respectively; $\theta(Q)$ is the partial coverage on the latter at equilibrium pressure p, and should be calculated from the Langmuir equation

$$\theta(Q) = \frac{b(Q)p}{1 + b(Q)\,p} = \frac{1}{1 + 1/b(Q)\,p} \tag{II.53}$$

In Eq. (II.53) the expression

$$b(Q) = b_0 \exp(Q/RT)$$

holds as before.

The expression of θ according to Eq. (II.52) is mathematically the expectation of $\theta(Q)$, the continuous random variable.

Let us now assume that, as suggested by Halsey and Taylor [33], $g(Q)$ follows the exponential distribution:

$$g(Q) = \frac{1}{Q_m} \exp(-Q/Q_m)$$

By the substitution of these expressions into Eq. (II.52) we obtain:

$$\theta = \frac{1}{Q_m} \int_0^{\infty} \exp(-Q/Q_m) \{1 + [\exp(-Q/RT]/[b_0 p]\}^{-1} dQ \approx$$

$$\approx \frac{1}{Q_m} \int_{-\infty}^{+\infty} \exp(-Q/Q_m) \{1 + [\exp(-Q/RT)]/[b_0 p]\}^{-1} dQ \qquad \text{(II.54)}$$

Extension of the lower limit of integration to $-\infty$ serves merely to facilitate the calculation of the definite integral. Halsey and Taylor have proved by numerical integration that the error involved is not significant (when Q is negative, with $Q_m \gg RT$, the integrand will approach zero).

By introducing the substitution

$$[\exp(-Q/RT)]/[b_0 p] \equiv \Phi$$

integral (II.54) takes the form:

$$\theta = (b_0 p)^{(RT/Q_m)} \frac{RT}{Q_m} \int_{-\infty}^{\infty} \frac{\Phi^{(RT/Q_m - 1)} \, d\Phi}{1 + \Phi}$$

whose solution is [34]:

$$\theta = (b_0 p)^{RT/Q_m} \frac{RT}{Q_m} \pi \operatorname{cosec} \left(\frac{RT}{Q_m} \pi \right) \qquad \text{(II.55)}$$

For small values of the argument of the cosec function, i.e. when $Q_m \gg \gg \pi RT$, Eq. (II.55) takes the simpler form:

$$\theta = (b_0 p)^{RT/Q_m} = cp^{1/n} \qquad \text{(II.56)}$$

where

$$c = b_0^{RT/Q_m} \quad \text{and} \quad n = \frac{Q_m}{RT}$$

According to the definition of the isosteric heat of adsorption:

$$Q_{st} = R \left[\frac{\partial \ln p}{\partial (1/T)} \right]_\theta = Q_m \ln \theta \qquad \text{(II.56a)}$$

Though the concept of surface heterogeneity is still rather a matter of contention, the logarithmic dependence of Q_{st} on θ according to Eq. (II.56a) (confirmed by certain observations) is usually too great to be attributed to adsorbate–adsorbate interactions, but may be explained in the majority of the cases, by the heterogeneity of the surface.

To decide whether experimental isotherm data can, in fact, be described by means of the Freundlich isotherm equation, Eq. (II.56) is preferably written in its logarithmic form:

$$\log\theta = \frac{RT}{Q_m} \log b_0 + \frac{RT}{Q_m} \log p \qquad \text{(II.57)}$$

If the equation is valid, then $\log \theta$ plotted against $\log p$ should lead to a straight line whose slope and intercept are both proportional to T. It can be proved by elementary calculation that the lines obtained from Eq. (II.57) for different temperatures intersect at

$$p = \frac{1}{b_0}, \quad \theta = 1$$

If θ is unknown then, similarly to the indirect method used before, the relation $\theta = \chi/\chi_m$ can be used and Eq. (II.57) is written in the following form:

$$\log \chi = \log \chi_m + \frac{RT}{\Phi_m} \log b_0 + \frac{RT}{Q_m} \log p \qquad \text{(II.58)}$$

χ_m can be determined most conveniently from the common intersection points of the lines represented by Eq. (II.58) for different temperatures, since at the intersection point $p = \dfrac{1}{b_0}$ and $\chi = \chi_m$.

The empirical expression of the Freundlich isotherm [Eq. (II.51)] might suggest that with increasing p, χ increases without limit. There was a time when it was concluded from the fact that the experimental isotherms failed to display this behaviour, that there was something theoretically wrong with the Freundlich isotherm. Indeed, form (II.56) of the equation shows that the theoretical isotherm reaches saturation at $p = 1/b_0$.

For the case of dissociative adsorption, essentially similar arguments lead to an isotherm equation formally identical with Eq. (II.56).

Example.

Frankenburg [35] and Davis [36] have proved with very detailed data for wide temperature (− 194 — +600 °C) and pressure (10^{-5}—30 torr) ranges that the chemisorption of hydrogen and nitrogen on tungsten powder can be described by the Freundlich isotherm. The difference noted between

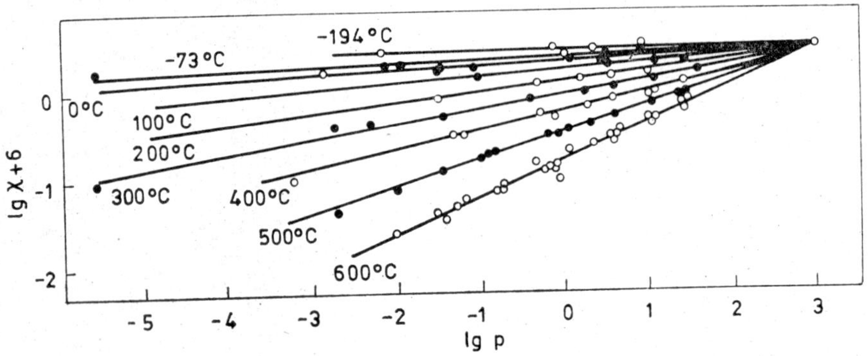

Fig. II.8. Hydrogen isotherms on tungsten powder in lg-lg scale

metal powders and vacuum-evaporated films is presumably due to the partial contamination of the powders.

Figure II.8 shows the hydrogen isotherms on a log-log scale. It can be seen that the experimentally obtained points fit straight lines, which, when extrapolated, intersect in the point S.

Since the specific surface of the powders was known, Frankenburg and Davis were able to demonstrate that the adsorbed quantity relating to point S corresponds to an adsorbed layer in which there is one H atom per W atom. More detailed investigations showed that the slopes of the lines in Fig. II.8 are proportional, not to T, but to $T/(1 - rT)$ where r is a constant. Such a temperature-dependence is shown by the Freundlich-type isotherm

$$\theta = (\sqrt{b_0 p})^{RT/[Q_m(1-RT)]}$$

derived by Halsey [37] for dissociative adsorption. The numerical values of the constants are:

$$1/b_0 = 3150 \text{ torr}, Q_m = 5700 \text{ cal mole}^{-1},$$

$$r = 0.75 \; 10^{-3} \text{K}^{-1}.$$

At coverages lower then $\theta = 8 \; 10^{-3}$ the above arguments no longer hold: on the log-log scale the isotherms at all temperatures were straight lines with slopes of 2; θ was thus proportional to $\sqrt{p}$, indicating the applicability of the Langmuir isotherm equation to dissociative adsorption at low coverages. This conclusion is supported by the changes in the heats of adsorption, which, in the case of $\theta \to 0$, will approach not $-\infty$, but a constant finite limiting value. It seems possible therefore that, in this case, the inhomogeneity can be attributed to secondary adsorbate–adsorbate interactions.

Qualitatively, the same is true for the adsorption of nitrogen on tungsten, the only difference being in the numerical values. Because of the slow establishment of the adsorption equilibrium, nitrogen isotherms on W cannot be measured at temperatures below 400 °C. With not too high coverages (up to about $\theta = 0.3$) the log-log isotherms are again straight lines with slopes of 2.

PARAGRAPH 2.2.3. THE TEMPKIN ISOTHERM

The extensive applicability of the Elovich rate relation (see Paragraph 6.3.2) appears to confirm the existence of a linear dependence upon θ not only with respect to $E_a^{\ddagger}$ and $E_d^{\ddagger}$ in chemisorption systems, but also, as a necessary consequence of the latter, with respect to the heat of adsorption:

$$Q_{st} = E_a^{\ddagger} - E_d^{\ddagger} = (E_{a,0}^{\ddagger} - E_{d,0}^{\ddagger}) + (\alpha + \beta)\,\theta = -Q_0(1 - \gamma\theta) \qquad \text{(II.59)}$$

where γ is a constant and Q_0 the heat of adsorption for $\theta = 0$.

The simplest way to conceive the linear dependence of the isosteric heat of adsorption on θ is to assume an *a priori* homogeneous surface where the decrease is caused by the increasing manifestation of the repulsive interactions as the coverage increases, or to assume that the number of adsorption sites having adsorption energies between Q and $(Q + \mathrm{d}Q)$ is constant, independent of Q, and that, at these sites, the *adsorption energy* decreases with increasing coverage of the surface according to the equation:

$$Q = Q_0 (1 - \gamma\theta) \tag{II.60}$$

In agreement with this concept of adsorption equilibrium let us write the probability density function $g(Q)$ in the form characteristic of a random variable with uniform distribution:

$$g(Q) = \frac{1}{Q_0 \gamma} = \text{const} \quad [Q_0(1 - \gamma) \leq Q \leq Q_0]$$

(see Paragraph 6.3.2) for which

$$\int_{Q_0(1-\gamma)}^{Q_0} g(Q)\,\mathrm{d}Q = \int_{Q_0(1-\gamma)}^{Q_0} \frac{\mathrm{d}Q}{Q_0\gamma} = 1$$

By repeating the sequence used in the derivation of the Freundlich isotherm with the above expression of $g(Q)$, we obtain:

$$\theta = \int \theta(Q)\, g(Q)\, \mathrm{d}Q = \frac{1}{Q_0\gamma} \int_{Q_0(1-\gamma)}^{Q_0} \frac{b\, p \exp(Q/RT)}{1 + b_0 p \exp(Q/RT)}\, \mathrm{d}Q = \frac{RT}{Q_0} \int_{\xi_l}^{\xi_u} \frac{\mathrm{d}\,\xi}{1+\xi}$$

via the substitutions

$$b_0 p \exp(Q/RT) \equiv \xi; \; b_0 p \exp[Q_0(1 - \gamma)/RT] \equiv \xi_l$$

and

$$b_0\, p \exp(Q_0/RT) \equiv \xi_u.$$

Integration and substitution of the limits of integration lead to

$$\theta = \frac{RT}{Q_0\gamma}\left[\ln \frac{1 + B_0\, p}{1 + B_0\, p \exp(-Q_0\, \gamma/RT)}\right] \tag{II.61}$$

where $B_0 \equiv \exp(Q_0/RT)$.

At average coverage values $(\theta_{\text{min}} \leq \theta \leq \theta_{\text{max}})$ it may be assumed that $B_0\, p \gg 1$, and at the same time Q_0 is sufficiently great for $B_0 p \exp(-Q_0\gamma/RT) \ll 1$. With this simplification Eq. (II.61) takes the following form:

$$\theta = \frac{RT}{Q_0\gamma} \ln(B_0\, p) \quad (\theta_{\text{min}} \leq \theta \leq \theta_{\text{max}}) \tag{II.62}$$

This is the so-called logarithmic or Tempkin isotherm.

For the isosteric heat of adsorption in fact

$$Q_{st} = R\left[\frac{\partial \ln p}{\partial(1/T)}\right]_\theta = -Q_0(1-\gamma\theta) \qquad (\theta_{min} \leqq \theta \leqq \theta_{max})$$

is obtained, in agreement with the statements in the Introduction. Below Q_{min} and above Q_{max} the dependence of Q_{st} upon θ is non-linear and the curve $Q_{st}(\theta)$ is less steep than $dQ_{st}/d\theta = Q_0\gamma$, which is characteristic of the linear section.

If the experimentally determined Q values are plotted vs. $\ln p$, according to Eq. (II.62), within the limits of its applicability, a straight line is obtained the slope of which is proportional to T. At too low and too high coverages deviations from linearity may occur even if Eq. (II.61) is otherwise valid.

It follows from Eq. (II.62) that, similarly as with the Freundlich isotherm, with increasing p values θ will increase without limit. It appears quite clearly from Eq. (II.61), however, which does not contain the simplification, that, here also, a saturation value exists ($\theta \to 1$). For dissociative adsorption Eq. (II.62) can be applied in the same form.

Example. The adsorption of N_2 and H_2 on iron can be described by the Tempkin isotherm [38]. This conclusion is supported not only by the shapes of the isotherms, but also by the direct experimental determination of the rate of chemisorption, whereby the constants of the Tempkin isotherm equation could be determined directly.

PARAGRAPH 2.2.4. OTHER ISOTHERM EQUATIONS

Derivations of the Freundlich and Tempkin isotherm equations proceeded from the quite general relation of Fowler [Eq. (II.52)], assuming that the surface fractions $g(Q)dQ$ between adsorption energies Q and $(Q + dQ)$ are homogeneous and that the partial equilibrium coverages on them can be calculated from Eq. (II.53). The type of isotherm equation obtained by this derivation is decided in any given case by the probability density function $g(Q)$. The same can be said of the differential heat of adsorption, whose changes with θ also depend on $g(Q)$ and further on the changes caused in $\theta(Q)$ by the individual gas doses:

$$-Q_{diff} = \frac{\int_{Q_{min}}^{Q_{max}} Q[\theta'(Q) - \theta(Q)]\, g(Q)\, dQ}{\int_{Q_{min}}^{Q_{max}} [\theta'(Q) - \theta(Q)]\, g(Q)\, dQ} \qquad \text{(II.63)}$$

where $\theta(Q)$ and $\theta'(Q)$ are the partial coverage functions at equilibrium pressures p and $(p + dp)$, respectively. If dp is sufficiently small, then from Eq.

(II.53), $[\theta'(Q) - \theta(Q)]$ may be replaced in both the numerator and denominator of the fraction:

$$\theta'(Q) - \theta(Q) = \frac{\partial \theta(Q)}{\partial p} \mathrm{d}p = \frac{b(Q)}{[1 + b(Q)\,p]^2} \mathrm{d}p$$

so that

$$-Q_{\text{diff}} = \frac{\int\limits_{Q_{\min}}^{Q_{\max}} Q \frac{b(Q)}{[1 + b(Q)\,p]^2} g(Q)\,\mathrm{d}Q}{\int\limits_{Q_{\min}}^{Q_{\max}} \frac{b(Q)}{[1 + b(Q)\,p]^2} g(Q)\,\mathrm{d}Q} \tag{II.64}$$

The earlier Fowler relation (II.52) is valid for θ as before.

It is clear from the above expression for Q_{diff} that there is no direct relationship between $g(Q)$ and $Q_{\text{diff}}(\theta)$.

In the derivation of a new isotherm equation, in principle the only approach beyond objection is that which starts from a distribution of the adsorption sites according to their adsorption energies and arrives by trial and error at a function $g(Q)$ which describes Q_{diff} and $\theta(p)$ in agreement with experiment. One difficulty in the application of this method is the lack of direct information whereby the validity of function $\theta(Q)$ can be checked. All things considered, the Fowler relation (II.52) is quite general, so that in principle any sorption equilibrium can be described with an appropriately chosen $g(Q)$ function. In addition, discrete distributions with respect to $g(Q)$ may furnish further practical results, since — as shown by experience — when the Langmuir isotherm parts are multiplied by appropriate weighting factors and summed, they often correctly reflect the dependence of θ on the equilibrium pressure.

The above examples were relatively simple cases in which there is such a considerable difference between the thermal effects of physical adsorption and chemisorption that by an appropriate choice of the experimental temperature (and, of course, the adsorption system), physical adsorption can be made quite insignificant. If there is no appreciable difference between the thermal effects of physisorption and chemisorption, and further if the experimental conditions favour physisorption (high specific surface, low temperature, high gas pressure, etc.), the two types of sorption may occur simultaneously. Quite frequently, but by no means necessarily, the physically sorbed layer may develop on top of the chemisorbed layer. Many such observations have been described in the literature [39].

SECTION 2.3. ADSORPTION OF GAS MIXTURES

It happens quite often in heterogeneous catalysis that the simultaneous adsorption of different molecules or atoms has to be considered. Poisoning, selective catalysis, and in a wider sense of the word, the promoter effect are all based on this phenomenon. Experience gathered in the past ten years with respect to adsorbed layers (see Paragraph 8.1.3) has proved that the simultaneous presence of several adsorbates of different types may cause, even on single crystal surfaces, such diverse manifestations of geometrical and energy effects and interactions that to account for them would be a task far beyond our present knowledge. This is a great obstacle in finding a kinetic description for heterogeneous catalytic processes, since these are always introduced by the sorption of one or two components, and it follows from what has been said above that there is an obvious link between the sorption kinetics and sorption equilibrium.

Starting from the Langmuir-type rate equations for the description of the rate of adsorption of mixtures, the isotherm equations first deduced by Markham and Benton [40] are obtained. There is no need to stress that the same simplifications and, consequently, the same restrictions are valid for these equations as for the Langmuir isotherm. A single detail should be pointed out, however: interaction between the components is manifested only in their independent ability to occupy the sites.

For the non-dissociative adsorption of two components, assuming that the layer is in a statistically random arrangement, equating Eqs (II.133), (II.134) and (II.135), (II.136) for the ad- and desorption rates:

$$\frac{\sigma_1 p_1}{\sqrt{2\pi m_1 kT}}(1-\theta_1-\theta_2)\exp(-E^{\ddagger}_{\mathrm{a},1}/RT) = K_1\theta_1 \exp(-E^{\ddagger}_{\mathrm{d},1}/RT)$$

$$\frac{\sigma_2 p_2}{\sqrt{2\pi m_2 kT}}(1-\theta_1-\theta_2)\exp(-E^{\ddagger}_{\mathrm{a},2}/RT) = K_2\theta_2 \exp(-E^{\ddagger}_{\mathrm{d},2}/RT)$$

or

$$b_1p_1(1-\theta_1-\theta_2) = \theta_1 \quad \text{and}$$

$$b_2p_2(1-\theta_1-\theta_2) = \theta_2 \qquad \text{(II.65)}$$

where

$$b_i = \frac{\sigma_i}{\sqrt{2\pi m_i kT}}\exp[-(E^{\ddagger}_{\mathrm{a},i}-E^{\ddagger}_{\mathrm{d},i})/RT]\frac{1}{K_i} \qquad (i = 1, 2)$$

The solution of the system of linear algebraic equations (II.65) is:

$$\theta_1 = \frac{b_1 p_1}{1+b_1p_1+b_2p_2} \quad \text{and} \quad \theta = \frac{b_2 p_2}{1+b_1p_1+b_2p_2} \qquad \text{(II.66)}$$

Generalization of equations (II.66) for n-components is quite evident:

$$\theta_i = \frac{b_i p_i}{1+\sum_i b_i p_i} \qquad (i = 1, 2 \ldots \mathrm{n})$$

If certain of the components are dissociatively adsorbed, e.g. the kth component, then

$$b_k p_k(1-\sum_i \theta_i)^2 = \theta_k^2 \quad \text{that is}$$

$$\sqrt{b_k p_k}(1-\sum_i \theta_i) = \theta_k$$

For this component, therefore, it holds that

$$\theta_k = \frac{\sqrt{b_k p_k}}{1 + \sqrt{b_k p_k} + \sum_{i \neq k} b_i p_i}$$

while for the others:

$$\theta_j = \frac{b_j p_j}{1 + \sqrt{b_k p_k} + \sum_{i \neq k} b_i p_i}$$

In fact, the analytical solution of equations of type (II.65) causes difficulty only if it is impossible to transform them by some simple method into a system of linear algebraic equations. In the theory of catalytic reactions proceeding according to the Langmuir–Hinshelwood mechanism, for want of a better method, surface coverages for each component are calculated in the above manner. Should the analytical tools prove to be insufficient for this purpose, recourse must be made to computer procedures, that is, to sub-routines for the numerical solution of non-linear algebraic equations.

An isotherm equation of a somewhat more general validity (but for the lack of an analytical solution implicit) is obtained by starting from the rate equations suggested by De Boer for homogeneous adsorbing surfaces [Eqs (II.165) and (II.166)]. Repeating the above sequence:

$$\frac{\theta_1}{1 - \theta_1 - \theta_2} = b_1' p_1 \, e^{-Q_1^0/RT} \, e^{-(c_1\theta_1 + c_2\theta_2)/RT} \tag{II.67}$$

$$\frac{\theta_2}{1 - \theta_1 - \theta_2} = b_2' p_2 \, e^{-Q_2^0/RT} \, e^{-(c_1\theta_1 + c_2\theta_2)/RT} \tag{II.68}$$

where the b_i' terms do not contain the Boltzmann factor formed with the help of the heat of adsorption.

Dividing Eq. (II.67) by Eq. (II.68) and putting b_1'/b_2' equal to unity (since its change is certainly not significant beside the exponential factor) we obtain:

$$\frac{\theta_1}{\theta_2} = \frac{p_1}{p_2} e^{(Q_2^0 - Q_1^0)/RT} \tag{II.69}$$

For adsorbates creating dipoles of identical orientations, as, for instance, in the case of the adsorption of C_2H_2 and C_2H_4 on nickel, the selectivity of the adsorption process can be estimated directly from Eq. (II.69). The values of the heats of chemisorption are -67 kcal mole^{-1} and -58 kcal mole^{-1}, respectively. Hence, at about 50 °C, provided the partial pressures are the same:

$$\frac{\theta_{C_2H_2}}{\theta_{C_2H_4}} = 10^6$$

Thus, the surface of a hydrogenating nickel catalyst is always covered with C_2H_2 until practically the very end of the hydrogenation reaction.

It seems evident, therefore, to explain selective hydrogenation by the selective chemisorption of the substrates [41].

Another characteristic example is the poisoning of the NH_3 iron catalyst by gases containing O_2 (even in the bound form, such as CO, CO_2, H_2O, etc.) [42].

In the first example no surface reaction could take place between the adsorbed C_2H_2 and C_2H_4. If the two adsorbed substrates are capable of reacting with one another on the surface, this greatly complicates the theoretical calculation of the stationary surface concentrations developing in the course of the reaction. This problem will not be dealt with here in detail; although some experimental work of this nature is known (see e.g. ref. [43]), a general solution of the problem has not yet been found.

To return to the question of activation energy and heat of adsorption, it follows clearly from Eqs (II.161)–(II.164) that surface contamination — provided this brings about a surface dipole layer of the same orientation as the adsorbate — will cause a decrease of the energy, i.e. of the absolute value of the heat of adsorption of the second component, the effective adsorbate. This explains the fact that on metals not fully reduced hydrogen is always adsorbed by activation, while on clean metal surfaces the value of the activation energy is practically zero (a few hundred cal mole^{-1}) (see Ref. [44], p. 147).

A fair number of examples are known in which the two components, by creating dipoles of opposite orientations on the adsorbent surface, considerably promote the adsorption of the partner. In this respect reference is made to the technical iron catalysts which contain Al_2O_3 and K_2O as promoters and on which, at temperatures around 400–500 °C, H_2 is adsorbed in the atomic state, due mainly to the effect of the promoters forming dipoles with their positive ends pointing away from the surface. In the chemisorbed state, nitrogen has a negative charge, so that the earlier-formed hydrogen layer has a favourable influence on nitrogen chemisorption [45]. In some cases this "promoter" effect can be so marked that a given iron sample is capable of chemisorbing a mixture of H_2 and N_2, whereas separately neither of the gases is sorbed.

On the whole it can be said that the theory of multicomponent adsorption has not developed, from either a theoretical or a practical aspect, to be equal to the important role which it ought to play in the kinetics of heterogeneous catalytic reactions. The above equations describing coverage with regard to the individual components may be objected to for the same reason as the analogous relations for single components. Although development in the field of mixed adsorption is likely to come not so much from new multicomponent adsorption research, as from other methods for the direct and indirect observation of the adsorbent surface, the establishment of the necessary conclusions as to adsorption from gas mixtures is an urgent task for the near future.

SECTION 2.4. DIFFICULTIES ENCOUNTERED IN THE INTERPRETATION OF MONOLAYER COVERAGE

Of the isotherm equation parameters, the sorbed quantity corresponding to monomolecular coverage (expressed in monomolecular numbers: $N_{a,\infty}$ [1] or [1 g^{-1}]; expressed in moles or cm^3 (S.T.P.): χ_m [mole g^{-1}] or [cm^3 (S.T.P.) g^{-1}], etc.) is the most important, since this is the parameter by means of which it is possible to determine the specific surface area from physisorption isotherms, and which also plays a significant role in surface determinations based on chemisorption data. This last problem involves the greatest number of contradictions, and we feel it necessary to make a few remarks in relation to this question.

Whenever it is possible to establish that the given sorption system strictly fulfills the demands of an isotherm equation based on some adsorption model, the experimental value of χ_m can be accepted as an informative datum, but in each case further investigation is needed to decide the actual layer structure which this value might represent.

On the faces of metal single crystals the number of atoms per unit surface is known exactly; although a number of authors have reported their observation of one surface atom corresponding to one sorbed atom in the saturated state of chemisorption, it might be advisable to adopt an attitude of cautious scepticism in this respect. Thanks to the investigations by LEED. we have currently at our disposal detailed layer structure data for several single crystal surfaces and various adsorbates, all of which indicate that the 1 : 1 correspondence is far from being a general phenomenon, and that the concept of monomolecular layer is, in fact, linked to the shape of the potential surface characterizing the adsorbent–adsorbate interaction and not to the structure of the adsorbent surface (which is otherwise related to the former). The simultaneous existence of more than one adsorbed layer produces a heterogeneous "surface", but this type of heterogeneity differs from the cases discussed earlier; due to lack of further information, therefore, the real meaning of monomolecular coverage and the equilibrium pressure at which such a layer will develop are somewhat questionable.

In the case of polycrystalline surfaces the concept of monomolecular coverage becomes arbitrary and, at times, meaningless. In extreme cases certain crystal faces may reach complete saturation without any significant sorption taking place on other faces. Most misunderstandings arise when sorption equilibrium can be described by one or other of the isotherms. This explains the practice suggested by Beeck [46] and fully accepted in chemisorption, when, for a given adsorption system state, $\theta = 1$ is assigned to some arbitrarily chosen equilibrium pressure (e.g. in the case of the system Ni/H_2, to a pressure of 0.1 torr). Only further research will clarify the errors caused, as, for example, in the calculation of thermodynamic functions, by some arbitrarily chosen monomolecular coverage based on one or another definition.

SECTION 2.5. PLURIMOLECULAR PHYSISORPTION

Below the critical temperature of the adsorbate, at equilibrium pressures near the tension p_0 of the liquid adsorbate, an adsorbed phase several molecular layers thick may develop on the surface of the adsorbent. Even on adsorbents without internal pore structure, the thickness of the film can, in certain cases, be so great that it is no longer possible to distinguish adsorption from condensation.

Great difficulties are encountered when an attempt is made to discuss the phenomenon of plurimolecular adsorption theoretically. The problem of the still unsolved theory of liquids [47, 48, 49] is further complicated by the fact that the field of the adsorbent must also be accounted for in the calculations. Because of the difficulties involved, the investigators engaged in finding a theoretical solution to the problem have been compelled to restrict themselves to more or less rough approximations, which will now be briefly described.

PARAGRAPH 2.5.1. THE BET THEORY AND ITS MODIFICATIONS

The theory of Brunauer, Emmett and Teller (the BET theory) [50] is one of the most successful attempts to describe plurimolecular adsorption. This, of course, is far from meaning that the theory is in every respect satisfactory; its assumptions are occasionally extremely rough, but nevertheless sufficient for the interpretation of some experimentally observed phenomena.

The most important practical application of the BET isotherm equation is the determination of the specific surface areas of adsorbents by means of an adsorption method. Fortunately, the results of surface determinations are not sensitively influenced by the deficiencies of the theory, so that surfaces determined according to BET are still the most reliable data of their kind.

The BET isotherm equation was originally derived by a kinetic method, but later Cassie [51] and Hill [52] arrived at the same result using statistical mechanics. The advantage of the statistical mechanical method is primarily its freedom from assumptions related to kinetics, while, at the same time, it expresses the constants and parameters with definite molecular data. We shall now present Hill's derivation of the BET isotherm equation.

The initial assumptions are as follows:

(1) In the first adsorbed layer B adsorption sites are available for localized adsorption per unit surface area of the adsorbent. The model does not consider the interaction between molecules in the first adsorbed layer.

(2) All molecules in the first adsorbed layer represent adsorption sites for the molecules which will form the second layer, and so on.

(3) The partition function (including the potential energy) of the second and subsequent layers is the same as in the liquid state $[\Omega_1 \exp(\bar{\varepsilon}_1/kT)]$, and, in general, differs from the partition function $\Omega_a \exp(\bar{\varepsilon}_a/kT)$ of the

first adsorbed layer. Horizontal interaction between molecules in the same layer is ignored in this theory.

In this way the adsorbed layer can be imagined as "piles" of different heights, built up from single molecules which do not exert an energetic influence on one another; the heights of the piles are determined by statistical laws such that in the state of equilibrium the free energy of the system shall have a minimum value. It follows from the assumptions that the molecules situated in the second, third, etc. layers form a "liquid phase" in which each chosen molecule has only two neighbours (one below and one above), in contrast to real liquids in which the number of nearest neighbours is something between 10–12.

Let us assume that a total of N_a molecules is adsorbed, of which X are in the first layer and $(N_a - X)$ in all the other layers.

The separate partition functions for the X molecules in the first layer and the $(N_a - X)$ in higher layers are then:

$$W_a = \frac{B!}{(B - X)!\,X!}\,[\Omega_a \exp(\bar{\varepsilon}_a/kT)]^X \tag{II.70}$$

$$W_1 = \frac{(N_a - 1)!}{(N_a - X)!\,(X - 1)!}\,[\Omega_1 \exp(\bar{\varepsilon}_1/kT)]^{N_a - X} \tag{II.71}$$

The factor $B!/(B - X)!X!$ is the number of distinguishable ways X identical molecules may be distributed among B adsorption sites, while $(N_a - 1)!/(N_a - X)!(X - 1)!$ is the number of possible arrangements of $(N_a - X)$ identical molecules in X piles if the "heights" of the piles are not limited (that is, the number of molecules from which the pile is built up is not restricted).

Neglecting 1 in comparison to N_a and X in Eq. (II.71), the complete partition function of the system is then

$$W = \sum^{n} W_a W_1 = \sum^{n}\left[\frac{N_a!\,B!}{(N_a - X)!\,(B - X)!\,(X!)^2}\,(\Omega_a e^{\bar{\varepsilon}_a/kT})^X\,(\Omega_1 e^{\bar{\varepsilon}_1/kT})^{N_a - X}\right] \tag{II.72}$$

where $n = N_a$ (if $N_a < B$) and $n = B$ (if $N_a > B$).

Using the usual approximation, $\ln W$ is made equal to the largest term of the sum. The value of X which gives this term is found from

$$\frac{\partial \ln (W_a W_1)}{\partial x} = 0 \tag{II.73}$$

that is for which $\ln W_a W_1$ has a maximum. It follows from Eq. (II.73), if the Stirling approximation is applied to the factorials, that

$$(N_a - X)(B - X) = \beta X^2 \tag{II.74}$$

where

$$\beta = \frac{\Omega_1}{\Omega_a} \exp[(\bar{\varepsilon}_1 - \bar{\varepsilon}_a)/kT]$$

The chemical potential of the adsorbed molecules is:

$$\frac{\mu_a}{kT} = -\frac{\partial \ln (W_a W_1)}{\partial N_a} = \ln \frac{N_a - X}{N_a} - \frac{\bar{\varepsilon}_1}{kT} - \ln \Omega_1 \qquad \text{(II.75)}$$

into which X should be substituted from Eq. (II.74).

The chemical potential of the molecules in the gas phase is:

$$\frac{\mu_g}{kT} = \frac{\alpha}{kT} + \ln p \qquad \text{(II.76)}$$

(the meaning of α is of no interest as far as the following argument is concerned).

By equating Eq. (II.75) with Eq. (II.76) we obtain

$$\ln \frac{N_a - X}{N_a} - \frac{\bar{\varepsilon}_1}{kT} - \ln \Omega_1 = \frac{\alpha}{kT} + \ln p \qquad \text{(II.77)}$$

At the same time, for the pure liquid, the equation

$$\frac{\mu_1}{kT} = -\frac{\bar{\varepsilon}_1}{kT} - \ln \Omega_1 = \frac{\alpha}{kT} + \ln p_0 \qquad \text{(II.78)}$$

holds, where μ_1 is the chemical potential of the liquid adsorbate. It follows from Eqs (II.77) and (II.78) that

$$x = \frac{p}{p_0} = \frac{N_a - X}{N_a} \qquad \text{(II.79)}$$

where $x = p/p_0$ is the relative pressure.

Finally, the BET isotherm equation is obtained by combining Eqs (II.74) and (II.79):

$$\theta = \frac{N_a}{B} = \frac{cx}{(1 - x)\,[(1 - x) + cx]} \qquad \text{(II.80)}$$

where $c = 1/\beta = \frac{\Omega_a}{\Omega_1} \exp\,[(\bar{\varepsilon}_a - \bar{\varepsilon}_1)/kT]$ the well-known BET constant.

In the relative pressure range between 0.05 and 0.35 the isotherm equation (II.80) gives a description of plurimolecular adsorption in agreement with experimental data. Its applicability to low relative pressures (between 0 and 0.05) is limited by the heterogeneity of the adsorbent surfaces in practical use*. At higher relative pressures (roughly above 0.35) the simplifications introduced for the "liquid phase" become less and less adequate, and it is not surprising that the equation fails to agree with experimental evidence at relative pressures higher than 0.35**.

* It is seen from Eq. (II.80) that for $x \to 0$ the BET equation is the same as the Langmuir equation, so that the errors of the latter will of necessity appear in the BET equation too.

** At relative pressures above 0.4 the BET equation indicates adsorbed quantities greater than the true amounts.

Modifications of the equation have aimed primarily at removing these deficiencies. Mention should be made of the attempts of Macmillan [53] and Walker [54] which kept the first condition of the BET model, but in the calculations accounted for the heterogeneity of the surface. Anderson [55] and Cook [56] went one step further in denying the equivalency of the second, third, etc. layers. They claim that only those layers of the adsorbed phase which are far from the adsorbent behave similarly to normal liquids. Hill [57] and later Halsey [58] tried to lend an exact formulation to the semi-empirical method of Anderson and Cook by accounting for horizontal interactions; these calculations, though theoretically interesting, have so far failed to produce any practically useful results.

In our opinion all attempts at developing the BET equation suffer from the common fault of keeping the condition of a localized first adsorbed layer in the Hill deduction. If this condition and its resulting consequences are dropped since they are in contradiction with physical reality, a realistic basis might be created for a revision of the BET theory and the pertaining isotherm equation. It is beyond all doubt that mono- and plurimolecular physical adsorption are characterized not by localized, but by delocalized adsorption layers; this has been confirmed by both theoretical arguments and convincing experimental data*. Though no derivation of the BET equation starting from the model of mobile layers is known, at present there is a derivation of the Langmuir isotherm equation for the limit $x \to 0$ [60], and there can be no doubt whatsoever that the derivation and development of the BET equation should be attempted by the same approach.

Cassel [61] criticised the BET equation from another aspect. He pointed out that by substituting the BET equation into the Gibbs equation [see Eq. (II.87)] a contradiction arises, namely:

$$\lim_{p \to p_0} \pi_f = \infty$$

which is an obvious impossibility. Cassel attributed this contradiction to the neglect of horizontal interactions. In our opinion Cassel's argument is not valid, since the transition $p \to p_0$ means that at saturation pressure the adsorbent surface is totally immersed in the liquid where (in fact, even before completion of the monolayer) the spreading pressure loses its physical meaning.

Although we consider Cassel's remark unjustified and difficult to understand, it had to be mentioned, for in fact it provided the incentive for the derivation of an isotherm equation which gives particularly good approximations at higher coverages, without being subject to the above-discussed contradiction.

The method by which Frenkel [62], Halsey [58] and Hill [63] succeeded in solving the task is fundamentally different from the others. The objective was to deduce an isotherm equation suitable, particularly at $\theta > 2$, for the

* Surfaces obtained by applying the BET equation are in good agreement, for example, with those calculated from liquid adsorption, a method in which close fitting and delocalized adsorption quite unequivocally exists [59].

description of the experimentally determined isotherm shapes. With such coverages the direct influence of the adsorbing surface can be ignored, while, at the same time, the adsorbed film several molecules thick increasingly approaches the character of a normal liquid. The main difference between the two is that the adsorbed film is subject to a different field of force from that prevailing in normal liquids. This fundamental argument of the theory is very similar to the well-known potential theory of Polányi [64].

The details of the derivation will not be given here, but the final result is stated simply as

$$\ln x = -\frac{\alpha}{T^3}$$

where α is characteristic of the interaction between the adsorbed film and the adsorbent; in the case of $\alpha \to 0$ the adsorption decreases. Substitution of the above equation into the Gibbs formula leads to:

$$\lim_{p \to p_0} \pi_f \neq \infty$$

Figure II.9 shows the adsorption isotherm of nitrogen (black dots) on anatase (Jura and Harkins [65]). In the vicinity of the equilibrium vapour pressure ($0.8 < x \leq 1.0$) the BET equation is reduced to the form $\theta = 1/(1 - x)$, (see ref. [79]; $X = B$). The curve marked "BET" was calculated from this equation, while the other full curves were obtained from the relation $\ln x = -k/\theta^3$ with different k values*. It can be seen that with the choice $k \approx 4$ excellent agreement between the calculated, and experimentally determined, values is obtained.

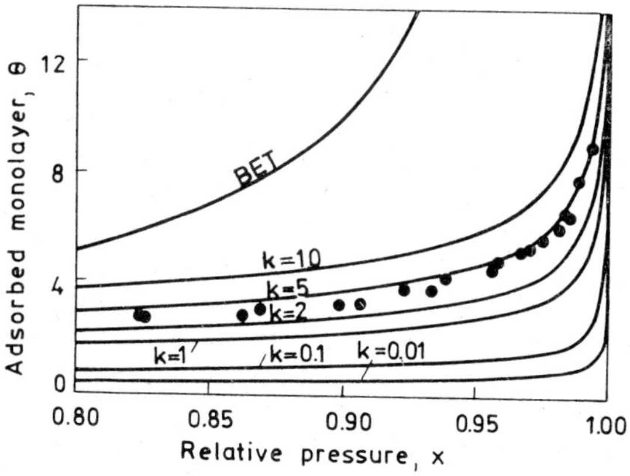

Fig. II.9. Adsorption of nitrogen on anatase. ● experimentally obtained points. The BET curve was calculated by means of $\theta = 1/(1 - x)$, the others from the equation $\ln x = -k/\theta^3$ for different k values

$$* \; k = \alpha \left(\frac{A_s}{N_{a,\infty}}\right)^3 = \alpha a_0^3$$

The equation gives good results mainly in the case of non-polar adsorbates. The measurements of Halsey [58] confirm the above examples: by choosing the exponents of Γ as 2.67 this equation can be successfully applied to the description of nitrogen isotherms even beyond the theoretical validity of the equation.

Harkins and Jura suggested a value of 2 as the exponent of Γ [66].

The study of the density of the adsorbed film, however, cast doubt on the theoretical reliability of the equation suggested by Frenkel, Halsey and Hill. Though Frenkel was fully aware of the fact that without an interaction between adsorbent and adsorbate there is no adsorption, he nevertheless assumed that the density of the adsorbed film is the same as that of normal liquids. On the other hand, Danforth, De Wries and others [67] were able to prove experimentally that the layer thickness of the adsorbed film has a fairly significant influence on the density of the film. Without going into details, it must be mentioned that Barrer and Robins [68] developed Frenkel's theory in this direction with success. Everything considered, the modifications resulted in the finding that at lower coverages it is advisable to use not 3.0, but values between 2.5 and 3.0 for the exponent of Γ.

CHAPTER 3. THERMODYNAMICS OF ADSORPTION

In the theory of the thermodynamics of adsorption it is expedient to distinguish two more or less independent trends:
(1) description based on surface excesses, i.e. on excess functions, when it is usual to define the excess, as compared to a reference system, with respect to extensive variables (volume, mass), and to thermodynamic functions (energy, enthalpy, entropy, etc.), including the quite general functional relations between them [69];
discussion of the "thermodynamics of solutions" [58] where the adsorbent ("solvent") plays the role of an inert material can be considered as a special case of (1) though here the interpretation of the terms "system" and "reference system" are somewhat different from those used in (1).
(2) discussion of the "thermodynamics of adsorption" when the boundary layer is considered a separate interface of finite thickness (and hence of defined volume) [70, 71]. This method of approach involves the error of interpreting the size of the adsorbent surface as an independent variable, which leads to difficulties in all cases in which the size of the surface can be changed only together with the mass of the adsorbent.
In our subsequent discussions the most up-to-date method of approach will be used; the description based on the excesses [(1)], and as suggested by Schay [72] the most important thermodynamic expressions will be derived in this connection. In the discussion of adsorption phenomena on solid adsorbents, however, the "thermodynamics of solutions" is the most widely used, and thus the claim to demonstrate the modifications of the excess functions in the hypothetical case when the adsorbent is inert cannot be renounced.

SECTION 3.1. DEFINITION OF THE EXCESS FUNCTIONS

Let us consider an open thermodynamic system consisting of the adsorbent α (of mass m^{α}) and a single sorbing gas component. In accordance with the demands of thermodynamic equilibrium the n moles of the latter are distributed between the gas and the sorbed phase. Let us give the symbol u to the internal energy (i.e. adsorbent + adsorbed phase + gas phase), s to the entropy, v to the volume of the total system and p to the equilibrium pressure in the gas phase. Instead of the entire accessible surface, A_s, of the adsorbent ($A_s \equiv a_s m^{\alpha}$, where a_s is the specific surface area) m^{α} is selected as the independent variable, since A_s cannot change independently of m^α if, conforming to general practice, a_s is taken as a constant.

Using these symbols* and restrictions we may write, as suggested by Gibbs [70], for any arbitrary infinitesimal change of the internal energy:

$$du = Tds - pdv + (\mu^\alpha + \gamma a_s)\, dm^\alpha + \mu dn$$

The third term on the right refers exclusively to the adsorbent; the use of the mass m^α instead of the "moles" of the adsorbent n^α is justified by the fact that for real adsorbents only the mass has a physical meaning. Thus μ^α is the specific free enthalpy of the compact adsorbent, and γ is the excess of the latter per unit surface, which is not necessarily identical with the surface tension of the solid. The adsorbent itself is characterized only by the value of γ for vacuum (γ^α) as an average value; γ is determined by the parameters of the actual interface (e.g. by the quantity of sorbed substance). μ is identical with the chemical potential of the gas component. u is a first-order homogeneous function of the extensive variables and can therefore be written directly in the integral form:

$$u = Ts - pv + (\mu^\alpha + \gamma a_s) m^\alpha + \mu n$$

from which it follows that for an infinitesimal reversible change of the intensive variables the Gibbs–Duhem-type relation:

$$sdT - vdp + m^\alpha(d\mu^\alpha + a_s d\gamma) + nd\mu = 0$$

is valid.

The enthalpy h, the free energy f and the free enthalpy g of the entire system** follow from the above expression for u, and taking their definitions into account:

$$h \equiv u + pv = Ts + (\mu^\alpha + \gamma a_s)\, m^\alpha + \mu n$$

$$f \equiv u - Ts = -pv + (\mu^\alpha + \gamma a_s) m^\alpha + \mu n$$

$$g \equiv f + pv = (\mu^\alpha + \gamma a_s) m^\alpha + \mu n$$

Interpretation of surface excesses requires the definition of a reference system consisting of an appropriately chosen quantity of adsorbent and of a separate gas phase, both having the same temperature and equilibrium pressure as the true system, with the sole difference that there is no interaction between them. Hence, the thermodynamic functions of the reference system can be obtained simply by addition of the corresponding contributions of the solid adsorbent and the gas phase. The surface excesses can be

* As regards the symbols used in this Chapter, in contrast to the practice of the preceding and following Chapters, the symbols referring to the state will be given not as the right lower but as the right upper index to avoid several indices in the case of extensive variables and thermodynamic quantities. A similar use of symbols in the other Chapters might have caused confusion with the exponents.

** In the thermodynamic discussion of adsorption, h and g are usually interpreted in a manner different from this, inasmuch as $\gamma a_s m^\alpha$, the capillary work, is subtracted from h and g.

obtained as the differences of the thermodynamic functions belonging to the real and the reference systems.

For the gas phase of the reference system the molar expression of the thermodynamic functions $U^{\mathrm{g}}_{\mathrm{r}}$, $H^{\mathrm{g}}_{\mathrm{r}}$, $F^{\mathrm{g}}_{\mathrm{r}}$ and $G^{\mathrm{g}}_{\mathrm{r}}$ * are as follows:

$$U^{\mathrm{g}}_{\mathrm{r}} \equiv TS^{\mathrm{g}}_{\mathrm{r}} - pV^{\mathrm{g}}_{\mathrm{r}} + \mu$$
$$H^{\mathrm{g}}_{\mathrm{r}} \equiv U^{\mathrm{g}}_{\mathrm{r}} + pV^{\mathrm{g}}_{\mathrm{r}} = TS^{\mathrm{g}}_{\mathrm{r}} + \mu$$
$$F^{\mathrm{g}}_{\mathrm{r}} \equiv U^{\mathrm{g}}_{\mathrm{r}} - TS^{\mathrm{g}}_{\mathrm{r}} = pV^{\mathrm{g}}_{\mathrm{r}} + \mu$$
$$G^{\mathrm{g}}_{\mathrm{r}} \equiv F^{\mathrm{g}}_{\mathrm{r}} + pV^{\mathrm{g}}_{\mathrm{r}} = \mu$$

These relations must be supplemented with the statement that when the complete differentials of these functions are formed, the Gibbs–Duhem equation

$$S^{\mathrm{g}}_{\mathrm{r}}\mathrm{d}T - V^{\mathrm{g}}_{\mathrm{r}}\mathrm{d}p + \mathrm{d}\mu = 0$$

holds here too.

As regards the adsorbent of the reference system, γ^{α} is, in fact, the surface tension of the adsorbent *in vacuo*. Choice of the reference system, however, requires a pressure p on the adsorbent. To eliminate this contradiction it is assumed that pressure p is exerted by an inert gas (e.g. He) which will have no influence on the value of γ^{α}. In addition, a_{s}, the specific surface area of the adsorbent is considered to be independent of the temperature.

After these preliminaries we can now write the specific thermodynamic functions for unit mass of the adsorbent:

$$u^{\alpha}_{\mathrm{r}} \equiv Ts^{\alpha}_{\mathrm{r}} - pv^{\alpha}_{\mathrm{r}} + \gamma^{\alpha}a_{\mathrm{s}} + \mu^{\alpha}$$
$$h^{\alpha}_{\mathrm{r}} \equiv u^{\alpha}_{\mathrm{r}} + pv^{\alpha}_{\mathrm{r}} = Ts^{\alpha}_{\mathrm{r}} + \gamma^{\alpha}a_{\mathrm{s}} + \mu^{\alpha}$$
$$f^{\alpha}_{\mathrm{r}} \equiv u^{\alpha}_{\mathrm{r}} - Ts^{\alpha}_{\mathrm{r}} = -pv^{\alpha}_{\mathrm{r}} + \gamma^{\alpha}a_{\mathrm{s}} + \mu^{\alpha}$$
$$g^{\alpha}_{\mathrm{r}} \equiv f^{\alpha}_{\mathrm{r}} + pv^{\alpha}_{\mathrm{r}} = \gamma^{\alpha}a_{\mathrm{s}} + \mu^{\alpha}$$

Here again the Gibbs–Duhem-type relation following from u^{α} is valid.

The definition of surface excesses now requires only a statement as to the composition of the reference system, that is, the number of moles of gas and the mass of the adsorbent. It seems most obvious to choose an adsorbent of mass m^{α}, and a quantity of gas (n^{g}) exerting an equilibrium pressure p in the gas phase (v^{g}) of the real system.

Since the volume v of the real system comprises the volume of the adsorbent v^{α} and the volume of the gas phase v^{g}:

$$v = v^{\alpha} + v^{\mathrm{g}}$$

and it follows that

$$m^{\alpha} = \frac{v^{\alpha}}{v^{\alpha}_{\mathrm{r}}} \quad \text{and} \quad n^{\mathrm{g}} = \frac{v^{\mathrm{g}}}{v^{\mathrm{g}}_{\mathrm{r}}}$$

v^{g} is obviously the same as the dead space measured with He, while v^{α} is simply the difference between v and v^{g}.

* The index r denotes the reference state.

It follows immediately from these considerations that the excess volume v^σ is zero:

$$v^\sigma = v - v^\alpha - v^{\mathrm{g}} = 0$$

This statement might sound surprising for more than one reason, since it is generally known that the adsorbent swells during the adsorption process and in certain cases (e.g. in plurimolecular adsorption) the volume of the adsorbed phase cannot be neglected. It must be borne in mind, however, that this result of $v^\sigma = 0$ is a necessary consequence of the equation defining v.

For the surface excess of adsorbate:

$$n^\sigma = n - n^{\mathrm{g}} = n - v^{\mathrm{g}}c = n - \frac{pv^{\mathrm{g}}}{RT}$$

where c is the equilibrium concentration of the gas component.

If the degree of adsorption is not too low and the equilibrium gas pressure not too high, n^σ will be identical with the sorbed amount n^{a}:

$$n^\sigma \equiv n^{\mathrm{a}}$$

The surface excess of internal energy will be

$$u^\sigma = u - m^\alpha u_{\mathrm{r}}^\alpha - n^{\mathrm{g}} U_{\mathrm{r}}^{\mathrm{g}} \tag{II.81}$$

and h^σ, f^σ, g^σ and s^σ are defined by entirely analogous equations.

As a natural consequence of the convention $v^\sigma = 0$:

$$u^\sigma \equiv h^\sigma \quad \text{and} \quad f^\sigma \equiv g^\sigma$$

Substituting u_{r}^α and U_{r}^α into Eq. (II.81) we obtain:

$$u^\sigma \equiv h^\sigma = T(s - m^\alpha s_{\mathrm{r}}^\alpha - n^{\mathrm{g}} S_{\mathrm{r}}^{\mathrm{g}}) - p(v - m^\alpha v_{\mathrm{r}}^\alpha - m^{\mathrm{g}} V_{\mathrm{r}}^{\mathrm{g}}) +$$

$$+ (\gamma - \gamma^\alpha)a_{\mathrm{s}} m^\alpha + \mu(n - n^{\mathrm{g}}) = Ts^\sigma + (\gamma - \gamma^\alpha)a_{\mathrm{s}} m^\alpha + \mu n^\sigma \tag{II.82}$$

Thus,

$$f^\sigma \equiv g^\sigma \equiv u^\sigma - Ts^\sigma$$

leads to

$$f^\sigma \equiv g^\sigma = (\gamma - \gamma^\alpha)a_{\mathrm{s}} m^\alpha + \mu n^\sigma \tag{II.83}$$

The differentials of u^σ and f^σ are:

$$\mathrm{d}u^\sigma \equiv \mathrm{d}h^\sigma = T\mathrm{d}s^\sigma + (\gamma - \gamma^\alpha)a_{\mathrm{s}}\mathrm{d}m^\alpha + \mu \mathrm{d}n^\sigma \tag{II.84}$$

and

$$\mathrm{d}f^\sigma \equiv \mathrm{d}g^\sigma = -s^\sigma \mathrm{d}T + (\gamma - \gamma^\alpha)a_{\mathrm{s}}\mathrm{d}m^\alpha + \mu \mathrm{d}n^\sigma \tag{II.85}$$

Complementing these equations with the function expressing the mutual relation between the arbitrary infinitesimal changes of the intensive variables:

$$s^\sigma \mathrm{d}T + a_{\mathrm{s}} m^\alpha \mathrm{d}(\gamma - \gamma^\alpha) + n^\sigma \mathrm{d}\mu = 0 \tag{II.86}$$

where n^σ (i.e. n^{a}) is often referred to unit mass or unit surface of the adsorbent (in the latter case $n^\sigma/a_{\mathrm{s}} m^\alpha = \Gamma^\sigma$ or $n^{\mathrm{a}}/A_{\mathrm{s}} = \Gamma$). The thermodynamic excess functions themselves are more often referred to one mole of sorbed substance, which means a simple division by n^σ (or n^{a}) or, when forming the partial molar (differential) quantities (see later), a simple differentiation.

It is one of the important properties of excess functions that for constant mass of adsorbent dm^α will be zero in the differential relations (II.84) and (II.85) corresponding to Eqs (II.82) and (II.83), suggesting that they do not include any data referring to the adsorbent. This might be one of the logical explanations for the fact that a number of authors [58] have believed that the thermodynamics of adsorption can be formulated without any explicit statement concerning the adsorbent [see also "solution thermodynamics" in point (1)]. In these arguments the role of the "inert" adsorbent would be limited to creating an external energy field without which adsorption ("the formation of solutions") is inconceivable.

Although experimental evidence proving the active participation of the adsorbent in the adsorption process does exist, let us try to envisage the consequences for the above system of equations of the assumption of an inert adsorbent.

It is easily seen that the internal energy of the real system u would be made up of the sum of the corresponding contributions of the adsorbent, the adsorbed layer and the gas phase as follows:

$$u = m^\alpha u_r^\alpha + n^a U^a + n^g U_r^g \qquad \text{(inert adsorbent)}$$

where U^a is the molar internal energy of the adsorbed layer. Substituting this expression for u into Eq. (II.81):

$$u^\sigma = n^a U^a = a^a \qquad \text{(inert adsorbent)}$$

Since the same procedure can be applied to h^σ, f^σ, g^σ, and s^σ too, it can be stated generally that in the case of an inert adsorbent the excess functions are identical with the thermodynamic functions of the adsorbed layer, when the quantity of the sorbed substance is n^a and the intensive thermodynamic variables have equilibrium values.

SECTION 3.2. THE GIBBS ADSORPTION (ISOTHERM) EQUATION

We find in all thermodynamic excess functions the difference

$$-\Delta\gamma = \gamma - \gamma_*^\alpha < 0$$

For mobile layers $\Delta\gamma$ is defined as the spreading pressure π_f, which has the same role in the two-dimensional equation of state of the adsorbed layer as the pressure in the equation of state of normal gases.

When the layers are localized the term "spreading pressure" has no meaning. According to Ref. [83], it is more correct to call the quantity $(\gamma - \gamma^\alpha)a_s m^\alpha$ quite generally (and thus independently of the state of the layer) the free energy excess of the surface.

From experimental data, $\Delta\gamma$ can be calculated by the following approach: substituting $dT = 0$ in Eq. (II.86) and rearranging:

$$\left(\frac{\partial \Delta\gamma}{\partial \mu}\right)_T = \frac{n^\sigma}{a_s m^\alpha} = \Gamma^\sigma$$

or, taking into account that, for ideal gases, $\mathrm{d}\mu_T = RT \,\mathrm{d} \ln p$:

$$\left(\frac{\partial \Delta\gamma}{\partial \ln p}\right)_T = RT\Gamma^\sigma = RT\Gamma \tag{II.87}$$

which is the same as the Gibbs adsorption equation.

Eq. (II.87) can be directly integrated:

$$\Delta\gamma \equiv \pi_\mathrm{f} = RT \int_0^p \Gamma^\sigma \,\mathrm{d} \ln p$$

supposing that the adsorption isotherm $\Gamma^\sigma(p)$ [i.e. $n^\sigma(p)$ or $n^\mathrm{a}(p)$] is known. The integral is always positive and increases monotonously, which means that γ decreases with increasing Γ^σ.

It follows from Eq. (II.83) that if $\Delta\gamma$ is known it is possible to calculate f^σ. Since the chemical potential of the gas component, μ, contains an arbitrary constant (the chemical potential of the gas component in the chosen standard state) the same is true for f^σ.

It follows from Eq. (II.85) that when m^α = constant and n^σ = constant:

$$\left(\frac{\partial f^\sigma}{\partial T}\right)_{m^\alpha, n^\alpha} = -s^\sigma$$

that is, by differentiating the function $[f^\sigma(T)]_{m^\alpha, n^\alpha}$ with respect to T the s^σ values for the different n^σ values can be calculated, as can u^σ, from Eq. (II.82). These calculations require nothing more than reliable isotherm data.

SECTION 3.3. DIFFERENTIAL QUANTITIES

As regards the thermodynamic functions, it is of fundamental importance to know the extent of the changes brought about in their values by unit increase of the quantity of substance sorbed, provided that, at the same time, the other parameters (m^α, T, p) remain constant. Change in n^σ must obviously be infinitesimal to leave the developed thermodynamic equilibrium state unaffected by this change.

It follows from Eq. (II.84) that:

$$\left(\frac{\partial u^\sigma}{\partial n^\sigma}\right)_{T, m^\alpha} \equiv \left(\frac{\partial h^\sigma}{\partial n^\sigma}\right)_{T, m^\alpha} = T\left(\frac{\partial s^\sigma}{\partial n^\sigma}\right)_{T, m^\alpha} + \mu$$

or, using simpler symbols:

$$\bar{U}^\sigma \equiv \bar{H}^\sigma = T\bar{S}^\sigma + \mu \tag{II.88}$$

Similarly, from Eq. (II.85):

$$\left(\frac{\partial f^\sigma}{\partial n^\sigma}\right)_{T, m^\alpha} \equiv \bar{F}^\sigma = \left(\frac{\partial g^\sigma}{\partial n^\sigma}\right)_{T, m^\alpha} \equiv \bar{G}^\sigma = \mu$$

in agreement with the requirements of thermodynamic equilibrium.

It holds for the chemical potential of the gas phase that:

$$\mu = H^{g} - TS^{g}$$

This can be substituted into Eq. (II.88), from which, by rearrangement:

$$\bar{U}^{\sigma} - H^{g} \equiv \bar{H}^{\sigma} - H^{g} \equiv \Delta H^{\sigma} = T(\bar{S}^{\sigma} - S^{g}) \equiv T\Delta S^{\sigma} \qquad \text{(II.89)}$$

In Eq. (II.89) ΔH^{σ} and ΔS^{σ} are identical with the virtual enthalpy and entropy change, respectively, accompanying the transfer of one mole of substance from the gas phase into the adsorbed phase at a constant temperature T, at a pressure p corresponding to adsorption equilibrium and at a sorbed quantity n^{σ}.

ΔH^{σ} is also called the isosteric heat of adsorption (Q_{st}) since its value can be obtained directly from the adsorption isosteres. It follows from Eq. (II.84) for m^{α} = const. and n^{σ} = const. that:

$$\mathrm{d}h^{\sigma} = T\mathrm{d}s^{\sigma} \quad (m^{\alpha}, n^{\sigma} = \text{const})$$

When this is differentiated with respect to n^{σ} the equation

$$\mathrm{d}\bar{H}^{\sigma} = T\mathrm{d}\bar{S}^{\sigma}$$

must hold for the differential excess quantities. Comparing this with the differential of Eq. (II.88):

$$\bar{S}^{\sigma}\mathrm{d}T + \mathrm{d}\mu = 0$$

The differential change of the chemical potential of the gas, however, is:

$$\mathrm{d}\mu = -S^{g}\mathrm{d}T + V^{g}\mathrm{d}p$$

which, when substituted into the above equation, leads to the Clausius–Clapeyron relation:

$$\left(\frac{\partial p}{\partial T}\right)_{m^{\alpha},n^{\sigma}} = -\frac{\Delta S^{\sigma}}{V^{g}} \qquad \text{(II.90)}$$

For an ideal gas $V^{g} = RT/p$, so that in place of Eq. (II.90):

$$\left[\frac{\partial \ln p}{\partial(1/T)}\right]_{m^{\alpha},n^{\sigma}} = \frac{\Delta H^{\sigma}}{R} = \frac{Q_{st}}{R} \qquad \text{(II.91)}$$

is obtained.

An analogous relation is obtained starting from Eq. (II.86) if it is stipulated that the intensive parameters of the system change in such a way that $(\gamma - \gamma^{\alpha})$ remains constant. Dividing Eq. (II.86) by n^{σ} and substituting $\mathrm{d}\mu$ by the above differential expression, and rearranging leads to

$$\frac{s^{\sigma}}{n^{\sigma}}\mathrm{d}T + \mathrm{d}\mu = \left(\frac{s^{\sigma}}{n^{\sigma}} - S^{g}\right)\mathrm{d}T + V^{g}\mathrm{d}p = 0$$

or

$$\left(\frac{\partial p}{\partial T}\right)_{\Delta\gamma} = -\frac{\dfrac{s^{\sigma}}{n^{\sigma}} - S^{g}}{V^{g}}$$

For ideal gases this can be expressed in a form similar to Eq. (II.91):

$$\left[\frac{\partial \ln p}{\partial(1/T)}\right]_{\Delta\gamma} = \frac{T\left(\dfrac{s^\sigma}{n^\sigma} - S^{\mathrm{g}}\right)}{R} \tag{II.92}$$

In contrast to Eq. (II.90), the expression $(\partial p/\partial T)_\gamma$ contains not the differential excess entropy $\bar{S}^\sigma$, but $\tilde{S}^\sigma \equiv s^\sigma/n^\sigma$, the integral molar excess entropy.

It may be worthwhile to point out again that for inert adsorbents:

$$\Delta H^\sigma \equiv \bar{H}^{\mathrm{a}} - H^{\mathrm{g}}$$

(inert adsorbent)

$$\Delta S^\sigma \equiv \bar{S} - S^{\mathrm{g}}$$

where S^{a} and H^{a} are the differential entropy and enthalpy, respectively, of the adsorbed layer. The quantity s^σ/n^σ in Eq. (II.92) is the same as the molar entropy of the layer:

$$\frac{s^\sigma}{n^\sigma} \equiv S^{\mathrm{a}} \qquad \text{(inert adsorbent)}$$

These relations are somewhat modified when the above-mentioned material transport from the gas phase into the adsorbed phase proceeds not under equilibrium conditions, but starting from some arbitrarily chosen standard state for the gas phase (e.g. from a state of unit pressure, or from one corresponding to the vapour pressure of the liquid adsorbate). If the molar enthalpy

$$^*H^{\mathrm{g}} = T\,^*s^{\mathrm{g}} + {}^*\mu$$

for the standard state of the gas component is subtracted from Eq. (II.88), the relation

$$\Delta^*H^\sigma = T\Delta^*S^\sigma + \Delta^*\mu \tag{II.93}$$

is obtained, where

$$\Delta^*H^\sigma = \bar{H}^\sigma - {}^*H^{\mathrm{g}}$$

$$\Delta^*S^\sigma = \bar{S}^\sigma - {}^*S^{\mathrm{g}}$$

$$\Delta^*\mu = \mu - {}^*\mu$$

Eq. (II.89) is obviously a special case of Eq. (II.93). Eq. (II.93) is equa to Eq. (II.89) when the equilibrium state is the same as the chosen standard state (i.e. when $\Delta^*\mu = 0$).

When the adsorbent is inert, then conforming with the above:

$$\Delta^*H^\sigma = \bar{H}^{\mathrm{a}} - {}^*H^{\mathrm{g}}$$

(inert adsorbent)

$$\Delta^*S^\sigma = \bar{S}^{\mathrm{a}} - {}^*S^{\mathrm{g}}$$

and even in the case of ideal gases:

$$\Delta^*\mu = RT \ln (p/{}^*p)$$

Using this expression for $\Delta^*\mu$ and rearranging, Eq. (II.93) may be written as:

$$\begin{aligned}\Delta^*\mu &= RT \ln (p/{}^*p) = \\ &= (\bar{H}^a - {}^*H^g) - T(\bar{S}^a - {}^*S^g) = \qquad \text{(II.94)} \\ &= (\bar{H}^a - {}^*H^g) - T(\bar{S}^a - {}^*S^g) \qquad \text{(inert adsorbent)}\end{aligned}$$

It should be mentioned here that if the saturated vapour of the liquid adsorbate is chosen as the standard state of the gas component, $-\Delta^*\mu$ will be equal to the Polányi adsorption potential, ε, [64]:

$$\varepsilon \equiv -\Delta^*\mu = -RT \ln x$$

where x is the relative pressure of the adsorbate. By means of calculations, the details of which will not be given here, it is possible to prove that in agreement with experimental findings the derivative of ε with respect to T along the adsorption isostere is equal to zero:

$$\left(\frac{\partial \varepsilon}{\partial T}\right)_{n^\sigma} = 0$$

Finally, it must be pointed out that in every case (e.g. at liquid interfaces) where the surface A_s can be varied independently of the adsorbent mass, the value of interface, A_s, may be written instead of $a_s m^\alpha$ in all the earlier equations without causing any fundamental change in their interpretation.

In the subsequent discussions, use will primarily be made of the equations relating to inert adsorbents, although we are fully aware of the fact that in the thermodynamic treatment of chemisorption systems it is extremely difficult to consider the adsorbent inert. As well as being the accepted convention, this procedure is justified by its descriptiveness, and not the least by the fact that no illustrative material can be found in the literature concerning the thermodynamics of chemisorption systems from any other standpoint. Although differences in interpretation are very significant in the majority of cases, this in no way affects the formality of the discussion.

CHAPTER 4. ENTHALPY CHANGES ACCOMPANYING ADSORPTION

SECTION 4.1. DEFINITIONS OF THE HEAT OF ADSORPTION

The parameter defined by the equation

$$Q_{\text{diff}} = \left(\frac{\partial u_{\text{a}}}{\partial n_{\text{a}}}\right)_T - U_{\text{g}} = \bar{U}_{\text{a}} - U_{\text{g}} \qquad \text{(II.95)}$$

is the differential heat of adsorption without external work ($p\mathrm{d}v = 0$). Since the differential internal energy of the layer, U_{a}, is a function of the quantity sorbed, the integral heat of adsorption liberated by the sorption of a quantity n_{a} of the substance is obtained from the integral of Eq. (II.95):

$$Q_{\text{int}} = \int_0^{n_{\text{a}}} Q_{\text{diff}}\,\mathrm{d}n_{\text{a}} = \int_0^{n_{\text{a}}} \left(\frac{\partial u_{\text{a}}}{\partial n_{\text{a}}}\right)_T \mathrm{d}n_{\text{a}} - n_{\text{a}} U_{\text{g}} =$$

$$= n_{\text{a}} \left(\frac{u_{\text{a}}}{n_{\text{a}}} - U_{\text{g}}\right) = n_{\text{a}}(U_{\text{a}} - U_{\text{g}}) \qquad \text{(II.96)}$$

where U_{a} is the integral molar internal energy of the layer.

For ideal gases the following relation exists between the differential and the earlier-defined isosteric heat of adsorption:

$$Q_{\text{st}} = \left(\frac{\partial u_{\text{a}}}{\partial n_{\text{a}}}\right)_T - (U_{\text{g}} + RT) = Q_{\text{diff}} - RT = \left(\frac{\partial h_{\text{a}}}{\partial n_{\text{a}}}\right)_T - H_{\text{g}} = \bar{H}_{\text{a}} - H_{\text{g}} \qquad \text{(II.97)}$$

since $Q_{\text{diff}} < 0$, $|Q_{\text{st}}| > |Q_{\text{diff}}|$.

If, contrary to the initial condition [Eq. (II.95)], the external work is not zero ($p\mathrm{d}v \neq 0$) we obtain the isothermal heat of adsorption which is related to the differential heat of adsorption by:

$$Q_{\text{is}} = Q_{\text{diff}} + v_{\text{g}} \left(\frac{\partial p}{\partial n_{\text{a}}}\right)_v \qquad \text{(II.98)}$$

where v_{g} is the volume of the gas phase. The differential quotient on the right hand side ($\partial p/\partial n_{\text{a}} < 0$) can be determined from the adsorption isotherms.

Recalling what was said in Chapter 1, the differential internal energy, U_{a}, can be considered in accordance with the equation:

$$\bar{U}_{\text{a}} = \bar{U}_{\text{a}}^{\text{therm}} + \bar{U}_{\text{a}}^{\text{pot}}$$

as being composed of a thermal term, and a second term characteristic of the adsorbent—adsorbate (and/or the adsorbate—adsorbate) interaction. Hence:

$$Q_{\text{diff}} = (\overline{U}_{\text{a}}^{\text{therm}} - U_{\text{g}}) + \overline{U}_{\text{a}}^{\text{pot}} \tag{II.99}$$

It might seem surprising that in the above relations the adsorbed layer is always characterized by differential parameters, such as the differential internal energy, $\overline{U}_{\text{a}}$, or the differential enthalpy, $\overline{H}_{\text{a}}$. This was done because $\overline{U}_{\text{a}}$ and $\overline{H}_{\text{a}}$ are generally functions of the surface coverage. This dependence on coverage is manifest mainly through the function $\overline{U}_{\text{a}}^{\text{pot}}$, which gives the potential energy for one mole of sorbed substance at the sorbed quantity n_{a} ("depth of the potential well"), preferably referred to the potential energy of one mole of free gas. However, $\overline{U}_{\text{a}}^{\text{pot}}$ has an indirect influence on $\overline{U}_{\text{a}}^{\text{therm}}$, since the number of degrees of freedom in the adsorbed phase is a function of, among other things, the mobility of the layer, and, as shown before, this is closely linked with the potential energy of the layer. These details obviously remain hidden in phenomenological thermodynamic descriptions, but are indispensable in the statistical mechanical discussion of the layer.

The dependence of $\overline{U}_{\text{a}}^{\text{pot}}$ on the coverage may be a result of the effects of many factors, usually summed up under the generic term of "surface heterogeneity". Heterogeneity might be an originally existing property of the adsorbent (caused by the various crystal faces, edges, peaks, etc.), but might also be caused by adsorbate—adsorbate interactions which change with the coverage (see also Paragraphs 2.2.4, 6.2.1 and Section 8.1).

Experimentally the heat of adsorption can be determined by direct calorimetric measurements. Adsorption calorimetry requires special equipment and a lot of experience, for the heat effects do not always lie in the range which can be easily measured. We shall not deal here with a description of the experimental technique. The practical question arises, however, as to which of the above-defined heats of adsorption is the one most closely approached by the calorimetrically determined heats of adsorption found in the literature.

In most cases there is external work during adsorption, the value for one mole varying between 0 and RT. In fact, depending on the experimental conditions, the calorimetrically determined heat of adsorption lies somewhere between the differential heat of adsorption, Q_{diff}, and the isothermal heat of adsorption, Q_{is}.

Because of this uncertainty only indirectly obtainable isosteric heats of adsorption are subject to strict comparison, since their values are hardly influenced by the experimental technique. In the calculation of these data it is essential that the isotherms [from which the $p = f(T)$, $\theta =$ const. isosteres are plotted] embrace the widest possible ranges of coverage, equilibrium pressure and temperature, and further that the θ vs. p, or χ vs. p data represent true equilibrium states.

In practical chemisorption the various definitions of the heats of adsorption are of no great importance, partly because of the difficulties of obtaining a reproducible measurement of the heat of chemisorption, and partly because of the fact that the heat effect due to external work is virtually negligible compared with the values of the heat of chemisorption, which may reach a multiple of -10 kcal mole^{-1}. Hence, since $\bar{U}_a^{therm} \approx U_g$:

$$Q_{st} \approx Q_{diff} \approx \bar{U}_a^{pot}$$

The meaning of $\bar{U}_a^{pot}$ acquires a descriptive meaning if an adsorption energy (Q), characteristic of the adsorbent—adsorbate interaction and depending on the prevailing coverage, is assigned to the "active centres" of the adsorbent; this energy is released as heat of adsorption when the centres in question become covered with the adsorbate (see also Paragraph 2.2.4). After what has been said in Chapter 1, it is perhaps superfluous to emphasize that Q is not a parameter of the adsorbent alone, but of the system adsorbent plus adsorbate. In the calculation of $\bar{U}_a^{pot}$ the reference state is the uncovered adsorbent plus the atom (or molecule) of the adsorbate at an (in principle) infinite distance from the adsorbent, and, since the potential energy of the system decreases in the course of the adsorption, $\bar{U}_a^{pot} < 0$. For reasons of expediency, and in the hope that this will cause no special confusion, since, as indicated for example by Eq. (II.63), there is no direct relationship between Q_{st}, Q_{diff}, $\bar{U}_a^{pot}$ and Q, in contrast with the currently used rule of signs in thermodynamics, the adsorption energy is defined as a *positive* parameter.

By means of the BET equation (II.80), an upper estimate still compatible with physical adsorption can be given for the maximum value of the heat of adsorption of the first layer (Q_a). Table II.4 contains data of this type for the most common adsorbates, together with the corresponding heats of evaporation Q_1.

The heats of adsorption determined for physical adsorption by various authors by means of adsorption methods are indeed between Q_1 and Q_a for the most varied adsorbents, permitting the conclusion that the value of the heat of adsorption is determined primarily by the nature of the substrate material. There is no distinction of this type between adsorbent and adsorbate in chemisorption processes, so that no classification similar to

TABLE II.4

Maximum values of heats of adsorption typical of physical adsorption

Gas	Q_1, cal mole^{-1}	$Q_{a,max}$ cal mole^{-1}
H_2	-220	$-2\ 000$
O_2	$-1\ 600$	$-5\ 000$
N_2	$-1\ 340$	$-5\ 000$
CO	$-1\ 440$	$-6\ 000$
CO_2	$-6\ 000$	$-9\ 000$
CH_4	$-2\ 180$	$-5\ 000$
C_2H_4	$-3\ 500$	$-8\ 000$
C_2H_2	$-5\ 740$	$-9\ 000$
NH_3	$-5\ 560$	$-9\ 000$
H_2O	$-10\ 570$	$-14\ 000$
Cl_2	$-4\ 400$	$-8\ 500$

that in Table II.4 can be made for the latter. Heats of chemisorption can, in fact, vary between very wide limits, even when the nature of the adsorbent and adsorbate are given, depending on the physical state of the adsorbent (single crystal face, polycrystalline surface, etc.), on the nature of the surface contaminants, on the temperature of the measurement and on the type of the adsorbed layer formed in the chemisorption as a result of these factors (see e.g. Paragraph 8.1.3).

It should be considered with regard to the physical state of the adsorbent, that for a given adsorbent and adsorbate the interaction energy will depend characteristically on the crystallographic orientation of the surface. If the adsorbent is present in powder form or has been prepared by vacuum evaporation, the frequencies of incidence of the different crystal faces will be different, and this is certain to have an effect on the value of the heat of adsorption.

The nature and quantity of the surface contamination influence the value of the heat of adsorption in a manner characteristic of the system under investigation. (For example, in the 1930's — 30 kcal mole^{-1} was measured for the heat of hydrogen chemisorption on Ni surfaces; this value is today much lower due to the far more efficient methods of surface cleaning.)

The effect of temperature can be studied only in relation to the properties and structure of the adsorbed layer (see Paragraph 8.1.3).

Instead of going into the details of the results of the experimental and theoretical studies on the heat of adsorption, we shall try, within the space available, to satisfy the demands of investigators engaged in the study of heterogeneous catalysis who wish to estimate the values of the heats of chemisorption from known physical parameters (primarily from the bond energies) of the adsorbent and the adsorbate. This might be required in the investigation of the physical reality of heats of adsorption and bond energies calculated via the kinetic parameters of catalytic studies.

SECTION 4.2. CALCULATION OF HEATS OF CHEMISORPTION

Until recently, calculations of heats of chemisorption mainly concerned carbon and metal adsorbents. The data obtained refer primarily to low coverages, since at higher coverages the absolute values of the heats of adsorption are considerably lower because of the adsorbate—adsorbate interactions.

For carbon and a simple gas, such as H_2, covalent bonds are formed during chemisorption, the energy of these bonds not differing significantly from the bond energies of simple carbon compounds. Hence the equation

$$2\ C + H_2 \rightarrow 2\ CH$$

can be considered an outline description of the chemisorption and, if it is assumed for the time being that the chemisorption has no effect on the bonds between the carbon atoms, the heat of adsorption at low coverages will be

$$Q_0 = E_{HH} - 2\ E_{CH} \tag{II.100}$$

where the E values are the corresponding bond energies.

Similarly, in the chemisorption of O_2, Cl_2 and NH_3 (the latter is chemisorbed in the form of NH_2 and H radicals):

$$Q_0 = E_{OO} - 2\ E_{CO} \tag{II.101}$$

$$Q_0 = E_{ClCl} - 2\ E_{Cl} \tag{II.102}$$

$$Q_0 = E_{NH} - E_{CN} - E_{CH} \tag{II.103}$$

The empirically determined and calculated heats of adsorption on graphite are shown in Table II.5.

TABLE II.5

Calculated and experimentally determined heat of chemisorption on carbon, in kcal mole^{-1}

Adsorbate	Q_0		Reference
	calculated	observed	
H_2	−71	−50	Barrer [74]
O_2	−173	−97	Bull, Hall, Garner [75]
Cl_2	−75	−31	Keyes, Marshall [76]
NH_3	−32	−17	Keyes, Marshall [76]

The calculated absolute values are all higher than those determined experimentally; this can be explained partly by the contamination of the carbon surface and partly by the inadequacy of the calculation method (which does not consider, for instance, the fact that some of the C—C bonds must also be split during the adsorption process).

In the case of metals the nature of the bond can be decided from the value of the dipole moment at low coverages (see Section 1.4).

For ionic layers the calculation is as follows:

(1) The first step of the adsorption: transfer of one electron from the infinitely distant adsorbate atom to the surface of the adsorbent. The heat of adsorption liberated by this process is:

$$(eV_i - e\Phi)$$

where $e\Phi$ is the work function and eV_i the ionization energy of the adsorbate.

(2) In the second step the positively charged ion interacts with the assumed infinitely large metal surface. The energy of interaction [Eq. (II.24)] is:

$$-e^2/4z_0 \quad (n_i = 1)$$

In this way the heat of adsorption of the overall process will be:

$$Q_0 = eV_i - e\Phi - \frac{N_A e^2}{4\ z_0} \tag{II.103}$$

where N_A is the Avogadro constant, and
$e\Phi$ and eV_i refer to gram atomic quantities.

Calculated heats of adsorption for Na, K and Cs on W are given in Table II.6 [77, 78].

The discrepancies are due partly to the neglect of the van der Waals type interactions between the ion and the metal and of the interaction due to repulsive forces, and partly to the uncertainty in z_0.

TABLE II.6

Calculated and experimentally determined heats of chemisorption, in cal g-atom^{-1}

System	$e\Phi$	eV_i	$N_A e^2/4z_0$	Q_0 (calc)	Q_0 (observed)
Na on W	104,000	118,000	44,500	−30,500	−32,000
K on W	104,000	99,600	35,900	−40,300	—
Cs on W	104,000	89,400	31,100	−45,700	−64,000

When gases are chemisorbed on a metal surface they often produce covalent adsorbed layers. The method of calculation is then similar to that applied to carbon-type adsorbents [79].

Let us consider the H_2—W system, for instance, where the chemisorption can be described by the following equation:

$$2\,W + H_2 \rightarrow 2\,WH \qquad \text{(II.104)}$$

so that at zero coverage the differential heat of adsorption will be:

$$Q_0 = E_{HH} - 2\,E_{WH} \qquad \text{(II.105)}$$

Splitting of the W—W bonds is again ignored in Eq. (II.105), but this is justified by the very high rate of the chemisorption process.

The bond energy E_{WH} in Eq. (II.105) is not known. According to Pauling it can be calculated from the relation:

$$E_{WH} = \frac{1}{2}(E_{WW} + E_{HH}) + 23.06(\omega_W - \omega_H)^2 \qquad \text{(II.106)}$$

where ω_W and ω_H are the electronegativities of the respective atoms, and $23.06(\omega_W - \omega_H)^2$ is the increment due to the partly ionic nature of the bond.

To a first approximation

$$\omega_W - \omega_H = \mu \qquad \text{(II.107)}$$

where μ is the dipole moment of the W—H bond at $\theta = 0$. Using Eqs (II.106) and (II.107), Eq. (II.105) may now be replaced by:

$$Q_0 = -E_{WW} - 46.12\,\mu^2 \qquad \text{(II.108)}$$

The bond energy E_{WW} can be determined from the heat of evaporation, λ. It is approximately true for face- and body-centred lattices that:

$$E_{Me-Me} \approx \frac{z_0}{12}\lambda \qquad \text{(II.109)}$$

$(\omega_W - \omega_H)$ can be obtained either from μ at the coverage $\theta = 0$, or from Pauling's electronegativity tables. Both methods lead to a value of $(\omega_W - \omega_H) = -0.6$ D.

Table II.7 lists the calculated and experimentally determined initial heats of adsorption of hydrogen on certain metals [80]. (For want of experimental data the values in brackets were estimated.)

TABLE II.7

Calculated and experimentally determined heats of chemisorption of hydrogen, in kcal mole^{-1}

Metal (*Me*)	E_{Me-Me}	μ, D	Q_0 (calc)	Q_0 (observed)
Ta	(30.7)	(−0.6)	−47.3	−45
W	33.8	−0.6	−50.4	−45
Cr	14.9	(−0.6)	−31.5	−45
Ni	16.4	−0.2	−18.2	−31
Fe	16.1	(−0.2)	−17.9	−32
Rh	(21.7)	(−0.2)	−23.5	−28
Pt	20.3	+0.6	−36.9	(−30)

This method of calculation can be applied to other gases too. The corresponding equations for O_2 and N_2 are as follows:

$$2\,Me + O_2 \rightarrow 2\,Me = O$$

$$2\,Me + N_2 \rightarrow 2\,Me \equiv N$$

Assuming that relations similar to Eq. (II.106) are also valid for double and triple bonds, then:

$$Q_{0,O_2} = -E_{Me=Me} - 46.12\,\mu^2 \qquad \text{(II.110)}$$

$$Q_{0,N_2} = -E_{Me\equiv Me} - 46.12\,\mu^2 \qquad \text{(II.111)}$$

The bond energies $E_{Me=Me}$ and $E_{Me\equiv Me}$ are assumed to be twice and three times, respectively, the bond energy E_{Me-Me}.

Some experimentally determined and theoretically calculated heats of adsorption are shown in Table II.8. Estimated dipole moments are again in brackets.

TABLE II.8

Calculated and experimentally determined heats of chemisorption of oxygen and nitrogen, in kcal mole^{-1}

System	$E_{Me=Me}$ or $E_{Me\equiv Me}$	μ, D	Q_0 (calc)	Q_0 (observed)	Reference
O_2 on W	67.6	(−0.78)	−95.6	−155	Roberts [81]
O_2 on Ni	32.8	(−0.78)	−60.8	−130	Beeck [43]
N_2 on W	100.4	(−0.61)	−117.4	−95	Beeck [43]
N_2 on Ta	92.1	(−0.61)	−109.1	−140	Beeck [82]
N_2 on Fe	48.3	(−0.61)	−65.3	−40	Beeck [43]

For oxygen layers the calculated heat of chemisorption is always lower than that determined experimentally. This discrepancy is probably caused by the fact that the dipole moments of the layers are known only for $\theta = 1$, and because of the mutual depolarizing effect of the dipoles this is in all probability considerably lower than the value for $\theta = 0$.

If Eqs (II.110) and (II.111) are accepted as being valid, then the dipole moments can be calculated from the experimentally determined heats of adsorption. It is then possible to calculate (according to Pauling [73]) the percentage ionic character of the surface bond from the expression $(1 - e^{-0.25\mu^2})$.

Ethylene (and probably other olefines too) is dissociatively adsorbed on metal surfaces [83, 84]. The dissociation products are H atoms and (in the case of ethylene) an adsorbed "acetylene" complex:

$$4\,Me + C_2H_4 \rightarrow Me_2C_2H_2 + 2\,MeH$$

The C_2H_2 radical is probably bound in the following manner to the metal:

$$\begin{array}{cc} \mathrm{CH}{=} & \mathrm{CH} \\ | & | \\ Me & Me \end{array}$$

so that the heat of chemisorption will be:

$$Q_0 = 2\,E_{\mathrm{CH}} - 2\,E_{Me-Me} - E_{\mathrm{HH}} - E_{\mathrm{CC}} - 46.12(\mu^2_{Me\mathrm{C}} + \mu^2_{Me\mathrm{H}}) \quad \text{(II.112)}$$

The momentum $\mu_{Me\mathrm{C}}$ can be calculated from the difference between the contact potentials of the pure and ethylene-covered surfaces. These data are known for Ni and W [19], though there is a significant difference here between the experimentally determined and calculated values of the heat of adsorption.

The way in which the above expressions are written may suggest that what was calculated was in fact the heat effects of gas phase reactions. It is extremely surprising that in a number of cases when a reaction actually occurring in the gas phase is equivalent to chemisorption, such as is the case for

$$W + \frac{3}{2} O_2 \rightarrow WO_3 \quad (\Delta H = -290 \text{ kcal mole}^{-1})$$

or for

$$Ni + 4\,CO \rightarrow Ni(CO)_4 \quad (\Delta H = -141 \text{ kcal mole}^{-1})$$

the heats of reaction for one O atom and one CO molecule agree almost within the limits of experimental error with the heats of adsorption of -193 kcal mole^{-1} measured for the W–O_2 system and -42 kcal mole^{-1} for the Ni–CO system [85]. This draws attention to the fact that in certain cases the bulk properties of the adsorbent lose their importance and the adsorption centres display the character of individual atoms.

SECTION 4.3. HEATS OF CHEMISORPTION ON DIFFERENT METALS

One of the most interesting problems which arises in the study of chemisorption and catalysis is whether the behaviour of the active centres of the metal used as the catalyst is determined by the bulk parameters of the metal or by the properties of individual atoms. Since surface atoms possess properties different from those of both individual and bulk atoms, such a rigid approach is no doubt wrong and misleading. Nevertheless, it cannot be denied that, depending on the effects to which the system is exposed, the way in which the surface atoms react will be determined in some cases by the properties of the individual atoms, and in others by those of the bulk. These principles, formulated in the various theories, appear throughout the whole of chemisorption and catalysis.

TABLE II.9

Percentage *d*-character of transition metals in the metallic state

Sc	Ti	V	Cr	Mn	Fe	Co	Ni	Cu
20	27	35	39	40.1	39.5	39.7	40	36
Y	Zr	Nb	Mo	Te	Ru	Rh	Pd	Ag
19	31	39	43	46	50	50	46	36
La	Hf	Ta	W	Re	Os	Ir	Pt	Au
19	29	39	43	46	49	49	44	—

The regularity in the values of the heats of chemisorption observed for transition metals induced Trapnell and Beeck [86] to seek a correlation between this parameter, which characterizes the "strength" of the chemisorption bond, and the percentage *d*-character in the Pauling theory of metals [87] (see Section 8.1).

Trapnell and Beeck claim that the chemisorption bond is identical with the covalent bond between the unpaired electrons of the atomic d-orbitals and the adsorbate (see Paragraph 8.1.1). It stands to reason that both the bond strength and the absolute value of the heat of adsorption will be the higher the more atomic *d*-orbitals are available for covalent bonds. Since the percentage *d*-character (δ) in the Pauling theory expresses the participation of *d*-electrons in the formation of dsp hybrid bonds between the individual atoms, $(100 - \delta)\%$ might characterize the ability of the atomic *d*-orbitals to form bonds with foreign substances.

The values of the heats of adsorption of H_2, O_2, N_2, CO, CO_2, C_2H_4 and NH_3 on various metals lie in the following order:

$$\text{Ti, Ta} > \text{Nb} > \text{W, Cr} > \text{Mo} > \text{Fe} > \text{Mn} > \text{Ni, Co} > \text{Rh} > \text{Pt, Pd} > \text{Cu, Au}$$

From the δ values of Bond [88] (see Table II.9) and using the theory of Trapnell and Beeck the following order is obtained for the absolute values of the heats of adsorption:

$$\text{Ti} > \text{Ta, Cr, Nb} > \text{Co} > \text{Fe} > \text{Ni, Mn} > \text{Mo} > \text{W} > \text{Pt} > \text{Pd} > \text{Rh} > \\ > \text{Cu, Au}$$

This is undeniably similar in some ways to the first series, but nevertheless differs from it.

In the equations of the preceding Section for a given adsorbate, but various metals, both the bond strength E_{Me-Me} (closely related to the heat of sublimation of the metal) and the electronegativity of the metal ω_{Me} vary, but E_{Me-Me} is the dominant factor.

From this, Dowden [86] arrived at the conclusion that the heats of adsorption should follow the heats of sublimation of the various metals:

$$\text{W} > \text{Nb, Ta} > \text{Mo} > \text{Rh, Pt} > \text{Ti} > \text{Co, Ni} > \text{Fe,} \\ \text{Pd} > \text{Cr} > \text{Mn} > \text{Au} > \text{Cu}$$

This order differs considerably from the experimentally determined order.

Despite this unsuccessful attempt at correlation there can be no doubt that the empty d-orbitals (or more probably d-bands) play a special role in the formation of chemisorption bonds. Studies of the catalytic behaviour of metals have led to similar conclusions.

CHAPTER 5. ENTROPY CHANGES IN ADSORPTION

SECTION 5.1. DEFINITIONS OF THE ENTROPY OF ADSORPTION

The entropy functions of the adsorbed layer are defined in a similar manner to the internal energy and the enthalpy. For a sorbed quantity n_a, the differential entropy $\bar{S}_a[\equiv(\partial s_a/\partial n_a)_T]$ is related to the integral entropy of the layer by the equation:

$$s_a = \int_0^{n_a} \left(\frac{\partial s_a}{\partial n_a}\right)_T dn_a = \int_0^{n_a} \bar{S}_a dn_a \tag{II.113}$$

The definition of the integral molar entropy of the layer is the value of s_a divided by n_a:

$$S_a \equiv \frac{s_a}{n_a} \quad \text{i.e.} \quad s_a = n_a S_a \tag{II.114}$$

From the rearrangement of Eq. (II.113):

$$s_a = n_a \left[\frac{1}{n_a} \int_0^{n_a} \left(\frac{\partial s_a}{\partial n_a}\right)_T dn_a\right] = n_a S_a$$

S_a is thus the mean value of the integral of $\bar{S}_a$ in the interval $(0, n_a)$.

Through the relation

$$s_a = k \ln W$$

the integral entropy is related directly to the thermodynamic probability, W, of the system. Hence, the value of s_a (and particularly of S_a) permits conclusions as to the degrees of freedom of the adsorbed molecules.

There are several possibilities whereby S_a can be determined from experimental data.

(1) The most obvious procedure is to integrate $\bar{S}_a(n_a)$ in Eq. (II.113) with respect to n_a.

$\bar{S}_a$ can be given as follows [Eq. (II.94)]:

$$\bar{S}_a = {}^*S_g - R \ln (p/{}^*p) - \frac{{}^*H_g - H_a}{T} = S_g - \frac{{}^*H_g - H_a}{T}$$

where ${}^*S_g$ and S_g are the entropies of one mole of gas at a pressure ${}^*p$ and at the prevailing pressure p of adsorption equilibrium, respectively. (It will be seen immediately that the first part of the expression is better suited for practical purposes.)

Substituting this expression of $\bar{S}_a$ into Eq. (II.113) we obtain:

$$s_a = \int_0^{n_a} \left[{}^*S_g - R \ln (p/{}^*p) - \frac{{}^*H_g - H_a}{T} \right] dn_a =$$

$$= n_a\, {}^*S_g + n_a R \ln {}^*p - R \int_0^{n_a} \ln p \, dn_a - \frac{{}^*H_g}{T} n_a + \frac{1}{T} \int_0^{n_a} \bar{H}_a \, dn_a$$

and after division by n_a:

$$\frac{s_a}{n_a} = S_a = {}^*S_g + R \ln {}^*p - \frac{R}{n_a} \int_0^{n_a} \ln p \, dn_a + \frac{1}{n_a T} \int_0^{n_a} Q_{st} \, dn_a \quad \text{(II.115)}$$

In practice this method usually does not work, as $\ln p$ cannot be extrapolated to $n_a = 0$ with the accuracy required here.

(2) From a reliable adsorption isotherm determined down to fairly low coverages the two-dimensional spreading pressure is calculated at a given temperature from the Gibbs equation [87], since this can be relatively well extrapolated to zero pressure:

$$\pi_f = RT \int_0^p \Gamma d \ln p \qquad (T = \text{constant}) \qquad \text{(II.116)}$$

The $\ln p$ vs. $1/T$ (π_f = const) curves can be plotted from several isotherms for different equilibrium temperatures; their derivatives with respect to $1/T$ are:

$$\left[\frac{\partial \ln p}{\partial (1/T)} \right]_{\pi_f} = \frac{T(s_a/n_a - S_g)}{R}$$

and with the help of Eq. (II.92) $s_a/n_a = S_a$ can be calculated as a function of π_f (i.e. of n_a).

In more simple cases it is possible to measure π_f directly (e.g. adsorption of heptane on a mercury surface [89, 90]), which greatly simplifies the application of Eq. (II.116).

(3) For the characterization of the process of adsorption the difference in the thermodynamic (primarily the entropy) functions of two arbitrarily chosen standard states is often used (see Chapter 3). Let us consider the relation (II.94)

$$\Delta^*\mu = RT \ln (p/{}^*p) = (\bar{H}_a - {}^*H_g) - T(\bar{S}_a - {}^*S_g)$$

in which the standard state of the gas phase, independently of the temperature, is the state for which ${}^*p$ = 760 torr. As recommended by Kemball and Rideal [91], let us choose a standard state for the adsorbed phase in which the volume occupied by one molecule in the adsorbed phase is the same as that in three dimensions at atmospheric pressure. The volume v, occupied by one molecule in the adsorbed layer is equal to the product of the surface φ occupied by one molecule and the layer thickness δ:

$$v = \varphi\delta$$

Let us arbitrarily choose δ to be 6 Å. In three dimensions one molecule occupies a volume of $135.18\,T$ Å, and thus in the chosen standard state of the adsorbed phase, from the equation $135.18\,T = 6\varphi$, the surface occupied by one molecule is $\varphi = 22.53\,\mathrm{T}$ Å^2. For the spreading pressure $\pi_f = 0.0608$ dyne cm^{-1} *.

In cases in which this standard state of the layer is associated with almost immeasurably low equilibrium pressures p, it is more expedient to choose coverages $\theta = 0.1$ or $\theta = 0.5$ as the standard state of the layer.

At a given value of T and at the pressure p pertaining to the standard state of the layer, Eq. (II.94) cannot be solved for $\bar{H}_a$ and $\bar{S}_a$. The possibility still exists of choosing two neighbouring temperatures T_1 and T_2 and of assigning the values of $\bar{H}_a$ and $\bar{S}_a$ calculated from Eq. (II.94) to the arithmetic mean of T_1 and T_2.

In contrast to the molar entropy S_a of the layer, which can be obtained additively from the contributions of the degrees of freedom of the molecules, the differential entropy cannot be formed in this way (see the next Section). If $\bar{S}_a$ (or $\bar{H}_a$) is available in only a single state of the layer, the chosen standard state, this information is not sufficient for an unambiguous comparison of the experimentally determined and theoretical $\bar{S}_a$ values. For this it is absolutely necessary to know the function $\bar{S}_a$ in the widest possible range of coverage.

SECTION 5.2. MOBILITY OF THE ADSORBED LAYERS

A large number of direct and indirect studies bear witness to the fact that in the majority of cases the chemisorbed layers are localized. This statement, however, should not be taken literally, since there are many exceptions depending on the chemisorption system (adsorbent, adsorbate) and on the actual experimental conditions (temperature, gas pressure and, consequently, coverage). Before going into the details of this problem it might be worthwhile to consider some of the ideas which will be dealt with later in Paragraph 6.2.2.

As a result of the interaction between the adsorbent and the adsorbate, in the first phase of chemisorption excited molecules or molecular fragments are formed on the surface and, depending on the prevailing temperature, the value of the excitation energy, the nature of the potential energy surface, the rate of energy dissipation, etc., these exhibit mobility on the surface for various periods. In the meantime they are again desorbed or bound

* 1 atm = $1.013\ 10^6$ dyne cm^{-2}. The thickness of the layer $\delta = 6\ 10^{-8}$ cm and the pressure in the adsorbed layer is also "1 atm", and hence the spreading pressure for "1 atm" pressure will be:

$$\pi_f = 1.013\ 10^6\ 6\ 10^{-8} = 0.0608 \text{ dyne cm}^{-1}.$$

at the prevailing most active adsorption sites. If the excitation energy is low and/or the rate of energy dissipation high (e.g. at a low temperature), displacement may be limited, promoting the formation of statistically random localized layers. Layers of this type may be converted at higher temperatures into ordered, localized layers, or, at even higher temperatures, into mobile layers. The layer generally becomes mobile at temperatures lower than that at which desorption begins.

Some general conclusions follow immediately from this:

(1) the higher the heat of adsorption characterizing the adsorption interaction, and the lower the temperature, the higher is the probability of formation of random or ordered localized layers.

(2) The lower the energy of the adsorption interaction and the higher the temperature, the greater is the probability that mobile layers are formed. Since the adsorption sites are occupied in the order of decreasing interaction, above a certain coverage part of the layer will be mobile while the fraction bound at low coverage is still localized.

(3) Mobilization of the layer begins at a lower temperature than desorption. Exceptions to this rule are the dissociatively sorbed substances, for which molecular desorption generally occurs at a lower temperature than the migration of the atoms.

Almost all details of this outlined picture refer characteristically to the non-equilibrium state of chemisorption. What conception can be formed of the pattern of the adsorption equilibrium?

For activated adsorption the impression of equilibrium might be given by the state when the activation energy of adsorption is several times higher than the thermal energy of the adsorbate at the temperature T of the experiment, and therefore adsorption does not proceed at a noticeable rate.

If, for one reason or other, the surface is heterogeneous and several layers of different properties exist simultaneously, some of them might be in equilibrium with the gas phase (i.e. their ad- and desorption rates might be identical) without the same being true for the other layers. The highly favourable case might also occur when the entire system is in thermodynamic equilibrium with the gas phase, perhaps in such a way that though certain layers are not directly desorbed, their material content is capable of direct exchange with the material of the other layers.

We believe it necessary to stress this point, since thermodynamic calculations furnish reasonable results for equilibrium systems only. For systems not in equilibrium it is extremely difficult to decide upon the limit of applicability of thermodynamic methods.

Mobility can be studied by direct and indirect experimental methods. The spreading of the adsorbed layer, for instance, can be observed directly on the screen of a field emission microscope. In the same way, other methods suitable for the direct measurement of surface concentrations (mainly thermal and light emission methods) can also be used with success.

Of the indirect investigation methods, entropy changes associated with the adsorption process and the analysis of the heat of adsorption may provide useful information concerning the state of the adsorbed layers.

(1) Field emission microscope

In brief, the construction and principle of operation of the field emission microscope is as follows (Fig. II.10):

A tiny metal needle fixed to an electrically heated tungsten wire is placed into the middle of a spherical glass vacuum flask of appropriate size. The wall of the flask opposite the needle is coated with a conducting fluorescent layer. As a result of a potential difference of some thousand volts between the screen and the needle, electrons are emitted from the needle and cause light emission from the screen. In the vicinity of the emitting point the intensity of the field, E, is

$$E \approx \frac{V}{5\,r}$$

where V is the voltage applied and r is radius of curvature of the needle.

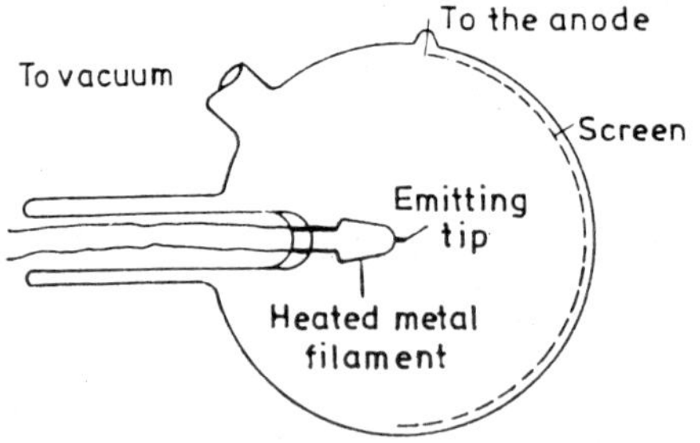

Fig. II.10. Diagram of the field emission microscope

At a radius of curvature $r \approx 10^{-5}$ cm and $V = 2-5$ kV, the field strength attains a value of $4\text{–}10 \cdot 10^{7}$ V cm^{-1}, which (by means of the tunnel effect) is sufficient to cause considerable electron emission even at room temperature ($\geq 10^{-9}$ A cm^{-2} screen surface). The kinetic energy of the electrons escaping from the emitting point is low, and thus (at least initially) their displacement is along the vectors of the electric field. Since the emitting point is a conductor (its surface is an equipotential surface), the vectors are normal to both the emitting point and the fluorescent screen, so that a highly magnified image of the tip of the needle appears on the screen. In the ideal case the linear magnification is determined by x/r, where x is the distance between the screen and the point. By an appropriate choice of the radii of curvature of the sphere and the emitting tip, a magnification of $10^{5}-10^{6}$ times can easily be achieved in this manner.

Since the development of monomolecular chemisorbed layers proceeds at a fairly high rate, even at very low pressures (estimating a sticking probability of 0.1 at 10^{-6} torr the time will be about 1 s, and at lower pressures correspondingly higher), the emission pattern of the uncovered surface will appear only at a pressure of about $10^{-9}-10^{-10}$ torr.

The mobility is studied by directing a gas jet from a gas source at an appropriate angle onto the emitting tip, and from the direct observation of the emission picture the migration of the adsorbed layer can be determined. To promote this, the accelerating voltage is switched off and, if necessary, the tip heated electrically to the required temperature.

Some characteristic field emission pictures are shown in Fig. II.11. These illustrate the mobility of oxygen layers formed at a low temperature on the surface of a tungsten tip.

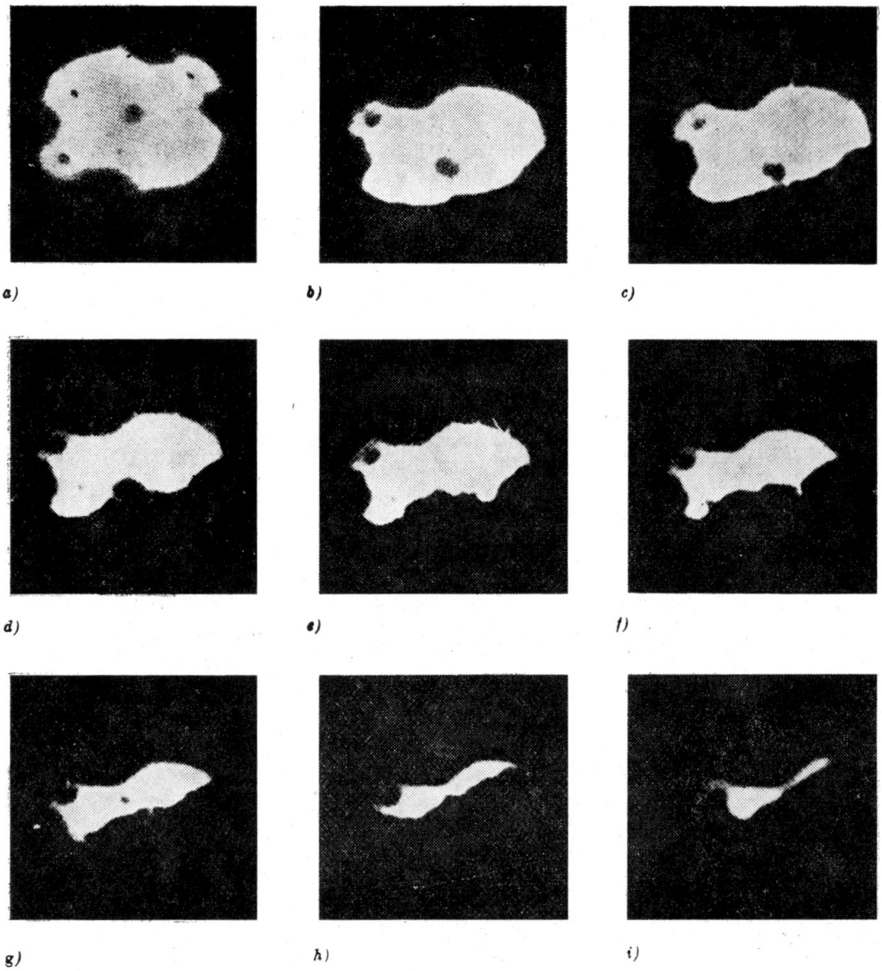

Fig. II.11.

a) Picture of the pure surface of the tungsten tip; b) the same when covered with a $\theta > 1$ oxygen layer at 4.2 K; c) the same after 0.43 s at 27 K; d) the same after 1.28 s at 27 K; e) the same after 2.28 s at 27 K; f) the same after 3.57 s at 27 K; g) the same after 5.28 s at 27 K; h) the same after 9.30 s at 27 K; i) the same after 12.6 s at 27 K; the light spot on the tip corresponds to a still uncovered surface

Figure II.11a shows a picture of the pure tungsten tip. Since crystal faces of different indices appear at various sites on the tip, the brightness of the field emission picture is not uniform, crystal faces having higher work functions being darker. (For instance, the bright spot in the left upper corner of the figure was caused by emission from a (001) face, and the dark spot in the middle by that from a (011) face.

When a plurimolecular oxygen layer is formed on certain parts of the surface at a temperature of about 27 K by applying an oxygen gas beam, the series of pictures taken at short intervals (Figs II.11b—II.11i) shows quite clearly the broadening of the dark region as the oxygen covers (relatively quickly) the entire surface of the tip*.

At this temperature *the monomolecular layer* becomes localized; this is apparent, among other things, from the fact that when the quantity of oxygen applied to the tip is not sufficient to produce a complete coverage, the migration previously observed is missing, and the field emission pattern remains unchanged for a considerable time. This type of diffusion can be observed with oxygen up to about 70 K.

These observations may be interpreted as follows:

Though the first (chemisorbed) layer is localized at 27 K, the physically adsorbed layers situated above it are delocalized. Thus, by surface migration the oxygen molecules can easily reach the edge of the first layer where they become chemisorbed and hence contribute to the spreading of the first layer. At temperatures above 70 K the average residence time of molecules in the physically adsorbed layers decreases to an extent when they are no longer able to cover a long path on the top of the first layer without being desorbed.

In the investigation of chemisorption the field emission microscope is used fairly widely, mainly for the examination of surface diffusion, as in the preceding example. In the theoretical description the diffusion is regarded as a two-dimensional migration, characterized by the rate of displacement of the layer boundary of the covered surface. If the path covered in a time t is X then

$$X = (2\,Dt)^{1/2}$$

where D is the diffusion constant. For D the equation

$$D = l^2 \nu_{\mathrm{d}} \exp\left(-E^{\ddagger}_{\mathrm{diff}}/RT\right)$$

is valid, where l is the distance between the deepest points of two potential wells, $\nu_{\mathrm{d}} \approx kT/h$, and $E^{\ddagger}_{\mathrm{diff}}$ is the activation energy of diffusion. Displacement of the layer boundary will only be uniform (and the field emission picture consequently sharp) if there are many deep traps; the boundary will otherwise have a diffuse structure.

So far, only the very simplest dehydrogenation reactions have been examined by means of the field emission microscope (the decomposition of ethylene and acetylene on Ir [93, 94]).

The field emission microscope does not provide a high enough resolution for the observation of individual atoms on the surface. The fact that the velocity vectors of the electrons emitted by the tip have a component parallel to the surface, causing a blurred picture, is in agreement with the Heisenberg uncertainty principle. Furthermore, even near 0 K, electrons

* The electron work function of the oxygen-covered surface is greater than that of the clean surface.

with energies corresponding to the Fermi level have kinetic energies of about 6 eV, with which statistically randomly oriented velocity vectors are associated. In the case of the tunnel effect the first to be emitted will be electrons having an energy corresponding to the Fermi level and, though in this case the velocity component normal to the surface dominates, the emitted electrons will preserve the other velocity components, too. Theoretically, from these two effects a maximum resolution of about 20 Å can be expected.

In 1951 Müller [95] developed the ion emission microscope, based on the principle of the ionization of appropriate gas atoms (He) in the inhomogeneous field of the tip, which causes acceleration of the ions and their collision with the fluorescent screen. Due to the differences in mass, resolution is improved down to the atomic region of 2–3 Å. By detecting the lattice defects of the emitting tip the method furnished some highly valuable results, mainly in metallurgy, but it is of tremendous importance in the fields of chemisorption and catalysis, too. For the first time in the history of chemisorption it became possible to observe individual adsorbed molecules and atoms [96].

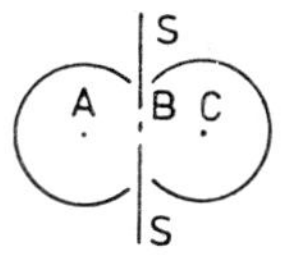

(2) Determination of the activation energy of surface migration

A useful means of measuring the activation energy of surface migration is provided by the thermal emission method, which has been successfully used for the study of adsorbed layers of foreign metals on metals. The principle of the method is as follows (Fig. II.12): one side of a tungsten filament is covered with, for example, thorium deposited by vacuum evaporation, and (thermal) electron emission is measured on both sides; changes in the emission give information on the rate of migration of the thorium layer. The method can only be applied when the work function is a linear function of surface coverage; in this case changes in emission permit unambiguous conclusions as to the changes in the surface concentration.

Figure II.13 shows idealized emission curves in which the electron emissions from opposite sides of the filament are plotted vs. time during the deposition and migration of the layer. In the course of the deposition of the thorium layer, emission on the facing side increases, but remains unchanged on the back side. On the other hand, when the layer begins to spread, the emission decreases on the facing side and increases on the back side, up to the point when the same limiting value is reached on both sides.

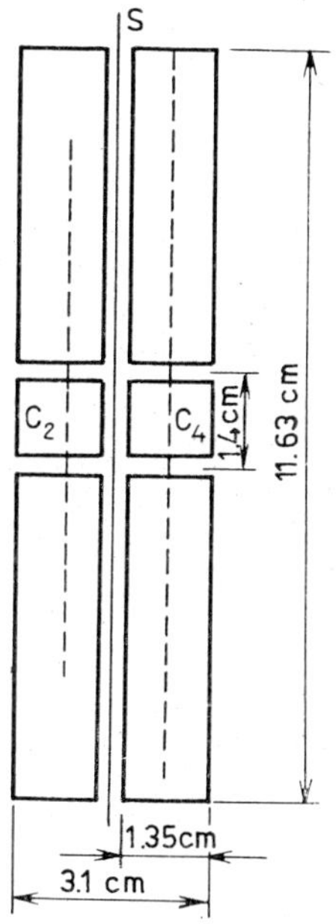

Fig. II.12. Determination of the work function by means of the thermal emission method

Brattain and Becker used essentially the same method to measure the rate of migration of thorium layers at 1535 and 1655 K [97], and found that the activation energy of the surface migration of thorium was 110 kcal mole^{-1}.

The mean value of the activation energy of desorption is 191 kcal mole^{-1} (see Paragraph 6.3.2), so that the activation energy of surface migration is

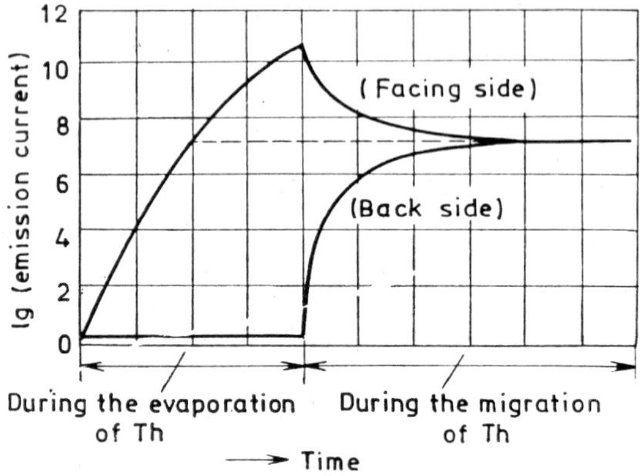

Fig. II.13. Idealized emission curves for the adsorption of Th on a W filament

57% of the activation energy of desorption. It is generally true that in non-activated adsorption processes the numerical value of the activation energy of surface migration is about one half that of the heat of adsorption.

Concentration changes due to migration on the surface of a metal filament have been followed in certain cases by means of the photoelectric (photoemission) method [23, 98]. By directing the beam of an ultraviolet light source onto various points of the filament surface, in the case of alkali metal coatings it was possible to estimate the coverage from the intensity of the photoelectric current.

PARAGRAPH 5.2.2. INDIRECT METHODS OF STUDYING MOBILITY

Because of the high electron work function, the thermal emission method is not suitable for studying the mobilities of gas layers on metal surfaces. In these cases the entropy change accompanying the adsorption, or the curves obtained for the dependence of the heat of adsorption on coverage, can, however, be applied for qualitative conclusions as to the state of the adsorbed layer.

(1) The entropy of adsorption

The function $S_a(\theta)$ [or, to a more limited extent, the function $\bar{S}_a(\theta)$] calculated from experimental data is, in principle, suitable for providing information on the degrees of freedom of molecules in the adsorbed layer.

For this the existence of certain degrees of freedom have to be *a priori* assumed (see Paragraph 2.1.1). It follows from the known principles of thermodynamics that the molar entropy of the layer is due to the sum of the contributions of the individual degrees of freedom:

$$S_a = S_a^{conf}(\theta') + S_a\ (\theta'') + S_a^{rot} + S_{vib}^{a} \qquad \text{(II.117)}$$

where S_a^{tr}, S_a^{rot} and S_a^{vib} are the molar translational, rotational and vibrational entropy contributions of the layer, and
S_a^{conf} is the configurational entropy of the layer.

In Eq. (II.117) the molar configurational entropy, S_a^{conf} is a function only of $\theta' = [1 - X(\theta)]\ \theta$, corresponding to the localized layer, while S_a^{tr} depends only on the partial coverage $\theta'' = X(\theta)\ \theta$, where $X(\theta)$ is the ratio of mobile to total (immobile plus mobile) adsorption $(0 \leq X(\theta) \leq 1)$. The coverage, θ, is the sum of these: $\theta = \theta' + \theta''$.

When the layers are either not completely mobile or not completely localized, and if a higher accuracy of the calculation is justified, in Eq. (II.117) $S_a^{rot} + S_a^{vib}$ should be substituted by:

$$X(\theta)\ (S_a^{rot} + S_a^{vib})_{mobile} + [1 - X(\theta)](S_a^{rot} + S_a^{vib})_{localized}$$

that is, the entropy contributions of the mobile and localized layers have to be calculated separately and added after weighting by the functions X and $(1 - X)$, respectively. This method is permissible for the rotational and vibrational entropies only, since these do not depend on n_a. Hayward and Trapnell considered (incorrectly) the configurational and translational entropies as being functions of a single coverage θ only [85].

Should the character of the entropy function calculated from Eq. (II.117) differ from the experimentally determined one, the number of otherwise possible degrees of freedom must either be increased or decreased until satisfactory agreement is reached. The unambiguity of the method is impaired by the fact that the function $X(\theta)$ can, in the end, be determined only from experimental data.

This way of writing S_a shows why experimental differential entropy [i.e. the function $\bar{S}_a(\theta)$] is less well suited for such an investigation. Let us multiply both sides of Eq. (II.117) by n_a and differentiate with respect to n_a:

$$\frac{\partial s_a}{\partial n_a} = \bar{S}_a = S_a + n_a \frac{\partial S_a}{\partial n_a} = S_a + n_a \frac{\partial}{\partial n_a} \cdot \{S_a^{conf}[(1 - X)\theta] + S_a^{tr}\, X\, \theta\}$$

$$\text{(II.118)}$$

After carrying out the indicated operations:

$$\bar{S}_a = S_a + n_a \left[\frac{\partial}{\partial X}(S_a^{conf} + S_a^{tr})\frac{\partial X}{\partial n_a} + \frac{\partial}{\partial \theta}(S_a^{conf} + S_n^{tr})\frac{\partial \theta}{\partial n_a}\right] \qquad \text{(II.119)}$$

Although all the functions in Eq. (II.119) with the exception of $X(\theta)$ and $\partial X/\partial n_a$ can be theoretically calculated, there is obviously no need to go into detailed reasoning to show that the fitting of the theoretical function $\bar{S}_a$ to the experimental results requires lengthy calculations without always leading to the desired result.

The theoretical entropies for the different degrees of freedom are as follows:

(1) The translational entropy of an ideal gas of molecular weight M at atmospheric pressure for three degrees of freedom is:

$$S_g^{tr} = R \ln (M^{3/2} T^{5/2}) - 2.30 \text{ (kcal mole}^{-1}\text{ degree}^{-1}) \quad (II.120)$$

Similarly the translational entropy of a two-dimensional ideal gas is [99]:

$$S_a^{tr} = R \ln (M T \varphi) + 65.8 \text{ (kcal mole}^{-1}\text{ degree}^{-1}) \quad (II.121)$$

where φ is the average surface available for one molecule, in cm^2.

The relationship between φ and the coverage θ can be established in the following manner:

According to one possible definition of the coverage, $\theta = n_a/n_{a,\infty}$. If the actual area of one molecule in the monomolecular mobile layer is $a_m(cm^2)$ and the surface of the adsorbent is $A_s(cm^2)$, then $n_{a,\infty} = A_s/N_A a_m$, and $n_a = A_s/N_A \varphi$, where N_A is the Avogadro number. Thus,

$$\theta = \frac{a_m}{\varphi} \quad \text{or} \quad \varphi = \frac{a_m}{\theta}$$

This expression for φ can then be substituted into Eq. (II.121).

Eq. (II.121) ignores the spatial requirements of the molecules, and consequently in the vicinity of $\theta = 1$ the equation is no longer accurate.

(2) Localized layers also possess a configurational entropy. For a homogeneous surface and non-dissociative adsorption the equation expressing the molar configurational entropy is [100]:

$$S_a^{conf} = -R\left[\ln \theta + \frac{1-\theta}{\theta} \ln (1-\theta)\right] \text{(kcal mole}^{-1}\text{ degree}^{-1}) \quad (II.122)$$

When the adsorption is dissociative, and further if the two fragments are identical and the layer is statistically randomly ordered, the value of S_a^{conf} is double that above. If the two adsorbed atoms remain near to each other [which is fairly frequent at low temperatures (see the next point)] the calculation will be more complicated.

(3) In the case of adsorbed polyatomic molecules the entropy contributions of the various normal vibrations can be calculated in the usual manner, with the condition that the frequencies of normal vibrations remain unchanged during the adsorption process. In quite a number of cases this simplification does not hold, because in the field of the adsorbent the molecules will be polarized, and this leads to a shift of the frequencies of the normal vibrations, and even to the elimination of their discrete character, as is the case for solutes.

The adsorption of a molecule or atom is associated with the transformation of at least one translational degree of freedom into a vibrational degree of freedom, which is related to the vibration of frequency ν of the centre of mass of the molecule normally to the surface. The entropy contribution of this vibration is:

$$S_{\mathrm{a}}^{\mathrm{vib}} = R\left[\frac{h\nu}{kT}\left(e^{h\nu/kT} - 1\right)^{-1} - \ln\left(1 - \mathrm{e}^{-h\nu/kT}\right)\right] (\text{cal mole}^{-1}\ \text{degree}^{-1}) \quad \text{(II.123)}$$

(4) The equation for the rotational entropy for n degrees of freedom is:

$$S_{\mathrm{a}}^{\mathrm{rot}} = R\left[\ln\left(\frac{1}{\pi_s}\right)\left\{\frac{8\,\pi^3(I_A^a\,I_B^b \ldots I_G^g)^{1/n}\,kT}{h^2}\right\}^{n/2} + n/2\right]$$

$$(\text{cal mole}^{-1}\ \text{degree}^{-1}) \quad \text{(II.124)}$$

where $a + b + \ldots + g = n$ and

$I_A, I_B \ldots$ are the moments of inertia with respect to the individual rotation axes.

Most of the literature concerning the entropies of adsorbed layers compares the differential entropy of a layer with theoretical entropies at a single coverage of the adsorbent chosen as the standard state. This procedure [see Eq. (II.118)] is theoretically inadmissible, since $\partial S_{\mathrm{a}}/\partial n_{\mathrm{a}} \neq 0$. At the moment we know of only two studies which are sufficiently detailed to enable the plotting of the functions $\bar{S}_{\mathrm{a}}(\theta)$ and $S_{\mathrm{a}}(\theta)$ from experimental data in the full range of coverage (0,1).

Example. Rideal and Sweett measured the chemisorption of H_2 on vacuum-evaporated Ni films [101] under very careful experimental conditions. The maximum temperature applied was 200 °C, and the lowest coverage $\theta = 0.005$. It was claimed that the chemisorption of H_2 on the films was undoubtedly reversible. The coverage of the surface was arbitrarily considered monomolecular at 1.45 10^{-4} mole sorbed substance per gramme of Ni.

Figure II.14 shows the isosteres plotted from the experimental data as log p vs. 1/T for various values of θ. According to Eq. (II.94):

$$\ln(p/{}^*p) = -\frac{[{}^*H_{\mathrm{g}} - \bar{H}_{\mathrm{a}}(\theta)]}{RT} + \frac{{}^*S_{\mathrm{g}} - \bar{S}_{\mathrm{a}}(\theta)}{R}$$

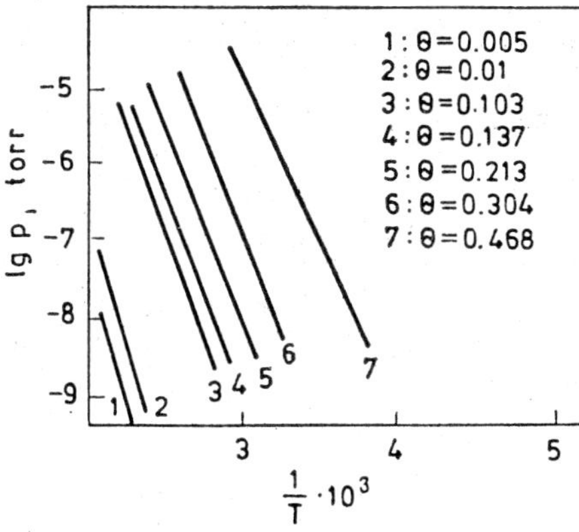

Fig. II.14. Adsorption isosteres of hydrogen on Ni film (after Rideal and Sweett)

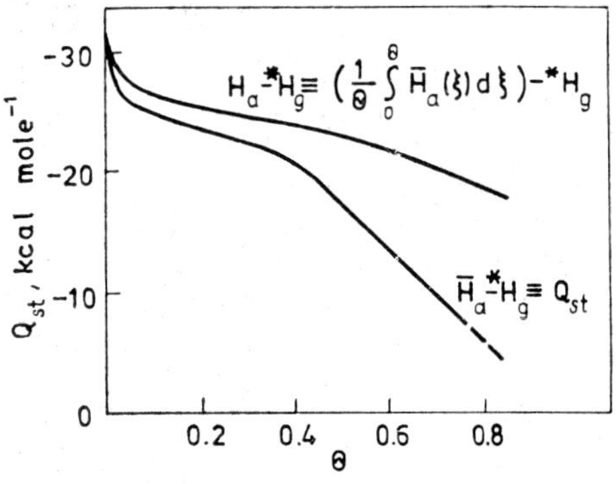

Fig. II.15. Isosteric heat of adsorption on Ni film (after Rideal and Sweett)

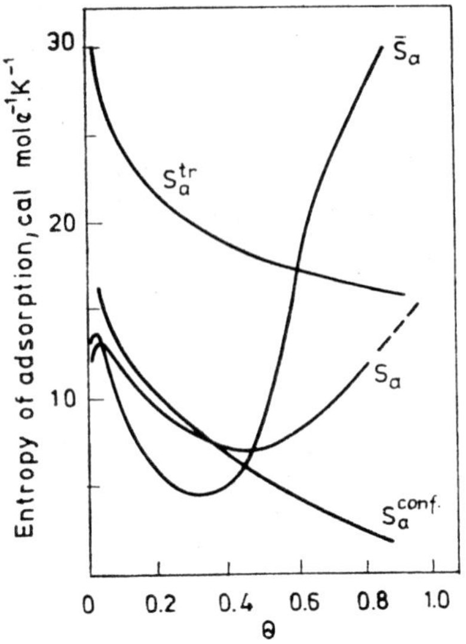

Fig. II.16. Entropy of the chemisorption of H on Ni film at 25 °C (after Rideal and Sweett)

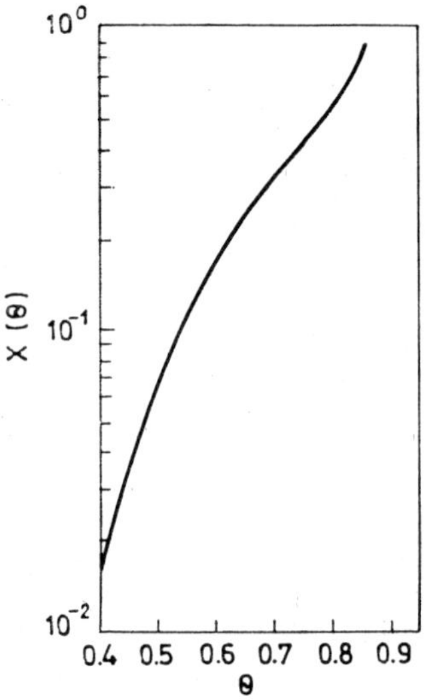

Fig. II.17. Mobility of H chemisorbed on Ni films

The intercepts of the isosteres $Y(\theta)$ are in direct relation to the differential entropy for the coverage θ:

$$Y(\theta) = \frac{[{}^*S_g - \bar{S}_a(\theta)]}{2.303\, R} + \log {}^*p$$

The linearity of the isosteres proves that in the temperature range of the experiment and at a fixed coverage $({}^*S_g - \bar{S}_a)$ and $({}^*H_g - \bar{H}_a)$ are constant, so that they can be carefully extrapolated.

The curves $(\bar{H}_a - {}^*H_g) = Q_{st}$ and $(H_a - {}^*H_g)$ vs. θ, obtained from the isosteres, are shown in Fig. II.15. Similarly, the curve $\bar{S}_a$ in Fig. II.16 shows the differential entropy obtained from the experimental data (the experimental points are not marked) while S_a is the integral molar entropy curve, obtained by the integration of $\bar{S}_a$ with respect to $\theta(\theta = n_a/n_{a,\infty}!)$, vs. θ at 25 °C. Of several otherwise possible adsorption models, Rideal and Sweett found the best agreement between experimental and theoretical $S_a(\theta)$ curves when they assumed a randomly ordered hydrogen layer and the chemisorption of 1 hydrogen atom for each adsorption site at $\theta = 1$. Using the data of Rideal and Sweett, Hayward and Trapnell attempted the calculation of the function $X(\theta)$. Taking into consideration that at 25 °C $S_a^{vib} \approx 0$, they arrived at the theoretically questionable (see the beginning of this Section) form of S_a:

$$S_a(\theta) = 2[1 - X(\theta)]\, S_a^{conf}(\theta) + X(\theta)\, S_a^{tr}(\theta)$$

Since it is possible to calculate $S_a^{conf}(\theta)$ and S_a^{tr} (see Fig. II.16), the function $X(\theta)$, with the help of which the calculated $S_a(\theta)$ most closely approaches the experimental $S_a(\theta)$ curve, can be determined from experimental data. Figure II. 17 shows $X(\theta)$ in a logarithmic scale (see Hayward and Trapnell, Ref. [85]). This, and the good agreement between the experimental value of S_a and S_a^{conf} in the range (0, 0.4) of the coverage, point to a localized layer up to a coverage of about $\theta = 0.4$, while at coverages higher than $\theta = 0.9$ the layer is already mostly delocalized. This appears in Fig. II.16 from the fact that S_a (experimental) $\to S_a^{tr}$ when $\theta \to 1$.

It follows from $S_a(\theta)$ (experimental) $< S_a^{conf}(\theta)$ in the range $0 < \theta < 0.2$–0.3 that there are centres with higher adsorption energies also on the surface (for the effect of the heterogeneity of the surface on the entropy of adsorption see Refs [102] and [103]).

From a theoretical point of view there is not much difference between this procedure and the calculation according to function $\bar{S}_a$. In the latter case $\partial X/\partial n_a$ and X should have been chosen so as to give the best possible agreement between the theoretical and experimental values of $\bar{S}_a$. X must obviously not differ from the numerical calculated integral of the curve $\partial X/\partial n_a$ for the initial condition $X(n_a) = 0$.

Example. Kisliuk studied the chemisorption of N_2 on a polycrystalline tungsten ribbon [104] at a temperature of about 1500 K, and found -119 kcal mole^{-1} for the isosteric heat of adsorption, independently of θ. The differential entropy, $\bar{S}_a$ vs. θ curve is shown in Fig. II.18. Assuming a localized atomic N layer, according to Eq. (II.118):

$$\bar{S}_a = 2\, S_a^{conf} + 2\, S_a^{vib} + 2\, n_a \frac{\partial S_a^{conf}}{\partial n_a} = 2\, S_a^{vib} - 2\, R \ln\left(\frac{\theta}{1-\theta}\right)$$

It is clear from Fig. II.18 that if the curve $-2\, R \ln(\theta/1-\theta)$ is shifted about 12 entropy units upwards, there is good agreement between the experimentally determined and the theoretical $\bar{S}_a$ functions, indicating that at about 1500 K S_a^{vib} for one gram atom of N_2 is approximately 6 cal mole^{-1}.

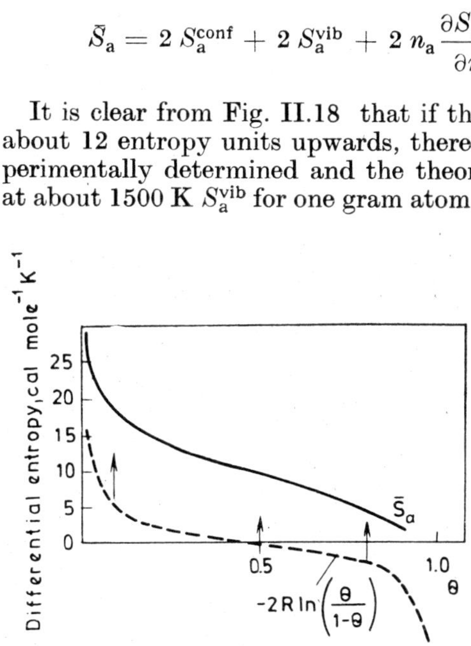

Fig. II.18. Differential entropy of nitrogen chemisorbed on W ribbon (after Kisliuk)

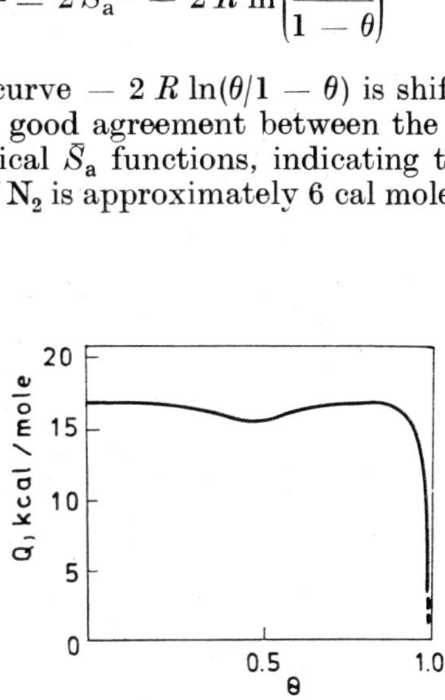

Fig. II.19. Calorimetric differential heat of adsorption of hydrogen on Ni film at -195 °C vs. coverage (after Wedler)

(2) The dependence of the heat of adsorption on θ, and the mobility

In certain cases it is possible to draw conclusions regarding the mobility from the Q_{st} vs. θ curves. Wedler [105] found, after repeated experiments, that at the temperature of liquid nitrogen the differential calorimetric heats of adsorption of H_2 and D_2 on vacuum-evaporated Ni films depends on the coverage as illustrated in Fig. II.19. Beeck [43] reported similar

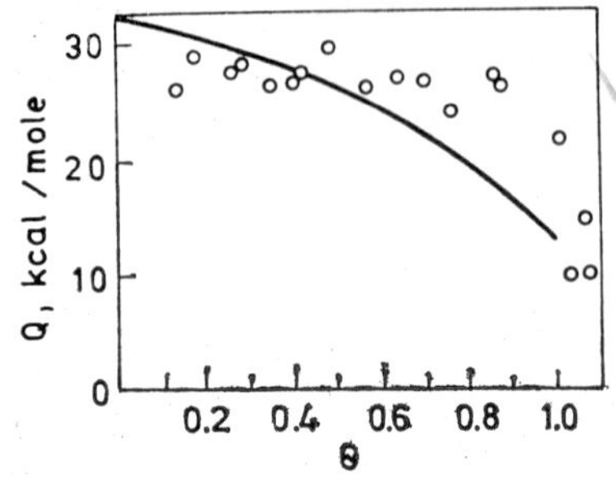

Fig. II.20. The Q vs. θ curve of hydrogen chemisorbed on iron film; the points are experimental data at —183 °C, the curve belongs to 23 °C

behaviour for the chemisorption of H_2 on vacuum-evaporated iron films. The isosteric heat of adsorption of H_2 on iron film is shown in Fig. II.20; the points refer to —183 °C, and the full line to 23 °C.

At first glance these findings might suggest that at a low temperature the adsorbent surface is homogeneous, while at higher temperatures this homogeneity ceases for some reason. Beeck [43] claims the far greater probability of a blocked or limited mobility of hydrogen atoms (or molecules) at low temperatures; they are thus not able to leave the immediate vicinity of collision (see Paragraph 2.2.4), so that independently of the adsorption energy, Q, the adsorbent surface is statistically uniformly covered. Hence, the function $\theta(Q)$ is constant in the entire investigated (Q_{min}, Q_{max}) range; that is, it depends on the prevailing "equilibrium" pressure, but not on Q.

This means, in other words, that the principle which postulates that with increasing coverage of the surface the adsorbed atoms will occupy the prevailing most active centres cannot be asserted because of the immobility.

Due to the mobility of the adsorbed atoms, this situation is fundamentally different at higher temperatures, as illustrated in Fig. II.20 by the curve for 23 °C.

It should follow from what has been said so far that the section of the Q_{st} curve for low temperatures up to about $\theta = 0.9$ ought to agree with Q_{int} ($\theta = 1$) calculated from the curve for 23 °C. The positions of the experimental data indeed suggests such an agreement, at least qualitatively.

This interesting relationship between the mobility, i.e. localization, of the adsorbed layers and the isosteric heat of adsorption supports the conception of the process of chemisorption (see Paragraph 6.2.1).

CHAPTER 6. RATE OF SORPTION

SECTION 6.1. HISTORY

The determination of the heat of adsorption on various porous and powdered adsorbents suggested, several decades ago, that the nature of the interaction between adsorbent and adsorbate may change depending on the experimental conditions: for example, for certain adsorption systems at the temperature of liquid air the absolute value of the heat of adsorption, characteristic for physical processes, is not more than a few kcal mole^{-1} [106], whereas at higher temperatures it attains several tens of kcal mole^{-1}, as in chemical processes [75, 76]. The shapes of experimentally determined isobars led to similar conclusions. Provided there is only a single type of adsorption, at constant pressure the quantity adsorbed should increase monotonously with temperature; however, it was found that in certain cases there may be a temporary increase of the adsorbed quantity in one or more, variously well-defined temperature ranges, resulting in the appearance of minima and maxima on the adsorption isobar. A characteristic example is given by the isobars of hydrogen on nickel powder in Fig. II.21, according to Benton and White [107]. In this figure the descending branches of the isobars after the maxima have slopes different from those obtained at lower temperatures.

Langmuir [108] and Taylor [109] drew attention to the contradiction-free interpretation of these findings by a general principle: physical adsorption with a low heat of adsorption is assumed at low temperatures, followed, at higher temperatures, by chemisorption, with a considerably higher heat of adsorption. The fact that the appearance of this type of chemisorption requires exceeding a lower temperature limit, was explained by Taylor in terms of the activation of the process, similarly to chemical reactions ("activated adsorption"), so that its rate is detectable only above this temperature limit. The slopes of the descending branches of the isobar before and after the maximum (see Fig. II.21) are roughly proportional to the heat of adsorption including the Boltzmann factor, $e^{Q/RT}$, and

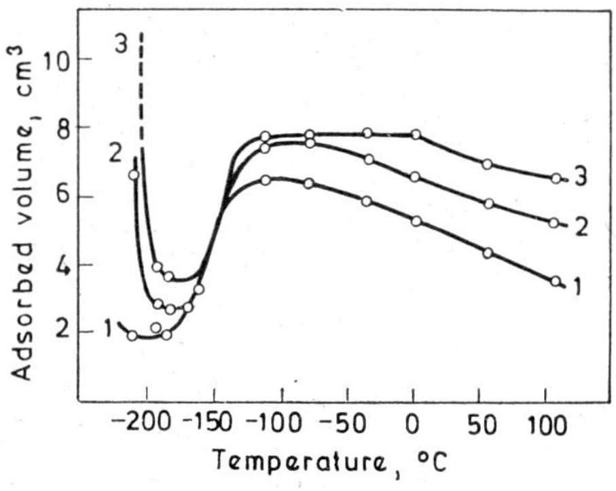

Fig. II.21. Adsorption isobars of hydrogen on Ni powder 1. 25 torr
2. 200 torr
3. 600 torr

14

since Q_{chem}, $Q_{phys} < 0$, and $|Q_{chem}| \gg |Q_{phys}|$, the slope relating to the higher heat of chemisorption is less steep.

Lennard–Jones [110] found an explanation for the fact of activated adsorption in the potential functions which characterize adsorbent — adsorbate interaction. In more complicated cases, these potential functions differ from those described in Chapter 1 (Fig. II.22). The most striking feature is the shift of the potential curve of the adsorbate $X - X$ in the direction of greater r distances, resulting in an intersection point P above the zero energy level. The distance of P from the zero level is the activation energy, $E_a^{\ddagger}$, of adsorption. Curve *1* corresponds to the case when only physical (repulsive and attractive) forces are present, while curve *2* describes (under otherwise identical conditions) the interaction of two X atoms with the surface. For the configuration corresponding to the intersection point of the two curves, two electron states of equal energy should be possible, but this is quantum-mechanically not permissible. Hence, this level is split as shown in the figure and the potential energy, in reality, decreases along the full curve. To the right of point P the internuclear distance $X - X$ is practically constant (physical adsorption), while to the left the two nuclei are separated and may, in principle, be removed to any distance from each other (dissociative chemisorption).

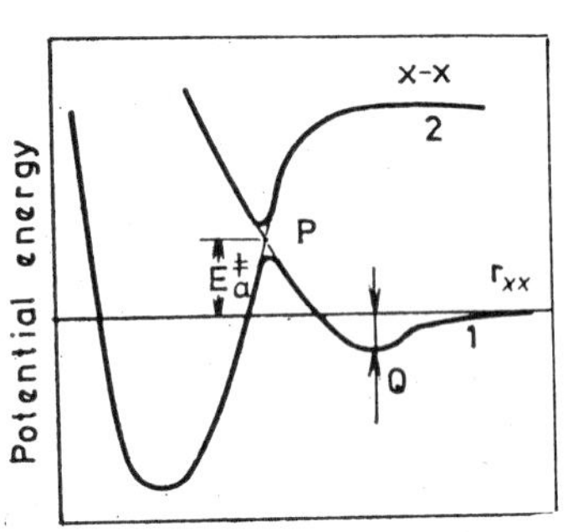

Fig. II.22. Potential energy change accompanying physical adsorption and chemisorption

It may be concluded from the relatively few data available on the adsorption of atoms, (mainly from recent experiments, see paragraph 8.1.1), that their adsorption does not require an activation energy. For di- or polyatomic molecules the formation of adsorbed layers of different structures (see Paragraph 8.1.1) can be conceived only by assuming fundamental changes in the electron configuration of the molecule prior to, or during, the formation of the final chemisorption bond. Rearrangement of the structure of the formed layers depends on many factors, including the availability of the necessary thermal energy at the given temperature.

The literature of the 1930's is extremely rich in experimental material illustrating the above point, and covers a very wide range of adsorbents (metals, various types of oxides, etc.) and adsorbates (H_2, O_2, N_2, H_2O, hydrocarbons, etc.) [111]. It might be advisable, however, to treat this with a certain criticism and reserve. It should first be pointed out that the majority of the slow processes observed on granular, porous adsorbents have very little to do with activated adsorption. The rates of very rapid (e.g. physical adsorption) processes, too, may decrease considerably if the material supply required by rapid sorption is not ensured by the transport processes (e.g. in gas mixtures the common gas diffusion ensuring material transport through the depleted zone at the boundary of the granule, and/or the internal diffusion which fulfils a similar role in the pore structure of the porous substance) [112, 113].

Inadequate cleaning of the adsorbent surface during the preparation for the adsorption process is another possible source of misunderstanding. The removal of surface contamination by means of reduction, adsorbate displacement, etc. is generally a slow process, the rate of which is not characteristic of the true interaction between adsorbent and adsorbate [114, 115, 116].

However, the reverse of this may also occur: a chemisorption process accompanied by its characteristic, relatively large thermal effect proceeds very rapidly at a low temperature without any notable activation energy]117]. If chemisorption is compared to chemical reactions, this finding no onger seems surprising.

SECTION 6.2. RATES OF ADSORPTION AND DESORPTION

PARAGRAPH 6.2.1. GEOMETRICAL CONSIDERATIONS

Definition and mathematical description of the rate of chemisorption is not an easy task, the less so since there is no detailed conception agreeing with the facts regarding chemisorption itself. Information on the surfaces of adsorbents and on the structures of the chemisorbed layers formed on them has been obtained during the past 45 years by means of field emission microscopy, the flash desorption method (see Paragraph 6.3.4) and, in particular, low energy electron diffraction (LEED) (for a description see Ref. [118]); this has forced us to modify our ideas about the geometry of chemisorption in several respects, and sometimes even to change them fundamentally.

This is primarily true for the interpretation of sorption phenomena on surfaces with regular structures, and hence mainly on crystalline substances, but, to a certain extent, for those on amorphous adsorbents too.

Before the introduction of the LEED method it was generally assumed that chemisorption takes place on regularly arranged, "rigid" atoms on the crystal face, or in the gaps between them, and that this leads, depending on the nature of the potential energy surface characteristic of the adsorption interaction and on the prevailing temperature, to the formation of localized or delocalized layers. With only a few exceptions (see Balandin's multiplet theory) no distinction was made in this respect between the individual crystal faces, and the random occupation of adsorption sites was also assumed. It was general practice to consider the coverage attained when the adsorption process ceased (when the adsorption "equilibrium" was reached) as being unity, ignoring the fact that this "monomolecular" layer in certain cases constituted only a negligible fraction of the atoms adsorbed on the surface. The interpretation of monomolecular layers on polycrystalline surfaces encountered even greater difficulties.

If the cause of this is sought, it must be borne in mind that it is not obvious where chemisorbed atoms (or molecules) will be bound on single-crystal faces. Figure II.23 shows the (111) face of a face-centred cubic lattice, with three possible arrangements (A, B, C) of the adsorbed atoms.

In the case of polycrystalline substances, depending on the arrangements of the atoms of the crystal face, only some of the above possibilities can be realized [e.g. a bridge bond *B* can be formed on the (111), (100) and (110) faces of face-centred cubic lattices, but not on the (100) faces of body-centred cubic lattices]. Furthermore, a given arrangement

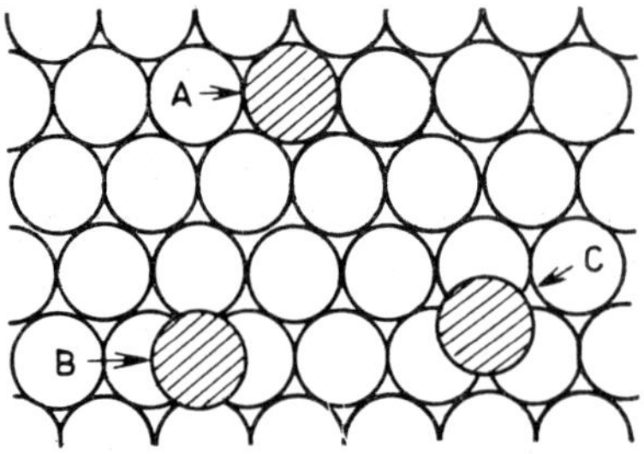

Fig. II.23. Model of classical adsorption on the (111) face of a face-centred cubic crystal. The adsorbate (hatched circles) may be adsorbed point-like (*A*), in a simple bridge bond (*B*) and in a triple bridge bond (*C*)

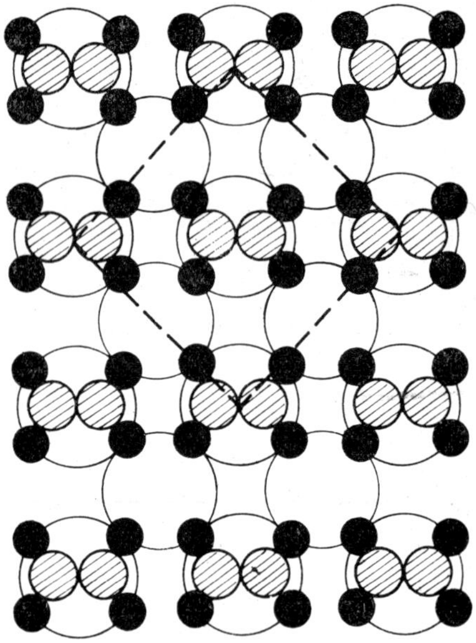

Fig. II.24. Two-dimensional structure of chemisorbed ethylene on the (100) face of Pt

type presupposes different bond strengths on the various crystal faces, depending on the number of nearest neighbours [for bonds of type *B* the number of neighbours on the (111) crystal face of the face-centred cubic lattice is 9, but on the (110) face only 7].

Classical methods of investigation are unsuited for the determination of the arrangements formed on the surfaces of polycrystalline materials. In principle, the LEED method opened the way for such determinations. Lander and Morrison [119], for instance, have demonstrated that on the (111) face of silicon chemisorbed iodine atoms are present in bonds of type *C*. Although it is not always possible to determine the position of the adsorbed atom, with the help of the LEED method it is possible to find out whether the chemisorbed atoms are in a regular arrangement on the surface (that is, whether all of them are present in the same type of bond, permitting the occupation of these regularly arranged potential adsorption sites in a statistically random manner, so that only in the vicinity of $\theta = 1$ does a regular layer structure develop), or whether adsorption is not accompanied by the formation of such a long-range order.

The first unexpected finding of the LEED method was the displacement of the atoms of the adsorbent by adsorbed atoms randomly occupying the regularly arranged adsorption sites [120]. (Since more than one atomic layer is involved in the development of the electron diffraction, it is possible to detect, albeit with weaker reflections, the adsorbent structure covered by the adsorbed atoms.) This pattern can be more or less reconciled with views on the adsorption process (see the statements of functions $f(\theta)$ and $f(\theta)$ in Paragraph 6.3.1).

The situation is entirely different with the LEED results of Davisson and Germer [121, 122], Germer, Scheibner and Hartman [123] and others [124, 125]: it was demonstrated that, in certain cases, the adsorbed atoms or molecules are arranged so as to bring about their own, regular, two-dimensional configuration on the surface, and moreover this configuration varies with the different degrees of coverage. As an illustration, Fig. II.24 shows the structure of a layer of C_2H_4 formed on the (100) face of platinum, but it must be added that not all the C_2-C_4 olefines and not all the investigated crystal faces of platinum are capable of producing a regular surface structure [247].

The result that, depending on the prevailing coverage, the adsorbed gas is capable of forming one or more regular two-dimensional structures on the surface of the adsorbent, either reminiscent of the adsorbent or quite unlike it, is of tremendous importance as regards adsorption and catalysis even though its kinetic consequences are not yet fully appreciated. These findings have certain consequences pertaining to energy: the existence of several different layer structures at a given temperature can only be due to the adsorption interactions changing with the coverage; if the temperature also changes, several sorbed layers can exist simultaneously even on a single crystal face; when two adsorbates are present, in addition to mutually influencing each other, the complex adsorbed layers, too, may have their own structures. The picture regarding the surface of polycrystalline substances is thus utterly confusing.

Perhaps an even more important discovery was that chemisorption is capable of altering the surface structure of the adsorbent itself; this is manifested in variations from the already-mentioned rearrangement of a few atoms to the growth of a complete new crystal face on the initial surface (see Paragraph 8.1.2).

In the light of our present knowledge we are able to formulate the sorption rate mathematically only for geometric conditions under which a statistically randomly ordered layer develops as a result of adsorption. This condition is fulfilled *a priori* in the case of delocalized layers, but for localized layers the conditions are probably extremely complex. To throw some light on this problem, let us briefly survey the various views concerning the elementary steps of adsorption.

PARAGRAPH 6.2.2. CONSIDERATIONS OF ENERGY

According to Langmuir the process of adsorption can be imagined as follows: if the molecule arriving from the gas phase possesses sufficient energy on the appropriate degree of freedom, and if the geometry of collision is suitable and the interaction energy is dissipated to the degrees of freedom of the adsorbent and/or the molecule itself within a sufficiently short time, then the molecule will adhere to the adsorption site with which it collides. If these conditions are not fulfilled, the collision will be elastic, and after a single vibrational period the molecule will return to the gas phase.

As far as the soundness of this concept is concerned, the assumption of collision and adherence on the same site appears doubtful. Although this may occur in special cases (see Paragraph 5.2.2), for the formation of a delocalized layer (which, in other ways too, cannot be reconciled with the above argument) the contradiction is obvious, and in fact the relevant literature makes no mention of such a restriction.

The situation is not quite so obvious when a localized adsorbed layer is finally formed. Langmuir's theory allows, in principle, for the development of a statistically randomly ordered localized adsorbed layer. If the layer has a regular structure, however, it must be assumed that after the elementary act of collision the particles possess sufficient mobility to occupy the most favourable adsorption sites. Since the interaction between adsorbent and adsorbate always results in variously excited molecules or molecular fragments, the possibility of displacement is, in principle, always ensured. The subsequent events are jointly determined by a number of factors.

Let us first examine qualitatively the influence of the potential energy surface. If the energies relating to the minima of the potential energy surface are commensurable with the translational energy (or some energy present with an appropriate degree of freedom) available from a thermal source, the layer will be delocalized. If the adsorbing surface is not homogeneous, but greater potential energy decreases may occur on certain adsorption sites or parts of the surface (e.g. in the case of polycrystalline surfaces) than on others, then the adsorbed molecules will first occupy the former, causing the depopulation of certain crystal faces and the simultaneous saturation of others. If the minima of the potential energy surface, which also reflect adsorbate—adsorbate interactions, display a regular arrangement, then their occupation may lead to layers with definite structures.

The other factor is the rate of cooling of "hot" molecules or, in the case of dissociative adsorption, of adsorbed atoms; this involves the time required for their energy to drop to the thermal level via the loss of the excitation energy. Besides the potential energy surface this is the other decisive factor, since it determines the path length of the surface diffusion. As the diffusion path length increases, the probability of the recombination of the adsorbed atoms, and thus of their desorption, also becomes greater.

Even this brief outline of the problem gives an idea of the tremendous variety of phenomena which can result with the different adsorbents and adsorbates. Perhaps the most important of all these is the realisation that collision and the final development of chemisorption or desorption are generally separated in space and time.

This latter was formulated by Ehrlich [126, 127] in a slightly modified way. Ehrlich assumes the existence of a precursor state (preceding chemisorption) in which the molecules are physically adsorbed, or are in a weakly chemisorbed state, and consequently possess sufficient mobility to change their positions. Their fate is either final chemisorption or, after a certain period in the precursor state, a return into the gas phase as a result of desorption. Whenever physisorption and chemisorption may take place simultaneously, or when two simultaneous types of chemisorption states may develop, Ehrlich's concept — though not so detailed as the earlier one of Robins [128] — may prove sound.

Currently, only for the "modernized" Langmuir concept, which permits the formation of delocalized layers, has it proved possible to formulate a detailed kinetical description. Extension of the theory to heterogeneous adsorbing surfaces is reasonable, and though, in principle, this has failed to give rise to new ideas, experiment has been enriched by a well-applicable rate relationship (see the Elovich equation). The discussion of adsorption from gas mixtures proceeds essentially on the same lines.

In principle the Robins concept is new, and it is unfortunate that so far only the desorption rate has been given a satisfactory mathematical description (see Section 7.5), and even then based on a simpler concept than that described above.

The Ehrlich method will be discussed in Section 7.4.

Wolkenstein and Pesev recently attempted to find a mathematical description for the rate of chemisorption on semiconductors [129]. Due to the lack of data the applicability of their equations cannot be judged.

PARAGRAPH 6.2.3. RATE EQUATIONS

According to the kinetic gas theory, at a pressure p (dyne cm^{-2}) the number of molecules colliding with unit surface of the adsorbent is $p/\sqrt{2\pi mkT}$, where m(g) is the mass of the molecule, k(erg/degree) the Boltzmann constant and T(K) the absolute temperature. If the probability of collision leading to chemisorption is s ("sticking probability"), then according to Langmuir the rate of adsorption is:

$$J_a = \frac{sp}{\sqrt{2\pi mKT}} \quad \text{(molecules cm}^{-2}\,\text{s}^{-1}) \qquad \text{(II.125)}$$

The sticking probability s is always smaller than 1, for the following reasons:

In agreement with what has been said before, only that fraction $\exp(-E_a^{\ddagger}/RT)$ of the molecules colliding with the surface can be chemisorbed, for which the thermal energy (or more accurately the thermal energy available with the degree of freedom necessary for chemisorption) is higher than $E_a^{\ddagger}$. If collision takes place in an unsuitable configuration, or following collision the excess energy of the molecule is unable to dissipate to its own degrees of freedom and/or the various degrees of freedom of the adsorbent, the collision will be elastic and after a single vibrational period the molecule will return to the gas phase. This is usually accounted for in the rate equations by the condensation coefficient (σ). Finally, $f(\theta)$ is the coverage-dependent probability that in the collision process a free adsorption site or part of the adsorbing surface is involved. Although collision with an occupied site may also lead to adsorption (adsorption energy is of about the same order of magnitude as intermolecular interactions), for this might be one way of obtaining plurimolecular adsorbed layers, under the experimental conditions of chemisorption studies (high temperature), or in the case of almost completely covered surfaces (when the molecule might be forced to travel a long distance on the top of the adsorbed layer in order to find a free adsorption site), the probability of binding is small.

With these points in mind the sticking probability is:

$$s = \sigma\, f(\theta) \exp\left(-E^{\ddagger}_{\mathrm{a}}/RT\right) \tag{II.126}$$

and after substitution the rate of adsorption can be written as follows:

$$J_{\mathrm{a}} = \frac{\sigma f(\theta) \exp\left(-E^{\ddagger}_{\mathrm{a}}/RT\right) p}{\sqrt{2\,\pi\, mkT}} \quad (\text{molecules cm}^{-2}\,\text{s}^{-1}) \tag{II.127}$$

In the majority of cases σ and $E^{\ddagger}_{\mathrm{a}}$ are not constants, but functions of the prevailing coverage. If the adsorbing surface is homogeneous, but the adsorption energy (and thus σ and E_{a}) nevertheless depends on the coverage, we usually speak of induced heterogeneity, which can be explained by the adsorbent—adsorbate interactions changing with coverage, that is, by the interaction between the adsorbed molecules becoming more dominant as the coverage increases. In this case the procedure is to consider σ and $E^{\ddagger}_{\mathrm{a}}$ in Eq. (II.127) as functions of θ.

More often we are dealing with *a priori* heterogeneous surfaces. If n_q is the probability density function of adsorption sites with adsorption energies $q (q > 0)$ and σ_q and $f(\theta)_q$ are their characteristic condensation coefficients and the function $f(\theta)$ respectively, then the adsorption rate will be:

$$J_{\mathrm{a}} = \frac{p}{\sqrt{2\,\pi\, mkT}} \int_0^{\infty} n_q\, \sigma_q\, f(\theta)_q \exp\left(-E^{\ddagger}_{\mathrm{a},q}/RT\right) \mathrm{d}q \tag{II.128}$$

So far only the assumption that $E^{\ddagger}_{\mathrm{a}}$ is a linear function of the coverage has led to rate equations of practical use (see later in the discussion of the Elovich equation). In principle, however, there is no objection to the computer application of the general rate relation (II.128), provided suitable models are available.

The desorption rate can be calculated as follows: Assuming that desorption from an occupied adsorption site can take place only if the necessary activation energy $E^{\ddagger}_{\mathrm{a}}$ is available at the involved degree of freedom, the probability of this at a temperature T is $\exp(-E^{\ddagger}_{\mathrm{d}}/RT)$ and the rate of desorption will be:

$$J_{\mathrm{d}} = k_{\mathrm{d},0}\, \bar{f}(\theta) \exp\left(-E^{\ddagger}_{\mathrm{d}}/RT\right) (\text{molecules cm}^{-2}\,\text{s}^{-1}) \tag{II.129}$$

where $k_{\mathrm{d},0}$ is the desorption rate constant and $\bar{f}(\theta)$ is the fraction of covered sites.

The following relationship exists between the activation energy of desorption, $E^{\ddagger}_{\mathrm{d}}$, the heat of adsorption, $Q (Q < 0)$, and the activation energy of adsorption, $E^{\ddagger}_{\mathrm{a}}$:

$$E_d^{\ddagger} = E_a^{\ddagger} - Q$$

showing that desorption is always an activated process.

Comparatively little is known of the mechanism of desorption, so that for want of a better concept it is assumed that $\bar{f}(\theta)$ can be calculated by simple statistical methods. As regards k_d and $E_d^{\ddagger}$, both are most probably functions of the coverage, $E_d^{\ddagger}$ since Q usually depends on θ, while k_d is related to the coverage-dependent vibrational frequency of the surface bond.

Analogous with what was said about the adsorption rate, here, too, it is expedient to distinguish between the two cases when desorption takes place from a homogeneous or a heterogeneous surface. The rate of desorption from heterogeneous surfaces can be derived in a manner similar to that given for the rate of adsorption.

In connection with the use of the relations mentioned so far (and of some which will be discussed later) it should be pointed out that under not specifically designed experimental conditions only the sorption rate produced by the difference between the adsorption and desorption rates:

$$\bar{J} = J_a - J_d \tag{II.130}$$

or even more probably, some consequence of this difference, (change in pressure, composition or weight, etc.), can be observed. It appears necessary to stress this point, since, with the exception of sorption measurements based on weight determinations, where there is a simple proportionality between the weight change per unit time and Eq. (II.130), the relationship between the observations and the sorption rate is generally not so direct.

This will be illustrated by the following simple example:

Let us assume that we wish to measure the sorption rate in an apparatus of constant volume by observing the pressure changes. Let m^{α} g of the non-porous adsorbent (specific surface area: a_s cm^2 g^{-1}) be in direct contact with the gas space of volume v_h (cm^3) in which the initial pressure is p_0 (torr). Further, let the adsorption system conform to the adsorption and desorption rate equations (II.127) and (II.129) by choosing $f(\theta) = 1 - \theta$ and $\bar{f}(\theta) = \theta$. After rearrangement Eq. (II.130) can be written in the more simple form

$$\bar{J} = J_a - J_d = k_a p(1 - \theta) - k_d \theta \quad (\text{mole s}^{-1})$$

where

$$k_a \equiv \frac{m_a^{\alpha} a_s \sigma \lambda \exp(-E_a^{\ddagger}/RT)}{N_A \sqrt{2\pi mkT}} \quad (\text{mole s}^{-1}\ \text{torr}^{-1})$$

$$k_d \equiv \frac{k_{d,0}\, m^{\alpha} a_s \exp(-E_d^{\ddagger}(RT)}{N_A} \quad (\text{mole s}^{-1})$$

λ (dyne cm^{-2} torr^{-1}) is the pressure conversion factor and N_A (mole^{-1}) is the Avogadro constant. Let n_g^0 (mole) and n_g (mole) be the quantity of adsorbate in the gas phase at $t = 0$ and time t, respectively, n_a the quantity of substance sorbed on the given sample at time t, and $n_{a,\infty}$ the quantity necessary for monomolecular coverage of the same sample.

It follows from the law of conservation of matter that

$$n_g^0 = n_g + n_a = \frac{pv_h}{RT} + n_a \tag{II.131}$$

and after differentiation with respect to t we obtain

$$-\frac{dp}{dt} = \frac{RT}{v_h}\frac{dn_a}{dt} = \frac{RT}{v_h}\bar{J}$$

Since $\theta = n_a/n_{a,\infty} = [n_g^0 - (pv_h/RT)]/n_{a,\infty}$ the above differential equation can be written in the form

$$-\frac{dp}{dt} = a_0 + a_1 p + a_2 p^2 \tag{II.132}$$

where

$$a_0 \equiv -\frac{RT}{v_d} k_d \frac{n_g^0}{n_{a,\infty}}; \qquad a_1 \equiv \frac{RT}{v_h} k_a \left(1 - \frac{n_g^0}{n_{a,\infty}}\right) + \frac{k_d}{n_a}$$

$$a_2 \equiv k_a/n_{a,\infty}$$

It can be clearly seen from Eq. (II.132) that the derivatives of the p vs. t curves are not in direct relationship to the original rate equation. By integration of the differential equation (II.132) with the condition $p(t=0) = p_0$, the quantity sorbed may be obtained from the rearranged Eq. (II.131) as a function of time. By comparing the theoretical equation p vs. t with the experimental data it is possible to calculate (e.g. by computer optimization) the constants a_0, a_1 and a_2, and if $n_{a,\infty}$ is known, the "rate coefficients" k_a and k_d, too.

With an appropriate experimental method it is possible, for a given phase or even the entire period of the measurement, that the resultant sorption rate is determined either only by adsorption (e.g. when observation begins) or only by desorption (e.g. during continuous removal of the gas). Several examples will be presented later to illustrate this point.

The above arguments can easily be generalized to describe the rate of adsorption of two (or more) components. From Eq. (II.127) the rate of adsorption of component "1" is

$$J_{a,1} = k_{a,1,0}\, f(\theta_1, \theta_2) \exp(-E^{\ddagger}_{a,1}/RT) p_1 \tag{II.133}$$

where

$$k_{a,1,0} = \frac{\sigma_1}{\sqrt{2\pi m_1 kT}}$$

$f(\theta_1, \theta_2)$ is the function of the fraction of empty sites vs. the coverage with respect to components "1" and "2"; and

$E^{\ddagger}_{a,1}$ is the activation energy of the adsorption of component "1".

Similarly, for component "2":

$$J_{a,2} = k_{a,2,0}\, f(\theta_1, \theta_2) \exp(E^{\ddagger}_{a,2}/RT)\, p_2 \tag{II.134}$$

According to Eq. (II.129) the rate of desorption is

$$J_{d,1} = k_{d,1,0}\,\bar{f}(\theta_1)\exp(-E^{\ddagger}_{d,1}/RT) \qquad \text{(II.135)}$$

where

$$J_{d,2} = k_{d,2,0}\,\bar{f}(\theta_2)\exp(-E^{\ddagger}_{d2}/RT) \qquad \text{(II.136)}$$

These equations are valid for the case when the desorption of components "1" and "2" is independent of each other, and their interaction during adsorption is limited to the necessity of accounting for the requirements of the second component in the calculation of the fraction of empty sites.

In the discussion of heterogeneous catalytic reactions the ad- and desorption rates are taken into consideration by means of such, or similar, functions, though these simple relationships are presumably even less suited for the description of multicomponent adsorption than the analogous Eqs (II.127) and (II.129). The corrections suggested by De Boer will be treated in Paragraph 6.3.2.

SECTION 6.3. FACTORS AFFECTING ADSORPTION AND DESORPTION RATES

Of the factors influencing the sorption rate, in general *a priori* assumptions based on chemical or physical evidence are applied to the functions $f(\theta)$ and $\bar{f}(\theta)$, and then checked for their agreement with experimental data. The activation energies and the condensation coefficient are parameters (or functions) to be determined experimentally, since, with the exception of the simplest cases, their values cannot be determined theoretically.

We shall next try to sum up the conclusions available, partly from theoretical and partly from experimental research, on the factors influencing the rate of sorption.

PARAGRAPH 6.3.1. THE FUNCTIONS $\bar{f}(\theta)$ AND $f(\theta)$

The dependence of the rate equations on θ appears in explicit form in the functions $\bar{f}(\theta)$ and $f(\theta)$. Let us consider two cases in this respect.

(1) The adsorbed molecule occupies a single adsorption site

If the molecule occupies a single adsorption site, then the probability of direct collision of the molecule with an empty adsorption site at coverage θ will be

$$f(\theta) = 1 - \theta \qquad \text{(II.137)}$$

irrespective of the localized or delocalized nature of the adsorbed layer developing.

Similarly, since desorption can take place only from an occupied site:

$$\bar{f}(\theta) = \theta \qquad \text{(II.138)}$$

When several components are sorbed under conditions otherwise identical with the above:

$$f(\theta) = 1 - \sum_i \theta_i \text{ and} \tag{II.139}$$

$$\bar{f}(\theta_i) = \theta_i \tag{II.140}$$

(2) The adsorbed molecule occupies more than one adsorption site

For a molecule to occupy more than one adsorption site requires that the molecule either undergoes complete or partial dissociation in the chemisorption process (type I), or that because of its size or for other (e.g. energy) considerations it prevents further adsorption at sites near the one it occupies (type II).

As regards the first case, it may be added that complete dissociation usually occurs when the breaking bond is a single one (dissociative adsorption). When a double or triple bond is split (e.g. if only the π bond is split, in a C = C double bond of $\sigma - \pi$ type, with the σ bond still keeping the atoms together), dissociation is in general only partial (associative adsorption).

In contrast to the case under (1), a distinction must be made here between localized and delocalized layers, and, moreover, within localized layers, between the random and ordered occupation of the sites.

For localized and statistically random layers of type I and for low coverages, $f(\theta)$ may be taken to a good approximation as being equal to $(1 - \theta)^2$. At higher coverages, however, the probability of adsorption will be higher than indicated by $(1 - \theta)^2$. It follows from statistical arguments that the probability of an adsorption site being empty is $(1 - \theta)$, and of a site next to a chosen empty site also being empty $\frac{z}{z - \theta}(1 - \theta)$, where z is the number of adsorption sites around the chosen site. Hence:

$$f(\theta) = \frac{z}{z - \theta}(1 - \theta)^2 \tag{II.141}$$

from which it follows for $\theta \to 0$ that $f(\theta) \to (1 - \theta)^2$, and at the same time for high θ values $f(\theta) \to \frac{z}{z - 1}(1 - \theta)^2$. Generalization to mixed adsorption is quite obvious.

The functions $f(\theta)$ of localized adsorbed layers of type I with definite structure are not yet known. It is not inconceivable, however, that for a polycrystalline adsorbent one or other of these functions more or less holds, since these adsorbents may contain in random distribution crystal faces of different indices perhaps with ordered layers. Nevertheless, it is to be expected that such surfaces will display *a priori* heterogeneity, due to the fact that, depending on the temperature, adsorption may play a preferred role on certain parts of the surface. This qualitative conclusion appears to be supported by the analysis of sticking probabilities recently determined experimentally on metal filaments and ribbons (see later). A particularly complex situation arises when several adsorbed layers with regular structures are built up on top of one another. It is an experimentally confirmed fact that when one or more components are adsorbed, the interlacing of the layers occurs in certain cases (see Paragraph 8.1.3). Accordingly, our ideas on the geometrical and energy conditions of mixed adsorption must be modified, which again throws a different light on the problem of heterogeneous catalysis.

For statistically random layers of type I, accurate calculations have led to the equation

$$\bar{f}(\theta) = \frac{(z-1)^2}{z(z-\theta)}\,\theta^2 \tag{II.142}$$

but for most practical purposes a sufficiently accurate approximation is obtained when $\bar{f}(\theta)$ is simply taken as being equal to θ^2. For mixed adsorption $\bar{f}(\theta)$ is thus $\bar{f}(\theta_\mathrm{i}) = \theta_\mathrm{i}^2$.

The resultant of the two opposing effects for delocalized layers of type I is (for details see Ref. [130]):

$$f(\theta) = (1-\theta)^2 \text{ and} \tag{II.143}$$

$$\bar{f}(\theta) = \theta^2 \tag{II.144}$$

Similarly, for adsorption from a gas mixture:

$$f(\theta) = (1 - \sum_i \theta_i)^2 \quad \text{and} \tag{II. 145}$$

$$\bar{f}(\theta_i) = \theta_i^2 \tag{II.146}$$

In localized layers of type II, $f(\theta)$ is a function not only of the crystal lattice structure, but also of the dimensions of the adsorbate molecule. Tonks derived the functions $f(\theta)$ for quadratic lattices [131]. The situation is complicated somewhat by the fact that, in accordance with what was said above, not only the size of the molecule can obstruct sorption on the adsorption sites neighbouring the occupied site, but also the changed adsorption interactions. Thus, it might be advisable to treat with some reservation the formulae derived by Tonks from a purely geometrical approach.

Since the adsorption processes creating layers truly of type II are non-dissociative, $\bar{f}(\theta) = \theta$ for both localized and delocalized layers.

PARAGRAPH 6.3.2. ACTIVATION ENERGY

Calculation of the activation energy of chemisorption from the relation which includes the sticking probability [Eq. (II.126)] is complicated by the fact that in the majority of cases the function $\sigma(\theta)$ is not sufficiently well-known, and $\sigma(\theta,\mathrm{T})$ perhaps even less so. Although an "activation energy" can almost always be calculated from the function $s(T)/f(\theta)$, all attempts to assign this to a defined adsorption state fails, not only because of what has been said above, but also because in the majority of cases $E_\mathrm{a}^\ddagger$ is not independent of the actual coverage.

Starting from the simple assumption that $E_\mathrm{a}^\ddagger$ and $E_\mathrm{d}^\ddagger$ are linear functions of θ, that is:

$$E_{\mathrm{a}}^{\ddagger} = E_{\mathrm{a},0}^{\ddagger} + \alpha\theta \tag{II.147}$$

$$E_{\mathrm{d}}^{\ddagger} = E_{\mathrm{d},0}^{\ddagger} - \beta\theta \tag{II.148}$$

Brunauer et al. [132] derived rate equations in excellent agreement with experimental findings (see later).

One of the features of the analogous empirical correlations (e.g. the Elovich equation [133]) is that they contain only an exponential θ-dependence, which is easy to understand if it is borne in mind that the rate equations from Eqs (II.147) and (II.148) always present an exponential dependence on θ, and further, that in most cases the influence of $f(\theta)$ or of $\bar{f}(\theta)$ can indeed be ignored.

There is an apparent contradiction in that direct measurements of the sorption rate do not seem to confirm the validity of the rate equation (II.127) or less convincingly than those in which the coverage-dependence of the activation energies is expressed explicitly, whereas in heterogeneous catalysis the first type of equations is almost generally used. However, if it is considered that, as a result of the elementary reactions increasing and decreasing the coverage, in steady-state or quasi-steady-state processes the surface coverage of the individual components is constant or nearly constant for a lengthy period of time, the objections seem to be somewhat less critical.

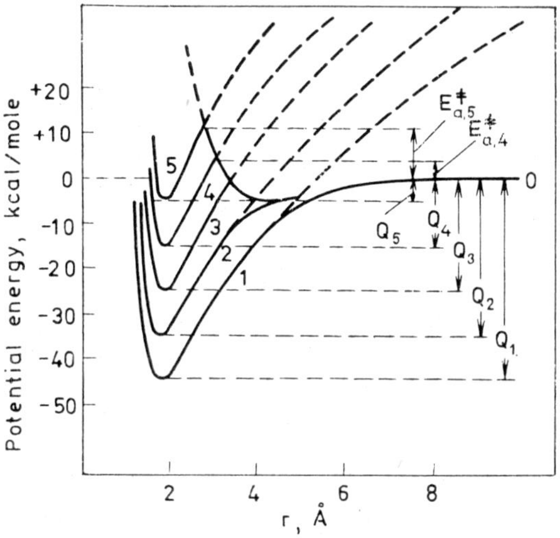

Fig. II.25. Potential curves typical of heterogeneously adsorbing surfaces

The dependence of $E_{\mathrm{a}}^{\ddagger}$ and $E_{\mathrm{d}}^{\ddagger}$ on the coverage is a result of the initially present, or induced, heterogeneity of the adsorbing surface. The conditions are outlined in Fig. II.25, showing the potential curves of an adsorbing system for a heterogeneous adsorbing surface. The potential curves are numbered in the order of decreasing absolute values of heats of adsorption and activation energies of desorption, i.e. in the order of increasing activation energies of adsorption. At high coverages the rate of adsorption decreases significantly, not only because of the function $f(\theta)$, but also because of the increase in the activation energy of adsorption, while that of desorption increases exponentially.

If the surface was initially heterogeneous, the potential curves should be considered as pertaining to the individual homogeneous parts of the surface. If the heterogeneity is induced, the gradual shift of the curves is a result of adsorbate—adsorbate interactions.

Elovich rate equations for one sorbed component

From the semi-empirical rule for the coverage-dependence of activation energies (and hence the differential heat of adsorption) Brunauer et al. derived their sorption rate relations (loc. cit.). Proceeding by the method outlined in the introduction, it is advisable to treat the two types of heterogeneity separately.

(1) The *a priori* homogeneous surface

Since the adsorption sites are *a priori* homogeneous, the corresponding rate equations will be:

$$J_a = \frac{\sigma p}{\sqrt{2\pi m kT}}(1-\theta)\exp(-E_a^{\ddagger}/RT) \tag{II.149}$$

and

$$J_d = k_{d,0}\,\theta\exp(-E_d^{\ddagger}/RT) \tag{II.150}$$

where $E_a^{\ddagger}$ and $E_d^{\ddagger}$ conform to Eqs (II.147) and (II.148).

At coverages neither too low nor too high the exponential dependence on θ dominates, in Eqs (II.149) and (II.150), so that:

$$J_a \sim \exp(-\theta/RT) \tag{II.151}$$

and

$$J_d \sim \exp(\theta/RT) \tag{II.152}$$

Formally, Eq. (II.151) is identical with the empirically postulated rate equation of Elovich (loc. cit.), which, since its presentation, has been considered the most useful equation describing the rate of chemisorption.

(2) The *a priori* heterogeneous surface

In order to be able to describe sorption rates on heterogeneous surfaces by the general relation of Eq. (II.128), explicit statements must be made concerning the functions in this equation.

For the activation energies, and consequently, for the isosteric heat of adsorption

$$-Q_{st} = Q_0(1-\gamma\theta) \tag{II.59}$$

to be a linear function of the coverage θ quite special functions $g(Q)$, $\sigma(Q)$, $f[\theta(Q)]$, $E_a^{\ddagger}(Q)$ and $E_d^{\ddagger}(Q)$ are required. It is another question, however, whether, beyond the experimental evidence relating to the behaviour of the differential heat of adsorption, this assumption of special functions can also be supported physically.

Equation (II.59) shows that on the surface the energies of adsorption lie somewhere between the values of Q_{min} and Q_{max}. If it is assumed that the adsorption sites fulfil the requirements of the probability density function

$$g(Q) = \frac{1}{Q_{max}-Q_{min}} \qquad (Q_{max} \leqq Q \leqq Q_{min}) \tag{II.153}$$

which describes the probability of uniform distribution with respect to Q, then the coverage θ [the expectation of $\theta(Q)$] can be given by the equation:

$$\theta = \int_{Q_{\min}}^{Q_{\max}} \theta(Q)\, g(Q)\, dQ = \frac{1}{Q_{\max} - Q_{\min}} \int_{Q_{\min}}^{Q_{\max}} \theta(Q)\, dQ$$

(see Paragraph 2.2.4). This integral can be related to Eq. (II.59) only by assuming that the adsorbate always occupies the most active sites available, that is if $\theta(Q)$ is in agreement with the following step function:

$$\theta(Q) \equiv 0 \quad \text{if} \quad Q_{\min} \leq Q \leq Q^*$$

and (II.154)

$$\theta(Q) \equiv 1 \quad \text{if} \quad Q^* \leq Q \leq Q_{\max}$$

Then:

$$\theta = \frac{1}{Q_{\max} - Q_{\min}} \int_{Q_{\min}}^{Q_{\max}} \theta(Q)\, dQ = \frac{1}{Q_{\max} - Q_{\min}} \int_{Q^*}^{Q_{\max}} dQ = \frac{Q_{\max} - Q^*}{Q_{\max} - Q_{\min}}$$

that is (omitting the upper index of Q^*):

$$Q = Q_{\max}\left[1 - \left(1 - \frac{Q_{\min}}{Q_{\max}}\right)\theta\right] = Q_0\,(1 - \gamma\theta) \text{ (see II.60)}$$

When $g(Q)$ is chosen in accordance with Eq. (II.153) and $\theta(Q)$ with Eq. (II.154) the above form of the adsorption energy [Paragraph 2.2.3, Eq. (II.60)] is obtained, and, as has been shown before, this leads to the isosteric heat of adsorption of Eq. (II.59).

Using the former expressions for $g(Q)$ and $f[\theta(Q)] \equiv 1 - \theta(Q)$, the integral in the general relation of Eq. (II.128) can be written, if $\sigma(Q)$ is considered to be independent of Q, as follows:

$$\int_0^{\infty} g(Q)\, f[\theta(Q)] \exp[-E_a^{\ddagger}(Q)/RT]\, dQ =$$

$$= \frac{1}{Q_{\max} - Q_{\min}} \int_{Q_{\min}}^{Q_{\max}} \exp[-E_a^{\ddagger}(Q)/RT] dQ$$

Taking into account that, according to Eq. (II.60), $dQ = -(Q_{\max} - Q_{\min})d\theta$:

$$\frac{1}{Q_{\max} - Q_{\min}} \int_{Q_{\min}}^{Q_{\max}} \exp[-E_a^{\ddagger}(Q)/RT]\, dQ =$$

$$= -\int_1^{\theta} \exp[-(E_{a,0}^{\ddagger} + \alpha\theta)/RT]\, d\theta = \int_{\theta}^{1} \exp[-(E_{a,0}^{\ddagger} + \alpha\theta)/RT] d\theta$$

from which the final form of the rate equation is:

$$J_a = \frac{\sigma p}{\sqrt{2\pi mkT}} \frac{RT}{\alpha} \exp(-E^{\ddagger}_{a,0}/RT)\,(e^{-\alpha\theta/RT} - e^{-\alpha/RT}) \tag{II.155}$$

Since in this case $f[\theta(Q)]$ and $\theta(Q)$ are complementary functions of each other, we obtain for J_d:

$$J_d = \frac{k_{d,0}}{Q_{max} - Q_{min}} \int_{Q_{min}}^{Q_{max}} \theta(Q) \exp[-E^{\ddagger}_d(Q)/RT]\,dQ =$$

$$= \frac{k_{d,0}}{Q_{max} - Q_{min}} \int_{Q}^{Q_{max}} \exp[-E^{\ddagger}_d(Q)/RT]\,dQ = \tag{II.156}$$

$$= k_{d,0} \int_0^{\theta} \exp[-(E^{\ddagger}_{d,0} - \beta\theta)/RT]\,d\theta = k_{d,0} \frac{RT}{\beta} \exp(-E^{\ddagger}_{d,0}/RT) \cdot [e^{\beta\theta/RT} - 1]$$

If, in the case of adsorption $\theta \ll 1$, $\exp(-\alpha\theta/RT)$ will be much greater than $\exp(-\alpha/RT)$, and similarly for desorption if $\theta > 0$, then $\exp(\beta\theta/RT) \gg 1$. Taking into consideration also that, for localized layers, $\left|\frac{d\theta}{dt}\right| = \frac{1}{n_s}\bar{J}$, Eqs (II.155) and (II.156) reduce to:

$$\frac{d\theta}{dt} = a \exp(-\alpha\theta/RT) \tag{II.157}$$

and

$$-\frac{d\theta}{dt} = b \exp(\beta\theta/RT) \tag{II.158}$$

with the added remark that the factors a and b contain not only the terms shown in Eqs (II.155) and (II.156), but also n_s (in the denumerator).

When the coverage θ is calculated from that prevailing at $t = 0$, then the initial condition for Eq. (II.157) is $\theta = 0$ when $t = 0$, so that the integral (II.157) takes the form:

$$\theta = \frac{RT}{\alpha} \ln\left(\frac{t + t_0}{t_0}\right) \quad \text{where } t_0 = \frac{RT}{\alpha a} \tag{II.159}$$

Similarly, integration of Eq. (II.158) with the initial condition $\theta = 1$ when $t = 0$, leads to

$$\theta = \frac{RT}{\beta} \ln\left(\frac{t'_0}{t + t''_0}\right) \quad \text{where } t'_0 = \frac{RT}{\beta b}$$

and

$$t''_0 = \frac{RT}{b} \exp(-\beta/RT) \tag{II.160}$$

The empirical Elovich equation was also derived for conditions other than those given above [134, 135].

In principle, the integrated Elovich equation (II.159) describes the rate of sorption (if at all) only at constant pressure p and for zero desorption. If there is a change in pressure during sorption, the procedure should be similar to that illustrated by the example in Paragraph 6.2.3. If the rate of desorption cannot be ignored in a given case, Eqs (II.157) and (II.158) must be integrated jointly, e.g. with the initial condition $\theta = 0$ when $t = 0$. Either or both of these two factors might be the reason that when the experimental data for higher coverages (p decreases and/or θ increases) are plotted in the θ vs. $\ln t$ coordinate system, not a straight line, but a curve (or "two" joined "lines") is obtained.

It might seem a contradiction that in the definition of $\theta(Q)$ the mobility of the adsorbed atoms was stressed, while later, the layer was considered to be localized. This, however, is in wide agreement with the views of Robins (loc. cit.) on the process of sorption. We have repeatedly emphasized that the mobility of "hot" adsorbed atoms formed in the process of chemisorption depends on the value of the excitation energy, and also on the depth of the still unoccupied potential wells. At first, the deepest potential wells are filled with relatively hot adsorbed atoms, and as soon as these are full, in accordance with the rate of energy dissipation, the less deep wells begin to trap the more or less cooled adsorbed atoms. Mobility is a property of excited molecules only, whereas adsorbed atoms cooled to the thermal level are probably localized.

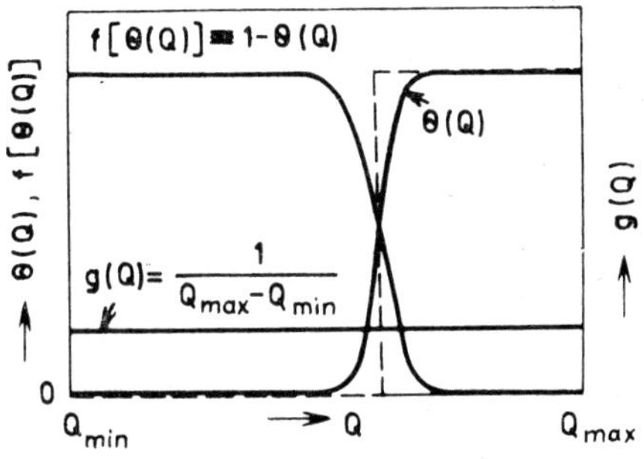

Fig. II.26. $\theta(Q)$ vs. Q in accordance with Eq. (II.53) (continuous curve) and its approximation for the calculation (dashed curve)

Finally, considering the conditions (if any) under which the special definitions of $g(Q)$ and $\theta(Q)$ might be attributed a physical meaning, let us examine Fig. II.26, where the continuous line is an attempt to illustrate $g(Q)$ and $\theta(Q)$ according to Eq. (II.53), and the dashed line is the limiting case on which the derivation is based [Eq. (II.154)]. It appears quite clearly that this seemingly arbitrary definition of the functions in question is one of the simplest approaches, yet does not contradict the requirements.

Elovich type equations for the description of the rate of adsorption of mixtures

By generalizing Eqs (II.147) and (II.148) for two components De Boer [44] derived equations for the description of the rate of adsorption of mixtures on homogeneous adsorbent surfaces.

For two adsorbates which produce dipoles of the same orientation on the surface during adsorption, generalization of Eqs (II.147) and (II.148) leads to the following relationships for the activation energies and heats of adsorption:

$$E_{a,1}^{\ddagger} = E_{a,1}^{\ddagger 0} + \alpha_1\theta_1 + \alpha_2\theta_2 \quad \text{(II.161)}$$

$$E_{a,2}^{\ddagger} = E_{a,2}^{\ddagger 0} + \alpha_1\theta_1 + \alpha_2\theta_2 \quad \text{(II.162)}$$

$$Q_1 = Q_1^0 + c_1\theta_1 + c_2\theta_2 \quad \text{(II.163)}$$

$$Q_2 = Q_2^0 + c_1\theta_1 + c_2\theta_2 \quad \text{(II.164)}$$

where $E_{a,i}^{\ddagger 0}$ and Q_i^0 are the initial values (pertaining to $\theta \rightarrow 0$) of the activation energy and the heat of adsorption, respectively, and α_i and c_i are positive constants.

In this way, the rate of adsorption of the compound of type "i" ($i = 1,2$) will be:

$$J_{a,i} = \frac{\sigma_i P_i}{\sqrt{2\pi m_i kT}}(1 - \theta_1 - \theta_2)\exp(-E_{a,i}^{\ddagger}/RT)\ (\text{molecules cm}^{-2}\text{s}^{-1}) \quad \text{(II.165)}$$

and the rate of desorption:

$$J_{d,i} = k_{d,i}\,\theta_i \exp[-E_{a,i}^{\ddagger} - Q_i)/RT]\ (\text{molecules cm}^{-2}\,\text{s}^{-1}) \quad \text{(II.166)}$$

By means of rearrangements similar to those in (1) we obtain:

$$J_{a,i} \sim \exp[-(\alpha_1\theta_1 + \alpha_2\theta_2)/RT] \quad \text{(II.167)}$$

and

$$J_{d,i} \sim \exp[(c_1\theta_1 + c_2\theta_2)/RT] \quad \text{(II.168)}$$

in agreement with Eqs (II.151) and (II.152).

If the dipole layers produced by the two adsorbates are of opposite orientation, one or other of the α_i-s or c_i-s of identical indices changes sign with the result that the adsorbates appear mutually to promote each other's adsorption.

Example. A number of chemisorption processes are known which take place according to Eq. (II.159). Let us consider as an example the chemisorption of H_2 on $2MnO \cdot Cr_2O_3$, investigated by Taylor and Thon [134] (Fig. II.27). If the experimentally determined initial rate [the constant a in Eq. (II.157)] is greater than the a value calculated from the intercept and slope of the plot of θ vs. $\ln[(t+t_0)/t_0]$, it follows that the initial process takes place at a high rate and does not follow the exponential law. In other instances there will be a break in the straight line of θ vs. $\ln[(t+t_0)/t_0]$ at certain definite coverages, as shown in Fig II.28 for the chemisorption of H_2 on ZnO. One of the possible explanations of the phenomenon involves the existence of two different types of adsorption sites, with different $E_a^{\ddagger}$ and α values on the surface. Another—perhaps more likely explanation—assumes that the adsorption process in fact follows not Eq. (II.159), but the θ vs. $\sqrt{t}$ function characterizing diffusion processes (moving boundary layer). In such cases, the transformation of experimental data as required by Eq. (II.159) usually causes a break in the straight line.

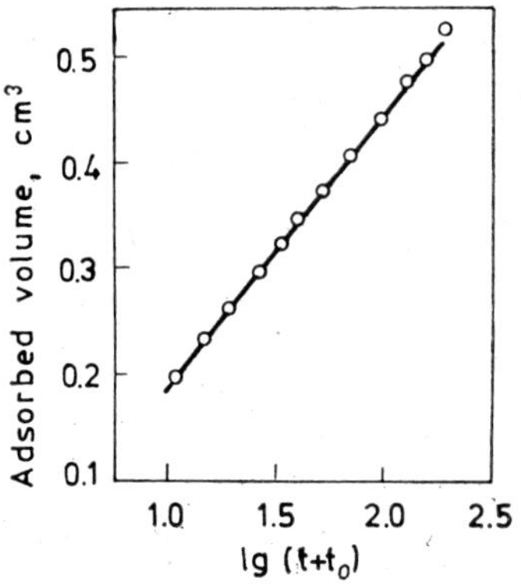

Fig. II.27. Adsorption of hydrogen on 2 MnO. Cr_2O_3 at 100 °C

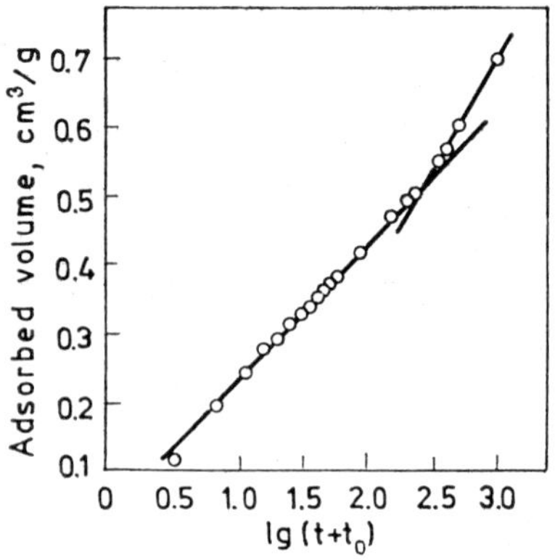

Fig. II.28. Adsorption of hydrogen on ZnO at 184 °C

The constant a in Eq. (II.159) often shows the temperature-dependence expected from $T^{-1/2}\exp(-E^{\ddagger}_{a,0}/RT)$ [Eq. (II.141)] or from $T^{1/2}\exp(-E^{\ddagger}_{a,0}/RT)$ [Eq. (II.155)], while, also in accordance with expectation, α is temperature-independent. The situation is similar with respect to the pressure-dependence of a. Provided the measurements are otherwise reliable, a different behaviour is usually due to some deficiency of the theory. The "explanations" occasionally appearing in the literature have little scientific value.

Equations of the type of Eq. (II.159) have provided a satisfactory practical interpretation in the following cases: chemisorption of H_2 on Pt, Ni and Fe; of CH_4 on Ni; of O_2 on CoO; and of N_2 on Fe. The very extensive study of the chemisorption of H_2 on Fe has confirmed the coverage-dependence of the activation energy of chemisorption as indicated by Eq. (II.147) [136].

Because of experimental difficulties encountered during the measurements the practical applicability of the desorption rate equation has not been investigated with sufficient thoroughness. It could be ascertained that k_d depends exponentially on θ, in agreement with the requirements of Eq. (II.160). The fact that the linear-dependence of $E^{\ddagger}_{d}$ on θ would also lead to such a rate relation means the loss of the unambiguity. A more thorough examination of the temperature-dependence of the desorption rate nevertheless usually permits a distinction between the two effects.

PARAGRAPH 6.3.3. THE CONDENSATION COEFFICIENT

The condensation coefficient is the least known parameter of the rate equation. Its value is undoubtedly influenced by the nature of the adsorbent and adsorbate, and by the interaction between these two which plays a primarily decisive role in the development of the properties of the adsorbed layer (localization, delocalization, and in connection with this: residual degrees of freedom, etc.). Experimental results suggest, however, that the "nature of the adsorbent" should be interpreted in a broader sense and should refer to an adsorbent of a given material nature in a given physical state (of crystallographic orientation and more or less unequivocally defined from the aspect of solid-state physics). Furthermore, the condensation coefficient is also affected by the "geometry" of the collision; this has been convincingly proved by, among others, studies of the dispersions of atomic gases on adsorbent surfaces [137]. In our opinion, the first points to be clarified are how far the interaction between the adsorbate and the adsorbent surface, leading to the binding of the molecule, can be described by the Boltzmann factor expressing the energy condition of a single molecule being sorbed, and, if this pattern is oversimplified, whether the result of ignoring certain factors does not appear in the condensation coefficient.

Calculation of the condensation coefficient by statistical mechanics has been described by Glasstone, Laidler and Eyring [138]. The theory of absolute reaction rates assumes that according to the stoichiometric equation

$$R + S \rightleftarrows RS^{\ddagger} \rightleftarrows RS \tag{II.169}$$

the localized, transition complex* $RS^{\ddagger}$, formed as a result of the interaction between the adsorbate R and the free site S, is in quasi-equilibrium with both the initial substances and the "product".

For delocalized transition complexes the free site S can be neglected as "initial substance", while in the case of dissociative adsorption the stoichiometric equation has to reflect the fact of dissociation:

$$R_2 + S_2 \rightleftarrows R_2S_2^{\ddagger} \rightleftarrows 2\,RS \tag{II.170}$$

If the concentration of the transition complex is $c^{\ddagger}$ (in molecules cm^{-2} units) and the frequency of the transition of the complex into the state of stable chemisorption is $\nu^{\ddagger}$, the rate of adsorption will be

$$J_a = \nu^{\ddagger} c^{\ddagger} \text{ (molecules cm}^{-2}\text{ s}^{-1}) \tag{II.171}$$

$c^{\ddagger}$ can be determined by assuming that the transition complex is in quasi-equilibrium with the initial substances. The equilibrium constant $K^{\ddagger}$ can be expressed from the corresponding concentrations and partition functions as follows:

for a localized transition complex: $K^{\ddagger} = \dfrac{c^{\ddagger}}{c_g c_s} = \dfrac{\Omega^{\ddagger}}{(\Omega_g/v)\,\Omega_s}$

for a delocalized transition complex: $K^{\ddagger} = \dfrac{c^{\ddagger}}{c_g} = \dfrac{\Omega^{\ddagger}}{\Omega_g/v}$

for a dissociative adsorption complex: $K^{\ddagger} = \dfrac{c^{\ddagger}}{c_g\, c_{s_2}} = \dfrac{\Omega^{\ddagger}}{(\Omega_g/v)\,\Omega_{s_2}}$

where

c_g (molecules cm^{-3}) is the concentration of the gas,
c_s (free sites per cm^2) is the concentration of free sites, and the
Ω terms are the corresponding partition functions.

* Also referred to as "activated complex" or "transition state".

By equating the two expressions for $K^{\ddagger}$ and then substituting $c^{\ddagger}$ into Eq (II.171) we obtain the rate of adsorption.

Certain transformations are necessary for these relations to fit into the physical pattern so that they also bear some resemblance to the rate equations of the earlier arguments. These transformations are:

(1) for an ideal gas: $c_g = p/kT$;

(2) when the zero-point energies are separated from the partition functions, the Boltzmann factor containing the activation energy is obtained: $\exp(-E_a^{\ddagger}/RT)$;

(3) since the vibrational frequency calculated from the characteristic temperature of the solid is high, at not too high temperatures the vibrations of the solid are not excited, and hence for the thermal partition function: $\Omega_s' \approx 1$;

(4) in the thermal $\Omega^{\ddagger\prime}$ the vibration normal to the surface has a low frequency ($\nu^{\ddagger}$), and consequently in its characteristic partition function, $1/[1-\exp(-h\nu^{\ddagger}/kT)]$, it is possible to apply the series expansion of the exponential expression up to the second term inclusive and to put the factor $kT/h\nu^{\ddagger}$ before the partition function. The residual partition function, $\omega^{\ddagger}$, thus contains only the contributions of the degrees of freedom left as follows:

(i) if the transition complex is localized, $\omega^{\ddagger}$ contains no translational and rotational contributions, but only a vibrational contribution, so that:

$$\frac{\omega^{\ddagger}}{(\Omega_g'/v)} = \frac{h^3}{(2\,\pi\,m\,kT)^{3/2}}\,\frac{1}{\omega}$$

where ω is the partition function corresponding to the rotational degrees of freedom of the gaseous substance;

(ii) in the case of mobile two-dimensional gaseous layers only one translational degree of freedom is lost (or transformed into a vibrational degree of freedom), and hence:

$$\frac{\omega^{\ddagger}}{(\Omega_g'/v)} = \frac{h}{\sqrt{2\,\pi\,m\,kT}}$$

(iii) if one rotational degree of freedom is also lost:

$$\frac{\omega^{\ddagger}}{(\Omega_g'/v)} = \frac{h}{\sqrt{2\,\pi\,m\,kT}}\left(\frac{h^2\,s}{8\,\pi^2\,IkT}\right)^{1/2}$$

where I is the moment of inertia of the molecule and s is the symmetry number;

(5) finally, for localized layers c_s and c_{s_2} can be simply calculated from the relations:

$$c_s = n_s f(\theta)$$

$$c_{s_2} = 1/2[z\,n_s\,f\,(\theta)]$$

where n_s is the number of adsorption sites (cm^{-2})
and z is the number of nearest neighbouring sites.

The relation for the rate of adsorption will contain $f(\theta)$ in the case of mobile layers too, for the intermediate complex can be formed only in connection with vacant sites.

Using these substitutions, the adsorption rate equations for the above four cases will be:
localized intermediate complex:

$$J_a = \frac{n_s h^2}{2\pi m kT\omega} \frac{p}{\sqrt{2\pi m kT}} f(\theta) \exp(-E_a^{\ddagger}/RT)$$

delocalized complex, no rotation lost:

$$J_a = \frac{p}{\sqrt{2\pi m kT}} f(\theta) \exp(-E_a^{\ddagger}/RT)$$

delocalized complex, with one rotational degree of freedom lost:

$$J_a = \left(\frac{h^2 s}{8\pi^2 IkT}\right)^{1/2} \frac{p}{\sqrt{2\pi mkT}} f(\theta) \exp(-E_a^{\ddagger}/RT)$$

dissociative adsorption complex:

$$J_a = \frac{n_s z h^2}{4\pi mkT\omega} \frac{p}{\sqrt{2\pi mkT}} f(\theta) \exp(-E_a^{\ddagger}/RT)$$

The factors before the collision numbers, $(p/\sqrt{2\pi mkT})$, correspond to the theoretical value of the condensation coefficient, σ.

For some of the above parameters only the order of magnitude of their values can be estimated; thus $n_s \approx 10^{13}-10^{15}$ cm^{-2}; $z = 4$.

It can be seen that for delocalized transition complexes (if the rotational degree of freedom is not lost in the adsorption process), $\sigma = 1$. Depending on the moment of inertia of the molecule, estimations for the loss of one rotational degree of freedom give $0.1 < \sigma < 0.6$. For localized transition complexes the value of σ is considerably lower: $10^{-4} < \sigma < 0.1$.

PARAGRAPH 6.3.4. THE RATE CONSTANT OF DESORPTION

Polányi and Wigner have derived a very simple equation for the rate of desorption of non-dissociatively adsorbed substances by assuming that a particle with sufficient activation energy will be desorbed during a single period of the vibration normal to the surface [7]. If the frequency of vibration is $\nu^{\ddagger}$, then:

$$J_d = \nu^{\ddagger} n_s(\theta) \exp(-E_d^{\ddagger}/RT) \quad \text{(molecules cm}^{-2}\text{ s}^{-1}) \qquad \text{(II.172)}$$

Comparison of Eqs (II.129) and (II.172) shows that:

$$k_{d,0} = \nu^{\ddagger} n_s$$

For chemical bonds the order of $\nu^{\ddagger}$ is 10^{13} s^{-1}, and further for several adsorbents $n_s \approx 10^{13} - 10^{15}\ \mathrm{cm}^{-2}$ and hence from these data $k_{d,0} \approx$ $\approx 10^{26} - 10^{28}\ \mathrm{cm}^{-2}\,\mathrm{s}^{-1}$.

It has been pointed out in the discussion of the condensation coefficient that according to the theory of absolute reaction rates the transition complex is in a state of quasi-equilibrium not only with the initial substances, but also with the products. Using a similar approach to that applied to the discussion of the adsorption rate, we may try to express the equilibrium constant $K^\ddagger$ from the right hand sides of the stoichiometric equations (II.169)—(II.170) in the hope of arriving at statistical mechanical relations for the rate of desorption.

Desorption without association:

$$K^\ddagger = c^\ddagger/c_a = \Omega^\ddagger/\Omega_a$$

Desorption with association:

$$K^\ddagger = c^\ddagger/c_a^2 = \Omega^\ddagger/\Omega_a^2$$

where $\Omega_a^\ddagger$ is the partition function of the adsorbed molecules.

Using the simplifications introduced in 6.3.3. in accordance with their meaning, the following equations are obtained for the rate of desorption:

Desorption without association:

$$J_d = n_s \frac{kT\,\omega^\ddagger}{h\,\Omega_a'} \bar{f}(\theta) \exp(-E_d^\ddagger/RT)$$

Desorption with association:

$$J_d = n_s \frac{kT\,\omega^\ddagger}{h\,\Omega_a'^2} \bar{f}(\theta) \exp(-E_d^\ddagger/RT)$$

where Ω_a' is the thermal partition function of the adsorbed molecules.

The theoretical values of $k_{d,0}$ are obtained from the factors in front of the function $\bar{f}(\theta)$.

The following two cases are worth distinguishing in connection with desorption without association:

(1) If the intermediate complex and the adsorbed molecules have the same degrees of freedom, and in addition the frequency of the vibration of the adsorbed molecule perpendicular to the surface is high (so that the thermal partial partition function for this degree of freedom is unity) $\omega^\ddagger$ is equal to Ω_a' and hence:

$$k_{d,0} = n_s \frac{kT}{h} \qquad \text{(II.173)}$$

(2) It may happen that the intermediate complex is less strongly bound to the surface than the adsorbed molecule, thereby possessing more degrees of freedom than the latter, and consequently $\omega^\ddagger/\Omega_a'$ is greater than unity. In special cases (e.g. when the intermediate complex, but not the adsorbed molecule, has translational and rotational degrees of freedom) the value of the factor $\omega^\ddagger/\Omega_a'$ may be even 10^3—10^4.

A similar situation arises when desorption is preceded by the association of the molecular fragments:

(1) when both the intermediate complex and the adsorbed radicals are localized, then $\omega^{\ddagger} = \Omega'_{\mathrm{a}} = 1$, and hence the theoretical value of $k_{\mathrm{d},0}$ agrees with that of Eq. (II.173);

(2) when the intermediate complex is delocalized, but the adsorbed radicals are localized, then (although $\Omega'_{\mathrm{a}} = 1$) because $\omega^{\ddagger} > 1$ the value of $k_{\mathrm{d},0}$ is higher than in the first case;

(3) when both the intermediate complex and the adsorbed radicals are delocalized it is difficult to predict the value of $k_{\mathrm{d},0}$.

CHAPTER 7. EXPERIMENTALLY DETERMINED RATES OF ADSORPTION AND DESORPTION

Experimental results on sorption kinetics can be roughly classified into three major groups according to the data of their origin, the experimental method, and, to a certain extent, the value of the conclusions which they permit. Studies of no interest as regards heterogeneous catalysis, and those which have failed to contribute significantly to the development of sorption kinetics, will be ignored.

One of the three main fields of research covers rate measurements on porous oxides, and, to a lesser degree, on metal powders. Interpretation of the experimental results for these highly complex phenomena is hampered by, among other things, the superposition of experimental errors due to inadequate preparation of the adsorbent, the porous structure of the oxides, etc. To avoid such errors, an increasing number of measurements are being carried out on metal filaments and ribbons, as well as on vacuum-evaporated films. In our view, experiments with metal filaments and ribbons are of outstanding importance, because they have drawn attention to a number of facts which might necessitate the revision of our concepts concerning the rate of chemisorption.

SECTION 7.1. CHEMISORPTION ON OXIDES

One of the main features of oxides used as adsorbents is that even after the most careful cleaning by chemical and physical methods they may still contain several types of adsorption centres of entirely different behaviour: metal ions, oxygen ions and various lattice defects. Consequently, depending on the chemical and physical properties of the oxide, on the adsorbate and on the temperature of the experiment, a multitude of different interactions may occur. Difficulties may also be encountered due to the fact that the perfect removal of chemisorbed "impurities" from oxides with high specific surface areas is an extremely difficult task, and the heating and evacuation used for this purpose lead, in the majority of cases, to irreversible structural changes (shrinkage, recrystallization, formation of compounds, loss of water, etc.). In the presence of surface impurities new interactions may arise, perhaps not even characteristic of the pure substance itself, while the pore structure may cause complex transport phenomena which completely mask the true chemisorption.

All this indicates that without additional tests (determination of the pore structure, infrared spectroscopy, ESR, conductivity measurement, etc.), it is almost impossible to draw conclusions about the processes taking place simply from a measurement of the quantity sorbed as a function of time, that is, from the results of the determination of the sorption rate. We are usually faced with not a single chemisorption phenomenon, but several simultaneous processes, some of which (for example slow surface rearrangement, surface compound formation) cannot be detected by means of the sorption of gas, but only by the direct (e.g. spectroscopic) observation of the sorbed layer.

As regards the experimental technique, that is, the method of gaining information, the rate at which the gas disappears is generally measured in an apparatus of constant volume, but the rate of isotope exchange is often determined. Garner et al. [139] suggested a special desorption technique which would hinder re-adsorption by the constant removal of the desorbed gas. Most of these methods are today obsolete, since they usually fail to ensure harmony between the methods of measurement and evaluation. Measurements performed at constant volume generally ignore simultaneous changes in the overall pressure. With Garner's method it is difficult to avoid re-adsorption. The most acceptable results as regards methodology are furnished by adsorption balances and by those other methods which permit the direct observation of the sorption rate on a small quantity of the adsorbent, for example of concentration changes in a flowing gas mixture [113].

However, the most general objections can be made about the evaluation of data obtained by the different methods. Almost without exception, authors ignore the influence of transport phenomena related to the pore structure on the observed sorption rate. In some cases the Elovich equation [see Eq. (II.159)] is used for the description of the rate of the process. If the experimental data fail to give a straight line in the transformation required by the equation, it is assumed that "several" processes take place simultaneously, whereas it may be that the rate of the process cannot be described by the Elovich equation (see Paragraph 6.3.2).

With more or less general validity it may be said of the sorption rates measured on oxides by means of various methods that at "low" temperatures rapid, non-activated processes take place, the majority of them of a physical nature, but, depending on the adsorbent and the adsorbate, chemisorption also may occur. At "high" temperatures, slow, activated processes gradually gain dominance, leading to partial or complete degradation of the adsorbate molecules, to the formation of surface compounds of various stabilities, or to a change in the nature of the surface bonding, etc. In chemisorption processes on oxides at high temperatures it is characteristic that not a single, but several species usually cover the surface of the adsorbent simultaneously.

In support of what has been said so far we shall quote in some detail only a few examples of certain fairly well-known systems, while measurements carried out with other oxide adsorbents are given in the form of references.

Several features of chemisorption on oxides can be well observed in the aluminium oxide—carbon dioxide system [140, 141, 142]. Under the usual conditions of preparation, chemisorbed —OH groups are present on the surface of Al_2O_3 of stoichiometric composition; these decompose on heat-treatment and form water. Decomposition is completed around 1100 °C. At temperatures between room temperature and 400 °C and an initial carbon dioxide pressure of about 3 torr, rapid sorption of short duration can be observed, which later slows down considerably [141] (the experimental data for 400 °C are plotted as p vs. log t in Fig. II.29). In the simultaneously measured infrared spectra there are absorption bands at wavenumbers of 1234, 1460, 1652 and 2365 cm^{-1}. The band at 2365 cm^{-1} develops very rapidly (in about 2 minutes at 20 °C) and disappears on evacuation (again at 20 °C) in 5 minutes. This band can be assigned to physically adsorbed CO_2. The other bands develop very slowly at 20 °C, and evacuation after adsorption does not affect their intensities. The band at 1460 cm^{-1} can be assigned to the CO_3^{2-} ion, situated in a symmetric environment (e.g. in micropores of molecular dimensions). Little [143] claims that the bands at 1652 and 1234 cm^{-1} originate from the chemisorbed bidentate ion:

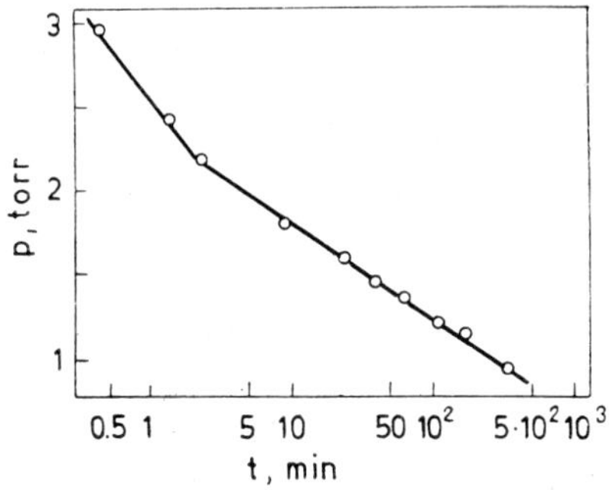

Fig. II.29. Rate of adsorption of CO_2 on Al_2O_3 at 400 °C (after Gregg)

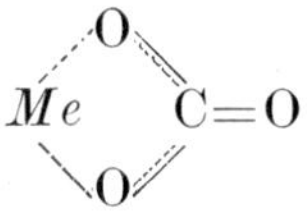

As the formation of such an ion requires a certain rearrangement of the surface structure, the process is undoubtedly activated and therefore slow.

It should be noted that the irreversibility of some of the processes is reflected in the adsorption "isotherms" showing a well-developed hysteresis. On samples pretreated at lower temperatures the slow development of bicarbonate $(HO—\underset{|}{\overset{}{C}}(—O)=O)^-$ groups can also be observed, possibly as a result of the transfer of the —OH group [140].

Adsorption of CO on Al_2O_3 surfaces leads to the development of similar surface structures, but the presence of formate groups can be demonstrated, indicating that the chemisorption of CO is accompanied by partial oxidation, in which the oxygen chemisorbed on the Al_2O_3 during heat-treatment may play some part [144]. Some of the oxidized CO is desorbed as CO_2 during stronger heat treatment.

The chemisorption system $ZnO—H_2$ has been the subject of thorough investigation for several decades. Figure II.30 shows the 60 torr isobar for H_2^* on ZnO pretreated at 340 °C [145]. There are two maxima (types *A* and *B*) in the figure. Simultaneous measurements of the electrical conductivity showed a monotonous increase from room temperature up to about 180 °C, after which the conductivity began to decrease. From the observation that the maxima of the conductivity and the isobar do not occur at the same temperature (see Fig. II.30), and further, that in the vicinity of 180 °C an uncertain inflexion point appears on the adsorption isobar, Narayana et al. assume the existence of a third, *C*-type adsorption.

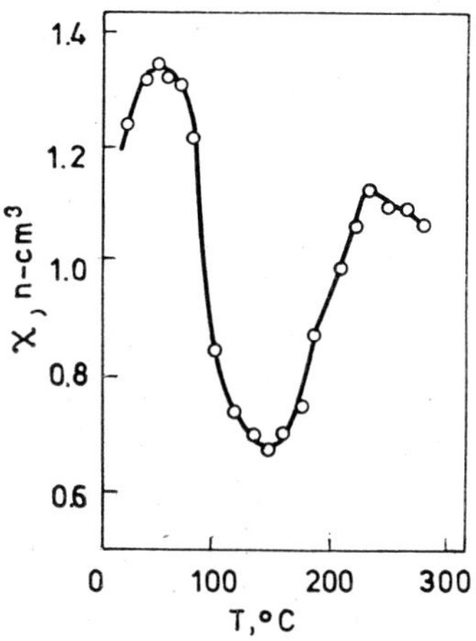

Fig. II.30. Adsorption isobar of hydrogen on ZnO; χ = sorbed quantity

At 275 °C, up to a partial pressure of 10 torr H_2, it is found that the conductivity depends on the partial pressure of H_2 according to the following relation:

$$\sigma = k(p_{H_2})^{1/4}$$

The rate curve for chemisorption at 30 °C (plotted as adsorbed quantity vs. log t) indicates a rapid and a slow process, while the simultaneously measured conductivity (which gives a straight line when plotted as log σ vs. log t) discloses the presence of only a single step with electrical consequences.

Eischens et al. [146] found two absorption bands, at 3510 and 1710 cm^{-1}, in the infrared spectrum of H_2 chemisorbed on ZnO at room temperature. The former can be assigned to the —OH group, the latter to the Zn—H surface group. Dent and Kokes [147] observed the ratio of the band intensities (with the exception of the highest coverages where there is a relative increase in the intensity of the —OH band) to be practically independent of p_{H_2}, while both bands disappear after a short period of evacuation. Eischens suggested that this layer (type *A* ?) should be attributed to the H_2 bound in a rapid irreversible process. It is not possible to observe the infrared absorption of the H_2 adsorbed in the slow process.

The semiconductivity drops dramatically when the chemisorbed layer formed at room temperature is brought into contact with O_2, while at the same time there is practically no change in the intensities of the infrared bands.

Tracer studies have shown that only the H_2 bound in the rapid process is active in the catalytic hydrogenation of ethylene.

Finally, it might be worthwhile to say a few words on the role of O_2. It has already been mentioned that O_2 chemisorbed at room temperature causes a drop in the semiconductivity. Other authors report the poisoning of the hydrogenation reaction by O_2, particularly when the H-layer has

* In fact the pressure is not strictly constant during the measurement.

been exposed to the effect of O_2 at higher temperatures. This point must be stressed, since Kokes claims that the simultaneous presence of three factors is needed to produce poisoning: high temperature, and the presence of both O_2 and H_2. This may suggest that the factor responsible for the poisoning is probably the water formed, and then firmly bound, by the interaction of the H-layer and O_2 at high temperature, and that this water hinders the development of the active H-layer necessary for hydrogenation.

By arranging the experimental observations the first fact to be noted is that at room temperature H adsorption corresponding to the maximum *A* of the isobar predominates, and hence the process is reversible, even at room temperature.

It soems evident to assume in this process an electron transfer (direct or indirect) to the localized vacant acceptor level of the ZnO:

$$(H_2)_g + (Zn^{2+}, O^{2-}) \rightleftarrows Zn^{2+} \ldots H + O^{2-} \ldots H$$

$$O^{2-} \ldots H \rightleftarrows O^{2-}eH^+$$

leading to a relatively weak proton-like bonding. As the acceptor levels are gradually filled, thermal ionization begins even at room temperature, lifting the localized electrons into the conduction band:

$$O^{2-}eH^+ \rightarrow OH^- + e$$

This process, of course, is accompanied by electrical consequences.

In the ZnO lattice the protons can be displaced at room temperature, and as the temperature increases such movements become even more likely. The migration of protons inside the lattice is probably a slow and activated process, which might be one of the reasons for the decreased sorption rate after the cessation of the rapid surface processes.

This train of thought leads to the conclusion that as regards their chemical properties the OH^- ions are not equivalent. Up to the time when the conduction electron occupying the acceptor level passes into the conduction band of the ZnO, the chemisorbed H atom remains active. At higher temperatures and under non-equilibrium conditions the surface coverage of this active hydrogen may be low, from which, however, it does not necessarily follow that, for example, the rate of olefine hydrogenation must also drop.

Dent and Kokes [148] suggest another alternative, but this ignores the conclusions resulting from the adsorption isobar and from the changes in conductivity. Essentially, it is as follows:

The introductory act of chemisorption is the rapid

$$(H_2)g + \underset{|}{-Zn}-\underset{|}{O} \rightleftharpoons \underset{\underset{H}{|}}{-Zn}-\underset{\underset{H}{|}}{O}-$$

process occurring on active centres of the type $\underset{|}{-Zn}-\underset{|}{O}-$ available in relatively limited numbers. (In agreement of what was said before this step should lead to electric consequences.) This process in followed by a slow proton migration according to the equations:

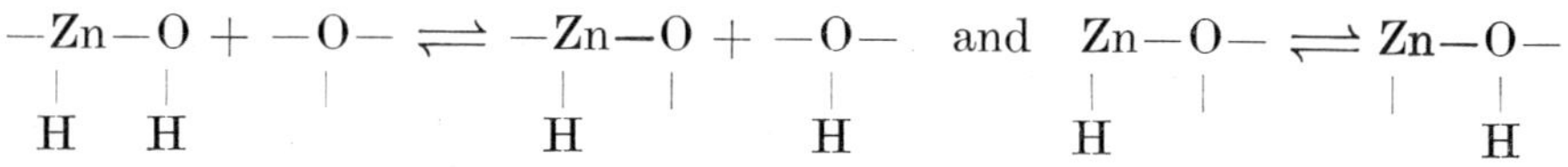

From this state H-desorption is presumably slow, for the recombination of H atoms, which migrated far away one from another, is quite improbable prior to desorption. It is reasonable to assume that OH^- ions have no hydrogenating activity, therefore, this reaction is due to H atoms chemisorbed on Zn^{2+}:

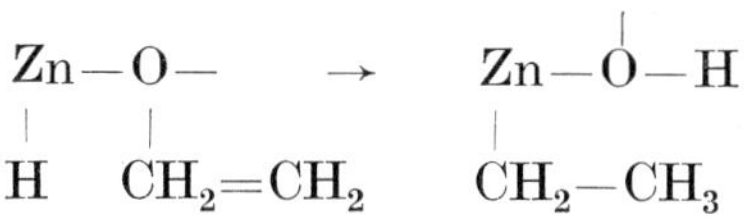

The last step, as suggested by Dent and Kokes, however, cannot be fully understood:

$$\underset{\underset{\text{Zn}-\text{O}}{\overset{\text{CH}_3}{|} \atop \overset{\text{CH}_2}{|}}}{} + -\underset{\underset{\text{H}}{|}}{\text{O}}- \rightarrow \underset{\underset{\text{Zn}-\text{O}}{\overset{\text{CH}_3}{|} \atop \overset{\text{CH}_2}{|} \overset{\text{H}}{|}}}{} + -\underset{|}{\text{O}}- \rightarrow \text{Zn}-\text{O}-+-\text{O}-+\text{C}_2\text{H}_6$$

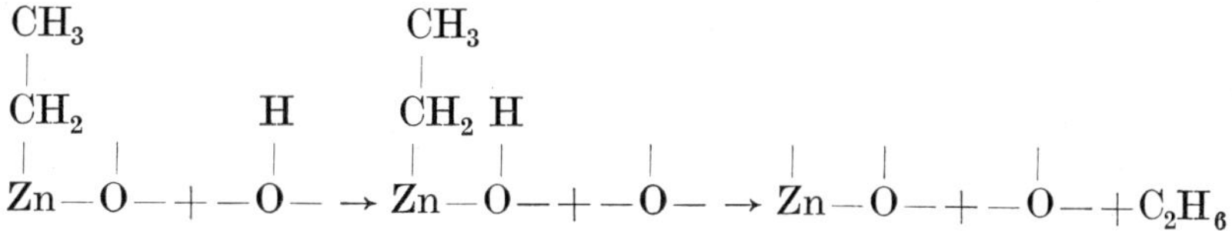

because here the OH^- ion would display its hydrogenating action.

The fundamental deficiency of both alternative suggestions is that they leave the reader in doubt about the nature of the bond of the H chemisorbed on the Zn^{2+} and O^{2-} ions. The —OH band observed at 3510 cm^{-1} in the infrared absorption spectrum is probably not identical with the absorption band at 3750 cm^{-1} found for isolated —OH groups on silica gel surfaces. The spectrum on Aerosil (Borello et al. [149]) also shows a well-developed band at 3500 cm^{-1}, but this is not assigned. The situation is similar with regard to the H band on Zn^{2+}.

The ESR investigations of Sedaka and Kwan [150] suggest the chemisorption of O_2 at room temperature according to the mechanism

$$O_2 + e \rightarrow O_2^-$$

which explains the drop in the number of conduction electrons, and hence the semiconductivity.

Beaufils [151] put forward an acceptable mechanism for the combustion of the H-layer, but proton migration, too, may result in the formation of water, split off, for example, from two neighbouring —OH groups at high temperatures as in the schematic reaction:

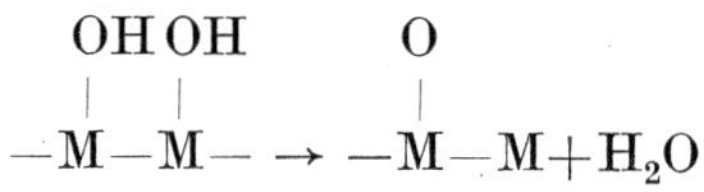

This example supports the arguments given in the introduction, that the determination of the rate of chemisorption on oxides in itself reveals little about the complex processes taking place on the surface of the adsorbent.

Information on other adsorption systems, without any claim to completeness, can be found in the references given in Table II.10.

TABLE II.10

Adsorbate	Adsorbent	References
H_2, D_2	Cr_2O_3, ZnO	[117], [152], [146]
H — D exchange	Co_3O_4	[153], [147], [154]
	NiO	[152]
	ZnO	[155]
^{16}O — ^{18}O exchange	ZnO	[156]
CO_2	$ZnO \cdot Cr_2O_3$	[7]
	$ZnO \cdot Cr_2O_3$	
	Al_2O_3	[140], [142], [141]
CO	Al_2O_3	[144]
	ZnO	[7]
O_2	ZnO	[157], [150], [158] [159], [160]
N_2O	NiO	[161], [162]

SECTION 7.2. EARLY INVESTIGATIONS OF THE RATE OF CHEMISORPTION ON POWDERED METALS AND METAL FILAMENTS

Besides measurements on oxides, the investigation of chemisorption on metal powders contributed considerably to the formulation of the concept of activated adsorption and to the clarification of the problems involved. For reasons which will be apparent later, with the exception of informative work on the chemisorptive properties of some industrially important catalysts [163], this trend of research has been abandoned.

There is an extensive literature relating to adsorption on the surface of liquid mercury, probably because the surface area of mercury is easy to measure and though this surface, and, thus, also the adsorbed quantity is generally small, indirect methods (e.g. based on the measurement of surface tension [90, 164, 165]) provide a fairly reliable basis for its calculation. Since these studies have no direct bearing on heterogeneous catalysis, they will not be discussed in detail.

The first studies of chemisorption on metal filaments date back to the 1930's, to the work of Taylor and Langmuir, and later Roberts [81]. Roberts used tungsten filaments heated in vacuo as the adsorbent and determined volumetrically (see Part VI, Paragraph 5.1.2) the adsorption of the two adsorbates, hydrogen and oxygen. He supplemented these experiments with the determination of the calorimetric heat of adsorption and of the accommodation coefficient of neon.

According to earlier investigations (Frankenburger et al. [166, 167]), the chemisorption of H_2 and N_2 is activated on tungsten powder which has been subjected to preliminary reduction at 750 °C and purification by evacuation. The hydrogen isobar has a minimum at 75 °C and a maximum at 150 °C.

Fundamentally different results were obtained when H_2 was adsorbed on a tungsten filament. Via the continuous measurement of the accommodation coefficient of neon during adsorption, Roberts et al. concluded that the adsorption process takes place at room temperature (just as it also does at the temperature of liquid air) in a few minutes, without being followed by more protracted, slow adsorption.

In order to decide whether this rapid process is physical adsorption or chemisorption, hydrogen was applied in several consecutive small doses to the surface of the filament and, in the meantime, the heat of adsorption of each small dose was determined.* Addition of the gas was continued till it was possible to measure the residual pressure in the system. (The data in Table II.11 illustrate the method.) Since the heat of adsorption of hydrogen in the case of physical adsorption is at most -2 kcal mole^{-1}, the data in Table II.11 suggest that the rapid process is a chemisorption and not a physical adsorption process.

TABLE II.11

The adsorption of hydrogen on a tungsten filament

Dose	Number of adsorbed H_2 molecules 10^{-14}	Heat of adsorption kcal mole^{-1}
First experiment		
1st	1.19	—34.2
2nd	1.18	—28.5
3rd	1.17	—20.3
4th	0.84	—17.5
After 4. dose pressure became measurable		
Second experiment		
1st	0.99	—
2nd	0.98	—34.2
3rd	0.97	—31.1
4th	0.96	—25.3
5th	0.51	—22.7
After 5. dose pressure became measurable		

Number of adsorbed molecules in the first experiment:

$4.38\ 10^{14}$

Number of adsorbed molecules in the second experiment:

$4.41\ 10^{14}$

Interesting conclusions can be drawn from the quantity of chemisorbed gas, too. Johnson [168] found that on the surface of tungsten filaments

* In these tests the filament itself was used as the calorimeter. Changes in filament resistance indicated the degree of heating, from which, provided the heat capacity was known, and by making the corrections usual in calorimetry, it was possible to calculate the heat of adsorption.

aged by heat-treatment the (110) and (100) faces dominate. On these surfaces the number of tungsten atoms per cm^2 is 14.24 10^{14} and 10.07 10^{14}, respectively, i.e. on the average 12.15 10^{14} (assuming that the frequency of incidence of both faces is the same). The apparent surface area, of the filament in this experiment was 0.55 cm^2, and the roughness factor (the ratio of real to geometric surface areas) of the aged tungsten filaments was 1.4 (Taylor and Langmuir [77]). From these data the number of surface atoms of the filament was 9.3 10^{14}, while, from several experiments, the average number of adsorbed hydrogen atoms was obtained as 8.6 10^{14}.

In the light of these arguments and of the experimental data there is very reason to assume that the adsorption of H_2 on tungsten surfaces proceeds according to the "reaction" equation

$$2\,W + H_2 \rightarrow 2\,WH$$

where W is a surface atom.

It follows unambiguously from the above that rapid chemisorption covers practically the entire filament surface with a monoatomic hydrogen layer.

In his hydrogen chemisorption experiments [81] Roberts used the apparatus shown in Fig. II.31.

An 18 cm long tungsten filament 0.0066 mm in diameter was placed into tube A. Neon at 0.1 torr was circulated above the filament by means of a diffusion pump. Oxygen traces in the H_2 under investigation and in the circulating Ne were removed by traps B, B, which were filled with activated carbon and cooled with liquid air. Mercury vapour was removed by similar traps between the MacLeod manometer and the apparatus.

The filament, which had been subjected to preliminary heating and evacuation, was reheated for a short time to a temperature above 2000 °C under constant neon circulation in order to remove surface contamination. After cooling, the filament was connected to a Wheatstone bridge and a current passed through it so that the temperature of the filament (T_2, determined by measurement of the resistance) was about 20 °C higher than the ambient temperature, namely, the carefully controlled temperature T_1 of the oil-bath. From the continuous recording of the resistance of the filament and from the current intensity, the Joule heat q (cal $cm^{-2}\,s^{-1}$) removed from the surface by the circulating neon could be accurately determined*. With the Joule heat available the accommodation coefficient, α, can be calculated from the equation [169]:

$$q = 1.74\ 10^{-4}\frac{\alpha p}{\sqrt{MT_1}}(T_2 - T_1)$$

where M is the molecular mass of neon,
and p is the partial pressure of neon in dyne cm^{-2}.**

* Under these conditions the heat losses due to convection and radiation, as well as those at the terminals of the metal filament, are practically negligible.

** The mean free path of neon must be greater than the diameter of the filament.

If the accommodation coefficient was measured after the application of a hydrogen dose of about 10^{-4} torr partial pressure (in the experiments described in Ref. [81]) the results presented in Fig. II.32 were obtained.

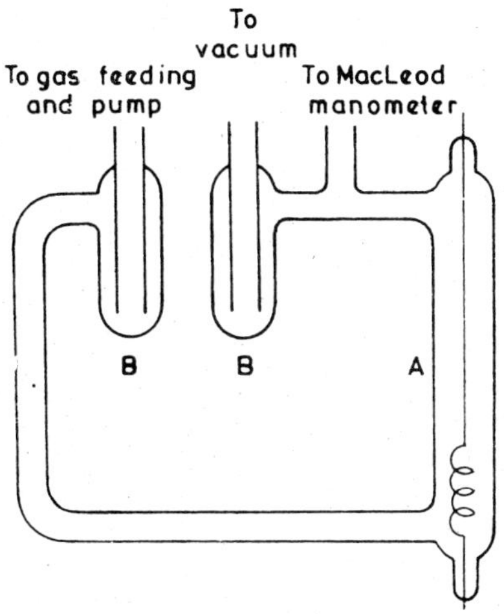

Fig. II.31. Apparatus for the continuous measurement of the accommodation coefficient

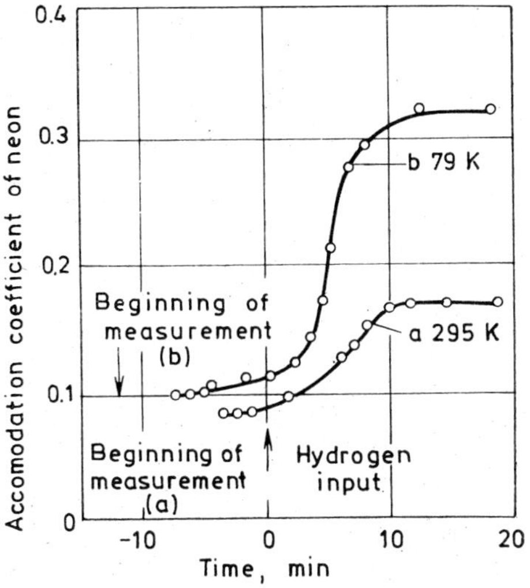

Fig. II.32. Adsorption of hydrogen on tungsten filament

The results obtained with oxygen were quite similar. On the same metal filament 4.9 10^{14} molecules of oxygen were adsorbed at room temperature with a heat of adsorption between -155 and -212 kcal mole^{-1}. The process is again described by the equation

$$2\ W + O_2 \rightarrow 2\ WO$$

Apart from slight modification the chemisorption of oxygen was measured in an essentially similar apparatus*. The modification involved, among other things, the maintenance of a steady partial oxygen pressure of about 2—5 10^{-9} torr in the vicinity of the filament surface by a constant oxygen input. At a higher pressure the adsorption rate would have been immeasurably rapid, and at a non-steady pressure not the true chemisorption rate, but the quantity of oxygen transported by diffusion and chemisorbed on the filament surface, would have been determined.

The results of an experiment at 2.3 10^{-9} torr partial oxygen pressure

* It was not possible to purify the by gas the same method as above, since oxygen would have been removed from the gas.

are shown in Fig. II.33. Roberts calculated the sorption rate from the $\alpha(t)$ curve by assuming a linear relationship between θ and α:

$$\theta = \mathrm{const}(\alpha - \alpha_0)$$

and hence the derivative of the $\alpha(t)$ curve with respect to time was proportional to the sorption rate.

Today, we can no longer completely agree with Roberts' method of evaluation, because the sticking probabilities calculated from his data are higher than the true values.

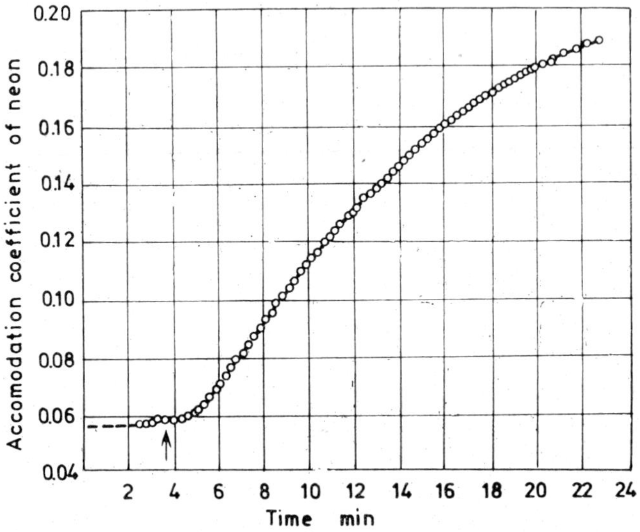

Fig. II.33. Adsorption of oxygen on a tungsten filament
↑ : beginning of experiment

Besides the accommodation coefficient, the change in the work function is also a true reflection of the state of the metal surface. Taylor and Langmuir were the first to apply this principle to the study of the rate of sorption of Cs vapour on a tungsten filament [77]. Since then, measurement of the work function has become one of the most important tools in the study of sorbed layers, so that it is worthwhile to survey briefly the theoretical considerations involved.

The work function, $e\Phi$, is the work necessary to remove an electron from the highest filled conducting level *in vacuo*. Its value depends on the crystallographic orientation of the surface (thus, on a polycrystalline surface some average value of the work functions of the single crystal faces can be observed), and also on the state of the surface. Since adsorption often leads to the development of surface dipoles, the presence of a foreign substance either raises, (if the negative ends of the dipoles are oriented away from the surface, i.e. if the layer is electronegative), or reduces, the value of the work function (when the layer is electropositive). In the course of physical adsorption, or when a surface layer with covalent bonding is formed, there is practically no change in the work function, so that from the sign of $e\Delta\Phi$ qualitative statements can immediately be made as to the state of the adsorbed layer. Moreover, in favourable cases, information is also gained which relates to the quantity of substance sorbed. If the double layer formed by adsorption is considered a capacitor, then the potential change in the layer will be:

$$\Delta V = 4\pi\, n_s\, \theta\, dq = 3.76\ 10^{-15}\, n_s\, \theta\mu \tag{II.175}$$

where n_s is the number of adsorption sites per unit surface area

- θ is the coverage
- d is the distance of the centre of the adsorbed atom from the surface
- q is the charge and
- μ is the dipole moment in D units.

However, ΔV agrees with the change of the work function, $\Delta\Phi$. When $\Delta\Phi$ and $n_s\theta$ are known, μ can be determined from Eq. (II.175). If μ is independent of θ, $\Delta\Phi$ will be directly proportional to θ.

The greatest importance of the experimental determination of the work function is its application to the study of chemisorbed layers on metals [filaments, tips (field emission and ion emission microscopy) foils, films].

There are several possibilities of measuring $\Delta\Phi$. The most important of these are the thermal emission, photoelectric and contact potential methods, and field emission microscopy.

The thermal emission method is based on the finding that the intensity of the current produced by electrons escaping from unit surface area of a solid of temperature T can be given by the relation:

$$i = C(1-\bar{r})\, T^2 \exp(-e\Phi/kT) \tag{II.176}$$

where C is the universal constant 120 amp cm^{-2} K^{-1}. Plotting $\ln(i/T^2)$ vs. $1/T$ in accordance with Eq. (II.176) we obtain the so-called Richardson line (the correction function depends only slightly on T), with slope $e\Phi/k$.

This method is not suited to the investigation of electronegative layers, because measurable saturation currents occur only at high temperatures where the structure of the adsorbed layer already undergoes extensive changes. For electropositive alkali metals, however, i is already satisfactorily measurable at around 150 – 200 °C.

For cold solids, between the frequency of light, ν, and the energy E of the escaping photoelectrons the following approximate proportionality holds:

$$E \sim (\nu - \nu_0) \tag{II.177}$$

The following relation exists between the limiting frequency, ν_0 (which is the lowest frequency at which an emission current still appears) and the work function:

$$e\Phi = h\nu_0$$

where h is Planck's constant.

If the work function exceeds $4-5$ eV, the frequency ν_0 will be in the far ultraviolet where accurate measurements are fairly difficult to carry out. In addition to the alkali metal layers with their low work functions, the method can be satisfactorily applied to vacuum-evaporated Cu and Ni layers too.

The principle of the method of measuring the contact potential is as follows: when two metals of a different nature, but at the same temperature, are in contact with each other in a common vacuo, electrons will transfer from the metal in whose filled conduction band the electron energy (and hence the Fermi level) is higher, into the other metal till this process is stopped by the created electrostatic potential difference. At equilibrium the condition

$$\Phi_1 + \Delta V_1 = \Phi_2 \tag{II.178}$$

must be fulfilled, where

Φ_1 is the work function of the metal with the higher electron energy;

Φ_2 is the work function of the other metal, and

ΔV_1 is the contact potential difference.

Let us assume that as a result of adsorption the work function of metal "1" changes to Φ_1' (let us suppose that metal "2" is a noble metal, and thus Φ_2 does not change). At equilibrium:

$$\Phi_1' + \Delta V_1' = \Phi_2 \tag{II.179}$$

By subtracting Eq. (II.178) from Eq. (II.179) we obtain:

$$\Phi_1' - \Phi_1 = \Delta\Phi = \Delta V_1 - \Delta V_1' \tag{II.180}$$

that is, the change in the contact potential difference is related to the change in work function due to the effect of adsorption. For the measurement of contact potential differences the most widely accepted technique is that of using a vibrating reed electrometer (see Paragraph 8.1.3).

Due to the different work functions of the faces of various crystallographic indices, in field emission microscopy it is possible to obtain a 10^5-10^6 times magnified picture of a metal tip (i.e. of the materials sorbed on it) having a radius of curvature of 1000–3000 Å.

The method was described briefly in Paragraph 5.2.1, and will now be supplemented by some theoretical considerations.

To leave the metal, electrons with the Fermi energy have to overcome a potential barrier $e\Phi$. When the necessary energy originates from a thermal source, we are dealing with thermal emission, of which it has already been mentioned that the emission current is proportional to $\exp(-e\Phi/kT)$. In field emission, by the application of an external potential the width of the potential barrier can be reduced to such an extent that by means of the tunnel effect cold emission will take place. This latter can be described with the help of the Fowler-Nordheim equation [170]:

$$i = a\, V^2 \exp(-b\, \Phi^{3/2}/V) \quad \text{(ampere)} \tag{II.181}$$

where

$$a \equiv 1.54\ 10^{-6}\ A/\Phi(5r)^2$$

$$b \equiv 6.8\ 10^7\ v(y)5r$$

A is the area of the emitting surface
V is the potential difference between the tip and the screen
r is the radius of curvature of the emitting tip
$v(y)$ is the elliptic integral of the variable
$y = 3.79\ 10^{-4}\ (V/5r\Phi)$, for which, in most cases, the inequality $0.8 < v(y) < 1.0$ is fulfilled.

The Φ value calculated from the Fowler-Nordheim equation is just as much a mean value as those calculated by any of the other methods. It can be proved that if the work function of face k is Φ_k, then

$$\Phi^{3/2} = \sum_k a_k \Phi_k^{3/2} \qquad \text{(II.182)}$$

However, it is possible to measure the work functions of individual crystal faces, for example, by using the apparatus constructed by Müller [171] based on the principle of leading the electron current through the sample opening of the screen into a Faraday-cage or an electron multiplier, whereby it becomes directly measurable. Such a technique might be necessary when for certain crystallographic faces the change in the work function is of opposite sign to the change on others.

In more exact studies the need may arise for the more accurate description of the shape of the potential barrier, and consequently of the probability of electron transfer, by means of the Wentzel-Kramers-Brillouin method [170], which may modify the exponential factor in the Fowler-Nordheim equation.

The field emission microscope is suitable for the measurement of relative adsorption and desorption rates when, in accordance with Eq. (II.175), the change in the work function, $e\Delta\Phi$, the change in the coverage, $\Delta\theta$, and the change in the brightness of the investigated crystal face on the screen of the microscope, or in Müller's arrangement, the change in the electron current, are proportional. Up to not too high coverages this condition is usually fulfilled. (For instance, Schmidt and Gomer [272] observed a linear decrease of the work function under the effect of metallic potassium adsorbed on the surface of the tungsten tip from 4.5 eV at coverage $\theta = 0$ to 2 eV at $\theta = 0.6$.)

Investigation of the rate of adsorption of metal vapours by means of the above methods revealed very high sorption rates, up to a coverage of $\theta = 0.98$ the sticking probability, s, being 1 [77]. At higher coverages a second, mobile, layer may also develop and, starting from this precursor state, the metal atoms, by surface migration, reach the still unoccupied vacant adsorption sites.

SECTION 7.3. EXPERIMENTS ON VACUUM-EVAPORATED METAL FILMS

The various experiments on vacuum-evaporated metal films, extending to the thorough experimental determination of adsorption, rate of adsorption, surface structure, the value of the work function, catalytic activity, etc., and continued in up-to-date research, have added important details to our earlier and current knowledge obtained with metal filaments and ribbons. Because of the greater surface area of films, the danger of contamination is not so marked, enabling the use of less efficient gas purification methods (traps cooled not with liquid air, but with dry ice are satisfactory), and thus of readily condensing adsorbates (e.g. NH_3, C_2H_4, etc.). There are no problems concerning the investigation of the structure of a film by X-ray diffraction and the subsequent measurement of the work function (e.g. by the photoelectric method). Heats of chemisorption, though fairly low because of the adsorbent surface of not more than a few cm^2, are still of an order which can be measured satisfactorily with sensitive resistance thermometers. The situation is similar as regards the rates of catalytic reactions.

In connection with the investigation of the rate of chemisorption on films, it seems a more or less generally valid conclusion that if chemisorption does take place at a given temperature, it usually proceeds at a high rate; at higher coverages, however, these rapid processes are associated with a slow activated process(es). Without any claim to completeness, the results of some typical studies by Allen and Mitchell [172], Beeck et al. [43, 173, 174], Kemball [175, 176] and Trapnell [177] are summarized in Table II.12.

TABLE II.12

Rates of chemisorption on metal films

Gas	Non-activated chemisorption	Activated chemi-sorption	No chemisorption below 0 °C
H_2	W, Ta, Mo, Ti, Zr, Fe, Ni, Pd, Rh, Pt, Ba	—	Cu, Ag, Au, K, Zn, Cd, Al, In, Pb
CO	the same as for H_2	Al	Zn, Cd, In, Sn, Pb, Ag, K
C_2H_4	the same as for H_2 + Cu, Au	Al	the same as for CO
C_2H_2	the same as for H_2 + Cu, Au	Al	the same as for CO, but for K
O_2	all metals with the exception of Au	—	Au
N_2	W, Ta, Mo, Ti, Zr	Fe	the same as for H_2 + Ni, Pd, Rh, Pt
CH_4	—	Fe, Co, Ni, Pd	—

It can be seen from the Table that the number of processes whose activated nature has been proved is small. In this respect the Fe—N_2 system is a classical example, just as is the chemisorption of saturated hydrocarbons. It seems an evident suggestion that the necessity of splitting the $\sigma - \pi$ bond of the N_2 molecule, or of the C—H σ-bond in the case of saturated hydrocarbons, hinders chemisorption. In fact, the situation may be more complicated, since even the vacuum-evaporated layers have polycrystalline structures (though not to the same degree as powders, and, moreover, in the presence of gas additives evaporation may lead to the preponderance of certain crystallographic faces [178]). Hence, it is difficult to decide which of the crystal faces are involved in the slow process and which in the rapid process, and for what reason. This question can only be answered by tests on single crystals or by field emission microscopy, when the behaviour of several crystal faces can be observed simultaneously and directly (see Paragraph 8.1.3).

It is difficult to find an interpretation for the data in the third column. One possible explanation for the absence of chemisorption might be found in the high activation energies of these processes, which can therefore take place at a measurable rate only at temperatures considerably above 0 °C. It is possible, however, that they are endothermic processes, so that they encounter thermodynamic obstacles.

Those of the slow processes which follow the rapid chemisorption and which take place in the presence of O_2 can unequivocally be attributed to further slow oxidation, while in other cases adsorption or the formation of some compound might be suspected. Beeck, Ritchie and Wheeler [179] found, for instance, that in the H_2 — Ni system the quantity of H_2 adsorbed in the slow process is independent of the sintering temperature of the film (i.e. of the size of the surface) and is proportional to the quantity of deposited metal; facts typical of absorption. More thorough investigations, however, have proved that there is hardly any difference in the ratio of the quantities of H_2 absorbed by the rapid and slow processes, despite a decrease in the absolute surface area to about 1/15 of the initial value [180]. All this seems to confirm that the slow process is also a surface phenomenon.

SECTION 7.4. MODERN INVESTIGATIONS ON METAL FILAMENTS AND RIBBONS

There is usually a difference between the sticking probabilities of a given adsorbate on polycrystalline filaments and on ribbons made from metals of approximately the same quality (see later). This can be explained, among other reasons, by the fact that on the surface in contact with the gas phase the probabilities of incidence of faces of different crystallographic indices are not the same. Electron diffraction and X-ray tests on tungsten and molybdenum ribbons have shown, for instance, that the frequencies of incidence of the (311), (411) and (100) faces are the highest [104], [181]. These results illustrate the extremely important fact that, without crystallographic information regarding the filament or ribbon, it is difficult to draw comparisons between the studies of different authors on otherwise chemically identical systems. The successful interpretation of experimental results obtained on identical metal filaments or ribbons with different adsorbates demands, as an indispensable condition, a knowledge of the surface structure.

In connection with the properties of surface layers (Paragraph 8.1.3), mention is made of the possibility of the simultaneous presence of several chemisorbed layers, generally possessing structure, on well defined crystal faces. Presumably each of them is the result of different (and therefore sensitively temperature-dependent) interactions resulting in chemisorbed layers of different structures. At low temperatures (perhaps on already formed chemisorbed layers) physical adsorption, too, necessarily occurs; as from a precursor state, the molecules may pass from this into the chemisorbed state, without this fact being detectable, e.g. by a change in the gas pressure.

A kinetic description allowing for all these conditions must therefore:

(1) take into consideration the differences in adsorption rate on the various crystallographic faces, including the special effects of the structure of the adsorbed layer and the different temperature-dependences of the rates of sorption on the various crystal faces;

(2) be capable of interpreting, in accordance with the points listed in (1), the sticking probability curve ($s \leftrightarrow \theta$) observed experimentally on a polycrystalline substance and its temperature-dependence;

(3) account for the desorption of simultaneously present layers of different properties formed on crystal faces with different and/or identical indices, and the temperature-dependence of the desorption.

With this, however, the subject is far from being exhausted, since, in addition, such a description must be capable of explaining, more or less independently of the assumed kinetics, the surface potentials formed on the various crystal faces as a result of adsorption and in relationship to the other properties of the layers.

In the past 10 or 15 years experimental work on metal filaments has centred round the measurement of the s vs. θ curves with the highest possible reliability.

Figures II.34 and II.35 (Becker and Hartman [182], Pasternak and Wiesendanger [183]) show the changes in sticking probability vs. coverage and temperature for N_2 adsorbed on tungsten and H_2 adsorbed on molybdenum. Apart from minor differences, all known systems reflect the same characteristics as those illustrated in the figures. For instance, the lower the temperature the higher the coverage up to which extends the section of the curve characterized by a constant value of s. The temperature-dependence of the s value extrapolated to $\theta = 0$ is, in general, not a typical property. The sticking probability of H_2 on molybdenum is constant in the temperature range 225—400 °C, $s \approx 0.35$ [183]. Proceeding towards higher coverages the curves suddenly begin to fall. A later flattening off of the curve appears only for N_2.

Sticking probabilities extrapolated to zero coverage for the same chemisorption system and temperature vary within wide limits. For the system $W - N_2$, $0.03 < s_0 < 0.55$, while for $W - CO$, $0.18 < s_0 < 0.62$. $s_0 \approx 0.3$ holds approximately, however, for almost all cases (we have seen that for metal vapours $s_0 \approx 1$).

Based on the abundant experimental material at their disposal, Hayward and Trapnell [85] made a detailed analysis of the reasons (e.g. the thermal energy of the molecule colliding with the surface, orientation of collision, angle of approach, momentum of collision and finally the site of collision) which might be responsible for s_0 values other than 1 for H_2. They arrived at the conclusion that at temperatures around 2500 K the site of collision is the only determining factor. Thus, at certain sites $s_0 = 1$, while at others $s_0 = 0$. Hence, in systems in which the sticking probability is independent of the temperature, s_0 is equal to the active fraction of the surface.

The conclusions arrived at by Hayward and Trapnell can probably not be generalized to chemisorption at low temperatures. It is far more probable that on various crystal faces the sticking probability may have values other than 1 and 0. With this assumption Smith [184] was able to produce an excellent description of the shapes and temperature-dependence of sticking probability curves. The features of the $W-N_2$ system shown in Fig. II.34 can be interpreted on this basis by assuming that the initial high rate is the contribution of many rapidly sorbing faces, while the protracted "tail" is the contribution of faces exhibiting low rates.

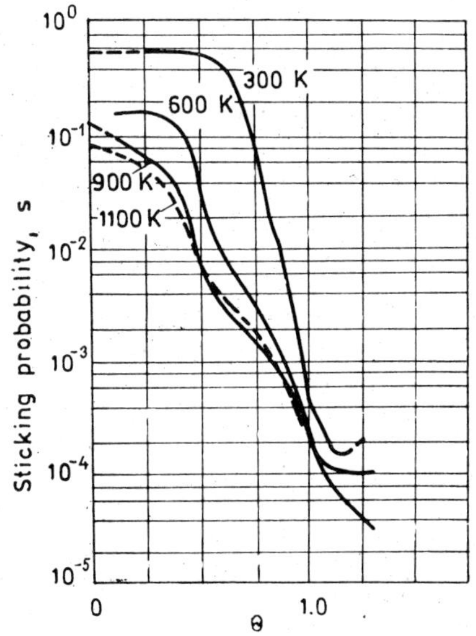

Fig. II.34. Dependence of the sticking probability of N_2 on tungsten upon the coverage (after Becker and Hartmann)

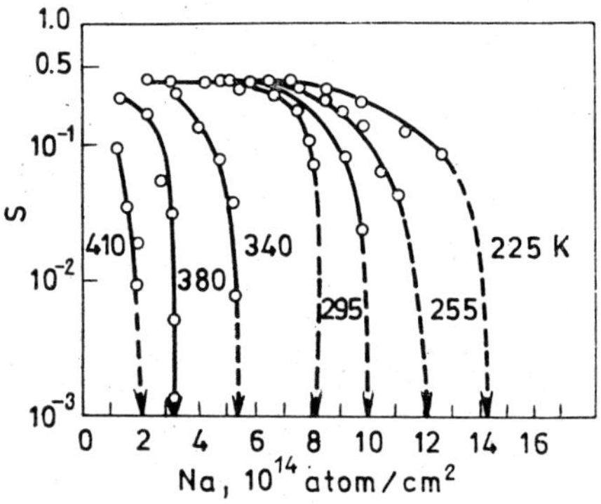

Fig. II. 35. Sticking probability (s) of H chemisorbed on Mo vs. sorbed quantity (after Pasternak and Wiesendanger)

According to Robins et al. [128] the differences between individual crystal faces may be caused by differences in the number of metal atoms neighbouring the chosen adsorption site. Experiments with the field emission microscope have provided evidence (see also Paragraph 8.1.3) that the (110) face of tungsten does not adsorb N_2; this can be explained by the conspicuously low surface atomic density of this face.

Due to the characteristic interaction energy, the atomic density of the surface may play a decisive role not only in the dissociation of the colliding molecule, but also in the lengths of the paths which the excited molecular fragments are capable of covering on the surface before becoming bound in an appropriate potential well corresponding to an adsorption site. There is, as yet, no direct evidence of a higher dissociation rate of faces with higher atomic densities. It is more likely that surfaces broken up by edges or steps fulfil a particular role in this respect, but the probability of bonding is certainly higher on faces of high atomic density. All things considered, differences in sticking probability are determined by the relative rates of dissociation and bonding; the formation of many fragments capable of adsorption on a certain face is of no use if the diffusion path on this face is long, so that sufficient time is available for recombination and subsequent desorption of the molecule, or for the migration of the excited molecular fragments to another crystal face. Hence, the populations of faces with adsorbed atoms are determined primarily not by the differences in bond energies, but by the kinetics of adsorption and by surface diffusion.

The site of the potential well (whether it is situated above or below the local surface) and the electronegativity of the adsorbed atom play important roles in the final surface charge density and thereby in the sign and value of the change in work function.

Ehrlich (loc.cit.) claims that the desorption or final chemisorption of the molecule in the precursor (physically adsorbed or weakly chemisorbed) state depends on the shape of the potential curves characterizing the interactions (see notes to Fig. II.22). If the activation energy of desorption from the precursor state is $E_{\mathrm{d}}^{\ddagger}$ and that of the transition into the final chemisorbed state is $E_{\mathrm{a}}^{\ddagger}$ then the sticking probability is:

$$s = \frac{\sigma f(\theta) \exp(-E_{\mathrm{a}}^{\ddagger}/RT)}{\exp(-E_{\mathrm{d}}^{\ddagger}/RT) + \sigma f(\theta) \exp(-E_{\mathrm{a}}^{\ddagger}/RT)} =$$

$$= \{1 + \exp[(E_{\mathrm{a}}^{\ddagger} - E_{\mathrm{d}}^{\ddagger})/RT]\,[\sigma f(\theta)]^{-1}\}^{-1}$$

Assuming a mobile transition complex $0.1 < \sigma < 1$; for uncovered surfaces $f(\theta) = 1$, so that for $s_0 \approx 1$ it is necessary that

$$\exp[(E_{\mathrm{a}}^{\ddagger} - E_{\mathrm{d}}^{\ddagger})/RT] < 0.1\sigma$$

For this condition to be satisfied, it is necessary that $E_{\mathrm{a}}^{\ddagger}$ be lower than $E_{\mathrm{d}}^{\ddagger}$ by about 1.5 kcal mole^{-1} at $\sigma = 1$, and about 3 kcal mole^{-1} at $\sigma = 0.1$. It also appears from the above relation that when $E_{\mathrm{d}}^{\ddagger}$ is greater than $E_{\mathrm{a}}^{\ddagger}$ at sufficiently low temperatures all molecules must be chemisorbed.

SECTION 7.5. RESULTS OF THE EXPERIMENTAL DETERMINATION OF DESORPTION RATE

It was mentioned in the preceding discussion of sorption rates that under appropriate experimental conditions (e.g. continuous evacuation of the gas space) it is possible to measure desorption rate separately from adsorption. When the merits of the Garner method (loc. cit) were pointed out, the difficulties in preventing re-adsorption on porous adsorbents were also stressed. Further, Garner's thermodynamic arguments that in non-activated processes the activation energy of desorption agrees with the heat of adsorption might also be erroneous, as the displacement of the chemisorbed atoms (immediately after adsorption or at the higher temperature necessary to cause desorption) cannot be excluded, and there is no guarantee that desorption will take place with a substance of the same state and from the same crystal face as in the adsorption.

The greatest value of the desorption tests — particularly since the introduction of the method of flash desorption — is the possibility they offer for the detailed investigation of simultaneously existing adsorbed layers on the adsorbent surface.

The principle of the method is as follows: the metal filament is brought at low temperature and at a certain partial pressure of the adsorbate into the state of adsorption equilibrium, and then heated according to a given temperature programme; the appearance of the desorption products in the gas space is recorded. When the gas is constantly withdrawn pressure peaks are obtained in the different temperature ranges, but when the gas space is evacuated only prior to the heating of the filament, an integral effect of the desorption peaks appears.

The properties, detected by flash desorption, of chemisorbed layers on various adsorbents will be dealt with in detail in Paragraph 8.1.3, and here only those results will be discussed which are related to the kinetics of the desorption process.

The most detailed investigations of this type were performed on the $W - N_2$ system [128], [182], [185], [186]. It was found that at room temperature chemisorbed nitrogen forms two distinct phases, α and β. At lower temperatures an additional third phase, γ, also appears. It was earlier assumed that the atomic β phase is desorbed by second-order kinetics, but Oguri demonstrated the splitting of the β peak into two (β_1 and β_2) peaks, the β_1 desorption being described by first-order, and the β_2 by second-order kinetics [187]. According to Madey and Yates the desorption of the β_2 peak follows a fractional order greater than one [188].

As regards the energy of adsorption, the bond energy of the α phase is 20 kcal mole^{-1} (desorbed by first-order kinetics), while those of the β_1 and β_2 are 81 and about 85 kcal mole^{-1}, respectively.

The atomic bonding of the β phases is supported by the complete isotopic mixing of ^{14}N and ^{15}N prior to desorption.

What has been said so far might suggest that the β_1 and β_2 phases develop on two different crystal faces. Though this possibility may also exist, there is evidence of the common presence of both phases, e.g. on the (111) face.

Starting from the work of Robins et al. [128], let us consider how to formulate, in agreement with the experimental results, the kinetics of desorption on a single crystal face.

Robins assumes that in the $W - N_2$ system the formation and decomposition of diatomic molecular complexes on the surface is a relatively rapid equilibrium process by comparison with the true desorption. If c_a (mole cm^{-2}) is the concentration of adsorbed atoms and c_m (mole cm^{-2}) that of the diatomic molecular complexes, the rate of formation of the latter is:

$$\overrightarrow{J}_m = k_1 c_a^2 \exp[-(Q_d^\ddagger + Q_m^\ddagger)/kT] \quad (\text{mole cm}^{-2}\,\text{s}^{-1})$$

where $Q_d^\ddagger$ is the activation energy of the surface diffusion
$Q_m^\ddagger$ the activation energy of the complex formation,
and the rate of decomposition is:

$$\overleftarrow{J}_m = k_2 c_m \exp(-Q_D^\ddagger/kT) \quad (\text{mole cm}^{-2}\,\text{s}^{-1})$$

where $Q_D^\ddagger$ is the activation energy of the dissociation.

At equilibrium $\overrightarrow{J}_m = \overleftarrow{J}_m$, and hence the equilibrium surface concentration of the molecular complexes is:

$$c_m = \frac{k_1}{k_2} c_a^2 \exp[-(Q_d^\ddagger + Q_m^\ddagger - Q_D^\ddagger)/kT] \approx K c_a^2 \exp(-Q_d^\ddagger/kT) \quad (\text{II.183})$$

[since according to experimental results the difference between the activation energies of desorption of the adsorbed atom and of the molecular complex, $(Q_m^\ddagger - Q_D^\ddagger)$, is much smaller than $Q_d^\ddagger$].

Calculated in moles of N_2, the total quantity of sorbed N_2, c_t (mole cm^{-2}), on a given crystal face is:

$$c_t = c_m + \frac{1}{2} c_a \quad (\text{II.184})$$

The derivative of this with respect to time, dc_t/dt, multiplied by the entire surface, A_s(cm^2), is the change in moles in the gas phase, and thus the rate of desorption:

$$-\frac{dn_t}{dt} = \left(\frac{dc_m}{dt} + \frac{1}{2}\frac{dc_a}{dt}\right) A_s$$

The individual rates are as follows:

$$-\frac{dc_m}{dt} = k_{dm} c_m \quad (\text{mole cm}^{-2}\,\text{s}^{-1}) \quad (\text{II.185})$$

and

$$-\frac{dc_a}{dt} = \frac{1}{2} k_{da} c_a^2 \quad (\text{mole cm}^{-2}\,\text{s}^{-1}) \quad (\text{II.186})$$

If the physical parameters characterizing the amounts of adsorbed atoms and molecular complexes (i.e. $Q_d^\ddagger$ and K) and the total sorbed quantity, c_t, are known, it is possible to determine the values of c_a and c_m from Eqs (II.183) and (II. 184). In principle it might also be possible to determine k_{dm} and k_{da} (and even $Q_d^\ddagger$ and K) from the flash desorption curves, since after the following rearrangements:

$$-\frac{dc_m}{dt} = \frac{dc_m}{dT}\frac{dT}{dt} \quad \text{and} \quad -\frac{dc_a}{dt} = \frac{dc_a}{dT}\frac{dT}{dt}$$

where dT/dt is some function typical of the heating programme for the filament (e.g. for linear programming $T = \alpha t + T_0$), dn_t/dT is directly proportional to the flash curve obtained under conditions of efficient evacuation, so that it can be adjusted to the latter by means of an appropriate computer programme [189]. Inefficient evacuation rate causes distortion of the flash curves, because the I signal of the ionization manometer is not determined by the desorption rate, $\mathrm{d}n_t/\mathrm{d}t$, alone, but is also influenced by the background signal, $h(t)$, of the gas molecules scattered from the walls of the vessel, so that:

$$I = \frac{\mathrm{d}n_t}{\mathrm{d}t} + \beta h(t)$$

With intensive evacuation $\beta \to 0$.

The computing difficulties involved might be relatively easily overcome. The real trouble is that the flash curves reflect not the desorption of a single surface, but the overall desorption of several surfaces. Not only does this multiply the number of unknown parameters; it is also necessary to know something about the possible rearrangement of the layers on the individual crystal faces during desorption. For want of the necessary information the kinetic description of this rearrangement cannot be attempted. It seems that, in the main, the application of field emission microscopy and the flash methods from the kinetic aspect promise new developments in this field. Apart from a few exceptions, (see, for example, Ref. [171]), the currently known work does not attempt a quantitative approach to the work function per crystal face, since there is no simple way to overcome the technical difficulties. In addition, the not completely clear-cut correlation between the work function and the corresponding change in coverage [see Eq. (II. 175)] introduces a further theoretical complication.

However, Eqs (II.185) and (II.186) allow certain qualitative conclusions. Experiments have shown that when a filament covered with N_2 is heated, a first-order desorption takes place first, producing the β_1 peak. It seems probable that this corresponds to the desorption of the molecular complexes in accordance with Eq. (II.185). If the temperature at which the associative desorption of the adsorbed atoms begins is considerably higher than the value pertaining to the β_1 peak, the β_2 peak will not be detectable, because due to the rapid surface rearrangement all the chemisorbed N_2 will leave the surface of the filament in the form of a complex. Partial overlapping of the β_1 and β_2 peaks indicates the absence of any significant difference between the two species with respect to the activation energy of desorption (in fact, the bond energies are 81 and 85 kcal mole^{-1}, respectively). For this reason the order of desorption has a value between 1 and 2, depending upon the relative participation of the molecular complexes and of (the associatively desorbed) adsorbed atoms in the overall desorption. It is to be expected that this ratio will not remain constant, but as the atomic desorption becomes predominant the order will increase. The kinetic factors obviously have an influence on the sizes of the peaks (on the quantity of substance desorbed).

Similar qualitative conclusions can be drawn for other systems too (see Paragraph 8.1.3).

CHAPTER 8. MECHANISM OF CHEMISORPTION

The majority of heterogeneous catalysts are metals or their semiconducting oxides, and recent decades have seen the dramatic growth of our knowledge of these catalyst types. If a short summary were to be given in the introduction to this Chapter as to the nature of this knowledge, mention would be made primarily of observations regarding the conditions under which chemisorbed layers are formed, and on the structures and reactivities of the layers. One of the central problems of heterogeneous catalysis is to throw light on the "activation" of the reacting substances which can be realized in the homogeneous phase by utilizing thermal or other energies, or with homogeneous catalysts of the same role, etc. Indirect test methods, including mainly the kinetic tests themselves provide relatively little information on this profound transformation, which decides the "fates" of the molecules of the initial substances. For this reason the use of the most up-to-date physical methods to study chemisorbed intermediates formed in heterogeneous catalysis has become an important task of reaction kinetics. The discipline of chemisorption can therefore hardly claim that these results are due to progress in the clarification of the "mechanism of chemisorption", since the knowledge acquired in the above manner contributes towards the elucidation of the mechanisms of heterogeneous catalytic reactions.

It is generally known that homogeneous chemical reactions may proceed according to molecular, radical or ionic mechanisms. In heterogeneous catalysis it is very difficult to make a similar distinction. Moreover, however surprising it may sound, it is even difficult to define what we mean by adsorbed molecule, ion or radical and in what way and to what degree the properties of these latter differ from those of molecules, ions or radicals in the gas phase. The intention of this Chapter is to discuss the information available in this field, linking this information with the various theories found satisfactory for the description of the electric properties of solids (metals and non-metals). Although extension of the metal and semiconductor theories to the field of chemisorption and catalysis has only partly justified expectations, in our opinion it will be necessary to deal next with the role and importance of these theories in the interpretation of chemisorption phenomena, for the reader to be able to assess the scientific value of the various conclusions.

SECTION 8.1. CHEMISORPTION ON METALS

Elaboration of the various metal theories and their experimental verification is due to solid-state physicists who were, at most, only slightly interested in the relationship between metal theory and chemisorption phenomena. It remained up to those actively engaged in the field of chemisorption to exploit these relationships and to draw the appropriate conclusions from them.

The simplest model of the electron structure of metals, the free electron model, (see Part I, Section 1.4) can be satisfactorily applied to the description and interpretation of a number of properties of metals, (e.g. the contribution of electrons to the specific heat, the magnetic susceptibility, the Hall coefficient), and primarily alkali metals, but it is not able to explain the general or more particular phenomena of chemisorption. The relevant experimental data are lacking and the theory is, in fact, not suitable for this purpose.

The band theory of metals introduced by Bloch [190] takes into consideration that the conduction electrons move within the metal crystal in the periodic potential energy field of the positive charge carrier atomic cores, shadowed by the conducting electrons. The solution of the Schrödinger wave equation:

$$\Psi_{\mathbf{k}}(\mathbf{r}) = e^{i\mathbf{kr}}\, u_{\mathbf{k}}(\mathbf{r})$$

is called the Bloch function or the Bloch crystal orbital (see Part I, Section 1.2), where

$\mathbf{k}$ is the wavenumber vector
$\mathbf{r}$ is the position vector
and $u_{\mathbf{k}}(\mathbf{r})$ is a function having the same periodicity as the potential symmetry in the crystal lattice.

Introduction of the periodic nature of the lattice and the resulting periodic potential energy field into the solution of the Schrödinger energy eigenvalue equation has resulted in far-reaching consequences. It has become clear, for instance, that in contrast to the conclusions drawn from the free electron model, the conduction electron cannot possess arbitrary energy*: the allowed energy values are concentrated in bands containing very many individual levels produced by the splitting of the atomic levels. These bands are separated by fairly wide forbidden bands. The degree of splitting, that is, the width of the bands in energy units, depends not only on the number of atoms making up the metal crystal, but also on the degree of interaction between the electron wave function of the individual atoms. The overlap integral

$$S_{ij} = \int \Psi_i \Psi_j \, dV$$

provides information on the value of this interaction.

* The energy of the free electron is also quantized, but similarly to the translational energy of the ideal gas the individual levels are very near to each other, so that distribution can be considered practically continuous.

With regard to the subsequent discussions it is of major importance that there is a relatively weak interaction between the electrons of an inner, incomplete shell (e.g. the d-shell of the iron group of the transition metals), and thus the corresponding bands of the metal crystal are fairly sharp. Moreover, it can be assumed that in certain cases of electrons in deep-lying shells (e.g. the second and third-row transition metals) the electron wave function is, to a good approximation, localized to the atom, so that the term band has no meaning here [191].

It is also obvious that if there are unfilled electron states in the isolated atoms, then in the bulk crystals there must also be energy bands not completely occupied by electrons.

These findings have been extremely useful, not only in explaining the outstanding thermal and magnetic properties of transition metals, but also — as will be shown later — in promoting a better understanding of the nature of chemisorption bonding between metals and gases (vapours).

It might be worthwhile to supplement the above discussion with numerical data. In the model of the first-row transition metals used by Slater [192, 193] the width of the 4s-band is about 10 eV, which partly overlaps with a narrower 3d-band about 5 eV wide (see Fig. II.36). 10 electrons are needed to complete the 3d-band, with 2 electrons for the 4s-band. Hence, the band width per electron is about 0.5 eV in the 3d-band and 5 eV in the 4s-band, in agreement with what was said above. Primarily the electrons of the 4s-band participate in current conduction.

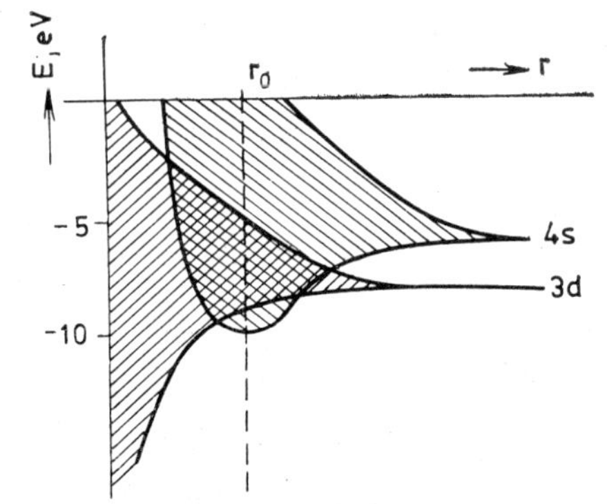

Fig. II.36. Band structure of the iron-group of transition metals according to Slater's model

The free-electron theory makes no distinction between the electrons responsible for the bonding (thereby stabilizing the lattice of positive atomic cores) and those participating in the conduction. While distinction between valence and conduction electrons (i.e. defect electrons) is a central problem in the band theory of semiconductors (see later), the band theory of metals is less concerned with this question. Some decades ago Pauling [194] tried to find a solution to this problem.

From the regularities and irregularities in the melting points and other physical parameters of metals, as expected from the periodic system of elements, Pauling distinguished three types of d-orbitals in the metal crystal: metallic d-orbitals, involved in electron conduction; valence d-orbitals forming hybrid (dsp) bonds between metal atoms; and finally, localized atomic d-orbitals (see also the concepts suggested by De Boer and Verwey), which play no role in either current conduction or the bonding of metal atoms. In transition metals these atomic d-orbitals are partly incomplete, this corresponding in a sense to the incomplete d-bands of the band theory.

The question now arises as to how these characteristics of metals, and particularly transition metals, observed for the bulk substances, will be manifested on the metal surface, and what other effects causing "unsaturation" of the metal surface may occur.

Surface atoms, similarly to the atoms within the crystal, possess incomplete d-bands or (as stated by Pauling) atomic d-orbitals, which, when a covalent bond is formed may be partly or completely filled with the electrons of the adsorbate molecule.

For a metal crystal of finite dimensions, the electron energy states typical of the bulk crystal are associated with electron states pertaining to electrons localized on the crystal surface (surface states, see Part I, Chapter 5). Here, too, the allowed levels merge into more or less wide bands which, in certain cases, may again be incomplete. In principle, the existence of surface states must be considered for all solids.

Since the initial work of Tamm [195] several researchers have embarked on the study of surface states using more complex and more realistic models [196]. However, these attempts have so far failed to provide quantitative results directly applicable to the theory of chemisorption bonding. It should perhaps be pointed out that according to the conclusions drawn from the model of Goodwin and Artmann (see Part I, Chapter 5) the surface levels are, in certain cases, occupied by electrons with uncompensated spins, which is equivalent to the appearence of free surface valence. Koutecký and Tomasek arrived at a similar conclusion [197] from an approximate calculation of the surface states of the three-dimensional diamond lattice by means of the LCAO MO method.

This is the solid-state physical background into which it seems that some of the chemisorption phenomena can be fitted. It must be stressed, however, that the above survey has quite deliberately been restricted to the concepts and information of solid-state physics which can successfully be related to phenomena observed in the study of chemisorption. It is unfortunate that even in these cases only a qualitative explanation of the experimental results is provided by solid-state physics.

Chemisorption causes a change in the electron configuration of the metal adsorbent, which can be observed by means of magnetic, conductivity, etc. methods; the structure of the adsorbate molecule also undergoes a change and this can be detected by spectroscopic, chemical or other direct or indirect methods. The experimental material will be presented separately according to these two aspects.

PARAGRAPH 8.1.1. EXPERIMENTALLY DETECTABLE CHANGES IN THE ELECTRON CONFIGURATION OF THE METAL ADSORBENT

Observation of the work function during chemisorption enables the calculation of the dipole moments of the adsorbed molecules via Eq. (II.175).

Experience has shown that, with the exception of O_2, for most inorganic and organic gases and vapours the dipole moments extrapolated to coverage $\theta = 0$ are lower than 0.5 D units [198] ($1\ D = 10^{-18}$ electrostatic unit.cm), which excludes the possibility of an ionic bond (for alkali metal vapours adsorbed on tungsten $\mu \approx 9-12\ D$).

Magnetization measurements on metals during adsorption appear to support the covalent nature of chemisorption bonding.

According to the band theory of metals, the paramagnetic susceptibility, χ_p, is proportional to the effective electron mass, m_e^*. It has already been mentioned that the widths of the inner d-bands of transition metals are very small and that in some cases these bands might be incomplete. The band theory of metals postulates that a small band-width is associated with a large effective electron mass ($m_e^*/m_e \gg 1$) and hence with a high magnetic susceptibility. In the formulation of Pauling: the unpaired electrons on the atomic d-orbital are the carriers of the paramagnetism. If the paramagnetic susceptibility decreases during adsorption, this indicates a decrease in the number of unpaired electrons. Such a decrease might also be caused by pair formation (spin compensation) with an electron originating from the adsorbate, that is by covalent bonding of the metal and the adsorbate. In estimating the value of the expected effect it must be noted that only those surface atoms of non-porous metal crystals can participate in the bond which are accessible from the side of the gas phase.

In their study of the adsorption of dimethylsulphide on Pd, Dilke, Maxted and Eley observed that the paramagnetic susceptibility of the Pd fell by 6—8% [199]. This can be satisfactorily interpreted by assuming the total disappearance of the paramagnetic susceptibility of the surface Pd atoms as a result of chemisorption. Dimethylsulphide is a compound with a free electron pair, and in a manner typical of catalyst poisons tends to form predominantly donor-type bonds with metals having incomplete d-bands (see later).

Typical representatives of catalyst poisons (such as pyridine, thiophene, carbon disulphide, phosphine, arsine, etc.) also produce donor-type covalent layers on metals with vacant d-bands. NH_3 which also has a free electron pair is chemisorbed as a radical (see later).

It appears from what has been said so far that electrons occupying the d-bands of metals and interacting not at all or only slightly with the neighbouring metal atoms and the vacant d-bands themselves play an important role in the development of chemisorption bonding. It cannot be accidental either that we find the catalysts of simple gas reactions among the transition metals and the metals of group IB of the periodic system. Trapnell [200] tried to establish a statistical relationship between these two findings. He arrived at the conclusion that simple gases such as H_2, O_2, N_2, CO, C_2H_4 and C_2H_2 are capable of chemisorption practically only on transition metals. The following metals appeared to be particularly active: Ca, Ti, Fe, Zr, Mo, Ba, Ta and W. N_2 is not chemisorbed on Ni, Rd, Pd or Pt, neither is H_2 on Cu, and Ag is capable of bonding O_2 only.

It is striking that with the exception of Ca and Ba all the metals of the first group are transition metals: Ti and Fe are the second and sixth members of the first row, Zr and Mo the second and fourth members of the second row, and Ta and W the second and third members of the third row.

The electron configurations of the incomplete shells of their isolated atoms are as follows:

Ti $3d^2 4s^2$ Zr $4d^2 5s^2$ Ta $5d^3 6s^2$

Fe $3d^6 4s^2$ Mo $4d^5 5s$ W $5d^4 6s^2$

The d-band can accomodate a total of 10 electrons, so the highest d-band unsaturation is shown by these metals. This might be the qualitative explanation of the finding that all the above gases are bound by strong covalent bonds to the metals listed (see later). It is also interesting to note that with the exception of Fe and Mo these elements are hardly used, if at all, as catalysts. Though Ca and Ba are in no way transition metals (their isolated atoms have no partially filled d-orbitals), the electrical conductivities of their crystal indicate that there is a partial overlapping between the 4s and 5s-bands and the corresponding d-bands, which ensures that the d-band is not vacant. In this respect Ca and Ba in bulk metal form resemble the transition metals.

Some of the above gases (e.g. CO, C_2H_4 and C_4H_4) be bound both on Cu and Au. Copper and gold follow immediately after the transition periods, and their d-bands are complete. It might be that the energy of the chemisorption interaction is capable of meeting the energy requirements of the d $\rightarrow$ s excitation (for Au this is 3.25 eV, and for Cu about 3.0 eV), thereby causing a vacancy in the d-band. Because of the excitation energy of about 4 eV on Ag, the chemisorption of all gases with the exception of O_2 would be endothermic, so that it cannot take place for thermodynamic reasons.

Oxygen behaves in a quite peculiar manner, as with the exception of Au it can be sorbed on all the metals. This "chemisorption" is in fact the introductory step to common oxidation during which, in contrast to the cases discussed earlier, the metal-metal bonds also are split. Chemisorbed O_2 produces atomic or molecular, negatively-charged dipole layers.

Nitrogen is chemisorbed only on those metals of the first group with the least complete d-bands; this appears to be the condition for the splitting of the $N \equiv N$ bond, with a bond energy of about 7.4 eV.

Recent investigations have confirmed the concepts of De Boer and Verwey (loc. cit.) and also the somewhat different formulation of Pauling on the simultaneous existence of localized electron states concentrated in bands [201], and, in a sense, the ligand field theory has developed the same concept to a higher degree [202].

According to the ligand field theory, the 5 degenerate orbitals in transition metals (with 2 electrons per orbital, capable of containing a maximum of 10 electrons) split as a result of the interactions with the 6 nearest neighbours situated around a given atom of the face-centred crystal of the transition metal. The higher-energy e_g-orbitals correspond to the $d_{x^2-y^2}$ and d_{z^2}-orbitals in the axial directions of the Cartesian coordinate system, while the t_{2g}-orbitals of lower energies (i.e. the d_{xy},d_{yz} and d_{xz}-orbitals) are oriented between these axes. Accordingly, in the case of surface atoms, e.g. on a

(100) face, the e_g-orbitals are perpendicular to, while the t_{2g}-orbitals form an angle of 45° with the normal to the surface (Fig. II. 37).

In the light of the ligand field theory the chemisorption of H_2 can be conceived of as the interaction of the H_2 molecule with the partially incomplete e_g-orbitals of two metal atoms, leading to the formation of a covalent *Me*-H bond. The condition for the formation of the bond is the presence of a single electron in the e_g-orbital. This is achieved by an electron transfer between the e_g-orbitals and the s–p or t_{2g}-orbitals of the adsorbent atom. This transfer is the easier the more the bands situated near the Fermi level are filled with electrons. The condition is best satisfied by the transition metals, and might play a decisive role in the development of their surprisingly high activities in chemisorption and catalysis (see Dowden, loc. cit).

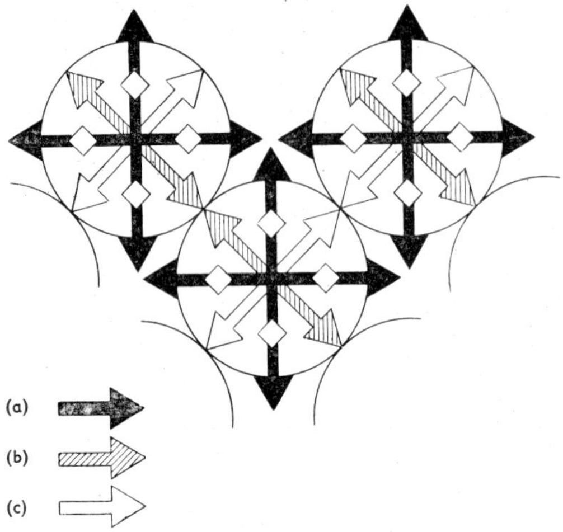

Fig. II.37. Diagram of the appearance of crystal orbitals on the (100) surface of a face-centred cubic crystal: (a) e_g-orbitals in the plane of the paper, (b) t_{2g}-orbitals in the plane of the paper, (c) t_{2g}-orbitals enclosing a 45° angle with the plane of the paper

A polycentric bond may be formed by the "wedge-like" sorption of the H atom (Fig. II.38) when there is a possibility of interaction between the H atom and the e_g-orbitals of five different metal atoms. Both types of bonding have been observed on platinum (see Paragraph 8.1.3).

Another interesting example of the assertion of the above principle is the associative adsorption of molecules with donor electron pairs [203]. In addition to typical representatives of "catalyst poisons" (such as thiophene, H_2S, phosphine, etc.), unsaturated hydrocarbons belong in this group. It must be assumed here that the e_g-orbital transfers its charge to the s–p or t_{2g}-band and behaves towards the electron pair of the donor molecule as an acceptor. If the adsorbed molecule possesses a vacant orbital of appropriate energy and symmetry, a π-bond will be formed with the t_{2g}-orbitals, so that, at the same time, the metal atom too becomes a donor with respect to the adsorbed molecule (back-donation). This may occur, for instance, in the chemisorption of olefins (Fig. II.39).

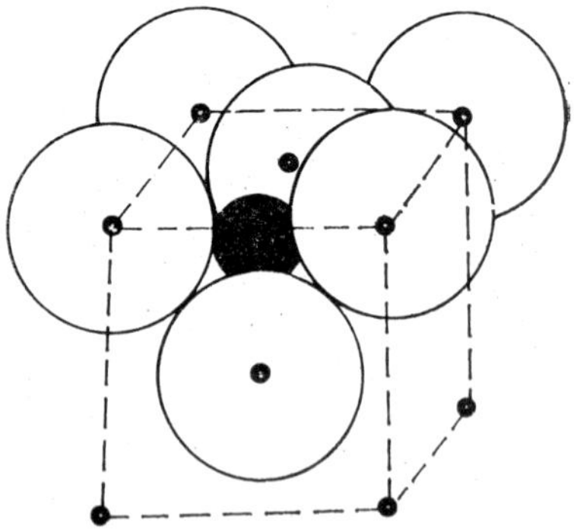

Fig. II.38. Wedge-like sorption of the hydrogen atom (●) on the (100) face of a face-centred cubic crystal

Thus, the associative chemisorption of olefins closely resembles the coordination of olefins with a coordinatively saturated metal atom. This interpretation has been found particularly useful in the catalytic hydrogenation of olefins [204]. LEED measurements fully support the possibility of the formation of a π-bond of this type (see Paragraph 8.1.3).

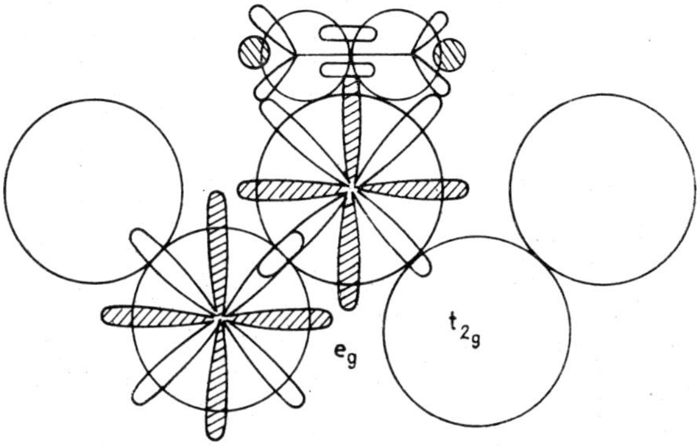

Fig. II.39. Location of π-adsorbed ethylene on the (100) face of a face-centred cubic crystal (for symbols, see text)

Nevertheless, spectroscopic evidence (see later) points to the fact that in certain cases the adsorption of olefins is accompanied by the splitting of the double bond, whereby the sp^2 carbon atom is hybridized into sp^3 symmetry, with the simultaneous formation of two σ-bonds with the t_{2g}-orbitals of two metal atoms of convenient geometry.

On the (100) face of transition metals, carbon monoxide may form a bridge bond with the t_{2g}-band of the metal (see Paragraph 8.1.3). It follows from the nature of the bond that in the state of complete coverage the ratio Ni : CO on the surface will be about 1 : 2. LEED studies have confirmed this assumption. Besides this, however, a weaker type of chemisorption is also observed (see the spectroscopic data), mainly on the (110) faces. This chemisorption presumably involves a σ-bonding with the e_g-orbital, while, at the same time, a donor effect, such as that mentioned in the discussion of the chemisorption of olefins, is also manifested. It should be pointed out that both types of bonds are well known for metal carbonyl complexes.

In the water-soluble complexes of transition metals with olefins it is primarily the electron configuration of the central metal ion which determines the stability of the complex [205]. Table II.13 contains experimentally determined equilibrium constants of reactions of the type:

$$[PtCl_4]^{2-} + \text{olefin} \rightleftharpoons [(\text{olefin})PtCl_3^{-}] + Cl^{-}$$

$$[Ag(H_2O)_n]^{+} + \text{olefin} \rightleftharpoons [Ag(H_2O)_m(\text{olefin})]^{+}$$

TABLE II.13

Equilibrium constants of the formation of olefin complexes

Olefine	K_{Ag} (25 °C)	K_{Pt} (30 °C)
$CH_2{=}CH{-}CH_2NH_3^+$	1.30	2 829
$CH_2{=}CH{-}CH_2OH$	22.91	13 011
trans$-CH_3CH{=}CH{-}CH_2NH_3^+$	1.28	450
trans$-CH_3CH{=}CH{-}CH_2OH$	5	3 008 (60 °C)

The d^8 electron configuration of Pt clearly ensures the formation of complexes of higher stabilities than does the d^{10} electron configuration of Ag. The few data available in this respect prove that the d^7 and d^6 electron configurations are associated with a lower stability of the complex. In the d^8 metal ion complexes the donor role of the metal is probably more marked and this stabilizes the complexes. The back-donation effect is less pronounced in the d^{10} case since the overlapping of the d-orbitals with the (—)-d-orbitals of the olefins is smaller and, at the same time, the ionization potential of the metal is higher. Similar aspects related to energy presumably hold in the formation of chemisorption bonds.

In the elucidation of the relationship between the electron configurations of solids and chemisorption bonding, great importance must be attributed to the experience gained from the chemisorption of organic radicals produced by means of thermal or catalytic decomposition of atoms and adsorbed substances. Despite obvious experimental problems it is difficult to understand the comparative dearth of literature as regards data of this kind.

Estimations available from the bond energies of H atoms chemisorbed on metals [212] seem to confirm that the d-bands do not play a predominant role in chemisorption. Atoms can be bound on any type of material (metals, semiconductors, ionic crystals), irrespective of the band structure [206, 207, 213].

The views resulting from the investigations of Kazanskii's team who carried out simultaneous chemisorption and ESR tests on H atoms, $CH_3^{\cdot}$ and $C_2H_5^{\cdot}$ radicals and high electron affinity molecules (O_2, SO_2) [208], can be summarized briefly as follows: it is worthwhile to distinguish two types of sorbed radicals, the first type including those formed by dissociation (dissociation might be caused by interaction with the adsorbent surface or by some external energy, e.g. radiation), while the second type includes those radicals which have become radical-ions by charge transfer between the molecule and the catalyst. In the first case, an equilibrium

$$\underset{\cdot\cdot}{R} \rightleftarrows \underset{\cdot}{R} + e$$

is assumed between weak one-electron and strong two-electron type surface bonds. In conductors and semiconductors there is a great probability that the one-electron bond is capable of trapping a second electron and thus changing into a strong two-electron bond. This might be one explanation for the fact that no free radicals have been observed on metallic or semiconducting adsorbents. This process cannot take place in the case of insulators because of the lack of free electrons and defects. At the same time this explains why free radicals bound by one-electron can exist only on the surface of insulators. Since one-electron bonding is relatively weak (not more than a few kcal mole^{-1}), at higher temperatures the radicals are liable to desorption and to act as initiators of homogeneous gas reactions.

These findings are, to a certain extent, supplemented and confirmed by the work of Selwood et al., who have investigated the behaviour of H_2 and certain hydrocarbons in relation to the magnetic properties of the metal adsorbent during adsorption [209]. Hydrogen and hydrocarbons probably undergo radical decomposition on metal surfaces (see also the result of $H-D$ exchange and the related studies of Kemball et al. [210, 211]), and the simple radicals (e.g. $H^{\cdot}$, $\cdot CH_3$, $\cdot CH_2$) and the more complex polyradicals (e.g. $\cdot CH_2-CH_2-\cdot CH-CH_3$) probably have a similar influence on the electron configuration of the metal as when directly chemisorbed as a radical. Selwood (loc. cit.) has shown that in the case of ethylene the decrease in magnetization agrees with the assumption that one of the $\sigma-\pi$ double bonds between the carbon atoms is split and two new carbon-metal bonds are formed with the unpaired electron of the metal, according to the principle of spin compensation (associative mechanism). In the case of ethane the decrease in magnetization indicates the formation of 4–6 new bonds, and in the case of cyclohexane a maximum of 4 new bonds (bath form).

Kemball (loc. cit.) investigated the $H-D$ exchange between the H of hydrocarbons and gaseous D_2 in the presence of metal catalysts, and observed simple or multiple exchange, depending on the hydrocarbon, the metal and the experimental conditions. For cases in which only multiple exchange occurs, it appears obvious to assume that the chemisorption of hydrocarbons involves the simultaneous breaking of several $C-H$ bonds and the bonding of the substrate molecule as a polyradical by very strong covalent bonding to the surface. Such chemisorbed material can be removed from the surface only by a chemical reaction. (The "displacement" effect of D_2 and H_2 can probably be explained by the change in the electron configuration of metal caused by the chemisorption of D_2 and H_2. By reacting with surface D and H atoms, the chemisorbed polyradical may be reconverted into the initial hydrocarbon, but partially containing D in place of H atoms.) In the case of ethylene the splitting of the $C-H$ bond is again observed [84] (dissociative mechanism), indicating that the mechanism typical of the chemisorption of paraffins also appears for olefins.

In future research concerned with the conditions required to produce chemisorption bonding it might be promising to take into account the electron density of the adsorbent surface. The calculations of Harrison [212] for the (110) crystal faces of aluminium and silicon point to a marked difference between the electron densities of metal and semiconductor surfaces. For Al the maximum deviation from the mean value is $\pm 38\%$ while in the case of Si the electrons are concentrated at the sites of bonds between the nearest neighbours, where the electron density can reach

120% of the mean value. With regard to interactions prior to the development of chemisorption bonding, these findings indicate qualitatively that on the (100) face of Al the overlapping of the wave functions is practically identical everywhere, in contrast to the situation on the semiconductor surface where the value changes from one site to another.

It has not yet been clarified whether the electrons occupying surface energy states contribute to the development of covalent bonds. It has been demonstrated in connection with the Goodwin-Artmann model (see Section 1.5) that an overlapping of the volume s and p–bands is a sufficient condition for the formation of surface states. This results in the subsequent hybridization of the s and p–wave functions of the surface atoms. Thus, in principle, it is conceivable that on a real adsorbent chemisorption causes the degradation and rehybridization of the (dsp) hybrid bonds of metal atoms.

There is a fairly marked discrepancy between the results of conductivity measurements performed during chemisorption on vacuum-evaporated films [214]. The conductivity of the metal is determined by the product of the number and mobility of the conducting electrons. Consequently, no clear-cut conclusion can be drawn from changes in conductivity during chemisorption as to the increase or decrease in the number of conducting electrons. According to the recent studies of Wedler [105], changes in conductivity due to chemisorption can almost always be attributed to some physical cause (diffuse scattering of electrons from atoms occupied by chemisorption, leading to reduced mobility). These findings seem to be supported by the decrease in conductivity which always occurs at the beginning, that is at low coverages, independently of the polarization of the chemisorbed layer (e.g. in the case of $Ni-H_2$, $Ni-CO$ and $Ni-N_2$ layers). Hence, however surprising it may sound, the well-known explanation that certain adsorbates increase, while others reduce, the number of electrons as a result of chemisorption, does not hold, at least for high-purity metal films. Assuming that on the surface covered by the chemisorbed layer the electrons are scattered diffusely (in all directions) from the occupied adsorption sites, whereby their mobility is changed, Wedler et al. were able, in a number of cases, to produce even a quantitative interpretation of the change in conductivity due to chemisorption.

PARAGRAPH 8.1.2. CHANGES IN THE SURFACE STRUCTURE OF THE METAL ADSORBENT AS A RESULT OF CHEMISORPTION

Surface layer structures, revealed by LEED investigations during the last few decades, will be treated in Paragraph 8.1.3. The discussion here is limited to evidences concerning the change in the geometrical structure of the adsorbent surface as a result of chemisorption.

It must first be recalled that the atoms of common metals possess a certain mobility even at room temperature. This has been common knowledge for a long time as far as mercury is concerned, and experiments on growing metal whiskers have provided convincing evidence on the mobility of other metals, too [215].

The phenomenon of thermal etching has also been known for some time. This can be observed at a high temperature for a given metal in the presence of characteristic gases, and sometimes even under the common microscope [216].

Prior to the LEED tests, nobody believed seriously in the possibility of a rearrangement of the surface. In 1962, Germer and Macrae [217] proved that the adsorption of H_2 on the (110) face of Ni causes a change in the arrangement of the atoms. This observation was based on the slight, hardly measurable scattering of the H atoms, when a change in the diffraction pattern has to be attributed to the rearrangement of the surface. Figure II.40 illustrates the ball model of the (110) face of an ideal face-centred

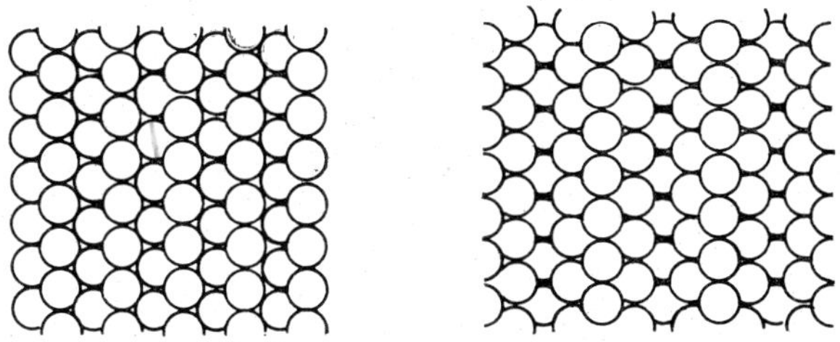

a) b)

Fig. II.40. a) Ball model of the (110) face of Ni; b) same surface showing rearrangement after the chemisorption of H. The H atoms are probably situated in the deepest "troughs" shown in this part of the figure

cubic lattice of Ni, and the distortion of this due to adsorption. It is quite clearly visible that in the second picture, too, the Ni atoms are ordered in rows in the direction of the (110) face, but the distance between the rows in the (100) direction is twice as great as in the first picture. The situation of the adsorbed H atoms on the surface is not specifically known. Since this phenomenon cannot be observed on the more compact (100) and (111) faces, it seems probable that rearrangement is promoted, among other factors, by the weaker adhesion of Ni atoms on the (110) face.

There is a certain rearrangement of the metal surface when O_2 is adsorbed, on Ni, but in this case the site of the O atoms incorporated into the surface can be determined, for their electron scattering is appreciably higher than that due to H atoms. To summarize the results in this field, with the help of the LEED method it was possible to reconstruct in order of decreasing surface atomic density two different surface structures on the (111) face, two on the (100) face and five on the (110) face [120]. The maximum coverage on the various faces is brought about by the following quantities: on the (111) face, 6 10^{14} atom cm^{-2}; on the (100) face, 8 10^{14} atom cm^{-2}; on the (100) face, 8.8 10^{14} atom cm^{-2}. 1/3, 1/3, 1/2 and 4/5 of the above maximal quantities (or only some of these) will be sorbed in the states corresponding to completely developed regular structures on the different crystal faces.

Results similar to those described above were obtained with other metals and semiconductors, too, using a great variety of adsorbate [120].

Surface rearrangement is not a necessary consequence of adsorption, or, at least, there is a temperature below which no rearrangement takes place. Germer and Macrae (loc. cit.) suggested certain factors which might cause rearrangement, but the problem has not yet been solved in all its details.

LEED studies may have a marked direct influence on the disciplines of chemisorption and catalysis. In addition to those aspects discussed above, the energy consequences of "geometrical observations" in chemisorption cannot be ignored, while in catalysis the conceptions of the reaction mechanism must definitely agree with the findings of the LEED studies. The importance of the existence of surface structure in heterogeneous catalysis has been confirmed by a large number of investigations and the latest results of biocatalysis serve to emphasize this.

As regards their direct usefulness in catalysis research, LEED measurements are able, primarily, to provide answers on the composition and purity of the surface.

As far as the composition of the surface phase is concerned, in recent decades an immense amount of experimental material has been collected on the catalytic activities of alloys without a knowledge of the true composition of the surface phase. Besides the LEED method, Auger electron spectroscopy tests promise to lead to substantial progress in this field [218].

Direct observation of stationary catalytic reactions by means of LEED studies may be of even greater significance. Attempts of this nature have so far been limited to the simplest oxidation reactions (H_2 and O_2 [219], and CO and O_2 [220]), but the catalytic reaction could not be maintained for longer periods during the measurements. This trend will probably gain ground in the future, although, for the operation of the electron gun, 10^{-4} torr is the upper limit, while at this, and higher, pressures electron scattering on the gas molecules causes a considerable reduction in sensitivity and resolution.

PARAGRAPH 8.1.3. CHANGES IN THE ELECTRON CONFIGURATION AND BONDING CONDITIONS OF CHEMISORBED MOLECULES

The configuration of chemisorbed molecules changes in the course of the chemisorption process. This change can be observed experimentally using appropriate methods. The most appropriate means of gaining information concerning the presence and mobility of adsorbed atoms and molecules or their fragments is via the change in the work function on the surface of the metal adsorbent, due to the adsorption process. The photoelectric method [221], the space-charge limited diode method [222] and the capacitor method [223] provide information on the change in the work function on the entire surface area of the sample, so that in the case of polycrystalline substances no "specific" conclusions can be drawn with respect to the individual crystal faces (see Section 7.2). This can be better achieved with the help of field (or ion) emission microscopy (see Paragraph 5.2.1 and Section

7.2), which provides a very great ($10^6 - 10^7$ times) magnification and permits direct visual observation of surfaces of different crystallographic indices and of the covering adsorbed layers on metal tips with very small radii of curvature. The method of low-energy electron scattering is being increasingly used for the investigation of the structures of surface layers (see Paragraphs 6.2.1 and 8.1.2), as it provides direct information on the structure of the surface layer (and of the underlying adsorbent surface) to a penetration depth related to the electron energy. Auger-type electron spectroscopy can be used in a similar manner [218] [224].

Useful information can be gained on the bonding strengths and states of adsorbed substances by means of flash desorption (thermal desorption) [225] and by bombardment with low-energy electrons [226]. The former method is suitable for a semi-quantitative estimation of the bonding strengths of adsorbed layers on metal filaments, films (and even on granulated adsorbents). The methods consist in heating the filament or film, etc., after adsorption equilibrium has been reached, to increasingly higher temperatures, when the desorption (observed either via a change in pressure or by means of the mass spectrometer) appears in a selective manner, depending on the bonding strengths of the various chemisorbed species (see Section 7.5). Electron bombardment has been in use for the cleaning of adsorbent surfaces and it has been found that from the nature (analyzed with the mass spectrometer) of the substances desorbed by electron bombardment in a "mild" process, using low-energy (from 10^{-1} to a maximum of a few 100 eV) electrons, conclusions can be drawn regarding the bonding of the adsorbed layer, the nature and structure of the adsorbate, etc.

In contrast to the earlier magnetic and conductivity measurements which are directly suited only for the observation of changes in the bulk properties, these methods provide information on the surface properties of the metal adsorbent and on the adsorbed layer itself.

In cases when prolonged heating, the presence of an inhomogeneous electric field, electron (or ion) emission or electron bombardment would result in the decomposition of the adsorbed molecules, the analysis of absorption (or emission or reflexion) spectra in the infrared or microwave ranges may promote our knowledge of the bonding conditions of chemisorbed molecules.

Properties of the adsorbed layers

Something of the properties of *hydrogen layers* on metal surfaces is revealed by the well established observation that metals capable of chemisorbing H_2 also act as catalysts of the H — D exchange*. This proves that chemisorption is accompanied by the splitting of the H—H bond and the formation of new metal-H bonds. Adsorption tests carried out on vacuum-evaporated nickel films [227] indicate that at —196° C and 0.1 torr equilibrium pressure one H atom is bound by one Ni atom. From the fact that even

* For metals the reverse is probably also true, while certain insulating or weakly semiconducting oxides in which some of the OH groups are capable of H-exchange with gaseous H_2 or hydrocarbons at relatively low temperatures, too, do not chemisorb hydrogen to a significant extent at temperatures below about 500 °C.

after repeated trials it was not possible to observe the infrared absorption spectrum of the hydrogen adsorbed on metals, Eischens and Plinskin [228] earlier arrived at the conclusion that the chemisorption of hydrogen on metals is associated with the formation of polycentred bonds. The small dimensions of the H atom (but particularly of the H atom deprived of its electron) make it quite obvious that during chemisorption hydrogen is capable of penetrating into the cavities in the metal lattice by interacting with several metal atoms at the same time. After improving their experimental technique, the same authors a few years later [229] observed the appearance of absorption bands at 4.86 and 4.76 μm. These two bands characteristically indicate the presence of two Pt—H bonds of different strengths; this was attributed by Pliskin and Eischens to topographical differences between the adsorbed H atoms (see Paragraph 8.1.1).

By means of the flash desorption method it has also been possible to observe several [230], but at least two [231], types of bonds of different strength for H_2 adsorbed on W filaments.

In a recent paper Wedler et al. [232] has demonstrated the presence of only three desorption maxima on this films of high purity (see Fig. II.41). The γ_1 peak at 90 K corresponds to the physical adsorption of H_2 (the absolute value of its heat of adsorption is less than 7 kcal mole^{-1}), while the β_1 and β_2 peaks represent the chemisorption of H_2. In fact, only the assignment of the β_2 peak to atomic H seems to be indisputable (the absolute value of its heat of adsorption is 18 kcal mole^{-1}), while for the time being it is not possible to identify accurately the β_1 peak. The parasite peak in the figure (and presumably the other peaks observed by several authors) is caused by the desorption of Ar contamination at 120 K.

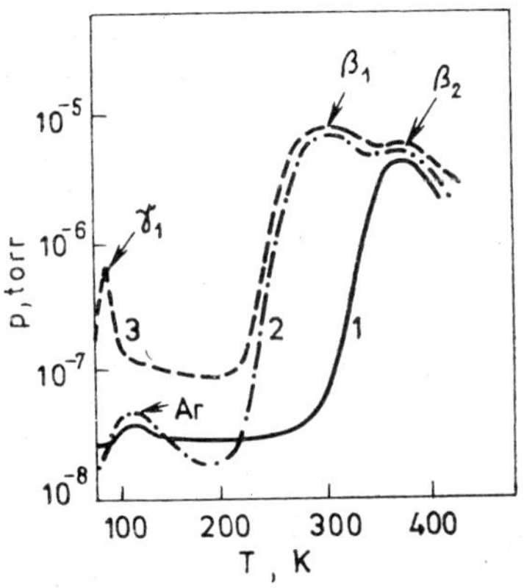

Fig. II.41. Differential desorption spectrum of H chemisorbed on Ni film (after Wedler)
1: $\theta_0 = 0.4$; β_2 peak
2: $\theta_0 = 1.0$; β_2 and β_1 peaks
3: $\theta_0 = 1.2$; β_2, β_1 and γ_1 peaks

When metals are cathodically polarized in electrolyte solutions, phenomena similar to those described above are observed [233].

Klopfer investigated the properties of H_2 sorbed on polycrystalline Mo by bombarding the surface with 150 eV electrons. Experiments at 420 K led to the conclusion of the presence of only a single sorption state of hydrogen [234], from which molecular H_2 and H^+ ions were desorbed as a result of electron bombardment. Prior to bombardment the surface coverage was 1.5 10^{14} particles per cm^2.

As already mentioned, the structure of H layers formed on metal surfaces cannot be studied directly by means of LEED measurements because of the very low electron scattering ability of the H atoms. This question has recently been studied from another aspect by Bauer [235], who found that adsorbed hydrogen atoms are also capable of electron scattering, but it is difficult to distinguish between this and the multiple scattering produced by the atoms of the adsorbent.

With a few exceptions, plurimolecular *oxygen layers* are formed on almost all metals even at low temperatures [120], [236], as a precursor of the oxidation which takes place at higher temperatures. Some decades ago, Roberts observed that after the formation of a monomolecular oxygen layer certain metals (e.g. tungsten) are capable of resisting further oxidation [81]. This can be explained partly by the high W—W bond strength and partly by steric reasons (for suggestions on the growth of oxide films see Ref. [237]).

Oxygen is bound to the metal surface by means of acceptor-type covalent bonds. In Pauling's estimation, at least 30—50% of the ***Me***—O bonds on various metals are of an ionic nature. The relatively high resistance to oxidation (and the high melting point) of tungsten explains the reason for its use in both field emission microscopy and flash desorption. The former has confirmed the existence of negatively charged metal-oxygen dipole layers [238].

In this case, too, flash desorption has revealed the presence of oxygen in several bonding states [231]. Figure II.42 shows the desorption of the α-layer at temperatures below 500 K, and that of the β-layer at around 1500 K. According to Hickmott, atomic oxygen leaves the surface of the heated filament at temperatures corresponding to both the α- and β-sections of the curve [231]. The electron bombardment experiments of Redhead give 62 and 110 kcal mole^{-1} as the desorption energies of the two types of chemisorbed phases [239]. From Klopfer's investigations (loc. cit.) it seems probable that there is an adsorbed phase of O_2 from which molecular O_2 is desorbed.

LEED studies also provide information on some other important details. As a supplement to what has been said in Paragraph 8.1.2 on the properties of the O-layers, Fig. II.43 (Germer [240]) illustrates the structure of the O-layer formed on the surface of a Ni single crystal at 350 °C, as a result of the sorption of 10^{-4} torr s^{-1} O_2. In this layer there is one O atom for every two Ni atoms. Further O_2 uptake can proceed only on top of the developed layer. As already mentioned, at lower coverages, too, layers develop with regular structures, but differing from that shown in the figure.

Infrared absorption spectroscopy is particularly suited to the investigation of the adsorption bond formed in the course of the *chemisorption of CO*. For the structure of the CO-metal bond the following speculative possibilities exist:

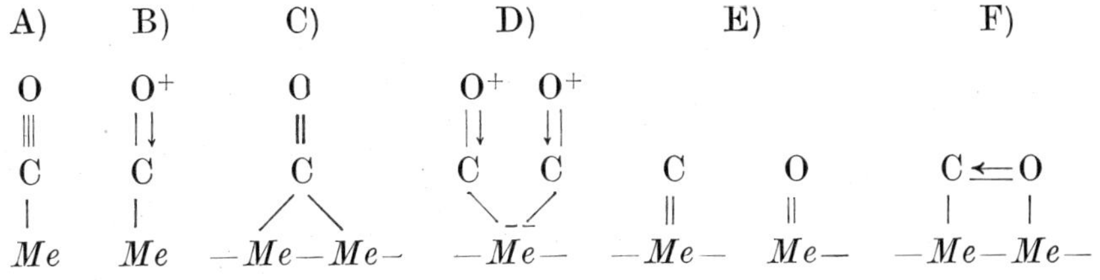

Eischens and Pliskin (loc.cit.) claim that CO is adsorbed on Pd on a Cabosil support in accordance with the two-site mechanism C), and at higher coverages in accordance with the one-site mechanism A). Spectroscopy showed only the linear form A) on Pt. It is very interesting that the nature

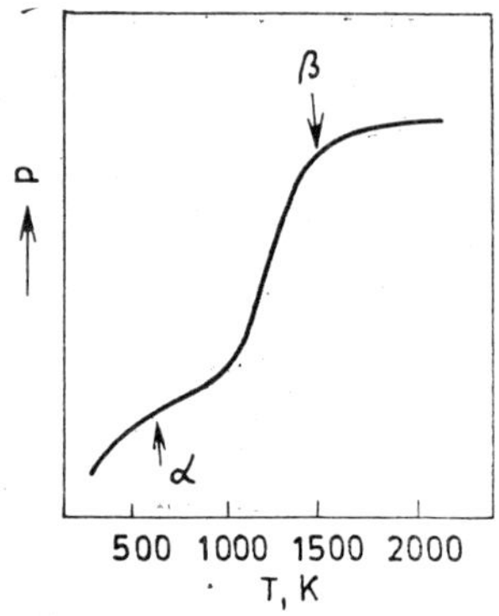

Fig. II.42. Integral flash curve of oxygen chemisorbed on a tungsten filament (after Hickmott)

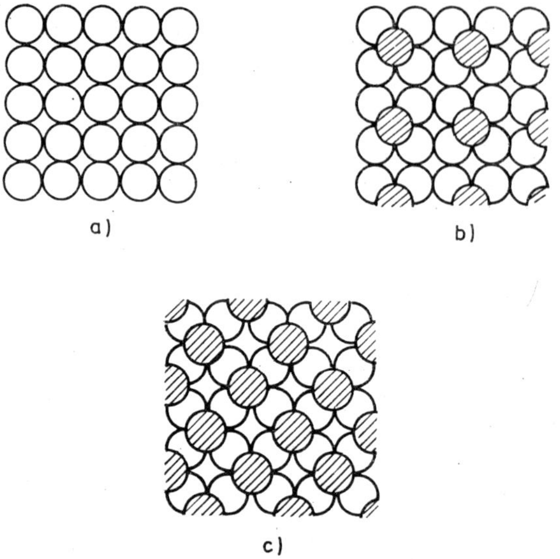

Fig. II.43. Surface structure of oxygen chemisorbed on the (100) face of Ni at different coverages
a) uncovered surface
b) surface structure at low coverage (● oxygen atom)
c) surface structure at complete oxygen coverage

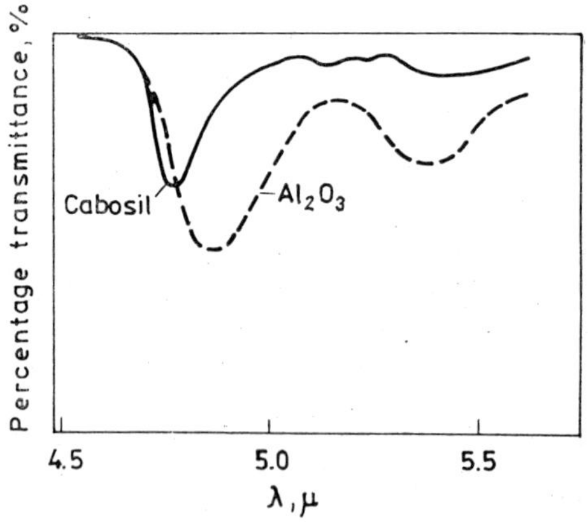

Fig. II.44. Absorption spectra of carbon monoxide chemisorbed on Pt on a Cabosil and on a Al_2O_3 support, respectively (after Eischens)

of the support influences the electron configuration of the supported metal, this effect occasionally being manifested in the infrared absorption spectrum of the substance chemisorbed on the metal. Figure II.44 (Eischens, loc. cit.) gives the infrared spectra of CO chemisorbed on Pt on Cabosil and Al_2O_3 supports. In the vicinity of 4.9 and 5.5 μm, there appear bands characteristic of structures A) and C), respectively. Closer examination of the 5.5 μm band intensities reveals the presence of about 50% of structure C) in the CO bound by Pt on the Al_2O_3

support, while on Pt on the Cabosil support this type of chemisorption is hardly represented. A qualitative explanation of this difference may be found in the transition of n-type conduction electrons from the Al_2O_3 to the Pt, thereby facilitating the development of multi-electronic bonding of type C).

Yang and Garland [241] observed the co-existence of the structures A), C) and D) on Rh on an Alon-C support. The relative quantities of the different types depended upon the coverage.

Experiments with the field emission microscope have confirmed structure E) for CO chemisorbed on tungsten at temperatures above 950 K; that is, dissociative chemisorption takes place under these conditions.

By means of electron bombardment experiments Menzel was able to distinguish three adsorption states for CO chemisorbed on W at temperatures around 100 K [226]. In addition to the "virgin" (physically adsorbed) CO layer there are layers symbolized α_1 and α_2. When these layers are bombarded with 100 eV electrons CO^+ is desorbed from the first, and O^+ from the second. The latter is identical with the "normal" α-layer observed earlier at room-temperature. The only difference between the "virgin" layer and the α_1 layer is in the directions of their dipole moments. Under the effect of thermal or electron energy both the "virgin" layer and the α_1-layer are converted into α_2, the structure of which appears to correspond to the A) form. At higher temperatures, above 450—500 K, a strongly bound β-phase emerges (bond energy about 50 kcal $mole^{-1}$) from which electron bombardment causes the desorption of O^+ ions. This suggests a structure for the β phase corresponding to C), though certain authors prefer structure F) [242]. The fact that no desorption of C^+ ions was observed from the β-CO layer is in favour of structure C).

From the many studies, performed using various methods, concerning the formation and properties of CO layers on W surfaces [212] the following facts appear. The β states formed at high temperatures are not uniform, since CO is bound by different energies to the different crystallographic surfaces. So far only the β-layers on the (110) face have been studied in detail: a particular (β_1) layer is preferentially formed. The more weakly bound α-layer is not homogeneous either, and contains vacant centres in the vicinity of which there are occupied centres in different states.

On metals capable of *chemisorbing* N_2 (see group I in Trapnell's earlier-cited classification) at least two types of bonds are again observed. For instance, at low temperatures chemisorption on iron films is rapid with a heat of adsorption of about -10 kcal $mole^{-1}$, followed at room temperature by a slow activated process. At low coverages the heat of adsorption reaches -70 kcal $mole^{-1}$ [43]. This latter chemisorption is certainly atomic, as this is the condition of NH_3 synthesis.

Flash desorption reveals three types, γ, α and β-type bonding on tungsten (Hickmott and Ehrlich, loc. cit.). Nitrogen of γ-type is completely desorbed at around 200 K, while the α-layer at between 1350 and 1800 K. Hickmott claims that even for the γ-layers it is not sufficient to consider bonds produced by dispersion forces alone (here too the heat of adsorption reaches -20 kcal g-$atom^{-1}$).

At temperatures above 1000 K the desorption of the β-layer follows kinetics of the second order.

Independent measurements [243] give 87 kcal mole^{-1} for the activation energy of desorption. All this indicates a β-layer of atomic N, while the good agreement of the activation energy of desorption points to a heat of adsorption almost independent of the crystallographic index of the surface.

One of the most interesting properties of the atomic N-layer is its inability to occupy the entire surface of tungsten; at room temperature the sorption amounts to only half the monomolecular layer. At room temperature, in addition to the β-layer an α-layer (less than 15% participation) also develops, mainly on the (111) faces. As the temperature is reduced, a γ-layer develops on areas not occupied by atomic N. This consists either of molecularly bound N_2 (the desorption of the layer proceeds according to first-order kinetics), or atom pairs which remain near each other because of their hindered mobility.

Ion emission microscopic investigations suggest the following explanation for the peculiar behaviour of the atomic nitrogen layers formed at room temperature. At room temperature all crystal faces are saturated with atomic N-layers with the exception of the (110) face and probably also the (130) and (100) faces. On the surfaces participating in the process of saturation the arrangement of the adsorbed atoms is such that the possibility of further adsorption is excluded on 4 neighbouring sites (Fig. II.45). Surfaces of this type, and the (110) faces which are quite unable to chemisorb at room temperature, still permit the chemisorption of layers in the γ-state at low temperatures, and these may then prevent the formation of the β-layer.

Fig. II. 45. Sites excluded from subsequent adsorption on a square lattice. The adsorbed atoms (○) prevent the adsorption of more atoms on the four nearest-neighbour adsorption sites (●)

A further interesting feature of the chemisorption of N_2 is the important role of the (110) surface (which is otherwise inactive at room temperature) in the kinetics of the chemisorption process: it acts as a reservoir (precursor state) from which the N_2 molecules may move to more active surfaces via surface migration. These conclusions are supported by the results of investigations on macroscopic samples [212].

The characteristic role of the (110) surface again underlines the truth of the statement that chemisorption phenomena cannot be judged on the basis of the chemical nature of the substrate alone, and that knowledge of the arrangement of the adsorbent atoms on the surface is an absolute necessity.

However, we have by no means exhausted the findings on the chemisorption of *N_2 and CO* which are extremely interesting from the aspect of catalysis. The adsorption tests of Nasini and Ricca [244] with CO and N_2 gas mixtures demonstrate that CO is capable of acting at room temperature in the same way as N_2 only at lower temperatures, viz. it may occupy the sites left vacant by N_2 and populate the (110) surfaces too. In this manner the CO and N_2 layers, so to speak, "interlace". On the sites left vacant by N_2 α CO-layers develop, and on the (110) surface β CO-layers. Considered from the point of view of heterogeneous catalysis, the situation is extremely complex. Even if the individual sorptions of both components are known, no definite conclusion can be drawn regarding their behaviour when they are sorbed together. Progress in this field can be expected only from the joint development of the theoretical and experimental methods.

On metals which catalyze the *decomposition of NH_3* this latter is certainly dissociatively chemisorbed; hence, despite the presence of the free electron pair of the N in NH_3, a dative bond with the metal will not result, but, in a manner similar to hydrocarbons, partial or total dissociation of the H must first take place, followed by the bonding of the radicals formed [176].

The structures of the various *hydrocarbon layers* might be better understood with the help of infrared spectroscopy. The experience gained here can be generalized for any olefin or paraffin hydrocarbon, and accordingly we shall next deal primarily with the relevant regularities.

Supplementing what has already been said concerning the *chemisorption of ethylene*, depending on the experimental conditions the spectrum of ethylene chemisorbed on Ni suggests both an associative (accompanied by the splitting of the C=C bond) and a dissociative (splitting of the C—H bond) mechanism. The most important factors to be taken into account are: temperature, partial pressure of hydrogen, and the condition of the surface, i.e. whether prior to the adsorption of ethylene it has been covered with hydrogen or not. Briefly, it may be said that on a metal surface covered with H, C_2H_4 will be associatively chemisorbed at room temperature. (When the chemisorbed ethylene, $\cdot CH_2-\cdot CH_2$, is exposed to the effect of H_2 at 4 torr partial pressure, it is also possible to observe the spectrum of adsorbed ethyl radicals, $\cdot CH_2-CH_3$.) If practically all the hydrogen is removed from the metal by evacuation at 350 °C for 30 minutes, C_2H_4 will be dissociatively chemisorbed on this metal surface. According to Eischens (see Eischens and Pliskin, loc. cit.) the surface complex formed is not of definite composition. In some cases the possibility of the ethylene losing all its hydrogen in the course of the chemisorption must not be ignored either. It is interesting that these partially or totally dehydrogenated surface complexes can be hydrogenated at 35 °C to form adsorbed alkyl radicals; these can be reconverted (in a more or less reversible manner) into the surface complexes as the partial pressure of H_2 is reduced. This hydrogenation-dehydrogenation process takes place with the participation of the hydrogen atoms of the ethylene itself, too, so much so that ethane also appears in the gas phase. This disproportionation of ethylene, which is obviously a characteristic feature of most hydrocarbons, has been studied by a number of authors [80], [83, 84] and [245].

The flash desorption of ethylene and acetylene from the surface of tungsten filaments [246] supports and supplements the above. Figure II.46 gives the H spectra of the flash desorption of monomolecular ethylene and acetylene layers produced on tungsten filaments at 95 K. The ordinate of the diagram shows the ion current of the mass spectrometer used for the recording of the H (this current is proportional to the partial pressure). The hydrogen peaks β_1 (at 300 K) and β_2 (at 450 K), originating from the decomposition of ethylene (see Wedler's data, loc. cit.), correspond to approximately identical hydrogen quantities. In the case of acetylene a single β_2 peak is observed at a higher temperature (higher thermal stability!).

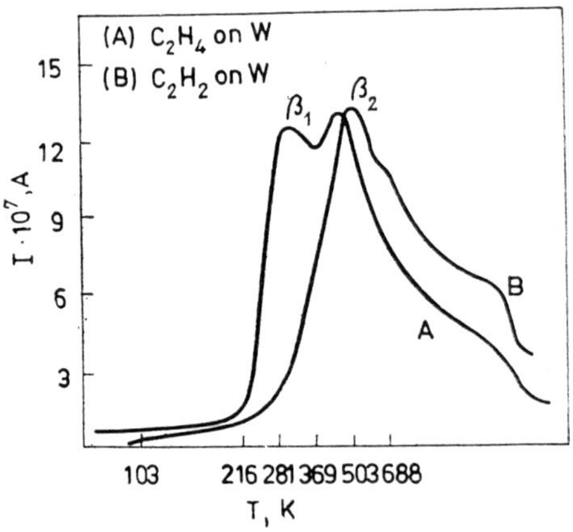

Fig. II.46. Differential flash curves of chemisorbed hydrogen originating from the thermal decomposition of ethylene (A) and acetylene (B) chemisorbed on a tungsten filament; I: ion current of the detecting mass spectrometer

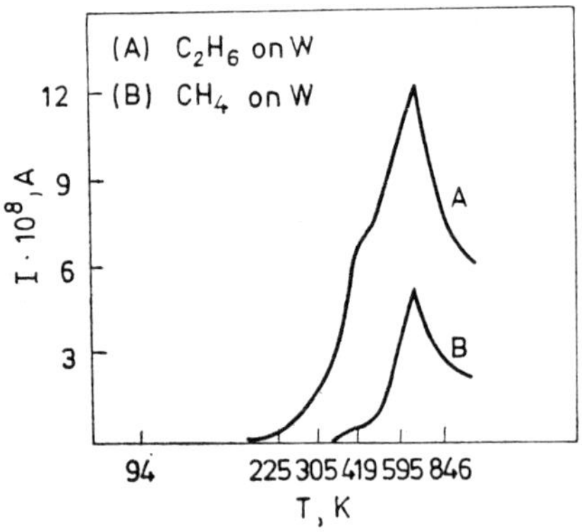

Fig. II.47. Differential flash desorption spectrum of hydrogen originating from the decomposition of chemisorbed ethane (A) and methane (B); I: ion current of the mass spectrometer

All this can be satisfactorily explained by the following schematic presentation of the decomposition process:

$$\cdot CH_2 - \cdot CH_2 \xrightarrow[> 200^\circ K]{} \cdot CH = \cdot CH + 2 \cdot H$$

$$\cdot CH = \cdot CH \xrightarrow[> 300^\circ K]{} \cdot\cdot(C_2) + 2 \cdot H$$

$$2 \cdot H \xrightarrow[\text{(rapid)}]{} (H_2)_g$$

At 3000 K carbon, too, is removed from the surface and forms CO with the oxygen chemisorbed on the glass wall.

Figure II.47 presents similar curves for *ethane* (curve A) and *methane* (curve B). Because of its higher thermal stability, methane begins to give off hydrogen only at temperatures higher by about 165 K. The shoulder near 420 °K on the curve (of the β_2-type peak) corresponds to the reaction

$$(CH_4)_g \rightarrow \cdot CH_3 + \cdot H$$

while the peak at 665 K is due to the subsequent decomposition of the chemisorbed methyl radical. This confirms the earlier finding, according to which the chemisorption of methane and ethane (and of saturated hydrocarbons in general) can proceed only by a dissociative mechanism. Thus, hydrogen chemisorbed on the metal surface hinders the chemisorption of saturated hydrocarbons. On surfaces free from hydrogen these substances are dehydrogenated, whereby surface complexes similar to those found in the chemisorption of unsaturated hydrocarbons are formed.

Disproportionation, as observed for ethylene, is also characteristic of the chemisorption of acetylene, the spectrum of which points to the presence of surface complexes similar to those found during the chemisorption of ethylene and ethane. For acetylene (and this property is also almost certainly typical of most hydrocarbons!) polymerization too occurs. Comparing the relative intensities of the bands at 3.38 and 3.42 μm, which are characteristic of the valence vibration of the asymmetric C—H bonds in the CH_3 and CH_2 groups, respectively, it appears that as a result of disproportionation and polymerization adsorbed butyl radicals, $\cdot CH_2-CH_2-CH_2-CH_3$, are present on the surface in a significant concentration. The sequence and other details of the processes leading to the formation of adsorbed oligomers are unknown. It is a fact, however, that the greater the preliminary hydrogen loss of the acetylene (resulting even in the formation of a "carbide") the higher the quantity of various oligomers formed on subsequent hydrogenation.

Kemball's often cited, very carefully performed kinetic studies appear to support even the details of the above conclusions. We wish to point out a single feature: according to Kemball [211], information may be obtained regarding the nature of the radicals chemisorbed on the surface of the metal by deuterating them at a low temperature (between −30 and 0 °C) with excess gaseous D_2. It is easily understood that the success of the experiment depends on whether the "exchange" stops at the point when, in all the C-metal and H-metal bonds, the metal is replaced by deuterium, or whether the hydrocarbons, which have already partly exchanged their hydrogen for deuterium and can be detected in the gas phase too, participate in a new exchange reaction.

Hydrocarbon layers of regular arrangements have recently been detected on the (100) and (111) crystal faces of Pt [247]. The hydrocarbons were various olefins, and when cooling started at 200 °C, layers began to be formed at about 130 °C. These surface structures decomposed when heated to 150 °C, without marked decomposition of the hydrocarbon. H_2 formation was observed only in the temperature range between 200 and 340 °C. All this points to these layers containing the hydrocarbons in a non-dissociative chemisorbed form. It is one of the most interesting results of these investigations that while almost every C_2-C_4 olefins produces a regular surface structure on the (111) face of Pt, with the exception of C_2H_4 no other olefins form a similar structure on the (100) face.

Although this explanation (within the framework of the ligand field theory) about the two possible bonds of olefins, the σ-double bond and the π-absorbed state (see Paragraph 8.1.1), does not contradict the LEED structures of these hydrocarbons (see Fig. II.24), the molecular orbital theory cannot be applied to the interpretation of the chemisorption of olefins on (111) faces. Nor is the theory suited to explain the chemisorption of acetylene derivatives on the (100) faces of face-centred metals by means of simultaneous bonding of two π-orbitals.

SECTION 8.2. CHEMISORPTION ON SEMICONDUCTORS

The electronic theory of chemisorption has proved to be more appropriate for the interpretation of chemisorption phenomena on elemental semiconductors and semiconductor oxides than for metals, where most results consist of qualitative analogies without adequate theoretical support. The reason for this is no doubt to be sought in the deficiencies of the current metal theories. Semiconductors have been at the centre of extensive theoretical and practical research, particularly during the last ten years, whereas there has been a noticeable slackening in the development of theories concerning metals. A fair number of conclusions following from the band theory of metals still lack experimental confirmation.

It must be pointed out, however, that the electronic theory of chemisorption on semiconductors has also developed from the band theory of semiconductors, and here, too, original ideas going beyond the disclosure of analogies have been put forward only in recent decades. These are not yet generally accepted in the theory of chemisorption, however, where results drawn from analogies still dominate. We shall see, for example, that chemisorbed particles differ from the "contaminants" of the bulk phase only in that the theory does not postulate an obligatory localization for the surface contaminants [248]. In the barrier-layer theory of semiconductors it is again quite easy to recognize the concepts with which rectification by semiconductor diodes is explained (see Ref. [249]). To summarize the impression which the reader (including the present author) with no special training in solid-state physics may gain from a study of the pertaining literature: it seems that the greater success of the electron theory of adsorption on semiconductors in the interpretation of chemisorption phenomena is mainly, if not entirely, due to the fact that it can rely on the more ample and theoretically more sound semiconductor theory.

It might be suggested that the theory of chemisorption and catalysis applied to semiconductors can usefully contribute only to the interpretation of experimental results obtained with elementary, and various other (mainly oxide-type), semiconductors. In our view this is not completely correct. In all experiments except for the relatively small number performed on metal filaments and vacuum-evaporated films under conditions of the highest possible laboratory sterility, so that the partial pressure of foreign gas impurities, and in particular that of oxygen, was less than 10^{-10}–10^{-11} torr, the presence of the oxide of the metal or the semiconductor must be assumed. In commercial-scale catalysis, where the above conditions of purity are even less likely to be achieved, the presence of oxides is practically certain. The question now arises of whether, at a more advanced stage of development than exists at present, the electronic theory of semiconductors will help lead to a concise and homogeneous explanation of the phenomena of chemisorption and catalysis. It must be said, unfortunately, that at present this goal is still quite distant.

By means of a short digression, the following indicates how the theory of chemisorption has reached its present state.

It was a long, rough road of almost fifty years which finally led to the recognition of the future important role of the electronic theory of semiconductors in the interpretation of the phenomena of chemisorption and catalysis. Although we do not believe it our task to present a complete description of this development, we do wish to point out certain milestones which have proved to be of fundamental importance.

The first indication that the "electronic factor" (by which the electronic structures of adsorbents, catalysts and interacting molecules should be understood) plays some role in catalysis can be found in the literature of the 1920's. This was followed about 10 years later — mainly on the initiative of Wagner and his team [250] — by extensive research aimed at establishing the influence on the conductivity of the nature and partial pressure of the gas in contact with oxide-type semiconductors. Starting from the assumption that the interaction between the gas and the semiconductor is manifested in electron transfer, and by applying the law of mass action, the concentration of charge carriers in the solid, and, hence, the conductivity were calculated. These studies already involved the band structure of semiconductors, the donor and acceptor levels, the position of the Fermi-level, etc. The procedure consisted essentially in substituting the bulk contaminants of the semiconductor, for which Fermi-statistics stipulates well-defined rules, by the "contaminants" of the crystal surface (electron and crystal defect traps formed on the surface by adsorption). In this way, apart from a few minor modifications the rules for bulk contaminants were accepted as being still valid. By and large we may agree with these conclusions, but it must be noted that the theory ignores the mobility of the charge carriers, just as it ignores the fact that this mobility (and thus also the conductivity) is different in the boundary layer from that in the bulk phase [251].

In the 1940's solid-state physicists began to realize that the electric double layers formed on solids and producing a space charge may play an important role in the understanding of surface problems. This led later to the development of the "barrier-layer theory". Rectification is due to the electric double layer formed where the metal and semiconductor are in contact, and results in the facilitation of the flow of the current in one direction and its hindrance in the other (see Kittel [loc. cit.]). As regards the relationship between metal and semiconductor the metal can be substituted by an adsorption "state" from where, provided the value of the electrochemical potential permits, electrons may pass into the semiconductor, or vice versa. There is a perfect analogy between the two phenomena. In the 1950's, mainly on the initiative of Soviet, American, French and German schools respectively [252], these ideas served as the detailed foundation of the barrier-layer theory of chemisorption.

As experimental techniques improved, support was given to the earlier conviction [253] that when the change in conductivity of the semiconductor due to chemisorption is evaluated, both the type of the semiconductor (intrinsic, n- or p-type) and the properties of the adsorbate must be taken into consideration. It will be seen later that in the case of n-type semiconductors when donor type molecules are adsorbed, and in the case of p-type semiconductors when acceptor molecules are adsorbed, the concentration of charge carriers (in the conduction bands the electrons, and in the valence band the defect electrons) increases so that this type of chemisorption is called cumulative chemisorption. In contrast, when an n-type semiconductor adsorbs donor-type molecules, the number of charge carriers decreases (depletive chemisorption).

Although solid-state physics has apparently gained more and more ground in the theory of chemisorption on semiconductors, there are many investigators with different opinions. Mention has already been made in Section 8.1 of the dangers arising from recognizing the existence of surface states while, at the same time, completely neglecting their role in surface phenomena. The same also holds for semiconductors of course. Quantum chemists are convinced that no significant progress can be expected in the theory of chemisorption without the elaboration of relatively reliable and rapid computerized methods for the calculation of surface states.

What has been said so far in a sense determines the content and grouping of the following Paragraphs in which we wish to present some of the details of the subject given in the title of this Section. This rough outline of the development, however, may have given the impression that the electronic theory of chemisorption on semiconductors proceeded from its beginning up to the present time without deviation, parallel to the growth of information on the band theory of semiconductors. In fact, there is also another trend, quite distinct from the electronic configuration of semiconductors, which tries to explain the chemisorption properties of oxide-type semiconductors starting from the presence of two types of active centres in oxides, the metal ion and the oxygen ion. In principle, this offers the possibility of attributing the properties of individual semiconductors to the different electronic configurations of the metal ions [152]. To mention only a single example: it is obvious from what has been said that such a theory cannot explain the difference between cumulative and depletive chemisorption. It is problematic, however, whether these objections also hold for insulators. It seems that such an approach may provide the correct answer, for instance, to the problem of chemisorption on molecular sieves containing several types of metal ions [254].

PARAGRAPH 8.2.1. BAND STRUCTURE OF SEMICONDUCTORS AND CHARGE TRANSFER DURING CHEMISORPTION

In theoretical considerations the particles adsorbed on the surface of the semiconductor are usually treated like the bulk contaminants, provided these surface impurities are also capable of exchanging electrons (or defect electrons) with the adsorbent. In semiconductors all energy states whose occupation is not forbidden by the band theory under the given conditions are available to the charge carriers. Population of the bands is determined by statistical rules which, in favourable cases, can be calculated with fair reliability by means of Fermi statistics.

In semiconductors the existence of three types of electron states are assumed.

There is nothing which distinguishes the electrons occupying the Bloch states from the conduction electrons of metals; they may change their positions within the crystal lattice almost freely and may occupy all lattice points with equal probability. Their energies are concentrated in well-defined energy bands (conduction bands), which are separated from each other by forbidden bands. In the case of a perfect semiconductor with intrinsic conduction properties, the band containing the lowest energy states, the valence band, is completely filled at the absolute zero point, while the conduction bands are totally empty. Under these conditions the semiconductor behaves as an insulator. Thermal excitation populates the conduction bands with electrons (and, at the same time, the valence bands with defect electrons). Conduction is attributed to these charge carriers.

In crystals of non-stoichiometric composition the bulk defects, called by the generic term "impurities", may bind electrons (or defect electrons) and become their localization centres. In the case of donor-type impurities, the energies of the localized electron states are somewhat lower than for those at the bottom of the conduction band, while in the case of acceptor-types the energies are somewhat higher than those of the upper part of the conduction band (see Figs I.10 and I.11). If it is assumed that the localized electron or defect electron is situated on a Bohr orbital, the order of magnitude of the ionization energy, $|E_i|$, can be estimated from the relation:

$$|E_i| = \frac{2\pi^2 m_e^* e^4}{n^2 h^2 \varepsilon^2} \qquad \text{(II.187)}$$

(after substitution of m_e^* and m_h^*, the effective electron and defect electron mass, respectively, and the dielectric constant of the medium, ε). With $n = 1$* for donor-type defects $|E_i| \approx 0.05$ eV, and for acceptor-types $|E_i| \approx 0.08$ eV. (This difference is due to the different effective mass of the electrons and the defect electrons.) The value of the ionization energy indicates that at low temperatures the electron (or defect) is not separated from the parent atom, but ionization occurs after the provision of relatively low energy. After ionization the delocalized electron or defect electron is capable of free movement within the conduction and valence bands of the semiconductor [255]. (For the calculation of the concentrations of delocalized electrons and defects electrons see Sections I.3.4–5–6.)

Finally, as has been repeatedly pointed out earlier, with real semiconductors of finite size the existence of surface states must also be considered. These energy states are identical with the energies of the two-dimensional Bloch states. The attribute "two-dimensional" is intended to indicate that

p* Investigation of the optical properties of these substances revealed long ago that hysical meaning can be ascribed to quantum numbers $n > 1$ too.

these energy states can be occupied only by electrons from the surface atoms of the lattice. Figuratively, it can be said that, in a sense, these electrons are reminiscent of the electrons localized on the bulk impurities of semiconductors, since their energies are lower than those of the electrons in the conduction band. As they belong not to a single surface atom, but to all surface atoms simultaneously, their energies widen out into a narrow band. However, the values of the energies are mostly unknown or only approximately known for a few substances, and until recently this hindered the correct interpretation of their role in the development of chemisorption bonding.

Semiconductors with intrinsic conduction properties (primarily elementary Ge and Si) are not, or only occasionally, used as catalysts, and then mainly in studies of theoretical interest (e.g. in the catalytic decomposition of silanes and germanes). The most interesting of the n- and p-type semiconductors are the various oxides, mainly n-type conducting ZnO with a metal excess (catalyst of methanol synthesis) and p-type conducting NiO and Cr_2O_3 with cation deficiencies (Cr_2O_3 is the catalyst or catalyst base material in several dehydrogenation and hydrogenation processes).

Our knowledge of the development of the energy band structure of semiconducting oxides may be summarized as follows: the electrons of the metal ions in oxides occupy lower energy states than those of the oxygen ion, so that to a first approximation, from the aspect of chemisorption bonding, it is more important to be acquainted with the energy state of the oxygen ion lattice. In addition, because of the greater distance between the metal atoms, the overlapping of the corresponding electron wave functions is practically negligible. The 2s and 2p electron states of the isolated O^{2-} ion are completely filled, and hence so too is the band formed by them within the oxide crystal. Above this band, and separated from it by a forbidden band of given width, is the conduction band formed by the splitting of the 3s and 3p states (in the case of oxides of stoichiometric composition the conduction band is unfilled).

For ZnO with a metal excess it may be assumed that the electron of the donor-type Zn atom occupies a localized electron state with an energy below the conduction band. When excited with approximately 0.05 eV, ionization may take place according to

$$Zn \rightarrow Zn^{+} + e$$

and the electron may pass into the initially vacant conduction band of ZnO. The situation is quite similar for NiO, where the electron transition according to

$$Ni^{3+} + e \rightarrow Ni^{2+}$$

causes the delocalization of the defect electron localized on the Ni^{3+} ion, and the defect electron passes into the valence band of NiO. The excitation energy in the second case is about 0.08 eV.

Similar behaviour has been assumed for the electron or defect electron originating from the adsorbate: the electron (or defect) initially occupying the electron state localized on the adsorbate may be delocalized by ionization and pass into the conduction or valence band of the semiconductor.

Ionization, or more precisely, ionization of all electrons or defect electrons occupying localized adsorption states, however, is not a necessary condition of the formation of chemisorption bonding between the substrate and the adsorbent, as suggested by certain investigators. The importance of this fact has been emphasized by Wolkenstein [256] since the beginning of the 1950's. The chemisorbed particle is thus capable of forming a non-ionic bond with the surface without actual charge transfer. Wolkenstein gave the symbol *CL* to this type of bond (*C* for the chemisorbed particle and *L* for lattice) and called it a "weak" form of chemisorption (see Chapter VI.7).

It is quite obvious that the electron wave function of the adsorbed neutral atom will be distorted in this case too, and will, so to speak, merge with the crystal lattice. According to Wolkenstein's terminology the "weak" bond becomes strong when the electron leaves this wave function, that is, it becomes delocalized (or when a defect electron becomes localized) and leaves an adsorbed ion behind. Wolkenstein introduced the symbols *CeL* (acceptor bond) and *CpL* (donor bond) for the designation of "strong" chemisorption bonding, depending upon whether it represents the localization of an electron or defect electron (or the delocalization of a defect electron or electron). As regards their nature, both bonding types may be ionic, or purely covalent, or any intermediate form. In the long run, this is determined by the distribution of the electrons or defects responsible for the bond formation between the adsorption centre and the adsorbed particle.

PARAGRAPH 8.2.2. THE BOUNDARY-LAYER THEORY OF CHEMISORPTION

Depending on the type (n- or p-type, symbolized as A_n and A_p) of the semiconductor and the properties of the adsorbate (acceptor or donor-type, symbols S_a and S_d) four adsorbent–adsorbate combinations are possible. In two ($A_p - S_a$ and $A_n - S_d$), adsorption causes a rise in the number of charge carriers, while in the other two ($A_n - S_a$ and $A_p - S_d$) adsorption results in a decrease in the number of charge carriers. The first is called cumulative, the second depletive chemisorption. A change in the number of charge carriers not only affects the conductivity of the semiconductor crystal, but may also have other interesting consequences worth closer scrutiny.

As already mentioned, because of its high electron affinity the O_2 molecule produces negatively-charged surface dipole layers on both metals and semiconductors. Depending on the type of the semiconductor, there are several possibilities for the withdrawal of the electron. In the case of n-type ZnO it is highly probable that chemisorption bonding is brought about by the contribution of those electrons which initially occupied the localized electron states and passed into the conduction band as a result of ionization, since the electrons of highest energy are the most easily withdrawn from the semiconductor. On the other hand, the conduction band of ZnO is populated by a relatively low number (of the order of 10^{18} cm^{-3}) of electrons. In principle, each Zn atom at an interstitial defect site furnishes one electron to the conduction band, so that the development of higher coverages (approximating to $\theta = 1$) is unlikely. In fact, even at low coverages the electron withdrawal must involve an increasingly deeper layer of the crystal. This layer, depleted of electrons and having a thickness of the order of 10^{-5}—10^{-6} cm, is called the boundary layer. By producing an appropriate potential barrier, the adsorbed layer, consisting of negatively-charged adsorbed ions, hinders electron supply from the bulk phase, causing a significant drop in conduction, particularly in the boundary layer. This fact is implied in the term "depletive adsorption" ($A_n - S_a$ type).

The one-dimensional Poisson equation can be applied to the electrostatics of the boundary layer. The relation between the volume charge density, ϱ, and the resulting potential V is:

$$\frac{d^2V}{dx^2} = \frac{4\pi\varrho}{\varepsilon} = \frac{4\pi}{\varepsilon} = N_d e$$

where x is the distance from the surface measured in the opposite direction to the normal of the surface
ε is the dielectric constant of the medium (semiconductor crystal)
N_d is the volume concentration of positive charges (ionized donors, i.e. Zn^+ ions), and
e is the electron charge.

For simple models the equation can easily be solved [257]. Let us assume in agreement with the above that the boundary layer is a Schottky type boundary layer in which the donor impurities are completely ionized. For the chemisorption of O_2 to have an acceptor nature, the electron energy (E_a) assigned to the adsorbed molecule must be lower than that of the Fermi level: $E_a < E_F$. With the supplementary condition that at a depth L of the boundary layer the potential is zero, $V(L) = 0$ (where L is defined by the equation for the conservation of charge: $N_d L = N_a$, N_a being the number of molecules adsorbed on unit surface), while the charge density on the surface is $-N_a e$, the solution of the Poisson equation will be:

$$V(x) = \frac{2\pi N_d e x^2}{\varepsilon} - \frac{4\pi N_a e x}{\varepsilon} + \frac{2\pi N_a^2 e}{N_d \varepsilon}$$

from which we obtain the value of the surface potential:

$$V(x = 0) = \frac{2\pi N_a^2 e}{N_d \varepsilon} \quad \text{(II.188)}$$

For the uncovered surface the energy of chemisorption of O_2 can be calculated as follows [see Eq. (II.104)]:

$$Q_0 = e\Phi - E_e + \Psi$$

where E_e is the electron affinity of the O_2 molecule
$e\Phi$ is the work function of the semiconductor, and

Ψ is the interaction energy between the O_2^- ion and the semiconductor adsorbent.

Adsorption causes a distortion of the band structure of the semiconductor in the boundary layer (E_F remains unchanged, while the electron energy and with this $e\Phi$, belonging to the lowest edge of conduction band, increases) and with this $e\Phi$ will increase quadratically with N_a (see Fig. III.30a). If V_s is the value of the surface potential at which $Q_\theta = (\Phi + V_s)e - E_a + \Phi_\theta = 0$, the maximum attainable coverage can be estimated from Eq. (II.188). With the numerical values $V_s \approx 1$ V (1/300 absolute potential units), $\varepsilon \approx 13$ and $N_d \approx 10^{18}$ cm^{-3}, we obtain $N_a \approx 10^{12}$ cm^{-2}. Since the number of adsorption centres on the surface of oxides is of the order of 10^{15} cm^{-2}, this value of N_a is roughly equivalent to a coverage of $\theta = 10^{-3}$.

It follows further from these arguments that Q_θ decreases quadratically with the coverage. For other models linear dependences have been derived [258].

As regards the value of the coverage, the theoretical conclusions have been qualitatively confirmed by experimental findings. (Chemisorption of O_2 on ZnO is indeed very slight. More marked chemisorption, resulting in 16% coverage, has been observed in a single instance [259], but this particular ZnO sample had an unusually high Zn content.) At the same time it has to be stressed that quantitative agreement has not been found between the calculated and experimentally determined values of θ; in the majority of cases the experimentally measured coverages were significantly higher than the calculated values. As pointed out by a number of authors, this discrepancy is probably due to the simplifications introduced into the chosen model. Of these, the complete ionization of the adsorbed atoms is, in particular, a rather doubtful condition. If this is omitted from consideration, it is not a necessity that the adsorption shall end at the coverage determined by the boundary-layer theory, because, as well as chemisorption resulting in adsorbed ions, adsorption without charge transfer may also take place. We shall return later to this same problem.

According to what has been said so far, chemisorption bonding between O_2 and ZnO is brought about by the mediation of the free electrons in the conduction band of the ZnO. At present, the literature contains no answers to such questions as to whether the adsorption of the adsorbed O_2^- ions is localized, and if so, in the neighbourhood of which bulk ions, or the question of the nature of the forces keeping the adsorbed ion on the surface. It should be added that the views on the origin of the bonding electrons are not in complete agreement either. It appears that the Soviet school attributes major importance in the formation of the bonding not to the free electrons but to the electrons occupying localized electron states (see Wolkenstein, [loc. cit.]).

In p-type NiO and Cu_2O the conduction band is empty, so that when O_2 is chemisorbed, electrons can be withdrawn only from the practically full valence band. In contrast to n-type conductors, the electron transition $O_2 + e \rightarrow O_2^-$ does not reduce, but increases the number of charge carriers, i.e. defect electrons, in the valence band (cumulative chemisorption), so that the chemisorption of O_2 on these semiconductors is accompanied by an increase in conductivity. It is further typical of cumulative chemisorption that no potential barrier develops on the surface of the adsorbent, because, due to the almost full valence band, sufficient bonding electrons are available in the immediate neighbourhood of the semiconductor surface to ensure complete monomolecular coverage. Chemisorption of O_2 on NiO and Cu_2O is of the $A_p - S_a$ type.

The chemisorption of H_2 on ZnO may serve as an example of $A_n - S_d$ type cumulative chemisorption. This leads to the formation of surface OH^- ions (see Section 7.1):

$$1/2\ H_2 + O^{2-} \rightarrow OH^- + e$$

The electrons of H_2, chemisorbed on NiO according to the previous mechanism, fill the holes of the valence band (depletive, $A_p - S_d$ type chemisorption). Since the number of holes in the valence band is of a similar order as that of free electrons in the conduction band of the ZnO, increasingly deeper layers of NiO will become involved in electron acceptance. Parallel to the filling of the holes in the valence band, the conductivity will decrease and, as with the chemisorption of O_2 on ZnO, the potential barrier will rise, at a certain point impeding the continuation of the chemisorption process.

PARAGRAPH 8.2.3. ENERGY CHANGES ASSOCIATED WITH ELECTRON TRANSITIONS

We have seen in the preceding Paragraph that in the charge transfer between semiconductor and adsorbed atoms sometimes the electrons of the conduction band, and, at other times, the electrons in the valence band of the semiconductor are involved. The possibility of electrons (or defect electrons) occupying localized states being withdrawn directly, without preliminary ionization, from the primary localization centres cannot *ab ovo* be excluded either. The energy condition of the new donor and acceptor levels created by the substrate will finally decide which of these possibilities applies for a given semiconductor and substrate. Much work has been devoted to the clarification of this important problem, mainly by the Soviet school, as this may finally lead to information regarding the transition of the adsorbed particle from one form of bonding to another, and also the equilibrium and the ratio of the adsorbed forms with and without charge. Utilization of the later arguments is hindered by the practical obstacle that the energies involved are still unknown, as suitable calculation methods are lacking.

Electron transitions accompanying adsorption are best illustrated by the diagram of the semiconductor energy bands. In Fig. II.48 the upper shaded part represents the conduction band (the electron energy at its lower edge is E_c), while the bottom part represents the valence band (at its upper end the energy is E_v). The width in energy of the forbidden band is E_g. The chemisorbed particles can be illustrated by the acceptor level (E_a) or the donor level (E_d). A "weakly" bound particle (Wolkenstein's terminology) (that is a particle which, in accordance with what has been said above, has not exchanged its charge with the adsorbent) may have simultaneous affinity for both the free electron and the defect. and it is then represented by an acceptor and a donor level. The positions of E_a and E_d are jointly determined by the properties of the semiconductor adsorbent and the particle.

Endothermic particle transition between the energy levels occurs as a result of thermal excitation. If the energy level E_a is filled by means of the substrate, in the "weak" adsorption bonding, trapping, or localizing an electron, this means that the "weak" bond is transformed into a "strong" acceptor bond. Electron transition can be realized in two ways: (i) the electron is transferred from the conduction band to the E_a level (for instance when O_2 is chemisorbed on ZnO):

$$(1)\ CL + eL \rightarrow CeL \qquad (v^- = E_a - E_c < 0)$$

or (ii) by means of excitation the electron passes from the valence band into the E_a level (for instance when O_2 is chemisorbed on NiO or Cu_2O):

$$(2)\ CL \rightarrow CeL + pL \qquad (v^+ = E_a - E_v > 0)$$

In a similar manner the substrate producing the donor level, E_d, might localize a defect, while the "weak" bond is converted into a "strong" one, by withdrawing and delocalizing an electron from the E_d level. This, too, might take place either by an electron in the E_d level recombining with a positive defect of the valence band (in other words, by "filling" the hole, as is the case when H_2 is adsorbed on NiO):

$$(3)\ CL + pL \rightarrow CpL \qquad (w^- = E_v - E_d < 0)$$

or by means of excitation when the electron is transferred into the conduction band (e.g. when H_2 is adsorbed on ZnO):

$$(4)\ CL \rightarrow CpL + eL \qquad (w^+ = E_c - E_d > 0)$$

The various electron transitions, with the accompanying explanations, speak for themselves. An additional comment seems fitting, however, as the figure permits some remarks concerning Wolkenstein's somewhat inconsistent "free valence" concept [261].

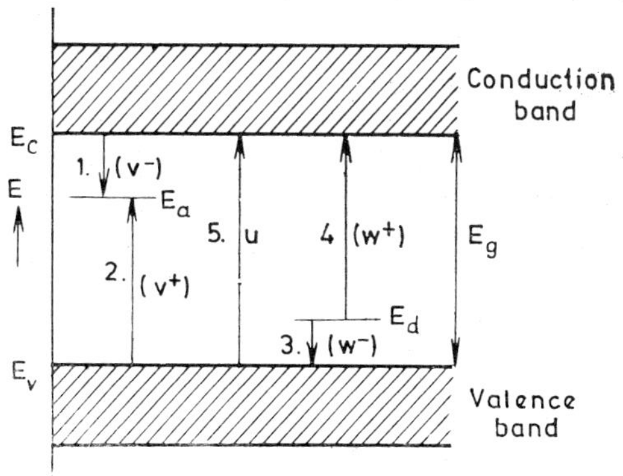

Fig. II.48. Diagram of electron transitions accompanying chemisorption on semiconductor type adsorbents

Wolkenstein claims that the free electrons and the free positive defects of the lattice fulfil the role of the "free valence" of the adsorbent. Thus, if an electron initially in the valence band is "lifted" into the conduction band by an electron transition according to (5) in Fig. II.48:

$$(5)\ L \to eL + pL \qquad (u = E_c - E_v > 0)$$

simultaneously a positive (free electron) and a negative (defect electron) "free valence" are created. The various defects of the crystal lattice must also be considered as "free valences". (This would mean that, for example, the Zn^+ ion in ZnO is equivalent to a free electron, i.e. to a positive "free valence", while the O^- ion is equivalent to a free positive defect, or, in other words, to a "free negative valence" among the normal structural components of the lattice.) Wolkenstein's free valence has nothing to do with the free valence mentioned briefly in connection with surface states. If we accept Wolkenstein's views, the former electron transition can be interpreted as the formation and disappearance of "free valences". For instance, in process (1) a "free valence" (eL) disappears, while process (2) produces a negative "free valence" (pL), etc.

The order of magnitude of the energy change accompanying each electron transition is determined by the width of the forbidden band, E_g. This may vary from a few tenths of an eV to several eV, and thus may represent a significant fraction of the overall interaction energy.

PARAGRAPH 8.2.4. EQUILIBRIUM BETWEEN DIFFERENT TYPES OF CHEMISORPTION

The lack of agreement between the coverages estimated from the initial primitive models and those experimentally observed was one of the deficiencies of the boundary-layer theory, or more generally, of the semiconductor theory for the interpretation of chemisorption phenomena. Boudart [252] and members of the Wolkenstein school soon recognized these contradictions as a consequence of the oversimplification of the statistical problem by the assumption that all "impurities" in the bulk phase, that is, those created in the course of adsorption, are present in their ionized states. The only correct and useful solution is obviously to account for all the electron states involved and to calculate their populations statistically. In other words: if N particles are distributed among k states so that in each state the number of particles is N_j ($j = 1, 2 \ldots k$), it is required to know the values of the corresponding "mole fractions", $x_j = N_j/N$ (where $N = \Sigma N_j$). Different authors have tried to find an answer to this question by means of various methods. One frequently used approach is, in essence, the application of the law of mass action to the chemisorbed "impurities". Quite equivalent to this are the relations derived from Fermi–Dirac statistics. We next show an example of the latter, as presented by Wolkenstein [261].

If equilibrium is established between the electron transitions (1) and (4) (see preceding Paragraph), some of the adsorbed N particles will be bound to the surface by "weak", and some by "strong", donor or acceptor bonds. Let N^0, N^- and N^+ represent the number of particles in the corresponding states per unit surface and let us introduce the following "mole fractions":

$$x^0 = \frac{N^0}{N} \quad x^- = \frac{N^-}{N} \quad \text{and} \quad x^+ = \frac{N^+}{N}$$

for which, of course:

$$x^0 + x^+ + x^- = 1$$

According to Fermi–Dirac statistics:

$$\frac{N^-}{N^0 + N^-} = \frac{1}{1 + \exp[(\varepsilon_s^- - |v^-|)/kT]} = \frac{1}{1 + \exp[(E_a - E_F)/kT]}$$

and

$$\frac{N^+}{N^0 + N^+} = \frac{1}{1 + \exp[(\varepsilon_s^+ - |w^-|)/kT]} = \frac{1}{1 + \exp[(E_F - E_d)/kT]}$$

The meanings of these symbols are given in Fig. II.49. From the last three equations we obtain for the unknown "mole fractions":

$$x^0 = \frac{1}{1 + 2A(\Delta u)\cosh\xi}; \quad x^- = \frac{A(\Delta u)e^{-\xi}}{1 + 2A(\Delta u)\cosh\xi};$$

$$x^+ = \frac{A(\Delta u)\,e^{\xi}}{1 + 2A(\Delta u)\cosh\xi}.$$

where

$$A(\Delta u) \equiv \exp(-\Delta u/kT) \quad \text{and} \quad \xi \equiv \frac{\varepsilon_s^+ - u^+}{kT}$$

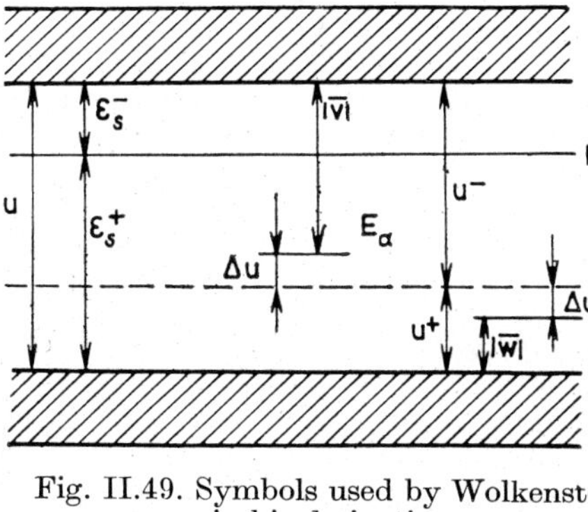

Fig. II.49. Symbols used by Wolkenstein in his derivation

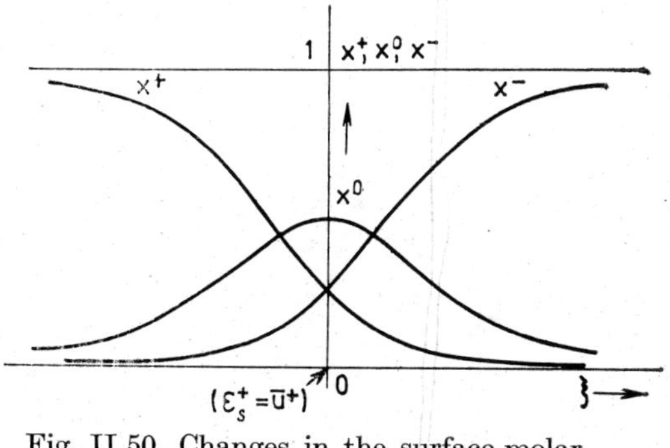

Fig. II.50. Changes in the surface molar fractions of particles in "weak" (x^0), strong "acceptor" (x^-), and "donor" (x^+), bonds vs. ξ characterizing the position of the Fermi level

Δu and u^+ are given by E_a and E_d, so that with ε_s^+ the "mole fractions" depend essentially only upon the energy of the Fermi level. The various functions have been plotted in Fig. II.50. It appears from this figure that as the Fermi energy (E_F) increases (and with it ε_s^+ and ξ), x^- monotonously increases while x^+ monotonously decreases, that is, an increasing fraction of the particles form acceptor bonds with the surface. The number of particles in "weak" chemisorption bonding is a maximum at $\xi = 0$, that is, when the Fermi energy is the arithmetic mean of E_a and E_d. In extreme cases the particles produce either only donor, or only acceptor, bonds.

The electron theory has thus led to the important result that the equilibrium between the simultaneously existing chemisorption bond types is essentially determined by the position of the Fermi level.

Using other initial conditions, other relations have also been derived. The very extensive literature on the subject has not yet been surveyed from uniform aspects.

Partial and complete coverages with respect to particles with and without charge ($\theta^0 = N^0/n_s$, where n_s is the number of adsorption sites on unit surface, θ^-, etc. and $\theta = \theta^0 + \theta^- + \theta^+$) are sometimes relatively simple (linear, or resembling the mathematical formulation of the Langmuir isotherm), and, at other times, highly complex functions of the total pressure p, and cannot be brought to an explicit form. The isotherms often contain many solid-state physical parameters, so that their overall effect is extremely involved and very difficult to check by experiment. This may explain the lack of reports in the literature relating to investigations aimed at establishing agreement between the experimentally determined sorption isotherm and all the parameters of the isotherm equation.

PARAGRAPH 8.2.5. VALIDITY AND LIMITS OF APPLICABILITY OF THE BOUNDARY-LAYER THEORY

All in all, the extension of the band theory of semiconductors to chemisorption has proved very fruitful. It has explained the why and how of the influence of electric properties on the adsorption activity of the semiconductor towards a given substrate, and also the effect of adsorption on the semiconductivity. Let us now examine how far the theory has been able to meet expectations.

We have seen that the sorption properties of the semiconductor is closely related to its conduction type, and to the quality of the substrate. The $A_n - S_d$ and $A_p - S_a$ adsorbent—substrate pairs have been described by the term cumulative sorption, a name intended to stress the creation of new free charge carriers in the semiconductor by means of sorption. Similarly, it has been shown that with the systems A_n—S_a and A_p—S_d only minor sorption can be expected because of the development of a potential barrier, and that sorption is accompanied by a decrease in conductivity. It is a logical conclusion worth adding that a correlation must exist between the conductivity in vacuum and the sorption ability. The correlation between conductivity in vacuum and the quantity adsorbed should therefore exhibit a synbate course for n-type semiconductors and acceptor gases (A_n—S_a) and an antibate course for p-type semiconductors and acceptor gases (A_p—S_a). Similar conclusions can be drawn for the other two pairings.

Finally, it should be mentioned that the type of conduction of a given semiconductor can be assessed from the position of the Fermi level. The nearer the Fermi level is to the conduction band and the farther from the valence band, the more the electron conduction (n-type) will gain preponderance over the defect conduction in the semiconductor, and vice versa. Hence, we may add to what has been said above that in those cases when the Fermi level is situated somewhere in the middle of the forbidden band and donor or acceptor properties do not dominate in the substrate either, quite strictly speaking the adsorption system cannot be included in any of the above four types. In these cases adsorbed atoms in a great variety of bonding forms may be present simultaneously on the surface of the adsorbent (see preceding Section).

The qualitative validity of these conclusions is also supported by experiment.

It has been demonstrated that carbon monoxide may be chemisorbed on semiconductor oxides by two different mechanisms: at low temperatures a reversibly rapid chemisorption takes place, resulting in weak bonds (Wolkenstein's "weak" bond?), while at temperatures above room temperature the chemisorption is slow and of the donor type:

$$CO + 2\,O^{2} \rightarrow CO_3^{2-} + 2\,e$$

In the case of p-type semiconductors the electrons pass into the valence band, while with n-type semiconductors they reach the conduction band. Theoretically it might be expected that following the chemisorption of CO the adsorbent surface will have a higher sorption power for acceptor gases, e.g. for O_2. Experience has provided extensive confirmation of this conclusion: on CuO [139] and ZnO containing Cr_2O_3 [139] additional oxygen sorption to almost exactly half the volume of the adsorbed CO was observed.

The similar interaction between CO_2 and oxygen ions

$$CO_2 + O^{2-} \rightarrow CO_3^{2-}$$

causes no saturation as regards oxygen, and moreover in these processes (and also in those with SO_2 for example) the semiconductor nature of the oxide is also less prominent.

On most oxides two types of mechanism are assumed for the chemisorption of O_2, too. On ZnO, for instance, at lower temperatures (room temperature and below) the adsorbed layer probably consists of O^-, and at higher temperatures, (200–300 °C and above), of O^{2-} ions [259]. The existence of O_2^- ions has been assumed, but not yet confirmed by experimental evidence.

The two types of CO and O_2 sorption are in good agreement with the boundary-layer theory: weak sorption causes (probably) no change in conductivity, and can therefore be treated in accordance with Wolkenstein's concept as a sorption producing a "weak" bond, with all its consequences, while the other types must be brought into agreement one by one with the requirements of the theory.

"Weak" chemisorption can be observed at a low temperature in the case of H_2 too. The "weakness" of the bond, however, is relative, for substances on which this type of chemisorption occurs catalyse the H—D exchange. At a higher temperature, as a result of strong donor-type chemisorption, adsorbed OH^- ions are formed on the surface of semiconductor oxides, by the same mechanism as described above. As also occurs in the case of CO, following chemisorption the surface appears to be unsaturated towards O_2. All this is in agreement with the demands of the boundary-layer theory. However, if the various oxides are compared on the basis of their activities in the H—D exchange, a somewhat peculiar result is obtained which, at first glance, seems to contradict the earlier concepts: the oxides which are the most active (e.g. Cr_2O_3, Co_3O_4, NiO, etc.) are those in which the metal ions possess an optimum number (not too few, but not too many either) of unpaired d-electrons (in Cr_2O_3 the electron configuration of the outermost shell of the Cr^{3+} ion is: $3d^3$; of cobalt in Co_3O_4: $3d^6$ or $3d^7$, of Ni^{2+} in NiO: $3d^8$ [152]).

A similar difficulty is encountered when we try to explain the relatively high activities of the oxides of transition metals in certain hydrogenation and dehydrogenation reactions. It is not impossible that a covalent bond is formed between the d-shell of the metal ion and the gas molecule, but nor is it unreasonable to assume that the oxide which catalyses hydrogenation is partly reduced to metal and that it is the metal which acts as catalyst.

These findings should be brought into line with the experimental data available on the stabilities of the complexes of olefins with metal ions. It is surely not accidental that there is a similarity in the activities of the oxides in the H—D exchange and in hydrogenation, as well as in the complex-forming tendency of the metal ion (see Paragraph 8.1.1).

All this is, in fact, a reversion to the earlier views expressed in 1937 by De Boer and Verwey, who claimed that the overlap integrals of the electrons on the d-orbitals of transition metal oxides are so small that it is not worthwhile to consider them as bands. This apparent contradiction is, in our opinion, no reason for major concern, because in the absence of many important details it is difficult to conceive behaviour which would not contradict the boundary-layer theory. We might add that the dispute: atomic localized bond versus band is at least as old as solid-state physics itself.

As regards sorbability, the elementary semiconductors, e.g. Si and Ge, present a greater problem. In contrast with the conclusions from the boundary-layer theory, Brennan et al. [262] observed an unexpectedly high chemisorption of O_2 on vacuum-evaporated Ge and Si films (at room temperature a monomolecular layer is formed almost instantaneously, and even at -195 °C coverages of around $\theta = 0.5$ develop), apparently quite independently of the properties of the semiconductor. Again in a quite irregular manner, these semiconductors are unable to chemisorb H_2.

The explanation of this phenomenon [263] is the fact that in elementary semiconductors during transition from extreme n-type to extreme p-type (e.g. as an effect of doping) the position of the Fermi level hardly changes (the change in the case of Si is 0.2 eV and even less for Ge). Hence, in these two characteristic types of semiconductors not the "electronic factor" but the chemical properties determine the chemisorption behaviour.

Occasional contradictions between the actual change in conductivity due to chemisorption and the expected change might be the result of (in addition to many errors which cannot be discussed here in detail, but the majority of which can be attributed to experimental errors) the current lack of knowledge regarding the mobilities of the charge carriers, and on the relationship between mobility and sorption. So far, little attention has been paid to the Coulomb attraction of the parent ion of opposite charge, or, more generally, to the shadowing effect of opposite charges, though there are definite references to this subject in the literature [264]. Some authors consider this the main deficiency of the band theory.

Application of the band theory of semiconductors may equally help in the understanding of heterogeneous catalysis.

There is a similar parallel between the electrical properties of semiconductors and their catalytic activities as exists between the conductivity and sorption properties. Depending upon the properties of the semiconductors and the substrates (where, by substrate S, we do not mean necessarily the initial substance of unchanged chemical composition, but possibly an intermediate or end-product formed from the former by chemical transformation), conductivity in vacuum and catalytic activity may exhibit the same or opposite changes. The changes agree when both reaction and semiconductor are n- or p-type (the reaction is n-type when the substrate, as defined above, has some influence on the reaction rate and is of the acceptor-type, or, using the earlier symbols, is of the S_a type, while a p-type reaction is assigned to the substrate S_d type). In mixed types (where the reaction is n-type and the semiconductor p-type, or vice-versa) changes in conductivity and catalytic activity are of opposite sign. This finding is in perfect agreement with what has been said above concerning the relationship between the conductivity and adsorption properties of adsorbents.

Let us consider some relevant experimental results.

According to Dell et all. [265] p-type conducting oxides are the most active catalysts of N_2O decomposition, since only they are capable of significant chemisorption of the intermediate of the n-type reaction, the adsorbed O^- ion. NiO doped with Al_2O_3 has a higher catalytic activity than NiO without an additive [266]. This experimental finding might be related to the decrease of the concentration of the defect electrons and to the associated higher adsorbed O^- ion coverage on the additive-containing NiO surface.

It has been observed, again with NiO, that the addition of Li^+ reduces the rate of oxidation of CO, (because of its size the Li^+ ion is capable of substituting the Ni^{2+} ion in the lattice), while the incorporation of Cr^{3+} or Al^{3+} ions raises the rate of oxidation of CO [267]. Keier, Roginskii and Sazonova arrived at a similar conclusion [268]. Analogous with the preceding example, it might be suggested that the lower activity caused by the Li^+ additive, just as before, is related to a reduced O^- coverage. However, Keier et al. also measured the sorption of CO and O_2 and found, contrary to expectation, a higher O_2 sorption and a reduced CO sorption on NiO with Li^+ additive. Investigation of the work function finally clarified the behaviour of Li^+ ions on the surface, where they act not as acceptor, but as donor-type defects.

With the help of the boundary-layer theory it is easy to interpret the experimental result that O_2 chemisorbed on a ZnO surface reduces the rate of decomposition of isopropanol [269]. The alcohol decomposition is an n-type reaction, and the decrease in the reaction rate can be explained by the competition of the alcohol and O_2 for the free electrons.

The above discussion is intended to show that a good start has been made in the elucidation of the relationship between catalytic activity and semiconductor properties, but this field still includes many theoretical and practical problems awaiting solution.

REFERENCES TO PART II

1. Roberts, J. K.: Proc. Roy. Soc. **A 152,** 445 (1935)
2. Taylor, H. S. and Williamson, A. T.: J. Am. Chem. Soc. **53,** 2168 (1931)
3. Trapnell, B. M. W.: Proc. Roy. Soc. **A 206,** 39 (1951)
4. Brunauer, S.: Physical Adsorption of Gases and Vapours. Univ. Press, Oxford, 1944 p. 69
5. Knözinger, H.: private communication
6. Wright, M. M. and Taylor, H. S.: Can. J. Res. **27,** 303 (1949)
7. See in Garner, W. E.: J. Chem. Soc. 1239 (1947)
8. Born, M. and Mayer, J. E.: Z. Physik **75,** 748 (1932)
9. Verwey, E. J. W. and De Boer, J. H.: Rec. Trav. Chim. **59,** 633 (1940)
10. Brunauer, S.: The Adsorption of Gases and Vapours Univ. Press, Oxford–New York, 1944, Chapter IV
11. Ross, S. and Olivier, J. P.: Adv. Chem. Ser. No. **33,** 309 (1961)
12. Morgenau, H.: Rev. Mod. Phys. **11,** 1 (1939)
13. Lennard-Jones, J. E. and Dent, B. M.: Trans. Faraday Soc. **24,** 92 (1928)
14. Avgul, N. N., Kiselev, A. V., Ligina, I. A. and Poskus, D. P.: Izvest. Akad. Nauk. USSR, **1959,** 1196
15. Polányi, M. and London, F.: Naturwiss. **18,** 1099 (1930)
16. Crowell, A. D.: J. Chem. Phys. **22,** 1397 (1954)
17. See e.g. De Boer, J. H. and Custers, J. F. H.: Z. phys. Chem. **B 25,** 225 (1934)
18. De Boer, J. H.: Adv. Colloid Sci. **3,** 44 (1950)
19. Mignolet, J. C. P.: Disc. Faraday Soc. No **8,** 105 (1950); J. Chem. Phys. **21,** 1928 (1953)
20. De Boer, J. H.: The Dynamical Character of Adsorption. Clarendon Press, Oxford, 1953 p. 146
21. De Boer, J. H. and Veenemans, C. F.: Physica **1,** 753 (1934)
22. Bosworth, R. C. L. and Rideal, E. K.: Proc. Roy. Soc. **A 162,** 1 (1937)
23. Bosworth, R. C. L.: Proc. Roy. Soc. **A 154,** 112 (1936)
24. Taylor, J. B. and Langmuir, I.: Phys. Rev. **44,** 423 (1933)
25. Langmuir, I.: Phys. Rev. **22,** 357 (1923)
26. Hill, T. L.: J. Chem. Phys. **16,** 181 (1948)
27. Cassel, H. M.: J. Phys. Chem. **48,** 195 (1944)
28. Hill, T. L.: J. Chem. Phys. **14,** 441 (1946)
29. De Boer, J. H.: loc. cit. (21), p. 170
30. Bard, Y. and Lapidus, L.: Catalysis Rev. **2**(**1**), 67–112 (1968)
31. Fowler, R. H.: Proc. Camb. Phil. Soc. **31,** 260 (1935)
32. Fowler, R. H.: Statistical Mechanics. Univ. Press, Cambridge, 1936
33. Halsey, G.: Adv. Cat. **4,** 259 (1952)
34. Gradshteyn, I. S. and Ryzhik, I. M.: Table of Integrals, Series and Products. Acad. Press, New York, 1956, p. 292
35. Frankenburg, W. G.: J. Am. Chem. Soc. **66,** 1827 (1944)
36. Davis, R. T.: J. Am. Chem. Soc. **68,** 1395 (1946)
37. Halsey, G.: Adv. Cat. **4,** 259 (1952)
38. Brunauer, S., Kowe, K. S. and Keenan, R. G.: J. Am. Chem. Soc. **64,** 751 (1942)
39. Rideal, E. K. and Trapnell, B. M. W.: Proc. Roy. Soc. **A 205,** 409 (1951)
40. Markham, E. C. and Benton, A. F.: J. Am. Chem. Soc. **53,** 497 (1931)
41. Bond, G. C. and Sheridan, J.: Trans. Faraday Soc. **48,** 651, 664 (1952)
42. Bokhoven, C.: Proc. 2nd Radio-Isotope Conf. Oxford, 1954 p. 53

43. Beeck, O.: Adv. Cat. **2**, 151 (150)
44. De Boer, J. H.: Adv. Cat. **8**, 141 (1956)
45. Bokhoven, C., Van Heerden, C., Wertrik, R. and Zwietering, P.: Catalysis. Reinhold Publ. Co. New York, 1955 Vol. 3, p. 312
46. Beeck, O.: Disc. Faraday Soc. **8**, 118 (1950)
47. Mayer, J. E.: J. Chem. Phys. **5**, 67, 75 (1937)
48. Kirkwood, J. G.: J. Chem. Phys. **3**, 300 (1935)
49. Born, M. and Green, H. S.: Proc. Roy. Soc. A **188**, 10 (1946); **A 189**, 103 (1947)
50. Brunauer, S., Emmett, P. H. and Teller, E.: J. Am. Chem. Soc. **60**, 309 (1938)
51. Cassie, A. B. D.: Trans. Faraday Soc. **41**, 450 (1945)
52. Hill, T. L.: J. Chem. Phys. **14**, 263 (1946); **17**, 772 (1949)
53. MacMillan, W. G.: J. Chem. Phys. **15**, 390 (1947)
54. Walker, W. C. and Zettlermoyer, A. C.: J. Phys. and Colloid Chem. **52**, 47, 58 (1948)
55. Anderson, R. B.: J. Am. Chem. Soc. **68**, 686 (1946)
56. Cook, M. A.: J. Am. Chem. Soc. **70**, 2925 (1948); **71**, 791 (1949)
57. Hill, T. L.: J. Chem. Phys. **15**, 767 (1947)
58. Coolidge, A. S.: J. Am. Chem. Soc. **48**, 1795 (1926); Halsey, G.: J. Chem. Phys. **16**, 931 (1947)
59. Schay, G. and Nagy, L.: J. Chimie Physique **58**, 49 (1961); Schay, G.: in Surface Area Determinations. IUPAC Publication, Butterworths, London, 1970 p. 273
60. Schay, G.: Acta Chim. Acad. Sci. Hung. **3**, 511 (1953)
61. Cassel, H. M.: J. Chem. Phys. **12**, 115 (1944); J. Phys. Chem. **48**, 195 (1944)
62. Frenkel, J.: Kinetic Theory of Liquids. Oxford Univ. Press, Oxford, 1946
63. Hill, T. L.: J. Chem. Phys. **550**, 668 (1949)
64. Polányi, M.: Verhandl. deutsch. phys. Ges. **15**, 55 (1916)
65. Jura, G. and Harkins, W. D.: J. Am. Chem. Soc. **66**, 1356 (1944)
66. Harkins, W. D. and Jura, G.: J. Am .Chem. Soc. **66**, 1362 (1944)
67. Brunauer, S.: loc. cit. (10), p. 419
68. Barrer, R. M. and Robins, H. B.: Trans. Faraday Soc. **47**, 773 (1951)
69. Defay, F., Prigogine, I., Bellemans, A. and Everett, D. H.: Surface Tension and Adsorption. Longmans, New York, 1966
70. Gibbs, J. W.: On the Equilibrium of Heterogeneous Substances. Collected Works, Longmans, New York, 1928 Vol. 1
71. Erikson, J. C.: Thermodynamics of Surface Phase Systems, Vols. I—V; Ark. Kemi **25**, 331, 343 (1965); Ark. Kemi **26**, 49, 117 (1966); Surface Sci. **14**, 221 (1969)
72. Schay, G.: J. Colloid and Interface Sci. **35**, 254 (1971)
73. Pauling, L.: The Nature of Chemical Bond. Cornell Univ. Press, New York, 1945
74. Barrer, R. M.: J. Chem. Soc. 1256 (1936)
75. Bull, H. I., Hall, H. M. and Garner, W. E.: J. Chem. Soc. 837 (1931)
76. Keyes, F. G. and Marshall, M. J.: J. Am. Chem. Soc. **49**, 156 (1927)
77. Taylor, J. B. and Langmuir, I.: Phys. Rev. **44**, 423 (1933)
78. Bosworth, R. C. L.: Proc. Roy. Soc. **A 162**, 32 (1937)
79. Eley, D. D.: Disc. Faraday Soc. **8**, 34 (1950)
80. Beeck, O.: Disc. Faraday Soc. **8**, 118 (1950)
81. Roberts, J. K.: Proc. Roy. Soc. **A 152**, 445 (1935)
82. Beeck, O., Cole, W. A. and Wheeler, A.: Disc. Faraday Soc. **8**, 314 (1950)
83. Trapnell, B. M. W.: Trans. Faraday Soc. **48**, 160 (1952)
84. Beeck, O.: Rev. Mod. Phys. **17**, 61 (1945)
85. Hayward, D. O. and Trapnell, B. M. W.: Chemisorption. Butterworths, London, 1964 p. 209
86. Farmer, M. F.: Leicester Chem. Rev. **No. 7**, 39 (1965)
87. Pauling, L.: Proc. Roy. Soc. **A 196**, 343 (1949)
88. Bond, G. C.: Catalysis by Metals. Acad. Press, London, 1962 p. 23
89. Cassel, H. M.: Z. Physik **26**, 862 (1925); **28**, 152 (1927); Z. Elektrochemie **37**, 642 (1931); J. Phys. Chem. **20**, 523 (1936)
90. Kemball, C. and Rideal, E. K.: Proc. Roy. Soc. **A 187**, 53 (1946)
91. Kemball, C. and Rideal, E. K.: Adv. Cat. **2, 233** (1950)

92. See e.g. Gomer, R.: Field Emission and Field Ionization. Harvard Univ. Press, Cambridge, 1961
93. Arthur, J. R. and Hansen, R. S.: J. Chem. Phys. **36,** 2062 (1962)
94. Arthur, J. R. and Hansen, R. S.: Ann. N. Y. Acad. Sci. **101,** 756 (1963)
95. Müller, E. W.: Z. Physik **131,** 136 (1951)
96. Müller, E. W.: Adv. Electron. Phys. **13,** 83 (1960)
97. Brattain, W. H. and Becker, J. A.: Phys. Rev. **43,** 428 (1933)
98. Bosworth, R. C. L.: Proc. Roy. Soc. **A 150,** 58 (1935)
99. Kemball, C.: Proc. Roy. Soc. **A 187,** 73 (1946)
100. Everett, D. H.: Proc. Chem. Soc. 38 (1957)
101. Rideal, E. K. and Sweett, F.: Proc. Roy. Soc. **A 257,** 291 (1960)
102. Drain, L. E. and Morrison, J. A.: Trans. Faraday Soc. **49,** 654 (1952)
103. Hill, T. L.: J. Chem. Phys. **17,** 762 (1949)
104. Kisliuk, P.: J. Chem. Phys. **30,** 174 (1959)
105. Wedler, G.: private communication
106. Dewar, J.: Proc. Roy. Soc. **A 74,** 122 (1904)
107. Benton, A. F. and White, T. A.: J. Am. Chem. Soc. **52,** 2325 (1930)
108. Langmuir, I.: J. Am. Chem. Soc. **40,** 1361 (1918)
109. Taylor, H. S.: J. Am. Chem. Soc. **53,** 578 (1913)
110. Lennard-Jones, J. E.: Trans. Faraday Soc. **28,** 333 (1932)
111. Schwab, G. M.: Catalysis. MacMillan, London, 1937
112. Ward, A. F. M.: Proc. Roy. Soc. **A 133,** 522 (1931)
113. Varga, K. and Fejes, P.: Acta Chim. Acad. Sci. Hung. **74,** 417 (1972)
114. Schuit, G. C. A. and De Boer, J. H.: Nature (London) **168,** 1040 (1951)
115. Roberts, J. K.: Proc. Roy. Soc. **A 152,** 445 (1935)
116. Trapnell, B. M. W.: Proc. Roy. Soc. **A 218,** 566 (1953)
117. Beebe, R. A. and Dowden, D. A.: J. Am. Chem. Soc. **60,** 2912 (1938)
118. See in Anderson, R. B. (Ed.): Experimental Methods of Catalytic Research. Academic Press, New York, 1968 p. 265
119. Lander, J. J. and Morrison, J.: J. Appl. Phys. **34,** 2298 (1963)
120. May, J. W.: Ind. Eng. Chem. **57,** 25 (1965)
121. Davisson, C. H. and Germer, L. M.: Phys. Rev. **30,** 737 (1927)
122. Germer, L. M.: Z. Physik **54,** 408 (1929)
123. Germer, L. M., Scheibner, E. J. and Hartman, C. D.: Phil. Mag., Ser. 8, **5,** 222 (1960)
124. Tucker, C. W.: Appl. Phys. Lett. **1,** 34 (1962)
125. Tucker, C. W.: Surface Sci. **2,** 516 (1964)
126. Ehrlich, G.: J. Phys. Chem. Solids **1,** 3 (1956)
127. Ehrlich, G.: J. Phys. Chem. **59,** 473 (1955)
128. Robins, J. L., Warburton, W. K. and Rhodin, T. N.: J. Chem. Phys. **46,** 665 (1967)
129. Wolkenstein, F. F. and Pesev, O.: Kinetika i Kataliz **6,** 95 (1965)
130. Laidler, K. J.: J. Phys. Chem. **57,** 318 (1953)
131. Tonks, L.: J. Chem. Phys. **8,** 477 (1940)
132. Brunauer, S., Love, K. S. and Keenan, R. G.: J. Am. Chem. Soc. **64,** 751 (1942)
133. Zeldovich, Ya.: Acta Physicochim. USSR **1,** 449 (1934); Roginskii, S. Z. and Zeldovich, Ya.: Acta Physicochim. USSR **1,** 554, 595 (1934); Elovich, S. Yu. and Zhabrova, G. M.: Zhur. Fiz. Khim. **13,** 1761 (1939)
134. Taylor, H. A. and Thon, N.: J. Am. Chem. Soc. **74,** 4169 (1952)
135. Low, M. J. D.: Chem. Rev. **60,** 267 (1960)
136. Porter, A. S. and Tompkins, F. C.: Proc. Roy. Soc. **A 217,** 529 (1952)
137. Logen, R. M. and Keck, J. C.: J. Chem. Phys. **49,** 860 (1968)
138. Glasstone, S., Laidler, K. J. and Eyring, H.: The Theory of Rate Processes. McGraw-Hill Co., New York, 1941
139. Garner, W. E. and Veal, F. J.: J. Chem. Soc. 1487 (1935); Dowden, D. A. and Garner, W. E.: J. Chem. Soc. 893 (1939); Garner, W. E. and Word, T.: J. Chem. Soc. 857 (1939); Word, T.: J. Chem. Soc. 1244 (1947); Garner, W. E., Gray, T. J. and Stone, F. S.: Proc. Roy. Soc. **A 197,** 294 (1949); Garner, W. E., Stone, F. S. and Tiley, P. F.: Proc. Roy. Soc. **A 211,** 472 (1952)
140. Parkyns, N. D.: J. Chem. Soc. 410 (1969)

141. Gregg, S. J. and Ramsay, J. D. F.: J. Phys. Chem. **73,** 1243 (1969)
142. Jakerson, V. I., Lafer, L. I., Ganiusevskii, V. Ya. and Rubinstein, A. M.: Izv. A. N. USSR **1,** 19 (1969)
143. See in Little, L. H.: Infrared Spectra of Adsorbed Species. Acad. Press, London, 1966 p. 74
144. See ref. (142). p. 26
145. Narayana, D., Subrahmanyam, V. S., Jagdish Lal, Mahmood Ali, M. and Kesavulu, V.: J. Phys. Chem. **74,** 779 (1970)
146. Eischens, R. P., Pliskin, W. A. and Low, M. J. D.: J. Catalysis **1,** 180 (1962)
147. Dent, A. L. and Kokes, R. J.: J. Phys. Chem. **73,** 3772 (1969)
148. Dent, A. L. and Kokes, R. J.: J. Phys. Chem. **73,** 3781 (1969)
149. Borello, E., Zecchina, A. and Morterra, C.: J. Phys. Chem. **71,** 2938 (1967)
150. Sedaka, Mi. and Kwan, T.: Bull. Chem. Soc. Japan **38,** 1414 (1965)
151. Beaufils, J. P.: Compt. Rend. **263,** 7 (1966)
152. Dowden, D. A., Mackenzie, N. and Trapnell, B. M. W.: Proc. Roy. Soc. **A 237,** 245 (1956)
153. Little, L. H.: Infrared Spectra of Adsorbed Species. Academic Press, New York, 1966 pp. 100–137
154. Garner, W. E. and Kingmen, F. E. T.: Trans. Faraday Soc. **27,** 322 (1931)
155. Smith, E. A. and Taylor, S.: J. Am. Chem. Soc. **60,** 362 (1938)
156. Barry, T. I. and Stone, F. S.: Proc. Roy. Soc. **A 255,** 124 (1960)
157. Winter, E. R. S.: J. Chem. Soc. 1522 (1954)
158. Molinari, E. and Parravano, G.: J. Am. Chem. Soc. **62,** 1393 (1940)
159. Aigneperse, J. and Teichner, S. J.: J. Catalysis **2,** 359 (1963)
160. Cimino, A., Molinari, E. and Cipollini, E.: Proc. 2nd International Congress on Catalysis, Vol. 1. (1961) p. 263
161. Winter, E. R. S.: Chemisorption. Butterworths, London, 1957 p. 189
162. Rudham, R. and Stone, F. S.: Chemisorption, Butterworths, London, 1957 p. 205
163. Scholten, J. J. F., Zwietering, P., Konvalinka, J. A. and De Boer, J. H.: Trans. Faraday Soc. **55,** 2166 (1959)
164. Cook, S. G.: Phys. Rev. **34,** 513 (1929)
165. Kemball, C.: Proc. Roy. Soc. **A 190,** 117 (1947)
166. Frankenburger, W. and Hodler, A.: Trans. Faraday Soc. **28,** 299 (1932)
167. Frankenburger, W. and Messner, G.: Z. phys. Chem. Bodenstein Verband, 593 (1931)
168. Johnson, R. P.: Phys. Rev. **54,** 459 (1938)
169. Miller, A. R.: The Adsorption of Gases on Solids. Cambridge Univ. Press, Cambridge, England, 1949 p. 9
170. Hansen, R. S. and Gardner, N. C.: Experimental Methods of Catalytic Research. Academic Press, New York, 1968 p. 169
171. Müller, E. W.: J. Appl. Phys. **26,** 732 (1955)
172. Allen, J. A. and Mitchell, J. W.: Disc. Faraday Soc. No. **8,** 309 (1950)
173. Beeck, O. and Wheeler, A.: J. Chem. Phys. **7,** 631 (1939)
174. Beeck, O., Smith, A. E. and Wheeler, A.: Proc. Roy. Soc. **A 177,** 62 (1940)
175. Kemball, C.: Proc. Roy. Soc. **A 217,** 376 (1953)
176. Wahba, M. and Kemball, C.: Trans. Faraday Soc. **49,** 1351 (1953)
177. Trapnell, B. M. W.: Proc. Roy. Soc. **A 218,** 566 (1953)
178. Harnsdorf, M.: Z. Naturforschung **A 23,** 1059 (1967)
179. Beeck, O., Ritchie, A. W. and Wheeler, A.: J. Coll. Sci. **3,** 505 (1948)
180. Gundry, P. M. and Tompkins, F. C.: Trans. Faraday Soc. **52,** 1609 (1956); **53,** 218 (1957)
181. Jones, P. L. and Pethica, B. A.: Proc. Roy. Soc. **A 256,** 454 (1960)
182. Becker, J. A. and Hartman, C. D.: J. Phys. Chem. **57,** 153 (1953)
183. Pasternak, R. A. and Wiesendanger, H. V. D.: J. Chem. Phys. **34,** 2062 (1961)
184. Smith, T.: J. Chem. Phys. **40,** 1805 (1964)
185. Ehrlich, G.: J. Chem. Phys. **34,** 29 (1961)
186. Redhead, P. A.: Proc. Symp. Electron Vacuum Physics, Hungary, 1962 p. 89
187. Oguri, T.: J. Phys. Soc. Japan **18,** 1280 (1963)
188. Madey, T. E. and Yates, J. T.: J. Chem. Phys. **44,** 1675 (1965)
189. McCarroll, B.: J. Appl. Phys. **40,** 1 (1969)

190. Bloch, F.: Z. Physik **52**, 555 (1928)
191. De Boer, J. H. and Verwey, E.: Report on a Conference on the Conduction of Electricity in Solids. Physical Society, London, 1937
192. Slater, J. C.: Phys. Rev. **49**, 537 (1936)
193. Fletcher, G. C.: Proc. Roy. Soc. **A 65**, 192 (1952)
194. Pauling, L.: Phys. Rev. **54**, 899 (1938)
195. Tamm, I.: Physik Z. USSR **1**, 733 (1932)
196. See e.g. Morrison, S. R.: Adv. Cat. **7**, 259 (1955)
197. Koutecký, J. and Tomasek, M.: Phys. Rev. **120**, 1212 (1960)
198. See e.g. Culver, R. V. and Tompkins, F. C.: Adv. Cat. **11**, 67 (1959)
199. Dilke, M. H., Maxted, E. D. and Eley, D. D.: Nature **161**, 804 (1948)
200. Trapnell, B. M. W.: Proc. Roy. Soc. **A 218**, 566 (1953)
201. Goodenough, J. B.: Phys. Rev. **120**, 67 (1960)
202. Bond, G. C.: Platinum Metals Rev. **10**, 87 (1966); Bond, G. C.: Disc. Faraday Soc. **41**, 200 (1966); Dowden, D. A.: J. Res. Inst. Catalysis, Hokkaido Univ. **14**, 1 (1966)
203. Rooney, J. J., Gault, F. C. and Kemball, C.: Proc. Chem. Soc. 407 (1960)
204. Bond, G. C.: The Role of Adsorbed State in Heterogeneous Catalysis. Faraday Society, London, 1966 p. 200
205. Hartley, F. R. and Venanzi, L. M.: J. Chem. Soc. 324, 328, 330–333 (1967)
206. Lapunskii, O. I.: Acta Phys. Chim. USSR **5**, 271 (1936)
207. Culver, R. V., Prichard, J. and Tompkins, F. C.: Proc. 2nd International Congress on Surface Activity. Butterworth, London, 1957, p. 243
208. Kazanskii, V. B. and Pariiskii, G. B.: Proc. 3rd Int. Congress on Catalysis, North-Holland, Amsterdam, 1964 p. 367
209. Moore, L. E. and Selwood, P. W.: J. Am. Chem. Soc. **78**, 697 (1956); Selwood, P. W.: J. Am. Chem. Soc. **79**, 3346 (1957)
210. Wright, P. G., Ashmore, P. G. and Kemball, C.: Trans. Faraday Soc. **54**, 1962 (1958)
211. Galwey, A. K. and Kemball, C.: Trans. Faraday, Soc. **55**, 1959 (1959)
212. Ehrlich, G.: Proc. 3rd Int. Congress on Catalysis. North-Holland, Amsterdam, 1964 p. 113
213. Hickmott, T. W. and Ehrlich, G.: J. Phys. Chem. Solids. **5**, 75 (1958)
214. See e.g. Suhrmann, R. and Schultz, K.: Z. phys. Chem. (Neue Folge) **1**, 69 (1954)
215. Brenner, S. S.: Science, **128**, 569 (1958); Hardy, H. K.: Progr. Metal Physics **6**, 45 (1956)
216. Blakely, J. M. and Mykura, H.: Acta Met. **9**, 595 (1961); Young, F. W. and Gwatmey, A. T.: J. Appl. Phys. **31**, 225 (1960)
217. Germer, L. M. and Macrae, A. U.: Proc. Nat. Acad. Sci. **48**, 997 (1962); Germer, L. M. and Macrae, A. U.: J. Appl. Phys. **33**, 2923 (1962)
218. Taylor, N. J.: J. Vacuum Sci. and Techn. **6**, 241 (1969)
219. Gwatmey, A. T. and Benton, A. F.: J. Chem. Phys. **8**, 569 (1940); Leidheiser, H. jr. and Gwatmey, A. T.: J. Am. Chem. Soc. **70**, 1200 (1948)
220. Park, R. L. and Farnsworth, H. E.: J. Chem. Phys. **40**, 2354 (1964)
221. See e.g. Suhrmann, R. and Sachtler, W. M. H.: Proc. Int. Symposium on Reactivity of Solids, 1954 p. 601
222. See e.g. Gysae, B. and Wagener, S.: Z. techn. Physik **19**, 264 (1938); Z. Physik **115**, 67 (1959); Culver, R. V. and Tompkins, F. C.: Adv. Cat. **11**, 67 (1959); Jones, P. L. and Pethica, B. A.: Proc. Roy. Soc. **A 256**, 454 (1960)
223. See e.g. Mignolet, J. C. P.: Disc. Faraday Soc. **8**, 326 (1950); Eberhagen, A., Jaeckel, R. and Strier, F.: Z. angew. Phys. **11**, 131 (1953); Delchar, T. A., Eberhagen, A. and Tompkins, F. C.: J. Sci. Instr. **40**, 105 (1963)
224. Hangstrum, H. D.: Trans. N. Y. Acad. Sci. **101**, 674 (1963); Hangstrum, H. D.: Phys. Rev. **96**, 325 (1954)
225. See e.g. Ehrlich, G.: Ann. Rev. Phys. Chem. **17**, 295 (1966)
226. See e.g. Menzel, D. and Gomer, R.: J. Chem. Phys. **41**, 3311 (1964); J. Phys. Chem. **41**, 3329 (1964)
227. Beeck, O. and Ritchie, A. W.: Disc. Faraday Soc. **8**, 159 (1950)
228. Eischens, R. P. and Pliskin, W. A.: Adv. Cat. **10**, 26 (1958)

229. Pliskin, W. A. and Eischens, R. P.: Z. phys. Chem. (Neue Folge) **24,** 11 (1960)
230. Völter, J.: private communication
231. Hickmott, T. W. and Ehrlich, G.: J. Phys. Chem. Solids **5,** 47 (1958)
232. Wedler, G., Fisch, G. and Papp, M.: Berichte der Bunsen-Gesellschaft für phys. Chem. **74,** (3), 186 (1970)
233. Telcs, I.: private communication
234. Klopfer, A.: Surface Science **20,** 129 (1970)
235. Bauer, E. G.: Coll. Intern. Centre. Natl. Res. Sci. No. 152 (1965)
236. Brennan, D., Hayward, D. O. and Trapnell, B. M. W.: Proc. Roy. Soc. **A 256,** 81 (1960)
237. Cabrera, N. and Mott, N. F.: Rept. Progr. Phys. **12,** 163 (1949); Lanyon, M. A. H. and Trapnell, B. M. W.: Proc. Roy. Soc. **A 227,** 387 (1955)
238. Hill, R. L., Jacobs, P. W. M. and Lodgem, G. W.: 4th Int. Conf. on Electron Microscopy. Springer Verlag, Berlin, 1960, p. 808
239. Redhead, P. A.: Can. J. Phys. **42,** 886 (1964)
240. Germer, L. M.: Adv. Cat. **13,** 191 (1962)
241. Yang, A. C. and Garland, C. W.: J. Phys. Chem. **61,** 1504 (1957)
242. See e.g. Madey, T. E., Tates, J. T. and Stern, R. C.: J. Chem. Phys. **42,** 1372 (1965)
243. Redhead, P. A.: Vacuum **12,** 203 (1962)
244. Nasini, A. G. and Ricca, F.: Trans. N. Y. Acad. Sci. **101,** 791 (1963)
245. Jenkins, G. I. and Rideal, E. K.: J. Chem. Soc. 2490 (1955)
246. Rye, R. R. and Hansen, R. S.: J. Chem. Phys. **50** (8) 3585 (1969)
247. Morgan, A. E. and Somorjai, G. A.: J. Chem. Phys. **51,** 3309 (1969)
248. Wolkenstein, F. F.: Adv. Cat. **12,** 189 (1960)
249. Kittel, C.: Introduction to Solid State Physics. John Wiley, New York, 1954 p. 386
250. See e.g. Wilson, A. H.: The Theory of Metals. Cambridge Univ. Press, Cambridge, England, 1954 p. 119; Parravano, G. and Boudart, M.: Adv. Cat. **7,** 47 (1955)
251. Schrieffer, J. R.: Phys. Rev. **97,** 641 (1955)
252. Wolkenstein, F. F.: The Electronic Theory of Catalysis on Semiconductors. Pergamon Press, New York, 1963; Aigrain, P., Dugas, C. and Germain, J. E.: Comp. Rend. **232,** 1100 (1951); Aigrain, P. and Dugas, C.: Z. Elektrochemie **56,** 363 (1952); Boudart, M.: J. Am. Chem. Soc. **74,** 1531, 3556 (1952); Weisz, P. B.: J. Chem. Phys. **20,** 1483 (1952); **21,** 1531 (1953); Hauffe, K.: Semiconductor Surface Physics. Univ. Pennsylvania Press, Philadelphia, 1957 pp. 259–282
253. Schottky, W.: Naturwissenschaften **26,** 843 (1938); Z. Physik **113,** 367 (1939); Z. Physik **118,** 539 (1942)
254. Dewar, M. J. S.: Bull. Soc. Chim. France **18,** C71 (1951)
255. Bardeen, J.: Phys. Rev. **75,** 865 (1949)
256. Wolkenstein, F. F.: Zhur. Fiz. Khim. **26,** 1462 (1952); **28,** 422 (1954); **29,** 485 (1955); Krusemeyer, M. J. and Thomas, D. G.: J. Phys. Chem. Solids **4,** 78 (1958)
257. See Aigrain, P., Dugas, C. and Germain, J. E. in ref. (252)
258. Germain, J. E.: J. Chem. Phys. **51,** 691 (1954)
259. Barry, T. I. and Stone, F. S.: Proc. Roy. Soc. **A 255,** 124 (1960)
260. Gray, T. J. and Savage, D. S.: Disc. Faraday Soc. **28,** 159 (1959)
261. Wolkenstein, F. F.: Zhur. Fiz. Khim. **32,** 2383 (1958)
262. Brennan, D., Hayward, D. O. and Trapnell, B. M. W.: J. Phys. Chem. Solids **14,** 117 (1960)
263. Allen, F. G. and Gobeli, G. W.: Phys. Rev. **127,** 150 (1962)
264. Gray, W. E. and Stone, F. S.: Proc. Roy. Soc. **A 197,** 294 (1949); Mott, N. F.: Nuovo Cimento, Suppl. **7,** 312 (1958)
265. Dell, R. M., Stone, F. S. and Tiley, P. F.: Trans. Faraday Soc. **49,** 201 (1953)
266. Rienäcker, G.: Chem. Technik **5,** 1 (1959)
267. Parravano, G.: J. Am. Chem. Soc. **75,** 1452 (1963)
268. Keier, N. P., Roginskii, S. Z. and Sazonova, I. S.: Dokl. Akad. Nauk USSR **106,** 859 (1956)
269. Miasnikov, I. A. and Psezetzkii, S. Ya.: Problemi Kinetiki i Kataliza (Problems of Kinetics and Catalysis) **8,** 165 (1955)

270. Machin, W. D. and Ross, S.: Proc. Roy. Soc. **A 265,** 455 (1962)
271. Ward, A. F. H.: Proc. Roy. Soc. **A 133,** 506 (1913)
272. Schmidt, L. and Gomer, R.: J. Chem. Phys. **42,** 3573 (1965)

ADDITIONAL REFERENCES

to Chapter 1. Adsorbent—adsorbate interaction

1. Tamm, P. W. and Schmidt, L. D.: "Interaction of hydrogen with (100) tungsten· Binding states". J. Chem. Phys. **51,** 5352 (1969)
2. Busby, M. R., Haygood, J. D. and Link, C. H.: "Classical model for gas-surface interaction". U. S. Clearinghouse Fed. Sci. Tech. Inform AD No. 708718 (1970)
3. Aldag, A. W. and Schmidt, L. D.: "Interaction of hydrogen with palladium". J. Catalysis **22,** 260 (1971)
4. Weinberg, W. H. and Merill, R. P.: "Crystal field surface orbital-bond-energy bond order (CFSO–BEBO) model for chemisorption. Application for hydrogen adsorption on a platinum (111) surface". Surface Sci. **33,** 493 (1972)
5. Gerschbacher, W. M. and Milfoed, F. J.: "Significance of many-body interactions in physical adsorption". J. Low. Temp. Phys. **9,** 189 (1972)
6. Primet, M., Basset. J. M., Mathieu, M. V. and Prettre, M.: "Infrared study of carbon monoxide adsorbed on platinum/aluminium oxide. Method for determining metal-adsorbate interactions". J. Catalysis **29,** 213 (1973)

To Chapter 2. Adsorption isotherms

7. Dong, B. W. and Mc. Quinstan, R. B.: "Diatomic adsorption on one-dimensional lattice spaces". J. Chem. Phys. **57,** 5013 (1972)
8. Podlovchenko, B. I. and Damaskin, D. B.: "Possible demarcation of adsorption isotherms based on repulsive interaction and surface inhomogenity". Elektrokhimija **8,** 297 (1972)
9. Yagud, B. Yu., Kefer, R. G. and Amirova, S. A.: "Use of infrared spectrophotometry for measuring adsorption isotherms at low adsorbate pressure". Zh. Fiz. Khim. **46(5),** 1237 (1972)
10. Misra, D. N.: "Jovanovich adsorption isotherms for heterogeneous surfaces". J. Colloid Interface Sci. **43,** 85 (1973)

To Chapter 3. Thermodynamics of adsorption

11. Bering, B. P., Makhashvili, N. I. and Serpinskii, V. V.: "Thermodynamics of adsorption equilibrium". Izv. Akad. Nauk SSSR, Ser. Khim. 2039 (1969)
12. Schay, G.: "Thermodynamics of the immersion wetting of solid adsorbents". Monatsh. Chem. **102,** 1419 (1971)
13. Dunken, H., Fritsche, H. G., Kadura, P., Kuennel, D., Mueller, H. and Opitz, C.: "Quantum chemistry of chemisorption". Z. Chem. **12,** 433 (1972)
14. Dubinin, M. M., Isirkiyan, A. A., Rakhmatkirev, G. U. and Serpinskii, V. V.: "Energy of adsorption of gases and vapors on microporous adsorbents". Izv. Akad. Nauk SSSR, Ser. Khim (6), 1269 (1972)
15. Pisani, C., Ricca, F. and Roetti, C.: "Calculated potential energies for the adsorption of rare gases on graphite". J. Phys. Chem. **77,** 657 (1973)

To Chapter 4. Enthalpy changes accompanying adsorption

16. Van Dongen, R. H. and Broekhoff, J. C. P.: "Isosteric heat of adsorption on homogeneous and patchwise heterogeneous surfaces". Surface Sci. **18,** 462 (1969)
17. Bonissent, A. and Mutafchiev, B.: "Adsorption and condensation on mixed layers II. Interpretation of heats of adsorption measurement". Surface Sci. **34,** 661 (1972)
18. Kubasov, A. A. and Smirnova, I. V.: "Heats of cyclohexene adsorption on aluminium oxide" Zh. Fiz. Khim. **46(5),** 1281 (1972)
19. Vochten, R., Petre, G. and Defay, R.: "Heat of reversible adsorption at the air-solution interface". J. Colloid Interface Sci. **43,** 310 (1973)

To Chapter 5. Entropy changes in adsorption

20. Miller, D. J. and Haneman, D.: "Electron paramagnetic resonance investigation of the surfaces at silicon-germanium alloys". Surface Sci. **33,** 477 (1972)
21. Weinberg, W. H.: "Entropy effects in the chemisorption of carbon monoxide on (100) palladium, nickel, and copper single crystal surfaces". J. Catalysis **29,** 173 (1973)

to Chapter 6. Rate of sorption

22. Maruthamuthu, P., Viswanathan, B., Swamy, C. S. and Srinivasan, V.: "Kinetics of chemisorption". Indian J. Chem. **8,** 1135 (1970)
23. Lyubitov, Yu. N. and Belenkii, V. Z.: "Kinetics of multicomponent adsorption". Dokl. Akad. Nauk SSSR **206(5),** 1162 (1972)
24. Varga, K. and Fejes, P.: "Rate of sorption on a single, spherical adsorbent particle III. Experimental confirmation of the rate equation". Acta Chim. Acad. Sci. Hung. **74,** 417 (1972)
25. Weber, B. and Cassuto, A.: "Adsorption state, adsorption rate and desorption rate of oxygen on rhenium, atomization and oxidation mechanism at high temperature and low pressure". Surface Sci. **36(1),** 81 (1973)

To Chapter 7. Experimental Results

26. Beckert, D., Michel, D. and Pfeifer, H.: "Nuclear magnetic resonance study of sorbed molecules". Bulg. Akad. Nauk **21,** 315 (1971)
27. Chen, T. S., Alloredge, G. P., De Wette, F. W. and Allen, R. E.: "Determination of surface heterogenity by gas adsorption". J. Chem. Phys. **55,** 3121 (1971)
28. Joyner, R. W., Lang, B. and Somorjai, G. A.: "Low pressure studies of dehydrocyclisation of n-heptane on platinum crystal surfaces using mass-spectrometry, Auger electron spectroscopy, and low energy electron diffraction". J. Catalysis **27,** 405 (1972)
29. Galkin, V. P., Golubev, V. B., Lunina, E. V. and Taskhai, A. N.: "Active centers of gallium oxide surface studied by an electron paramagnetic method". Zh. Fiz. Khim. **46,** 1312 (1972)
30. Schmidt, W. and Krautz, E.: "Field ion microscopic investigations of the interaction of methane with tungsten". Surface Sci. **32,** 349 (1972)
31. Egerton, T. A. and Stone, F. S.: "Adsorption of carbon monoxide by zeolite Y exchanged with different cations". J. Chem. Soc. Faraday Trans. I. **69,** 22 (1973)
32. Bendow, B. and See, Chen Ying: "Photon-induced desorption of adatoms from crystal surfaces". Phys. Rev. B**7,** 622 (1973)
33. Holland, B. W. and Woodruff, D. P.: "Missing spots in low energy electron diffraction". Surface Sci. **36,** 488 (1973)
34. Todireanu, S. and Hautecler, S.: "Study of mobile adsorption by cold neutron scattering". Phys. Lett. **A43,** 189 (1973)
35. Gottwald, B. A., Haul, R. and Roth, W.: "Studies of adsorption kinetics by molecular flow experiments". Vak.-Tech. **22,** 6 (1973)

To Chapter 8. Mechanism of chemisorption

36. Horiuti, I. and Toya, T.: "Chemisorbed hydrogen". Solid State Surface Sci. **1,** 1–86 (1969)
37. Broekhoff, J. C. P. and Van Dongen, R. H.: "Mobility and adsorption on homogeneous surfaces. Theoretical and experimental study". Phys. Chem. Aspects Adsorbents Catal. 63 (1970)
38. Panchenkov, G. M. and Tsabek, L. K.: "Kinetics of adsorption". Zh. Fiz. Khim. **44,** 318 (1970)
39. Anderson, J. B. and Bauer, B. G.: "Adsorption, kinetics, and surface structure in catalysis". Chemisorption React. Metal Films **2,** 1–61 (1971)
40. Ustinov, Y. K.: "Structure of chemisorbed layers on metals". Zh. Tekh. Fiz. **41,** 1472 (1971)
41. Deflin, M., Bavarez, M. and Bastich, J.: "EPR determination of the nature of the alumina sites involved in toluene chemisorption". C. R. Acad. Sci., Ser. C. **275,** 757 (1972)

42. Evseev, L. N., Plachenov, T. G. and Seballo, A. A.: "Kinetics of the adsorption of hydrocarbons on zeolites". Zh. Prikl. Khim. **45,** 2480 (1972)
43. Zolotarev, P. P.: "Theory of nonisothermal desorption kinetics". Zh. Fiz. Khim. **46,** 1104 (1972)
44. Tench, A. J., Lawson, T. and Kibblewhite, J. F. J.: Oxygen species adsorbed on oxides I. Formation and reactivity of $(O^-)_5$ on magnesium oxide". J. Chem. Soc. Faraday Trans. I. **68,** 1169 (1972)
45. Tench, A. J.: "Oxygen species adsorbed on oxides II. Formation of $(O_3^-)_5$ on magnesium oxide". J. Chem. Soc. Faraday Trans. I. **68,** 1181 (1972)
46. Kadura, P. and Opitz, E.: "Quantum chemical reference to very localized chemisorption in the system carbon monoxide on iron". Z. Phys. Chem. (Leipzig), **250,** 168 (1972)
47. Madey, T. E.: "Adsorption of oxygen on tungsten (100). Adsorption kinetics and electron stimulated desorption". Surface Sci. **33,** 355 (1972)
48. Kocirik, M. and Zikanova, A.: "Kinetics of adsorption in the case of the rectangular adsorption isotherms". Z. phys. Chem. (Leipzig) **250,** 250 (1972)
49. Lapujoulade, J. and Neil, K. S.: "Hydrogen adsorption on nickel (100)". Surface Sci. **35,** 288 (1973)
50. Gerei, S. V., Rozkhova, E. V. and Gorokhovatskii, Y. B.: "Propylene and oxygen chemisorption on cupric oxide and cuprous oxide catalysts". J. Catalysis **28,** 341 (1973)
51. Evseev, L. N. and Seballo, A. A.: "Kinetics of desorption and thermal desorption of hydrocarbons on NaY zeolite". Zh. Prikl. Khim. **46,** 130 (1973)

PART III

STRUCTURE AND ACTIVITY OF CATALYSTS

D. Kalló

INTRODUCTION

The problems which we wish to discuss in this Part have arisen from the almost banal recognition that although all heterogeneous catalysts are solids, only very few solids are heterogeneous catalysts. The basis of this distinction is whether or not a given solid possesses catalytic activity, or, in other words, whether its presence will or will not significantly alter the chemical conversions within an arbitrary system. Though, strictly speaking, this is the only appropriate formulation of catalytic action, in practice the question is usually: which of the chemical conversions of given components will be promoted by the catalyst, and why is a given solid an active catalyst of one conversion but, at the same time, inactive as regards other conversions? The reasons why a solid is an active catalyst of a certain process must be sought in the structure of the solid, since any chemical conversion can be defined for given components, while the structure of the solid used as the catalyst may vary within wide limits.

We shall first investigate the relationship between structure and activity from a phenomenological point of view, taking into consideration only simple geometrical and energy factors, and then study the effect of the electron system of the catalyst, that is, its internal bonding conditions.

A concise picture of catalytic reactions will first be presented in order to outline quite clearly the problems involved.

1. GENERAL DESCRIPTION OF CATALYTIC CONVERSIONS

During catalytic reactions matter is subject to the same types of qualitative changes as in any non-catalytic reaction. Because of this essential similarity our knowledge of non-catalytic reactions can be utilized in the discussion of catalytic reactions. Let us first of all recall some concepts of reaction mechanism which also occur in the study of the mechanism of action of catalysts. In this introduction, therefore, we shall sum up some of the more important well-known conclusions relating to non-catalytic reactions [1].

When we speak about chemical conversion we mean a process in the course of which the intramolecular or intermolecular bonding conditions undergo some change. The process itself consists of several steps, which can be separated from each other. Each step corresponds to an elementary act with interactions of different natures.

Let us assume that such an elementary step involves the following conversion:

$$AB + CD \rightarrow AC + BD$$

and let us investigate how it proceeds, as well as the characteristics which distinguish this step from any other.

First, AB and CD must be sufficiently near to each other for an interaction to take place between their electron systems. Secondly, the geometry of overlapping is not irrelevant; in the above case the desired orientation might, for example, be $\begin{matrix} A-B \\ C-D \end{matrix}$. The simple encounter of the molecules, however, is not sufficient, and the molecules must possess suitable energies for far-reaching changes to occur: the bonds between A and B, and C and D must loosen so that new bonds can be formed between A and C, and B and D. The bonds may be loosened by thermal excitation, for example, and as a result of appropriate collision. Essentially, a given energy is necessary for certain degrees of freedom. The value of this energy is determined by the barrier of the potential energy surface through which the system, reacting in the given elementary step, must pass. The potential energy surface is different for every reacting system and thus also for each single step, since it is determined partly by the potential curves of the bonds in the field-free state and partly by a series of modified potential curves due to interactions, that is, by the potential energy surface of the reaction.

Complete conversion proceeds through one or more such elementary steps. The mechanism of the reaction can be characterized by the successive elementary steps. The reaction mixture reaches its final state after passing through various energy barriers. The nature of these energy barriers, the requirements which the energy transfer has to fulfil, the probability of a single step taking place, and therefore the rate of the process of the given mechanism, depend on the mechanism of conversion, or, in other words, on the reaction path.

The rates for different reaction paths may be highly different, so that, in reality, a single mechanism usually dominates in the formation of a given product.

However, interaction and the rearrangement of bonds exclusively within, or between, the molecules subject to transformation is not the only means of chemical conversion. In addition to the reactants and to the products, a third type of substance might also be present, with which the components participating in the conversion process may enter into temporary chemical interaction, so that overall chemical conversion is realized through other intermediates and other potential paths. It may happen, in this way, that an easier path, which can be covered at a higher rate, is opened to the reaction, while reactions which are otherwise not possible without this interaction with a third substance may proceed at a reasonable speed. This third substance is the catalyst.

A simple example may serve to illustrate what has been said about the activation energy barrier. Figure III.1 shows the energy of the reacting system as a function of the reaction path (this method of illustration was first used by Polányi and has since been generally accepted; it is essentially a section of the potential energy surface with respect to the reaction path). The reactants which, as a result of activated "chemisorption", have got into a potential valley, E_{chem}, are converted into "chemisorbed" product(s) by taking up energy $E^{\ddagger}_{catalytic}$. The figure shows quite clearly the considerably lower energy barriers which the catalytic reaction has to pass compared to the same simple, non-catalytic conversion with the single input of activation energy $E^{\ddagger}_{homogeneous}$.

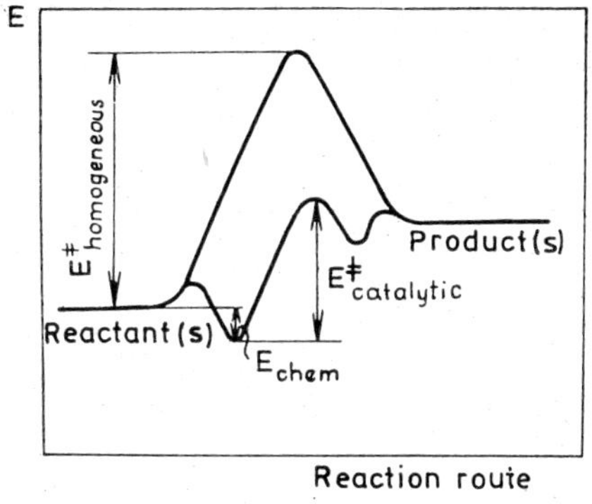

Fig. III.1. Energy diagram of catalytic conversion according to Polányi

Thus, we are dealing with a catalytic conversion when the introduction into the reacting mixture of a substance which is neither a reactant nor a product causes an increase in the conversion rate. This increase might appear relatively to the finite low rate of the non-catalytic reaction, or even when this rate is zero, that is, when, without catalyst, no conversion takes place.

In principle, the conversion rate could increase due to a higher concentration of the reactants on the catalyst as a result of physical adsorption. This is, perhaps, the oldest concept of the mechanism of the action of catalysts, but it has received no experimental confirmation, since in the presence of catalysts the conversion proceeds at a very much higher rate than in the homogeneous phase, and this can be explained only by the presence of unrealistically high surface concentrations.

Non-catalytic and catalytic processes have *in common* the fact that the natures of the transitory interactions are identical and that conversions take place decisively by the reaction paths of highest velocity; that is, there is always a preferential mechanism, a preferential sequence of the elementary steps.

The *difference* between catalytic and non-catalytic conversions is the active participation of a component which plays no part in the overall stoichiometry of the reaction. This component participates in certain phases of the reaction path in such a way that, at the end of conversion, it is again separated from the reacting system.

Thus, in the course of catalytic conversions there is at least one phase when certain components of the reacting mixture interact with the catalyst, followed by at least one phase in which converted substances joined to the catalyst undergo bonding rearrangement, and finally there is a phase in which the interaction between catalyst and reactants ceases and the catalyst is reconverted into its original form.

Consequently, in heterogeneous catalytic processes we distinguish the single steps of adsorption, surface reaction and desorption as successive part-processes of the overall conversion process. Considering this in molecular dimensions, in the course of the overall conversion process a certain area of the catalyst enters into the reaction, but it later becomes free again to offer the possibility of conversion to other reacting molecules. As regards the catalyst, this conversion in molecular dimensions is periodically repeated, whereby the reaction is manifested macroscopically. The periodic repetition of molecular conversions involving the catalyst is, to a certain extent, reminiscent of homogeneous non-catalytic chain reactions, where a chain carrier is formed at the end of the conversion cycle, enabling the initiation of at least one new cycle. The chain carrier bears a great resemblance to the active catalyst. In a homogeneous chain reaction, however, a chain carrier must be formed, in the chain-initiating step, and hence for perfect similarity, the catalyst ought to be activated or in some way made appropriate for promoting the reaction. This latter, though, is not a necessary condition. The chain carriers are always produced by new molecules, while the active part of the catalyst is always the same. Hence, a catalyst can be fouled, which cannot happen with chain carriers of course.

It follows from the above that the participation of the catalyst in the conversion process is fairly specific: of the thermodynamically possible conversions only one, or perhaps a few, will be promoted by the catalyst. The catalyst participates in the conversion process by interacting chemically with the reacting molecules, thereby enabling the rearrangement of bonds necessary for the reaction to take place. Of course, not all specific chemical interactions have to be considered, but only those which open a path to the reaction. This specific action of catalysts is called their selectivity. Selectivity, however, is not exclusive. The same catalyst is usually capable of accelerating chemically related processes or conversions of the same type in more or less the same way. It may also happen that widely different reactions are simultaneously accelerated. This might be due, for instance, to the identical function of the catalyst in some initial critical stage of the conversion process, when the same complex has to be formed, or to the participation of different types of active sites of the same catalyst in the different processes.

After this brief survey of the main characteristics of catalytic conversions, we shall consider some of the existing concepts.

2. DEFINITIONS

Since the reactants in catalytic processes enter into a temporary intimate interaction with the catalyst, they are often called *substrates*.

The *catalyst* is the component of the reacting system which, except for disturbing side-processes, such as poisoning, structural changes and deactivation, is the same before, and after conversion, by increasing the amount of which the process is accelerated.

Heterogeneous catalysts are catalytically active solids, whereas the fluid reactants and products form a separate phase, that is, the reactants and products are liquids and/or gases. In this case, catalytic conversion takes place in the boundary layer and is obviously detectable by the change in the composition of the fluid phase. An increase in the quantity of the catalyst involves an increase of its active surface, which can also be achieved by a higher dispersion of the contact. In this way, porous catalyst grains are obtained, which are sometimes agglomerates of very small particles.

The *product* of catalytic conversion is usually called the catalysate.

At a given composition, pressure and temperature the rate of the catalytic reaction is determined by the *activity* of the catalyst; vice versa: by the activity of a catalyst we mean the quantity of substance converted in accordance with some preferred reaction path on unit mass of catalyst (or, in the case of heterogeneous catalysts, on unit surface area of the catalyst) in unit time at some chosen constant composition, pressure and temperature.

The activity of the catalyst is sometimes less strictly defined, e.g. when not all the above parameters are fixed. For instance, to estimate the activities of catalysts the maximum attainable velocities of a given reaction are compared under identical composition and pressure conditions, irrespective of the temperature. Further, the maximum reaction rate obtained under optimum conditions on certain catalysts is occasionally considered a measure of activity. Again, in other cases, composition and pressure are kept identical and the temperature is determined at which the reaction rates are the same, etc. It should be noted, however, that since it is relatively more difficult to determine reaction rates, finite conversions are sometimes compared under standard experimental conditions, when the results are predominantly qualitative.

Selectivity is another important feature of catalysts, that is, the relative rates of conversions taking place on them. In a discussion of selectivity two cases must be distinguished: (a) simultaneous and (b) consecutive reactions. In these cases catalytic reactions are to be treated kinetically as the non-catalytic reactions, when changes in composition are considered as a function of time.

In the majority of practical cases both the activity and selectivity of the catalyst tend to change with time, due to the modification of its active surface during use.

As for solids in general, various structures may be formed on the surface of the catalyst, so that usually the entire surface cannot be considered as being homogeneously active. The formations on which the reaction takes place are called *active sites*. If all active sites possess the same activity, the catalyst is said to be of *homogeneous* activity; if not, it is of *heterogeneous* activity.

During catalytic conversion the first step of the interaction between the catalyst and the substrate leading to reaction is a *chemisorption* process on the active formations, followed by a *surface reaction* of one or several steps, after which the product is *desorbed* and the active formation is again free, thereby closing the *cycle* of catalytic conversion. In the individual steps via the active complexes *intermediates* of various stabilities are formed. For each step the reacting system has to pass through some energy barrier, the necessary energy being obtained from thermal energy, and thus it ought to be possible to determine the activation energy requirement of each step from the Arrhenius diagram of the rate constants of the various steps. Without going into the details of catalytic conversion and the kinetics of this consecutive reaction (see Part IV), we mention only that, usually, merely one of these consecutive steps determines the rate of the reaction, while the others can be considered as being in quasi-equilibrium. Thus, even in principle, only the rate constant of the rate-controlling step (true rate constant) can be experimentally determined. From the temperature-dependence of the true rate constant we obtain the activation energy of the rate-controlling step, which is also the activation energy of the overall catalytic conversion and is therefore called the *true activation energy*.

Rate equations correctly reflecting the mechanism of the process (which will be presented in Part IV) often take a form in which the rate constant of the rate-controlling step cannot be separated, and only a group of constants, the so called resulting rate constant (consisting of several rate and equilibrium constants), can be determined from them. The situation is similar as in many known non-catalytic complex reactions, in whose rate equations the group of characteristic constants of the individual steps is mathematically not separable. Such a group of constants contains, among others, the rate constant of the rate-controlling step. From their temperature-dependence, only the *resulting activation energy* can be determined.

Occasionally, however, the exact expression of the reaction rate is unknown and all we know is an empirical relationship which, at the temperatures under investigation, gives the reaction rate in the measured composition and pressure ranges in identical form. If these interpolation formulae have only a single temperature-dependent constant (the apparent rate constant), as, for example, in the case of an empirical kinetic expression of the first order, then it is possible to determine only the *apparent activation energy*. In the same way, some apparent activation energy is obtained from the temperature-dependence of the reaction rate on a given catalyst at identical composition and pressure.

There are differences between the values of the three types of activation energies.

Due to the manner of its derivation, the resulting activation energy necessarily comprises the algebraic sum of the enthalpy changes of the steps in quasi-equilibrium and of the true activation energy.

No statement of general validity can be given concerning the difference between apparent and resulting activation energies. Comparison of the empirical and exact rate equations may provide some information regarding this difference. It may sometimes occur that the forms of the two types of rate equations are so different that no mathematical parallel can be drawn between them. It is therefore surprising how the rate, or the corresponding constant of the empirical rate equation, can be plotted at all in the form of the Arrhenius diagram, when apparently this can be expected only for constants of exact rate equations of some entirely different form. Experience has shown that scattering caused by experimental errors is so high it tends to mask the discrepancies due to the incorrectness of the Arrhenius-type presentation: the various parameters characteristic of the rates of catalytic reactions might give a linear Arrhenius correlation without providing any information on the energy conditions of catalytic conversion.

3. OUTLINE OF THE SCOPE OF INVESTIGATION

This brief outline of the nature of catalytic action and of the consecutive steps of catalytic conversion permits the clarification of the most important concepts involved in heterogeneous catalysis, but the picture is incomplete, as it ignores the spatial separation and distribution of the heterogeneous catalytic system. This is, in fact, necessitated by the very nature of the heterogeneous system. However, a simple separation according to solid and fluid phases is not sufficient for a satisfactory distinction of the essential stages of the catalytic process. Instead of "phases", which are appropriate for simple physical distinction, it is preferable to use here the term "zones".

During heterogeneous catalysis several such zones which can be easily distinguished from each other, are formed. Thus, starting from the inside of the catalyst, we can consider as zone *1* the solid itself, that is the catalyst, followed by the surface layer, or, more accurately, the catalytically active sites (zone *2*) on which the reacting atoms of the substrate are situated (zone *3*); these are linked to those parts of the molecules subject to conversion which do not participate directly in the reaction (zone *4*). In the outermost region, zone *5*, only transport processes take place. This latter is also called the diffusion zone. Within each zone the conditions are defined by the same laws and relationships, while transition from one zone into another involves a marked change in these.

In the following we shall deal only with transformations taking place in zones *2* and *3*, that is to say, those in which the active surface of the catalyst and the group of reacting atoms of the substrate participate.

This point has to be stressed, since it is clear from the literature that conclusions have been drawn from experimental results for some zone other than that involved, and this has led to many misunderstandings. It has repeatedly happened, for instance, that though the modifications in the course of the reaction were not due to a change in zone *2* and/or *3*, conclusions were drawn from the results with respect to a relationship between the structure and activity of the catalyst. This means that in each case it is necessary to consider carefully whether, for example, a certain rate equation is indeed typical of the chemical process taking place on the catalyst surface, since only in this case can relevant conclusions be drawn from the values of the rate constant or of the activation energy. Although this may seem fairly obvious, it should nevertheless be stressed, as misunderstandings occur quite easily.

It should be added that the zones which develop in the course of catalysis interact with each other. In this way the bulk of the catalyst acts on the atoms of the active surface formations (see Part I, Chapter 5 and Part III, Section 2.3). It is quite obvious that zones *2* and *3* interact, while the role of the substituents in zone *4* must not be left out of consideration. It may suffice to draw attention to the spatial requirements of the substituents, not to speak of their role in influencing bond strengths (electron configurations) (see Part III, Section 2.2).

Occasionally, transport processes in the diffusion zone *5* may determine the macroscopically apparent catalytic conversion, as will be shown in

Part IV, Section 1.4. Further, the catalytic conversion can be influenced by the ratio of the reactant volume to the catalyst surface, if the reaction proceeds in the homogeneous phase, too.

All that has been said so far has been intended to illustrate the multitude of factors which may influence events on the active surface of the catalyst, and the macroscopic appearance of these events. It is futile, therefore, to expect the existence of a uniform theory by means of which the various phenomena of heterogeneous catalysis can all be explained. In addition, it is rather improbable that all of these numerous factors will simultaneously exert a significant effect.

Although our investigations concern phenomena in zones *2* and *3*, if necessary the effects of zone *1* and occasionally even of zone *4* will also be considered. Macroscopic factors of no interest from the aspect of catalytic action, such as those occurring in zone *5*, or the reactant volume/catalyst surface ratio, will be omitted from consideration and it will be assumed that these have no influence on the observed catalytic conversion.

In other words, an attempt will be made to present, in a fairly ordered manner, an interpretation of activity on the basis of catalyst structure, that is, to trace back catalyst activity to structural parameters. Of course, we shall not try to compile even an approximate survey of the tremendous amount of experimental data, but merely illustrate the validity of certain concepts by quoting some typical examples.

The phenomenological concepts will first be described, without going into the deeper causes of the interactions between catalyst and substrate. It will simply be accepted as a fact that there are certain forces acting between the catalyst and the substrate, and that the directions and strengths of these forces affect the catalytic conversion. The orientation of the interaction forces raises certain geometrical demands with respect to the fitting of the substrate to the catalyst, while their strength determines the shaping of essential energy conditions in the course of the catalytic conversion.

The deeper causes of the phenomena and the role of electron configurational factors will next be dealt with mainly with respect to the catalysts. In turn, metal, semiconductor, supported (metal + semiconductor), and insulator, catalyst will be discussed.

Mixed catalysts are beyond the scope of the present investigations. The objective of this book excludes discussion of homogeneous (e.g. complex) catalytic processes and irradiation reactions.

First, however, a brief survey will be presented of, perhaps, the oldest concept of the relationship between catalyst structure and activity, a concept called the theory of active sites, which, modified to some extent and supported more exactly, is still inserted in the various catalysis theories.

It was long claimed that as regards both chemisorption and heterogeneous catalytic processes the surface of the solid phase could be considered absolutely uniform and homogeneous from the aspect of the boundary layer phenomena. Experimental results and direct investigations cast doubts on the justification of this concept, and investigation of the chemisorption of hydrogen on a nickel catalyst finally led Taylor to introduce in 1923 the concept of active sites [2a]. His experimental findings convinced him that the surface atoms of the metal possess varying adsorption energies. Because of the role played by adsorption in catalysis the same is obviously true for catalysis too. This conclusion was supported by the observations of Armstrong and Hilditch [2b], who found that the quantity of poison which will cause the catalyst to lose its activity completely is far below the quantity required for monomolecular coverage, so that the reaction in question can take place only on a small fraction of the surface. At about the same time, on the basis of similar observations, Pease [2c] outlined a certain model-like conception of the active sites. He claimed that the atoms situated in sharp bends of the surface, mainly those on convex elements, peaks or ridges, tend to display a different behaviour. This theory was later confirmed experimentally by Schwab and Rudolph [3], who found that the activity per unit mass of the catalyst, a, is not directly proportional to the specific surface area, a_s, but as a_s increases, a is proportional to increasingly higher powers of a_s:

$$a = \alpha a_s^n$$

where $$n = 1.5-3.5$$

If $n = 1$ the reaction takes place mostly on the surface, as $a \sim a_s$. When $n = 2$, the activity is proportional to the length of the edges, as for cubes of density ϱ the specific edge length is

$$e = \frac{\varrho}{3} a_s^2,$$

that is, $a \sim a_s^2 \sim e$, and hence the reaction takes place at the edges. If $n = 3$, the activity is proportional to the number of peaks, as the specific peaknumber is

$$p = \frac{\varrho^2}{27} a_s^3$$

and thus $a \sim a_s^3 \sim p$, indicating that the reaction takes place on the peaks.

However, it is impossible to interpret the case when $n > 3$ in a similar manner, which is undoubtedly a deficiency of this concept.

Constable went beyond this concept when he suggested that the active sites are formed in fact by surface inhomogeneities, such as amorphous granules, or allotropic modifications, that is, formations with high energy content [4]. We shall return to the theoretical reasoning on which Constable's findings were based in the discussion of the compensation rule (see Section 2.3).

All these early hypotheses can be confirmed in certain cases by means of energetical considerations of catalysis, and moreover, provided our knowledge permits, they may be supported to some extent by the interpretation theory of catalysis (see e.g. Paragraph 4.2.3, the interpretation of the experiments of Schwab and Rudolph).

The active sites had to meet not only energy, but also geometrical, demands, however. Burk took the first step in this direction when he assumed in place of the hitherto accepted adsorption on a single site, the adsorption of dual-centre molecules on two sites [5], which helped to provide a satisfactory explanation for a number of experimental data. Balandin went considerably further and characterized the active sites by atomic ensembles in fixed positions in his multiplet theory [6], which will be described in detail later in this book. In agreement with other investigators, Balandin was later able to suggest the probable arrangement of atoms on the catalyst surface and in the reacting substrate when a suitable fit develops, including the various distances which might occur in such an arrangement. All this led, however, to the geometric aspects of catalysis, which forms the subject of the following Chapter.

CHAPTER 1. THE ROLE OF GEOMETRICAL FACTORS

SECTION 1.1. GEOMETRY OF CONVERSIONS ON THE SURFACE LATTICE PLANES OF CATALYSTS

We shall begin the description of catalysis theories with geometrical considerations which always arise when conversions take place on the surface. This subject also deserves priority because it helps to illustrate the process, so that, later, this directly understandable pattern has only to be supplemented with the appropriate energy content.

What will be discussed here is essentially the geometric part of the Balandin multiplet theory [7].

The theory was primarily based on hydrogenation reactions for which considerable experimental material was available, mainly from the extensive investigations of Zelinskii.

Evaluation of kinetic measurements led to the conclusion that surface reaction can take place, at most, in a monomolecular layer. The fact that the reaction rate hardly depends upon the length of the hydrocarbon chain indicates a definite orientation of the reacting molecules. Since the radius of action of chemical forces is very small, the atomic groups undergoing rearrangement must be situated very close to the atoms on the catalyst surface.

All this led to the conclusion that there must be a definite alignment between the groups of atoms participating in the reactions and the surface structure of the catalyst. Thus, if activated adsorption on two sites is required to initiate the catalytic reaction, the distance between the two chemisorbing atoms on the catalyst surface determines the distance between the two chemisorbed atoms of the substrate, and vice versa. For instance, in the case of the hydrogenation of ethylene Horiuchi and Polányi supposed a link between the two carbon atoms and the two neighbouring atoms, as shown in Fig. III.2a [8a]. According to Twigg and Rideal [8b], for every catalyst of the reaction the distance between two metal atoms must be in the range 2.4 – 2.8 Å otherwise the bond angle *Me*—C—C would differ greatly from the normal value of 109° [8c]. This rule is valid not only for metals, but also for metal sulphides, as in the latter, too, the carbon atoms of the double bond are joined to metal atoms (the bond electron is localized on the metal atom, see Part II, Chapter 8). In the hexagonal crystal

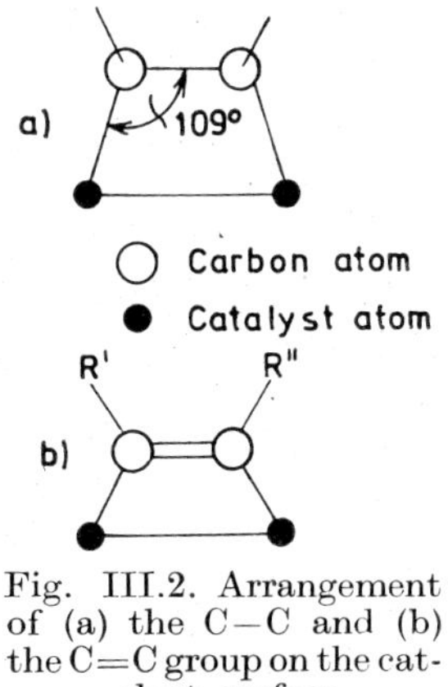

Fig. III.2. Arrangement of (a) the C—C and (b) the C=C group on the catalyst surface

lattice of nickel sulphide the Ni—Ni distance is 2.65 Å, and, thus, nickel sulphide is a good hydrogenation catalyst, whereas at temperatures above 150 °C it is converted by hydrogen into nickel subsulphide (Ni_3S_2), an inactive compound whose nickel atoms are situated farther from each other than the permissible value [9].

Certain stereochemical experiments support the disposal of the $C=C$ group as shown in Fig III.2b [10a].

When acetylene derivatives are hydrogenated to the corresponding olefins, it is always the cis-isomer which is formed, because, in the molecules linked to the surface, the bonds are in the same plane, so that the opening bond is directed towards appropriate atoms of the catalyst and consequently the substituents occupy cis-positions.

If compounds with double bonds are chemisorbed in a manner corresponding to Fig. III.2a, depending on the starting substance, hydrogenation leads to the following products:

$$
\begin{array}{ccc}
\begin{matrix} R & & R \\ \searrow & & \swarrow \\ C & = & C \\ \swarrow & & \searrow \\ Q & & Q \end{matrix}
& \longrightarrow &
\begin{matrix} R & & R \\ \searrow & & \swarrow \\ HC & - & CH \\ \swarrow & & \searrow \\ Q & & Q \end{matrix} \\
\text{cis} & & \text{meso}
\end{array}
$$

$$
\begin{array}{ccc}
\begin{matrix} R & & Q \\ \searrow & & \swarrow \\ C & = & C \\ \swarrow & & \searrow \\ Q & & R \end{matrix}
& \longrightarrow &
\begin{matrix} R & & Q \\ \searrow & & \swarrow \\ HC & - & CH \\ \swarrow & & \searrow \\ Q & & R \end{matrix} \\
\text{trans} & & \text{dl}
\end{array}
$$

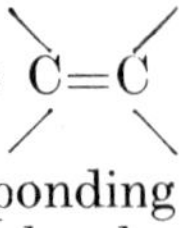

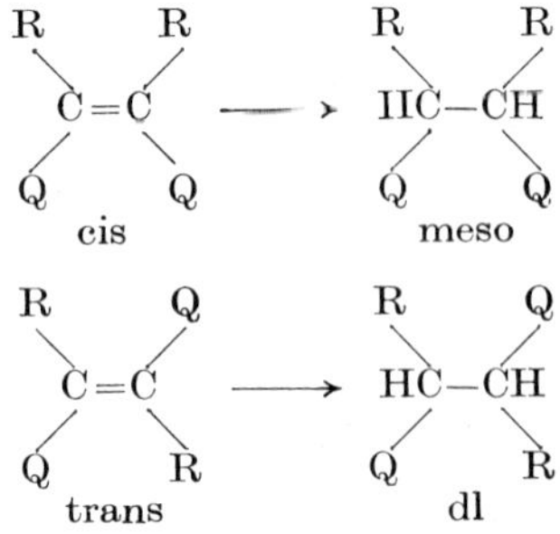

In fact, when starting from the cis-isomer the product has a meso-configuration, while under identical conditions the trans-isomer leads mainly to products of the dl-form.

These examples are striking illustrations of the geometry of the position of the substrate to be hydrogenated, but with respect to hydrogen the only assumption is that they are completely dissociated on neighbouring catalyst atoms. Balandin overcame this obvious deficiency of the geometric concept by introducing his multiplet theory, which also takes the hydrogen into consideration and, in addition, allows the interpretation of the roles of all partners in a great variety of catalytic processes.

Let us now consider the following atomic rearrangement:

$$ AB + CD \rightarrow AC + BD $$

where A, B, C and D are atoms to which further molecular parts might be linked. To a first approximation, catalytic conversion was depicted by the following diagram (· symbolizes the catalyst atom):

(a)
$$
\begin{matrix} A & C \\ | & | \\ B & D \end{matrix}
\quad \longrightarrow \quad
\begin{matrix} A & C \\ \vdots & \vdots \\ B & D \end{matrix}
\quad \longrightarrow \quad
\begin{matrix} A & - & C \\ B & - & D \end{matrix}
$$

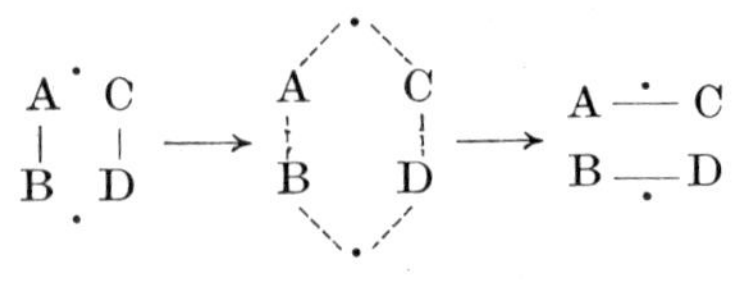

This scheme illustrates a reaction on a doublet. In the intermediate of the reaction the reacting atoms form bonds with the catalyst itself, while the initial bonds are loosened or even completely broken up.

Polányi suggested the following scheme for the same reaction (the vertical lines represent the "free valences" starting from the catalyst surface) [10b]:

(b)

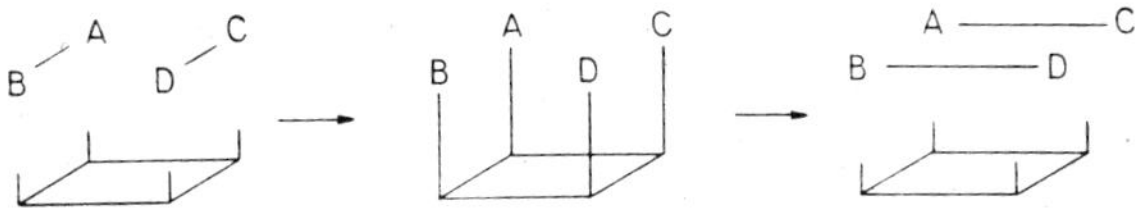

We shall see later that diagram (b) gives only a very formal picture of the real process. Nevertheless, this diagram reflects the real situation quite correctly from an important aspect: according to the diagram the reaction can be reversible, which was not indicated by the doublet model. If process (a) takes place from right to left, the position of the initial atomic groups on the catalyst will, in principle, be different from that expected from the reaction taking place from left to right. Since A, B, C and D represent any atoms, no such restriction can be made. However, if the initial state is represented according to Balandin by the following hypothetical symbol:

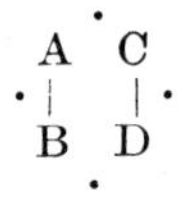

then there will be no contradiction. In other words, every cleaving and every forming bond is situated on a catalyst atom before any further rearrangement or conversion can take place. Thus, in the initial state the groups of the reacting atoms are situated in an arrangement corresponding to Fig. III.3, in the pits between the surface atoms of the catalyst. This assumption is rendered probable by energetical aspects, since, as the crystal lattice grows, the new lattice plane also begins to be built on these sites.

By mentioning these, and similar, concrete geometric concepts we do not wish to suggest that these concepts are necessary and of general validity. It is merely desired to illustrate that there must be a relationship between the lattice constant of the catalyst and the distance between the reacting atoms of the substrate, irrespective of the true arrangement in a given case. These examples are based only on the usual intramolecular bond picture, and the bond angle $Me-Me-C$, for instance, has been ignored.

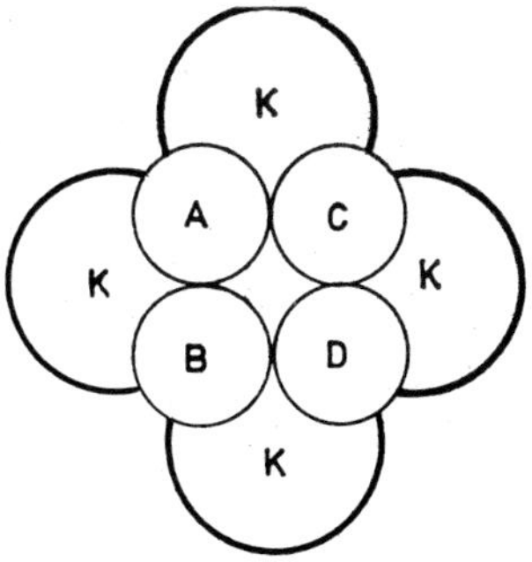

Fig. III.3. Arrangement of the reacting atoms (A, B, C, D) of the substrate on the surface atoms of the catalyst (K) according to Balandin [11]

According to the multiplet theory, the more common reactions in which the bonds are rearranged within an ensemble consisting of four atoms can be classified into the following groups [these concepts have been proved valid mainly for metal catalysts, though recent investigations [31] have shown their applicability to the metal atoms of metal oxides (see Section 2.2)]:

Dehydrogenation
of alcohols

$$\begin{array}{cc} >\!C\!-\!\!-\!\!O \\ |\quad\; | \\ H\quad H \end{array} \;\rightleftharpoons\; \begin{array}{cc} >\!C\!=\!\!O \\ \\ H\!-\!\!-\!H \end{array}$$

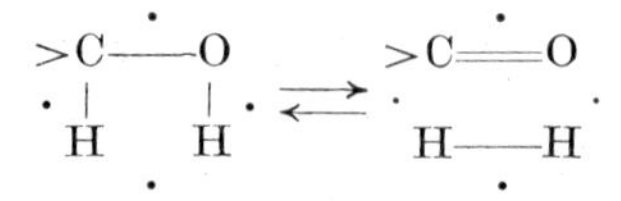

Dehydration of
alcohols

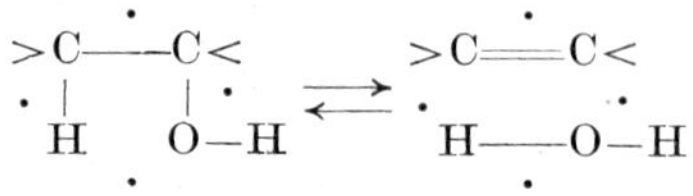

$$\begin{array}{cc} >\!C\!-\!\!-\!C\!< \\ |\quad\;\; | \\ H\quad O\!-\!H \end{array} \;\rightleftharpoons\; \begin{array}{cc} >\!C\!=\!\!C\!< \\ \\ H\!-\!\!-\!O\!-\!H \end{array}$$

Dehydrogenation
of hydrocarbons

$$\begin{array}{cc} >\!C\!-\!\!-\!\!O\!< \\ |\quad\; | \\ H\quad H \end{array} \;\rightleftharpoons\; \begin{array}{cc} >\!C\!=\!\!C\!< \\ \\ H\!-\!\!-\!H \end{array}$$

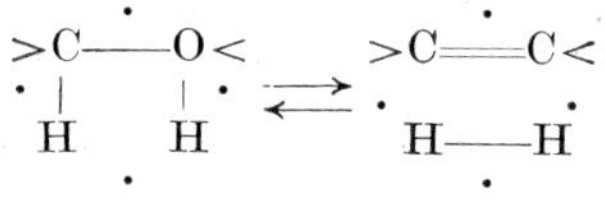

Oxidation of
olefines

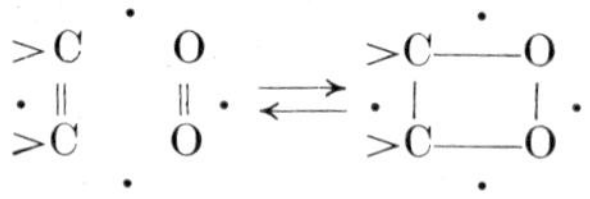

$$\begin{array}{cc} >\!C\quad O \\ \|\quad\; \| \\ >\!C\quad O \end{array} \;\rightleftharpoons\; \begin{array}{cc} >\!C\!-\!\!-\!\!O \\ |\quad\; | \\ >\!C\!-\!\!-\!\!O \end{array}$$

Decarboxylation

$$\begin{array}{cc} O\!=\!C\!-\!\!-\!\!O \\ |\quad\; | \\ >\!C\quad H \\ | \end{array} \;\rightleftharpoons\; \begin{array}{cc} O\!=\!C\!=\!\!O \\ \\ >\!C\!-\!\!-\!H\!\cdot \\ | \end{array}$$

Dealkylation

$$\begin{array}{cc} >\!C\!-\!\!-\!C\!< \\ |\quad\; | \\ >\!C\quad H \\ | \end{array} \;\rightleftharpoons\; \begin{array}{cc} >\!C\!=\!\!C\!< \\ \\ >\!C\!-\!\!-\!H \\ | \end{array}$$

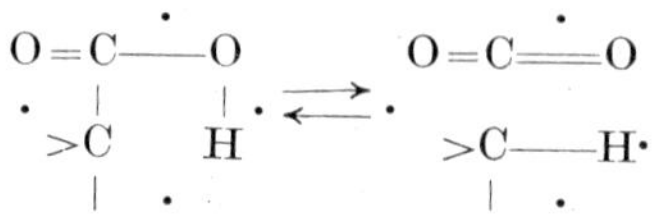

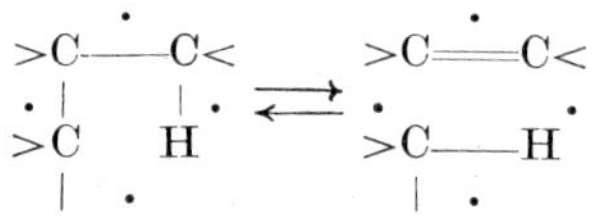

The atoms between which the bonds undergo rearrangement in the course of the reaction are called the index of the reaction group. The reaction equations are usually simply written with these indices, the opening bonds being represented by vertical, and the formed bonds by horizontal, lines:

$$\begin{array}{cc} A & C \\ | & | \\ B & D \end{array} \;\rightleftharpoons\; \begin{array}{cc} A\!-\!D \\ \\ B\!-\!D \end{array}$$

The intermediates of these reactions are the multiplet complexes. They are formed in several steps, and not so simply and directly as indicated by diagram (b). When alcohols are dehydrogenated the detailed reaction scheme is presumably made up of the following elementary steps [11]:

$$\begin{array}{cc} >\!C\!-\!\!-\!O \\ |\quad\; | \\ H\quad H \end{array} \;\rightleftharpoons\; \begin{array}{cc} >\!C\!-\!\!-\!O \\ |\quad\; | \\ H\quad H \end{array} \;\rightleftharpoons\; \begin{array}{cc} >\!C\!-\!\!-\!O \\ |\quad\; | \\ H\quad H \end{array} \;\rightleftharpoons\; \begin{array}{cc} >\!C\!=\!\!O \\ \\ H\!-\!\!-\!H \end{array}$$

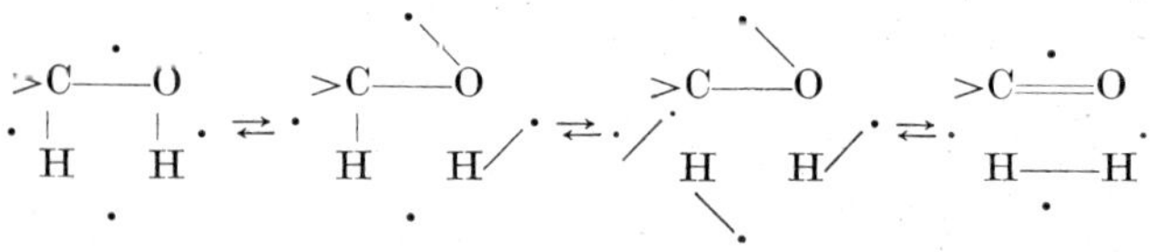

The formation of the "semi-hydrogenated product" observed in the process of radical hydrogenation and dehydrogenation of hydrocarbons, the dissociative adsorption of hydrogen, can also be considered as an intermediate step:

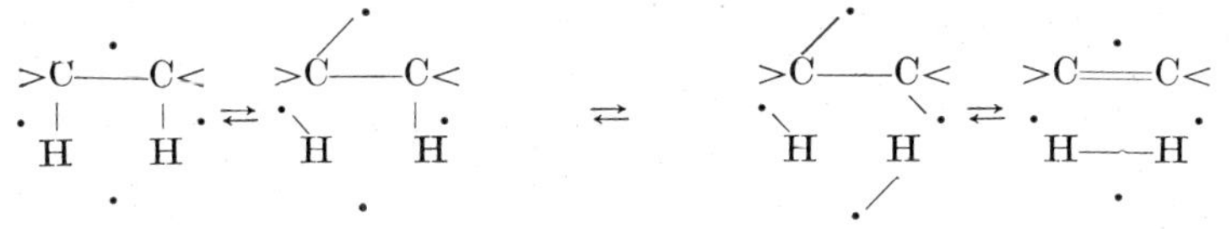

Semihydrogenated product

Dissociative hydrogen adsorption

Of course, this is not dissociative chemisorption, i.e. a complete cleavage of the bonds, as in the above concept of Polányi. What occurs is only a loosening of the bonds due to chemisorption, followed by their opening and rearrangement. The configuration obtained in step 2 corresponds perfectly to such an arrangement, as seen before in the position of the C=C group, when conditions were investigated only from the aspect of one of the reactants, the olefin.

It must be borne in mind, however, that this presentation of bonds is, of course, only schematic and for a correct picture an exact knowledge of the electron configuration is required.

From the geometrical aspect the considerably more concise multiplet theory explains even better the great importance of harmony between the structures of the catalyst and the substrate. It appears quite clearly from the above that the dehydrogenation and dehydration of alcohols require different orientations of the molecules on the catalyst surface. From this point of view the investigations of Rubinstein and Pribitkova are particularly interesting [12]. These authors found that the lattice constant of magnesium oxide catalyst is liable to a certain change, depending upon the conditions of preparation. The catalytic activity changes depending on the lattice constant: it was observed that when n-butan-l-ol is dehydrogenated the catalyst having the smaller lattice constant furnishes the optimum, while in the dehydration of the same compound the best results are obtained using the catalyst having the greater lattice constant. In the first case, the length of the cleaving O—H bond (1.01 Å) is smaller than the length of the C—O bond involved in the second reaction (1.45 Å). (This is presumably not the only explanation of the phenomenon, which will also be discussed in Section 1.3.)

So far, reactions of four atomic indices have been studied. The situation is different, however, for example when cyclohexane is dehydrogenated or benzene hydrogenated. When the entire ring participates simultaneously in the reaction, the six-membered ring presumably lies flat on the surface of the catalyst, wich, according to the multiplet theory, can be realized on a sextet (Fig. III.4). Simultaneity, of course, is the result of a simplification. In our case it has only been taken to mean that the C_6 ring stays put on the given site of the catalyst as long as the rearrangements of the bonds on the surface proceed. (The splitting-off of three H_2 molecules in the dehydrogenation of cyclohexane proceeds, in fact, stepwise as indicated by traces of cyclohexene [13].) This condition is encountered only in the case of those metals which have surface lattice planes with structures allowing the arrangement shown in Fig. III.4. This condition is fulfilled only by the (111) face of face-centred cubic catalysts and by the basal plane of hexagonal crystals when the atomic distances are in the range 1.20—1.40 Å. This generally valid conclusion is reached on the basis of a very great amount of experimental data.

Since the crystal lattice of rhenium satisfied the above requirement, it was expected that this metal would afford catalytic activity in the hydrogenation of the six-membered ring. This assumption was later confirmed by experiment. The same criteria are valid for alloys.

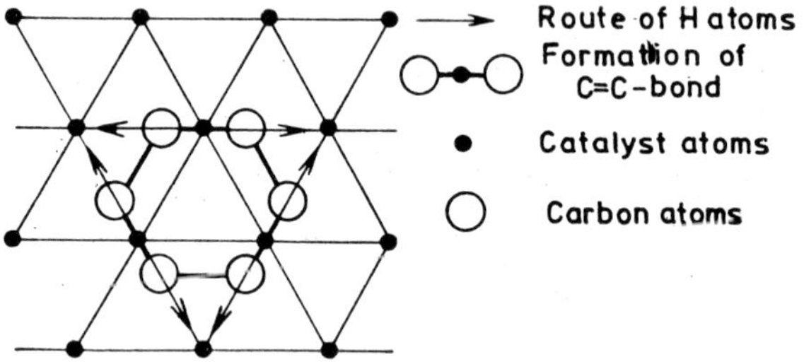

Fig. III.4. Adsorption of cyclohexane on a sextet

Metals on inert supports catalyse processes taking place according to the mechanism shown in Fig. III.4 only if the metal has reached a certain extent of agglomeration on the surface of the support, whereby a possibility is opened for the formation of a crystal face of desired orientation. In fact, magnetic susceptibility measurements on a supported nickel catalyst have proved that in such cases, too, the (111) octahedral face forms the boundary of the metal crystallite [14]. Rubinstein et al. noted the same when working with platinum catalysts on an activated carbon support [15]. Dehydrogenation of the six-membered ring may also proceed on quadruplets, however, and not only on sextets. Cyclohexane and decaline were dehydrogenated in the presence of nickel and chromium(III) oxide catalysts on an alumina support [16]. In the presence of nickel almost twice as many cyclohexane molecules were dehydrogenated as were decaline molecules. The molecules are probably situated with their planes on the surface (Fig. III.5) which explains why more of the first type find room on the same surface area. In the presence of chromium(III) oxide both hydrocarbons are converted at the same rate, indicating that the molecules rest on their edges (Fig. III.6). The activation energies are the same in each case on the same catalyst: 12.5 kcal mole^{-1} on nickel and 26 kcal mole^{-1} on chromium(III) oxide. The observed differences in reaction rate may be attributed to the different orientations of the molecules. (The remarkably low activation energy of the endothermic reaction is probably due to the removal of hydrogen in steps.)

With regard to the surface requirements of the substrate, some experiments referring to the poisoning of the catalyst should be mentioned. Platinum and nickel hydrogenation catalysts were deactivated to a different extent with sulphur. Rubinstein and Pribitkova found that if the quantity of sulphur is increased the catalyst gradually loses its activity, and only smaller and smaller molecules can be hydrogenated, presumably because the frequency of larger active areas on the surface decreases rapidly [17].

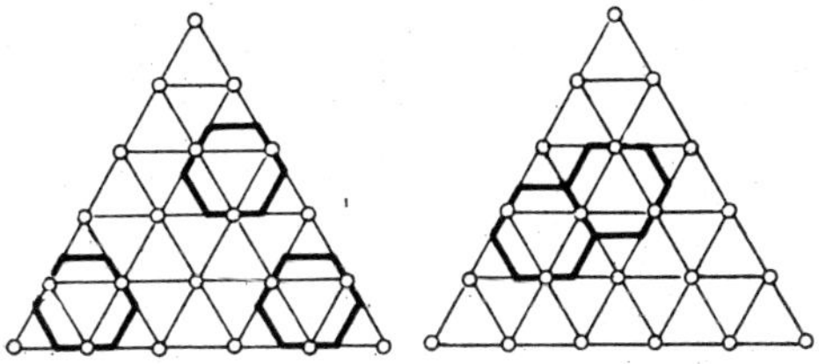

Fig. III.5. Planar arrangement of (a) cyclohexane and (b) decaline on the (111) face of platinum

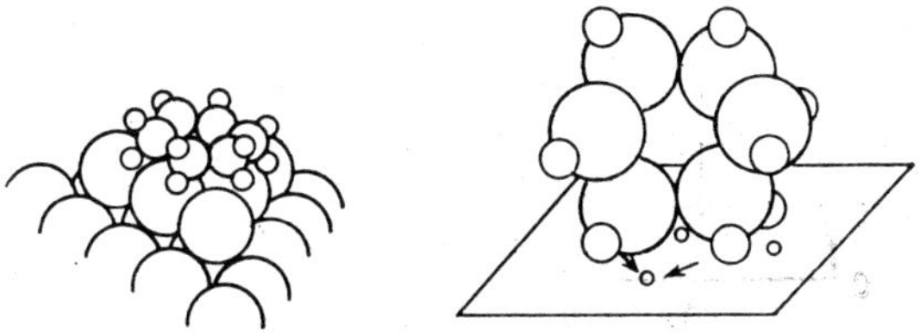

Fig. III.6. Two different orientations of cyclohexane (a) arrangement in the plane (b) arrangement along an edge

These geometric arguments have led to the result that a certain structural correspondence must exist between the surface of solid catalysts forming relatively large aggregates and the reacting substrate. This demands the formation of a crystallographic "face" consisting of four, six or even more atoms on the catalyst surface. These "faces" may be situated on the peaks of certain protrusions. It has been observed that, because of the very great spatial requirements of the rest of the molecule, ring hydrogenation of certain highly complex tripticene derivatives proceeds only on such peaks [18]. The size and direct environment of geometrically suitable active sites can, therefore, vary within wide limits, resulting, as will be shown later, in different activities.

SECTION 1.2. PROCESSES TAKING PLACE ON ADSORBED CATALYST ATOMS [19a, b]

In the discussion of geometric theories special attention must be devoted to the case when the process presumably takes place only on discrete catalyst atoms or atomic groups. According to the active ensembles theory of Kobozev, such a situation arises when the catalytically active component is applied in a very low concentration on the inert support, that is, when, with respect to catalyst atoms, the coverage of the surface is $\theta \approx 10^{-3}$*.

* Surface coverage may be defined as follows:

$$\theta = \frac{N\varphi}{a_s} \qquad \text{(III.1)}$$

where N is the number of catalyst atoms on 1 g of the support, g^{-1};
φ is the area covered by a single catalyst atom in the case of monomolecular coverage, cm^2;
a_s is the specific surface of the support, $cm^2\ g^{-1}$.

Changes in two parameters, the total activity and the specific activity referred to unit mass of the active substance, were investigated as functions of the coverage under these conditions. Figures III.7 and III.8 show two examples of the changes in the total activity (A) and in the specific activity ($a = A/\theta$)*. In the first case (Fig. III.7) A shows a maximum and a decreases monotonously, while, in the second case (Fig. III.8), both A and a pass through a maximum. In a number of cases of the second type it was found that

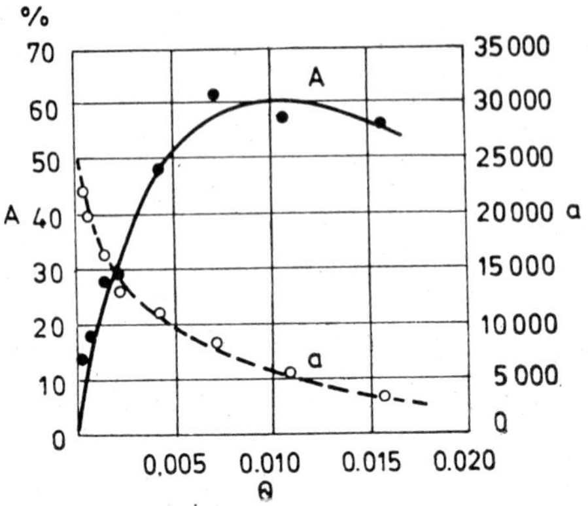

Fig. III.7. Oxidation of sulphur dioxide on a platinum-on-alumina catalyst

$$\frac{\theta_{A_{\max}}}{\theta_{a_{\max}}} = 1.5 \quad \text{or} \quad \frac{\theta_{A_{\max}}}{\theta_{a_{\max}}} = 2 \tag{III.2}$$

For low coverages of catalytically active components Kobozev interpreted the experimental results on the following assumption:

(1) Catalytic activity can be attributed to the catalyst in the atomic, precrystallite phase. The only role of the support is to enable the formation of this dispersed phase.

(2) As a result of the mosaic structure, the surface of the support is divided into discrete sections of identical sizes, which are covered by catalyst atoms in random distribution.

(3) Each process takes place only on those surface elements which contain a definite number, n, of adsorbed catalyst atoms. These are the active sites of the process.

Of these conditions, perhaps the first is the most liable to some reservation, but nevertheless their application allows the mathematical interpretation of the relationships plotted in Figs III.7 and III.8.

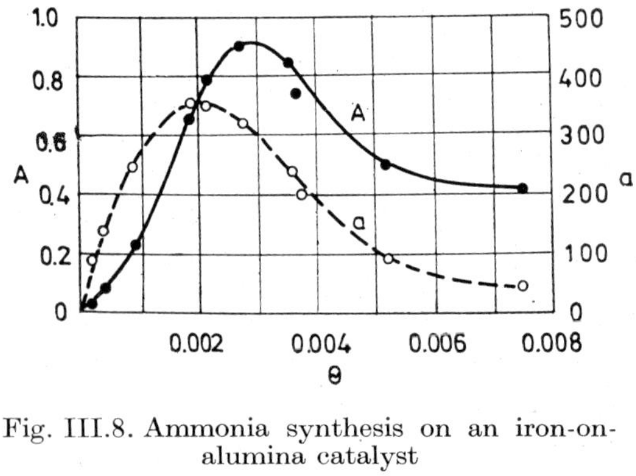

Fig. III.8. Ammonia synthesis on an iron-on-alumina catalyst

On the surface elements of the support there are various numbers of adsorbed catalyst atoms, irrespective of the coverage. The probability (W_n) of an ensemble of n atoms varies, however, depending upon the coverage:

* Total activity in this case means conversion relating to unit mass (or surface) of the support and taking place in unit time, in the same reaction, at the same pressure, temperature and concentration.

$$W_n = \frac{n_{av}^n}{n!}\exp(-n_{av}) \qquad \text{(III.3)}$$

if, on the average, n_{av} catalyst atoms pertain to each surface element. n_{av} is defined by the following equation:

$$n_{av} = \frac{N\Delta}{a_s} \qquad \text{(III.4)}$$

where Δ is the size of the surface element in cm²:

$$\Delta = p\varphi \qquad \text{(III.5)}$$

where p is the number of catalyst atoms on the surface element for complete coverage, and
φ is the surface covered by a single adsorbed atom.

Taking Eq. (III.1) into consideration we obtain:

$$n_{av} = p\frac{N\varphi}{a_s} = p\theta \qquad \text{(III.6)}$$

and by substituting Eq. (III.6) into Eq. (III.3):

$$W_n = \frac{(p\theta)^n}{n!}\exp(-p\theta) \qquad \text{(III.7)}$$

If the number of surface elements on a catalyst of unit mass is z, and the number of surface elements on which n catalyst atoms are situated is z_n, it follows from Eq. (III.7) that:

$$z_n = z\frac{(p\theta)^n}{n!}\exp(-p\theta) = \frac{a_s}{\Delta}\frac{(p\theta)^n}{n!}\exp(-p\theta) \qquad \text{(III.8)}$$

According to the third assumption of the theory, the catalytic activity is proportional to the number of those atomic ensembles in which a definite number, n, of atoms are situated:

$$A_n = \alpha_n z_n = \alpha_n \frac{a_s}{\Delta}\frac{(p\theta)^n}{n!}\exp(-p\theta) \qquad \text{(III.9)}$$

where α_n might be called the activity of a single active site (or of a single active surface element, which is precisely the same).

Specific activity, a_n, is obtained from Eq. (III.9):

$$a_n = \frac{A_n}{\theta} = \alpha_n \frac{a_s}{\Delta}\frac{p^n\theta^{n-1}}{n!}\exp(-p\theta) \qquad \text{(III.10)}$$

Eqs (III.9) and (III.10) are, in effect, the mathematical formulations of the Kobozev theory.

Eqs (III.9) and (III.10) have maxima at the following coverages:

$$\theta_{A_{max}} = \frac{n}{p} \qquad (III.11)$$

$$\theta_{a_{max}} = \frac{n-1}{p} \qquad (III.12)$$

which are known from the curves plotted from experimentally determined data (Fig. III.8), so that n and p can be calculated. From Eqs (III.11) and (III.12) we obtain, for the number of catalyst atoms in the active site:

$$n = \frac{\theta_{A_{max}}}{\theta_{A_{max}} - \theta_{a_{max}}} \qquad (III.13)$$

that is, if

$$\frac{\theta_{A_{max}}}{\theta_{a_{max}}} = 2, \quad n = 2$$

and if

$$\frac{\theta_{A_{max}}}{\theta_{a_{max}}} = 1.5, \quad n = 3$$

and if $n = 1$, $\alpha = f_a(\theta)$ has an extreme value at $\theta = 0$. From Eqs (III.11) and (III.12) it is possible to calculate the number of adsorbed catalyst atoms on a single surface element at complete coverage:

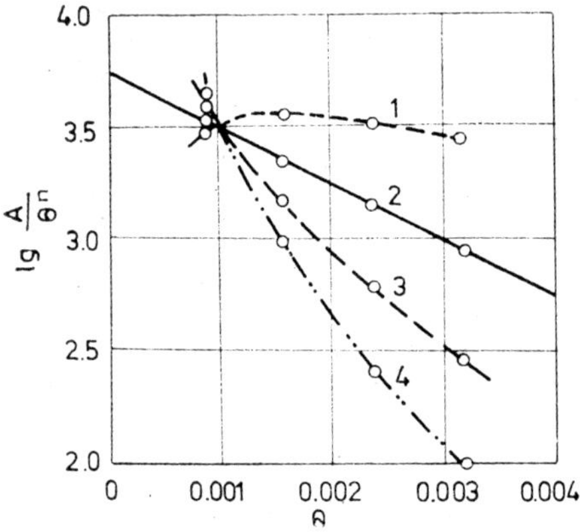

Fig. III. 9. Hydrogenation of 1,1-dimethyl-2-propine-1-ol on a palladium-on-carbon catalyst

1. $\lg \frac{A}{\theta} = f(\theta)$; $n = 1$

2. $\lg \frac{A}{\theta^2} = f(\theta)$; $n = 2$

3. $\lg \frac{A}{\theta^3} = f(\theta)$; $n = 3$

4. $\lg \frac{A}{\theta^4} = f(\theta)$; $n = 4$

$$p = \frac{1}{\theta_{A_{max}} - \theta_{a_{max}}} \qquad (III.14)$$

so that from Eq. (III.5) we obtain the size of the surface element, from which we can also calculate z, provided the specific surface is known. More accurate values are obtained when Eq. (III.10) is linearized by plotting $\lg(A_n/\theta^n)$ as a function of θ:

$$\lg \frac{A_n}{\theta^n} = \lg \left(\alpha_n \frac{a_s}{\Delta} \frac{p^n}{n!} \right) - \frac{p}{2.3} \theta \qquad (III.15)$$

Only when an appropriate value of n is substituted will a straight line be obtained (Fig. III.9), from the slope of which p can be calculated ($p = 600$).

In this manner, several experimental results were evaluated.

In the case of oxidation and reduction, as well as in the decomposition of hydrogen peroxide, it was found that the active sites are almost exclusively single metal atoms (Table III.1). It was assumed that during oxidation an intermediate of the structure 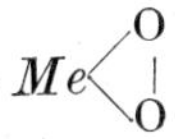 is formed [19c] (it follows from the electronic theory that the structure $Me{-}O{-}O{-}$ is more probable, see Part II, Sections 7.1 and 8.1).

TABLE III.1

One-atomic active centres

Reaction	Catalyst	Support	Active centre
$SO_2 + 1/2\ O_2 \rightarrow SO_3$	Pt	alumina gel, kieselguhr	$(Pt)_1$
	Pd	alumina gel, kieselguhr	$(Pd)_1$
$NH_3 + 5/4\ O_2 \rightarrow 3/2\ H_2O + NO$	Pt	alumina gel, kieselguhr	$(Pt)_1$
$2\ H_2 + O_2 \rightarrow 2\ H_2O$	Pt	alumina gel, kieselguhr	$(Pt)_1$
$SO_3^{2-} + 1/2\ O_2 \rightarrow SO_4^{2-}$	Fe^{2+}	activated carbon	$(Fe^{2+})_1$
	Cu^{2+}	activated carbon	$(Cu^{2+})_1$
	Ag^+	activated carbon	$(Ag^+)_1$
Reduction of nitrophenol	Pt	activated carbon	$(Pt)_1[(Pt)_5?]$
Reduction of picric acid	Pt	activated carbon	$(Pt)_1[(Pt)_5?]$
$H_2O_2 \rightarrow H_2O + 1/2\ O_2$	Pt	activated carbon	$(Pt)_1[(Pt)_5?]$
	Pt	alumina gel	$(Pt)_1[(Pt)_5?]$
	Pt	kieselguhr	$(Pt)_1[(Pt)_3?]$ $[(Pt)_7?]$
	Fe^{3+}	activated carbon	$(Fe^{3+})_1$
	Cu^{2+}	activated carbon	$(Cu^{2+})_1$
	Co^{2+}	kieselguhr	$(Co^{2+})_1[(Co^{2+})_3?]$ $[(Co^{2+})_7?]$
	Ni^{2+}	kieselguhr	$(Ni^{2+})_1[(Ni^{2+})_7?]$
	Ag^+	activated carbon	$(Ag^{2+})_1$
	haemin	activated carbon	$(haemin)_1$

There is, as yet, no interpretation for the ensembles in square brackets, though their presence can be concluded from Eq. (III.13), since the function $a = f_a(\theta)$ is of finite value at $\theta = 0$ and also has a maximum before $\theta_{A_{max}}$.

The oxidation of sulphur dioxide on a platinum catalyst can be conceived as proceeding according to the following mechanism [19d]: the first step is the activation of oxygen:

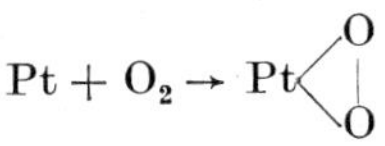

followed by the consecutive reaction of the activated oxygen atoms with two sulphur dioxide molecules:

$$\mathrm{Pt}\langle\begin{matrix}\mathrm{O}\\ |\\ \mathrm{O}\end{matrix} + SO_2 \rightarrow PtO + SO_3$$

$$PtO + SO_2 \rightarrow Pt + SO_3$$

that is, it seems highly probable that what happens is the reverse of that which occurs in the event of oxidation.

Further evaluation indicates that the hydrogenation of the $>C=C<$ bond and the dehydrogenation of the $\overset{|}{\underset{|}{HC}}-\overset{|}{\underset{|}{CH}}$ group take place on surface elements on which two catalyst atoms are situated (Table III. 2), while in ammonia synthesis the active ensemble is triatomic (Table III. 3).

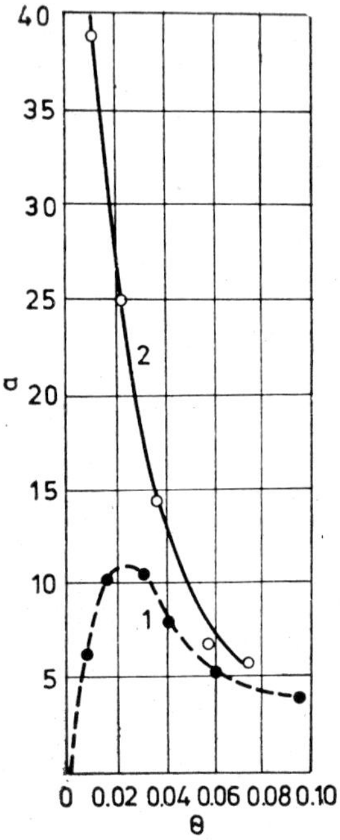

Fig. III. 10. Changes in the specific activity of a platinum catalyst on a kieselguhr support in the hydrogenation of maleic acid with two different initial catalyst compositions

For a reliable estimation of the number of atoms constituting the active ensemble it is necessary that probability (III.3) be valid for the distribution of adsorbed catalyst atoms on the surface. This is fulfilled for the statistical distribution of independent catalyst atoms, but as soon as an ensemble consisting of a definite number of atoms is involved, the distribution derived above refers to the distribution of these ensembles. For instance, if platinum is applied in the form of $PtCl_4$ on kieselguhr, after decomposition of the compound the probability of platinum atom ensembles will be given by (III.3). It appears from curve *1* in Fig. III.10 that when maleic acid is hydrogenated the specific activity will have a maximum, since the number of $(Pt)_2$ formations reaches a maximum at a certain coverage. On the other hand, when $K_2(Cl_3Pt-CH_2=CH-CH=CH_2-PtCl_3)H_2O$ is decomposed, the catalyst formed in this way will originally contain $(Pt)_2$ formations, the probability of which decreases monotonously with coverage, in the same way as in the case of monoatomic formations. This is illustrated by curve *2* in Fig. III.10.

The theory of catalytically active adsorbed atomic ensembles has also been confirmed in its generalized form.

If, according to Eq. (III.12), $\theta_{a_{max}}$ is substituted into Eq. (III.10) and Eq. (III.10) divided by the obtained expression, we arrive at the formula

$$\frac{a_n}{a_{n_{max}}} = \left[\frac{e}{n-1}\right]^{n-1} (p\theta)^{n-1} \exp(-p\theta) \qquad \text{(III.16)}$$

TABLE III.2

Two-atomic active centres

Reaction	Catalyst	Support	Active centre
$C_2H_4 + H_2 \rightarrow C_2H_6$	Pt	alumina gel, kieselguhr	$(Pt)_2$
	Ni	barium sulphate, calcium sulphate, kieselguhr, zinc oxide	$(Ni)_2$
$C_3H_6 + H_2 \rightarrow C_3H_8$	Pt	alumina gel, kieselguhr	$(Pt)_2$
$C_5H_{10} + H_2 \rightarrow C_5H_{12}$	Pt	alumina gel, kieselguhr	$(Pt)_2$
$C_4H_6 + 2\,H_2 \rightarrow C_4H_{10}$	Pt	alumina gel, kieselguhr	$(Pt)_2$
$C_7H_{14} + H_2 \rightarrow C_7H_{16}$	Pt	kieselguhr	$(Pt)_2$
$(CH_3)_2(HO)C{-}C{\equiv}CH + 2\,H_2 \rightarrow$ $\rightarrow (CH_3)_2(HO)C{-}CH_2CH_3$	Pd	activated carbon	$(Pd)_2$
HCCOOH ‖ HOOCCH + $H_2 \rightarrow (CH_2COOH)_2$	Pd	activated carbon, barium sulphate	$(Pd)_2$
HCCOOH ‖ HCCOOH + $H_2 \rightarrow (CH_2COOH)_2$			
$C_2H_5CH{=}CHCOOH + H_2 \rightarrow$ $\rightarrow C_4H_9COOH$	Pt	activated carbon	$(Pt)_2[(Pt)_6?]$
cyclohexene + $H_2 \rightarrow$ cyclohexane	Pt	nickel, kieselguhr	$(Pt)_2$
benzene + $3\,H_2 \rightarrow$ cyclohexane	Pt	kieselguhr	$(Pt)_2[(Pt)_6?]$
phenol + $3\,H_2 \rightarrow$ cyclohexanol	Pd	kieselguhr	$(Pd)_2$
toluene + $3\,H_2 \rightarrow$ methylcyclohexane	Pt	activated carbon	$(Pt)_2[(Pt)_6?]$
cyclohexane $\rightarrow$ benzene + $3\,H_2$	Pt	activated carbon	$(Pt)_2$
	Pd	activated carbon, magnesia	$(Pd)_2$

TABLE III.3

Three-atomic active centre

Reaction	Catalyst	Support	Active centre
$3\,H_2 + N_2 \rightarrow 2\,NH_3$	Fe	alumina gel, activated carbon, asbestos, crystalline iron	$(Fe)_3$

which leads to a quite general correlation between the relative specific activity values and $p\theta$, provided the n values agree. For reactions of the same type, and thus, when n is always the same, Eq. (III.16) is valid irrespective of the nature of the catalyst or of the support, the experimental conditions, the temperature, the substrate, etc. (see, for example, in Fig. III.11 the generalized correlation for $n = 2$, i.e. for diatomic active sites).

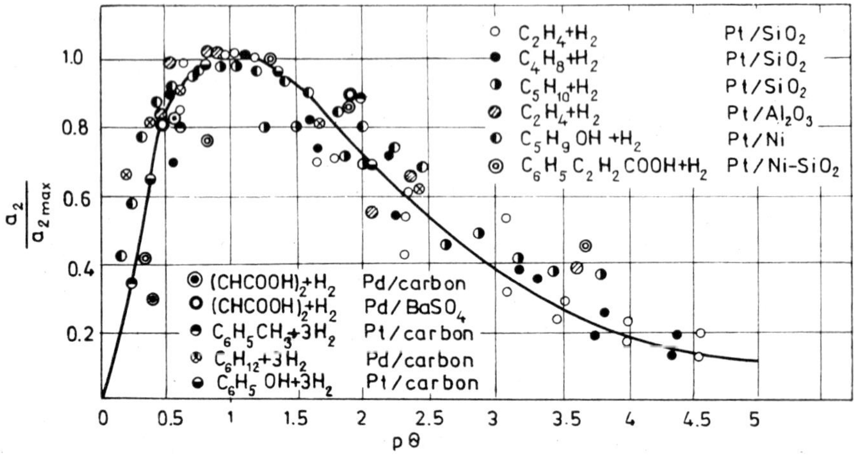

Fig. III.11. Generalized activation curve for two-atomic active centres

In an attempt to find an interpretation of poisoning phenomena it was assumed that the distribution of a given quantity of poison molecules on the support surface can be described by means of Eq. (III.3) just as well as the distribution of active catalyst atoms [19e]. It was also suggested that all surface elements with even a single poison molecule lose their activity, since physically adsorbed molecules migrate very rapidly on the surface of the support (see Part II, Section 5.2). This was followed by the determination of the probability, that is, the number of surface elements which remain vacant when a given poison quantity comes into contact with the surface. For the activity of these surface elements the above correlations are valid, i.e. in Eq. (III.8) z, and in Eq. (III.9) z_n have to be multiplied by a factor smaller than 1 depending upon the quantity of the poison. On a change of the quantity of the poison, the activity varied in accordance with the derived correlations, as confirmed by experiment. In this respect, too, the theory seemed to be correct.

SECTION 1.3. CRITICISM OF THE GEOMETRICAL THEORIES

Two extreme types of catalyst have been discussed. Crystalline catalysts on the continuous lattice planes of which the transformation takes place, and supported catalysts for which there is no possibility of the formation of crystalline agglomerates have been dealt with separately. It should be noted that these latter occur only very rarely in practice. The interpretation given in the preceding Section can be considered geometric, inasmuch as the relationships between the structure and activity of the catalyst were described without taking into consideration either energy or electronic parameters, but attempting solely the determination of the number of catalyst atoms within the catalytically active ensemble. The Kobozev theory in this sense might be called more a stoichiometric, than a truly geometrical, theory, since it allows no statement concerning the exact arrangement of the catalyst atoms participating in the reaction or the disposal of the substrate on them.

If, for instance, the hydrogenation of the C=C bond is simply compared on the basis of the number of catalyst atoms participating in the reaction in the case of the two types of catalysts, we arrive at different results. According to the Balandin multiplet theory, the reaction proceeds on 4 atoms in the lattice plane of the catalyst, while it follows from the Kobozev active ensembles theory that the reaction takes place on 2 supported catalyst atoms.

Kobozev explains this discrepancy by claiming that Balandin stresses the importance of a geometrical relationship between the converting molecules and the active sites, while in fact the nature of active atom ensembles is determined not by the structure of the reacting molecule, but by the structure of the catalytic process [19a]. This is the reason for the need of a triple atomic ensemble in ammonia synthesis, where three new bonds are formed; in oxidation the activation of oxygen plays the decisive role, but this proceeds on only a single catalyst atom.

In our view, the reason for this discrepancy has to be sought in the different mechanisms of the same conversion on catalysts of different types*. On a continuous metal surface this reaction proceeds, in fact, between neighbouring adsorbed hydrogen and olefin molecules [20], as proved by kinetic mechanism studies [20], while on two catalyst atoms only conversion according to the Rideal-Eley mechanism can take place, that is, only one of the reactants can be adsorbed. To the best of our knowledge, there are no kinetic experiments which support such a mechanism.

Nevertheless, it must be mentioned that Poltorak has attempted to apply the Kobozev mechanism to pure metal catalysts [21]. He assumed that on the surface of the metal the same type of "atomic phase" develops as is found in the case of supported catalysts. By means of statistical mechanical methods Poltorak derived correlations for the quantitative distribution of the crystalline phase—atomic phase for various crystallite sizes, surface tensions and temperatures. So far, however, these results have not been supported by the necessary catalytic experiments.

On the other hand, the following observation seems to confirm the different ranges of validity of the theories:

* It has been shown in Part II that the same molecule can be adsorbed in a different manner on different surfaces.

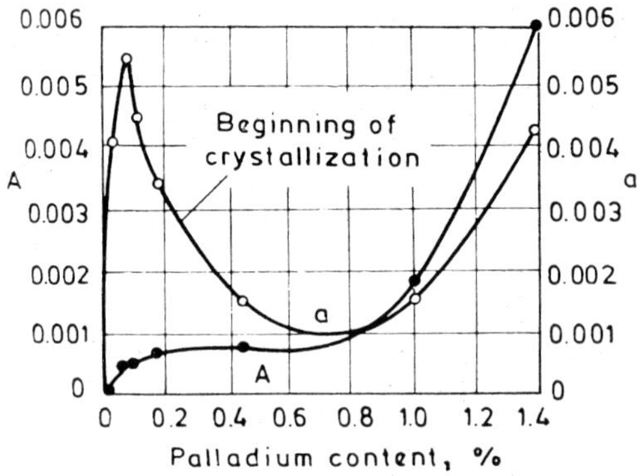

Fig. III.12. Hydrogenation of benzene on a palladium catalyst on a kieselguhr support [22]

The hydrogenation of benzene was studied in the presence of a palladium catalyst on a kieselguhr support (Fig. III.12) [22]. As long as the surface concentration of palladium was low, the curves of overall activity and specific activity had shapes corresponding to those expected from the theory of active atomic ensembles. As the palladium content increased, X-ray diffraction tests indicated the beginning of crystal formation on the surface, and a parallel steep rise of both curves. This increase cannot be explained by supposing the existence of active ensembles, whereas it can be interpreted perfectly by means of the multiplet theory, as the crystal surface on which the substrate can be fitted in the desired position may have occurred to an ever greater extent.

This example shows quite clearly that in one case one theory represents the true situation, while, in the other case, the other theory is more representative. Either because of their high mobility or as a result of their random distribution in the case of localized adsorption, on definite surface elements of the support the adsorbed catalyst atoms could conceivably adopt with a certain probability those formations which, according to the multiplet theory, are necessary for a given reaction to take place in both directions [which in Section 1.1 would be preferable according to the simplest scheme (a)]. In our view, even in this case the two theories cannot be simply merged without any restriction because the energy state and electron configuration of an independent adsorbed catalyst atom are quite different from those of atoms forming part of a continuous crystal surface. This fact alone would be sufficient for the assertion of entirely different geometries in the two cases.

It can be seen from the above that a geometrical concept alone is not enough to give a full picture of the mechanism of catalysis. It was pointed out at the end of Section 1.1 that the neighbourhoods of active formations meeting geometrical requirements, might differ from an energy viewpoint, and this difference must obviously be reflected in the activities. The different energy states of the catalyst surface are accompanied by changes of the atomic distances in the lattice. Under these conditions it is not enough to ascribe activity changes solely to geometrical factors, as was done with respect to the MgO catalyst according to [12] (p. 670).

An outline will now be given of the role and importance of the energy factors, in order to show how they modify the geometric concepts. A strong interaction between the catalyst and the substrate, that is, the formation of certain bonds between them, is a primary condition of catalytic interaction.

As regards adsorption on several sites, the bond strengths will obviously depend upon the spatial orientations, as well as on the nature, of the involved atoms. It is necessary to examine:

(1) how far the development of formations meeting geometrical requirements modifies the bond strengths (what results from the "deformation of the bond angles") and,

(2) what the result will be when, with identical geometric arrangements, the bond strength changes because of differences in the energy state of the catalyst surface, because of a change in the substrate and/or in the catalyst.

It has already been mentioned that the adsorption of the reacting molecule is possible only if the distances between the catalyst atoms in the surface lattice plane are within a certain range, which means that a certain deformation of the bond angles is still permissible. This, in fact, implies that certain "stresses" may appear which weaken the bonds between the substrate and the catalyst. In some cases it is possible to calculate the activation energy values devoted to "stressing" the molecule. Sherman and Eyring carried out such calculations [23] when determining the activation energies for the chemisorption of hydrogen on a fictive carbon surface in which there are different C—C atomic distances. It appears quite clearly from the theoretical values that the curve activation energy vs. C—C distances goes through a minimum, as has been shown in greater detail in Chapter 6 of Part II. The activation energy of the o → p hydrogen conversion was calculated in a similar manner for Pd—Au alloys [22]. The lattice constant of palladium is 3.88, and that of gold, 4.07 Å. The lattice is face-centred cubic in the entire alloy series. Considering the change in the lattice constant alone, a difference in activation energy of 0.8 kcal $mole^{-1}$ is obtained.

When the decomposition of carbon monoxide was studied with the help of the field emission microscope (for details of the method see Part II, 5.2.1 and Fig. II.10), it was found [24] that the various crystallographic faces possess different activities; this can be explained by the different atomic spacings in the various crystallographic faces.

More accurate measurements, suitable for quantitative evaluation, were performed with the same aim in mind [24]. The hydrogenation of ethylene was studied on a nickel single-crystal. There were differences amounting to, at most, a factor of ten between the activities of the faces of different orientations. This difference is too high to be attributed solely to differences in atomic distances. Thus, here we have to consider the manifestation of an effect of type (2): besides the "stressing" of the substrate molecule, another energy factor, namely, the different energy states of the surface catalyst atoms, is also involved.

Another example illustrates that despite more favourable geometrical conditions the catalytic activity can be lower. Destructive hydrogenation was studied on molybdenum disulphide and molybdenum dioxide catalysts [8c]. In both cases, the open surface lattice plane is formed by molybdenum atoms. According to calculations on the sulphide, the angle of the Mo—C—C bond within the Mo/C—C\Mo complex formed in the course of adsorption is 112°, hardly differing from the ideal 109°, while there is a considerably greater deviation on the oxide. In spite of this, the activity of the oxide catalyst was much higher than that of molybdenum disulphide.

The influence of energy factors of type (2) can be so decisive that, even though the geometric arrangement is entirely suitable, no noticeable catalytic conversion takes place. This is the case with copper for example. In the hydrogenation and dehydrogenation of C_6 cyclic hydrocarbons, copper is quite inactive, although on the atoms of the (111) face the substrate can be situated so as to give perfect fitting.

To summarize briefly the necessity of energy considerations, it may be said that appropriate geometrical factors are necessary, but not sufficient conditions for catalytic activity.

CHAPTER 2. ENERGY CONDITIONS DURING CATALYSIS

Having outlined the geometric concepts and mentioned the role of energy factors in them, we shall next discuss the latter in greater detail.

In the Introduction it was pointed out in connection with Fig. III.1 that a strong interaction, which might be called a chemical bond, must exist between the catalyst and the substrate. Only in this way can a conversion take place at all, or far more readily than in the homogeneous phase. Though the catalytic reaction may proceed in more steps than the homogeneous reaction the rate of the former will be the higher, for, in the catalytic reaction, the energy barriers between the individual steps are lower. Only an extremely large reactant volume/catalyst surface ratio can cause a higher reaction rate in the homogeneous *space* than in the field of the catalyst *surface*.

SECTION 2.1. CHANGES IN ENERGY LEVELS DURING CATALYTIC CONVERSIONS

Let us start from the state in which the distance between the reactant (R) and the catalyst (C) is so great that no interaction can arise between the two, while, further, each degree of freedom has the corresponding zero point energy value. In this case the energy level of the closed reacting system consisting of R and C will be taken as zero (see the height of $R+C$ in Fig. III.13).

For a bond to be formed between R and C their structures have to undergo certain modifications; in other words, R and C must be activated, as shown by examples in the Chapter on chemisorption in Part II. Because of its subordinate importance, there will be no discussion here of how physical adsorption might modify or complement these concepts of chemisorption. It will merely be pointed out that the energy of the system consisting of R and C must increase. In R and C the existing bonds between certain atoms must be loosened or even split for new bonds to be formed between R and C. In other words, in R and C certain degrees of freedom must possess higher energies than in the ground state. This, of course, might result in the simultaneous rise of energy levels of several other degrees of freedom; that is, the system will have an energy $E_{[RC]}{}^{\ddagger}$ (Fig. III.13) at which the desired energy values are ensured for the involved degrees of freedom.

Under appropriate conditions adsorption takes place: the loosened "excited" bonds of the activated state are replaced by new bonds between

R and C. The strengths of the new bonds are of the same order as the initial bonds in R and C, that is, R is chemisorbed. If all energies for each degree of freedom in the system are reduced to the ground level, the energy of the system differs from the initial value by the value of the heat of adsorption (in the example illustrated in Fig. III.13 it drops to $E_{[RC]}$), and the residual bonds in R and C are considerably weaker.

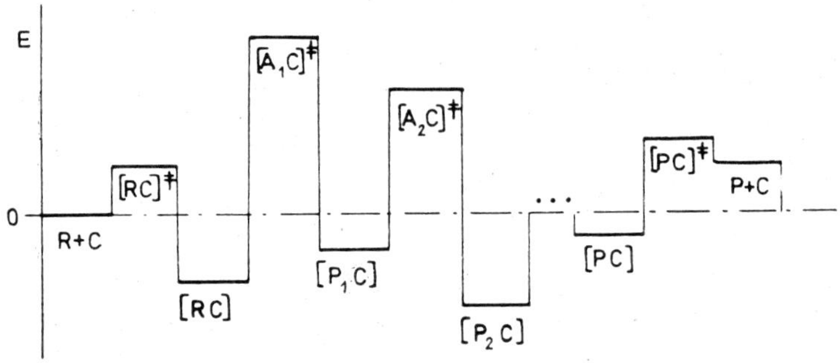

Fig. III.13. Energy scheme of conversion $R \rightarrow P$

As a result of the different adsorption interactions different bonds are weakened. Specific adsorption (see the Chapter on chemisorption in Part II), however, leads to the weakening of mainly those bonds whose rearrangement means the desired chemical reaction itself. Experience has fully confirmed this selectivity, since it is a well-known property of catalysts that they accelerate only a few, and usually only one, of the thermodynamically possible transformations.

The reaction takes place only when the bonds involved are further loosened by the input of more energy, which may cause even the complete elimination of the bonds. In this step an energy input must be provided, presumably for more than one degree of freedom, in the same way as shown for the activation of adsorption. In this manner the activated complex of catalytic reactions is formed.

If the overall transformation on the surface involves the rearrangement of more than one bond, in accordance with the Rice-Teller principle [11] several chemisorbed intermediates ($[P_1C]$, $[P_2C]$, . . .) may be formed through several activated complexes ($[A_1C]^{\ddagger}$, $[A_2C]^{\ddagger}$, . . .). In principle this resembles the case when chemisorption proceeds in several steps (see Part II) which might, of course, be the case when catalytic transformation is initiated (for simplicity we shall not go into these details). Desorption of the intermediates is excluded from our investigations, since these refer not to a consecutive reaction, but to the study of a single catalytic conversion.

After the rearrangement of the bonds and "cooling" of the system the level $[PC]$ is reached, which gives the zero-point energy value of the system for the adsorption of the product P.

For the desorption of the product the same arguments are valid (with appropriate modifications) as for the adsorption of the reactant: the bonds between P and C are excited ($[PC]^{\ddagger}$) by the input of the energy needed, for desorption, that is, for the activated state of desorption to be formed when the required bonds come into being within P and C. Finally, the distance between product and catalyst is so great that no interaction of any kind is possible and the corresponding zero-point energy pertains to every degree of freedom. The difference between the initial zero energy level and E_{P+C} gives the energy change of the reaction $R \rightarrow P$ at the absolute zero point (energy changes between the various steps of the reaction are obtained in the same way from the differences between the various zero-point energy levels).

Similar arguments can be applied to the reaction products, since the principle of microscopic reversibility must hold in catalysis too.

In this theoretical example an entirely general case was presented. It may happen that one or another of the steps takes place without an energy input. The chemisorption of radicals is a process requiring practically no activation; in hydrogenation the surface reaction usually proceeds with completely dissociated hydrogen molecules, when an excitation of the H—H bond is no longer necessary.

Both R and P may have components which do not interact with the catalyst (see the Rideal-Eley mechanism in Part IV), that is, in this case there are bonds within the reacting atomic ensemble which have not lost any of their strength as a result of interaction with the catalyst. All this, however, has no bearing on our original phenomenological concept of the system as a whole, but only means that to reach a certain level $E_{AC]}^{\ddagger}$ the same amount of energy is needed as would be required in the ho[mo-geneous phase. In this case the symbol means that the active state required in the sequence of steps of the catalytic conversion is reached by the activation of some component not interacting with the catalyst.

It is also quite obvious that the relative positions of the individual energy levels are different in different cases. The scheme of a real case differs even in principle from that illustrated in Fig. III.13, since, after each step, the system never reaches the energy level corresponding to zero degree, but an energy state corresponding to some finite temperature, and this also modifies the positions of the energy levels.

For a given catalytic reaction quantum chemical calculations might give the required answer. However, many difficulties are encountered here:

(1) The energy state of the catalyst surface is not accurately known (see Part I, Chapter 5 and Part III, Chapters 3 and 4), and nor are the bond strengths resulting from the interactions.

(2) For the time being it is not possible to calculate to what degree the individual bonds must be loosened for the formation of new bonds, nor the extent cf the energy often distributed over other degrees of freedom. (Experience in this respect is available only for homogeneous reactions, for which it has been determined what fractions of the energies of the formed bonds must be subtracted, depending on the type of the reaction, from the energies of the opening bonds [25]; that is, when, in addition, energy must be invested into other degrees of freedom [26].)

(3) Similarly, as regards the fate of the energies liberated in the individual exothermic steps it is not known whether they are used up entirely in the

activation of the subsequent step (whether the process is an "adiabatic" one), or are spread so that, for example, they are distributed over other degrees of freedom too (the process is "isothermal"). Such problems have been discussed in Part II, Paragraph 6.2.2.

Finally, a few words must be said concerning the relationships between the heights of the energy barriers and the rate of transformation. For homogeneous reactions a satisfactory answer has been provided in some cases by the theory of absolute reaction rates and, in principle, this method can be applied to the theoretical calculation of the adsorption rate (Part II, Paragraphs 6.3.3. and 6.3.4). Similar correlations have been derived for catalytic transformations, but, to the best of our knowledge, these have not been experimentally confirmed. The difficulties are obvious from what has been said so far, since, among other things, the exact changes of the energy levels during multistep processes are not known, while very little has been elucidated about the nature and structure of the intermediate complexes. It is also highly doubtful how far the equilibrium distribution valid for a large number of particles can be considered applicable to a small number of particles, such as the single activated complexes.

However, experience has shown that the concentrations of activated complexes requiring higher activation energies are lower and presumably the rates of conversions in which they are involved are also lower. It may sometimes happen, however, that the presumption of the theory of absolute reaction rates, according to which the activated complexes must be in equilibrium with the reactants, does not hold. In other words, when the "rate-controlling" process is not the decomposition, but the formation, of the activated complexes, the reacting system will stay the longer in the state preceding the step which requires activation energy, the higher the energy level of the activated state. Study of the energy conditions is therefore of paramount importance.

It is clear that there are many difficulties to be overcome in this case. Nevertheless, it appears probable that this will be the approach leading to the solution of the problem, that is, the clarification of the mechanism of action of catalysts (see the following Chapters).

It will next be shown that on the basis of these arguments and with several simplifying suppositions it is possible to solve some practical problems.

SECTION 2.2. CALCULATION OF ENERGY LEVELS ON CATALYSTS WITH HOMOGENEOUS SURFACES

It has already been mentioned that certain simplifications are permissible with respect to the energy diagram in Fig. III.13. For the sake of an easier understanding these will be given consecutive numbers and introduced wherever required in the discussion of the problem. Our arguments are based on the Balandin multiplet theory.

Thus, first of all,

(1) the chemisorption activation energies for R and P are taken as zero:

$$E_{[RC]^{\ddagger}} - E_{R+C} = 0$$

$$E_{[PC]^{\ddagger}} - E_{P+C} = 0$$

It has been seen that in the course of chemisorption the bond strengths in the substrate molecule are reduced. As a quantum chemical approximation of zero order, it is assumed that

(2) if the bond strength between the atoms A and B within a molecule was originally E_{AB}, and if, in the course of adsorption, the atoms A and B interact with the catalyst to give bonds of energies E_{AC}, and E_{BC}, then the energy of the bond $A-B$ in the chemisorbed molecule is [6]

$$E_{AB,C} = E_{AB} - E_{AC} - E_{BC} \tag{III.17}$$

that is, the energy level of the closed reacting system remains unchanged and, in strength, the initial bond $A - B$ has lost the sum of the strengths of the new bonds, $A - C$ and $B - C$ (see Fig. III.14). Thus,

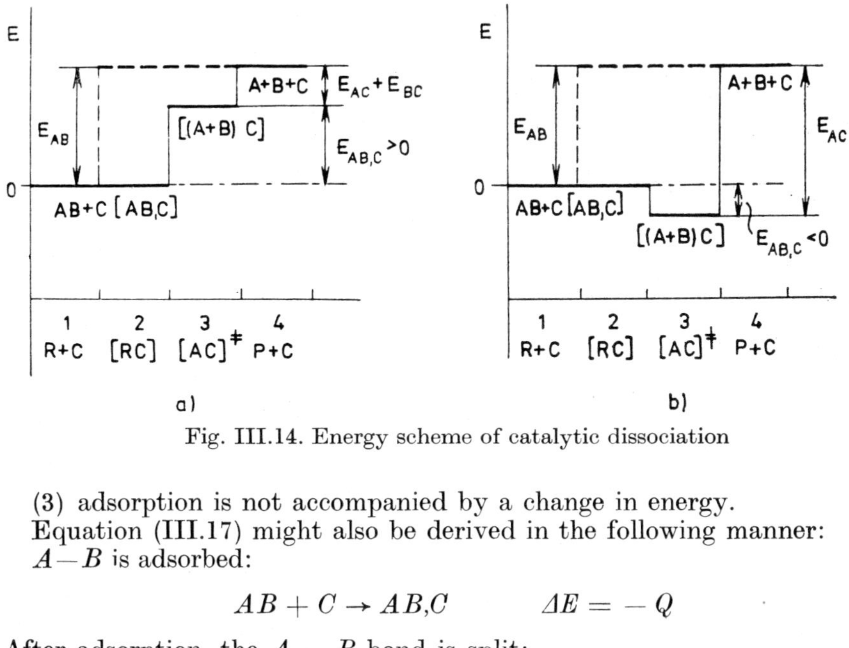

Fig. III.14. Energy scheme of catalytic dissociation

(3) adsorption is not accompanied by a change in energy.

Equation (III.17) might also be derived in the following manner:

$A-B$ is adsorbed:

$$AB + C \rightarrow AB,C \qquad \Delta E = -Q$$

After adsorption, the $A - B$ bond is split:

$$AB,C \rightarrow AC + BC \qquad \Delta E = E_{AB,K}$$

After dissociation on the surface, desorption takes place:

$$AC + BC \rightarrow A + B + C \qquad \Delta E = E_{AC} + E_{BC}$$

This is, in fact, the dissociation of $A - B$:

$$AB \rightarrow A + B \qquad \Delta E = E_{AB}$$

that is

$$E_{AB} = -Q + E_{AB,C} + E_{AB} + E_{BC}$$

which, if $-Q = 0$, leads after rearrangement to Eq. (III.17).

Now, in the light of points (1), (2) and (3), let us consider a simple dissociation.

The reaction can proceed in either of two ways: either in the homogeneous phase (along the dashed line in Figs III.14a and III. 14b), or catalytically (along the continuous line of Fig. III.14a if $E_{AB} > E_{AC} + E_{BC}$, and along the continuous line of Fig. III.14b if $E_{AB} < E_{AC} + E_{BC}$).

In homogeneous reactions the energy level of the reacting system must be raised by at least E_{AB}, that is, only those molecules are capable of dissociation which, at the given degree of freedom, possess a minimum energy E_{AB} [1]. In the case of thermal excitation and statistical equilibrium, the number of molecules in this state is proportional to the fraction exp $(-E_{AB}/RT)$ of the number of all the molecules, in accordance with the Arrhenius correlation. Essentially, therefore, E_{AB} is determined, when the activation energy of an elementary transformation is calculated.

It can be seen from Fig. III.14a that during catalytic transformation more than one step may require an energy input: the surface dissociation $E_{AB,C}$, and the desorption $(E_{AC} + E_{BC})$. Since these are consecutive processes, their kinetically measurable macroscopic rates will necessarily be the same. If the transformation of a single molecule is followed, however, it appears that the molecule must usually stay considerably longer in one state than in the other, since the individual energy differences are not the same. The transition requiring less energy proceeds at a considerably higher rate and, consequently, the activation energy obtained from the change of the experimentally determined rate constant gives the energy surplus needed for the greater change in energy. In other words: the activation energy will essentially be identical with the higher of $E_{AB,C}$ and $(E_{AC} + E_{BC})$, provided A and B are simultaneously desorbed and no homogeneous decomposition yet takes place.

In order to be able to perform calculations of practical use from these arguments, it is necessary to know the bond energies involved in the process. These can be determined from spectroscopic and thermochemical data. There is a very great scatter, however, between the results obtained by different methods [27], so that certain bond energies are fixed, conventionally, and other bond energies are calculated from these in such a way that the values obtained satisfy the system of thermochemical equations. Table III.4 contains data which, in this sense, are consistent*.

The left side of Table III.4 gives those zero-point bond energies which are valid for free molecules, that is to say when no external field (e.g. of the catalyst) acts on the molecule. From these average values and Eq. (III.17), the corresponding bond energies on nickel, $E_{AB,\mathrm{Ni}}$ can be calculated (see the right side of Table III.4).

* A similar table has been compiled starting from more accurate bond energies [28], but this is still far from being sufficiently complete to enable the calculation of the energy conditions of catalytic conversions.

TABLE III.4

Bond energy values for single σ bonds (kcal $mole^{-1}$)

	In the free state							On nickel						
	C	N	O	S	H	Cl	Br	C	N	O	S	H	Cl	Br
C	59							45						
N	49	20						26	−12					
O	70	61	35					4	−14	−83				
S	55	—	67	64				5	—	−35	−22			
H	87	84	110	88	103			27	15	−2	−8	−3		
Cl	67	38	49	66	103	58		−6	44	−76	−43	−16	−74	
Br	54	—	—	57	87	53	46	−3	—	—	−36	−16	−63	−54
Ni	7	16	59	43	53	66	50							

A few remarks seem to be appropriate with respect to these data.

As mentioned before, on the left side of Table III.4 we find average values which, in certain cases, may change to a certain extent, depending on the substituent and the state of the catalyst (e.g. the bond energy of $>C=O$ is 142 in formaldehyde, 149 in other aldehydes and 152 kcal $mole^{-1}$ in ketones).

Balandin suggested that

(4) substituents modify the energy of the $A - B$ bond in practically the same way as the bond energies of A and B with the catalyst. Since E_{AB} and $E_{AC} + E_{BC}$ have opposite signs in Eq. (III.17), the substituent effect is more or less compensated.

Irrespective of the position of the catalyst atom (which will be discussed in greater detail at the end of this Section), Table III.4 contains constant energy values for bonds with nickel, so that

(5) from an energy viewpoint, a homogeneous catalyst surface is indicated.

Unfortunately, it has not proved possible to find an example for catalytic dissociation which would prove how far simplifications (1)—(5) are acceptable, and which would provide evidence whether, in this way, the activation energy of the process can be determined, that is, the sequence of dissociation for the different bonds can be established in accordance with expectation. On the other hand, there are many examples to illustrate conversions in which the bonds between two atom-pairs are opened. In this case, the opening of the bonds is always part of a process in the course of which a bond rearrangement finally takes place between two atom-pairs:

$$AB + CD \rightarrow AC + BD$$

This T4MM-type transfer reaction can, of course, take place in the homogeneous phase too. According to our arguments, this would require the input of bond energies $E_{AB} + E_{CD}$ (Figs III.15a and b, the path marked with the dashed line). Should this exchange reaction in fact proceed by way of a bimolecular reaction, the experimentally measurable activation energy is always lower, as no complete cleavage of the bonds occurs in the intermediate complex. It may further happen that activation proceeds in several steps, so that in agreement with the above only the activation energy of the slowest step, that which requires the highest energy input, is obtained.

For the time being, however, let us consider the dashed line in Fig. III.15 as the energy path of the homogeneous reaction which is idealized to the same degree as the path marked with continuous straight lines for catalytic conversion, since both include the assumption that

(6) both bonds are activated at the same time by the complete opening of the two bonds.

Thus, catalytic conversion can be divided into the following steps:
for adsorption:

$$AB + CD + C \rightarrow AB,C + CD,C \qquad \Delta E = -Q_1$$

for surface dissociation:

$$AB,C + CD,C \rightarrow AC + BC + CC + DC$$

$$\Delta E = E_{AB,C} + E_{CD,C}$$

for association following dissociation:

$$AC + BC + CC + DC \rightarrow AC,C + BD,C$$

$$\Delta E = -E_{AC,C} - E_{BD,C}$$

and for the desorption of the product:

$$AC,C + BD,C \rightarrow AC + BD + C \qquad \Delta E = +Q_2$$

In agreement with supposition (3), the values of Q are regarded as zero, and the $E_{XY,C}$ values are calculated from Eq. (III.17). Under these conditions two steps of the conversion process may require an activation energy, namely the dissociation of the adsorbed reactants, the energy requirement of which is:

$$E_1 = E_{AB,C} + E_{CD,C} = E_{AB} + E_{CD} - (E_{AC} + E_{BC} + E_{CC} + E_{DC}) \qquad \text{(III.18)}$$

and the formation of the molecules of the adsorbed product which requires

$$E_2 = -E_{AC,C} - E_{BD,C} = -E_{AC} - E_{BD} + (E_{AC} + E_{BC} + E_{CC} + E_{DC}) \qquad \text{(III.19)}$$

The energy values depend upon the relative values of $(E_{AB} + E_{CD})$ and $(E_{AC} + E_{BC} + E_{CC} + E_{DC})$, and of $(E_{AC} + E_{BD})$ and $(E_{AC} + E_{BC} + E_{CC} + E_{DC})$. In Figs III.15a and b the energy diagrams of the same reaction $[(E_{AB} + E_{CD}) - (E_{AC} + E_{BD}) = \Delta E_r$ (identical)] are shown, in case (a) E_1 being greater, and E_2 less, than zero, while in case (b) both E_1 and E_2 are greater than zero. This shift is simply caused by an increase of $(E_{AC} + E_{BC} + E_{CC} + E_{DC})$ [compare Eqs (III.18) and (III.19)].

Next let us consider how far these energy arguments are supported by actual catalytic investigations.

From the data of Table III.4 with the help of the known correlations it is possible to calculate the differences between the various energy levels during nickel-catalyzed hydrogenolysis. In general, hydrogenolysis can be represented as follows:

$$AB + \text{HH} \rightarrow A\text{H} + B\text{H}$$

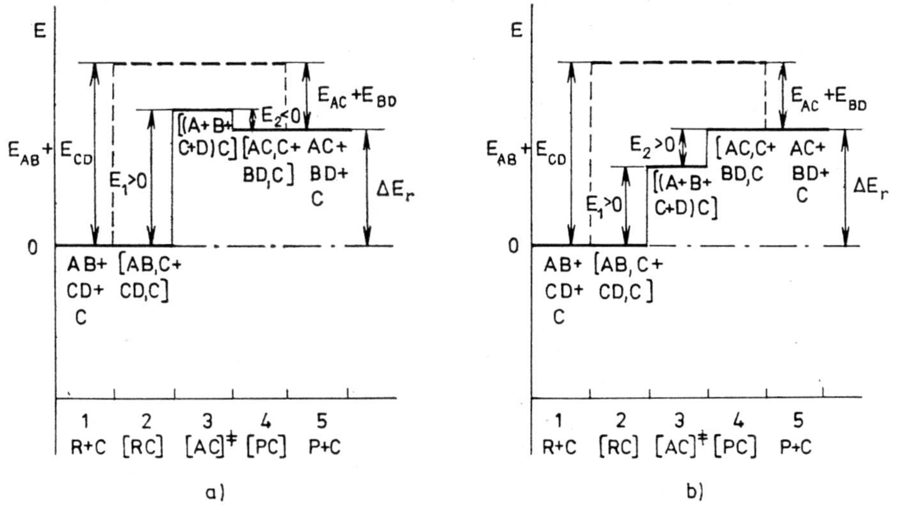

Fig. III.15. Idealized energy scheme of the catalytic reaction $AB + CD \rightarrow AC + BD$

where A and B are two atoms bound to each other within an organic molecule and H is the hydrogen atom.

The results of the calculations are given in Table III.5.

TABLE III.5

Differences in energy levels during hydrogenolysis (kcal)

A—B	C—C	C—N	C—O	C—Br	C—Cl	N—O
$E_{AB} + E_{\text{HH}}$	162	152	173	157	170	164
$E_{A\text{H}} + E_{B\text{H}}$	—174	—171	—197	—174	—190	—194
ΔE_r	—12	—19	—24	—17	—20	—30
$E_{AB,\text{Ni}} + E_{\text{HH,Ni}}$	42	23	1	—6	—9	—17
$-(E_{A\text{H,Ni}} + E_{B\text{H,Ni}})$	—54	—42	—25	—11	—11	—13

It appears from the Table that, in these examples, hydrogenolysis is accompanied by a decrease in energy (exothermic reaction, $\Delta E_r < 0$), and thus these reactions can be followed in the diagram of Fig. III.15 by proceeding from right to left.

The line ($E_{AB} + E_{HH}$) in Table III.5 gives values which, in the case of homogeneous reactions, correspond to the amount of energy required for complete dissociation of the reactant.

In the last two lines of the Table are the energy increments which may occur during catalytic hydrogenolysis: 1. $E_{AB,Ni} + E_{HH,Ni}$ the energy required to split the bond in the adsorbed reactant, 2. $-(E_{AH,Ni} + E_{BH,Ni})$ the energy needed for the formation of the adsorbed molecule from A, B, H and H.

Let us now consider the sequence provided by the experimental results.

The hydrogenolysis of p-nitrochlorobenzene was studied in the presence of nickel at increasing temperatures, under otherwise identical conditions [29]. (In such cases the bonds of the chemisorbed reactant are activated by the input of thermal energy, which means that, as the temperature is increased, transformations requiring increasingly higher activation energies will also take place.)

It was observed that first the N—O, then the C—Cl, later the C—N, and finally the C—C bond underwent hydrocracking (the double bonds were always saturated very quickly). Thus, it was not the heat of the reaction (the values of ΔE_r), or the energy required for homogeneous dissociation which determined the sequence of reactivities, but the order determined by ($E_{AB,Ni} + E_{HH,Ni}$).

The stepwise degradation of several hundred compounds was investigated in a similar manner [7]. This rich experimental material fully confirmed the sequence expected from the ($E_{AB,Ni} + E_{HH,Ni}$) values of Table III.5.

It should be added that though the data in Table III.5 were calculated from zero-point energies (see Table III.4), this sequence does not change as the temperature is raised; when the bonds are thermally excited the energy levels are higher, and the energy necessary for the complete cleavage of the bond is lower. According to quantum chemical calculations, however, this decrease is proportional to the bond strength, and the conclusions regarding the sequence are still valid.

One very important problem remains in need of further clarification: how is it possible that the sequence of reactivity can be extended to the negative values of ($E_{AB,Ni} + E_{HH,Ni}$), when no difference whatsoever can be observed between the rates of elementary steps accompanied by a decrease of energy? The reason for this must be sought among the numerous simplifications used during evaluation. One explanation might be the fact that, in Eq. (III. 17), a smaller value of E_{XC} than shown in Table III.4 has to be subtracted from the initial bond energy, [in this case it follows from Eq. (III.19) that $-(E_{AH,Ni} + E_{BH,Ni})$ would always have a considerably higher negative value than that shown in Table III.5]. Consequently, the bonds which are eliminated during the reaction are loosened on the catalyst to a lesser extent.

With very few exceptions the conclusions drawn from the bond energies have been found valid not only for hydrogenolysis, but also for a series of similar 4-atomic rearrangements when again several reaction paths are simultaneously available [7].

Hence, as a result of the simplifications (1) — (6), a quantitatively false, but self-consistent, picture is obtained of the energy conditions of the catalytic conversions under investigation.

In the exothermic conversions ($\Delta E_r < 0$, see Table III. 5) discussed, it has been seen that only the energy requirements of the surface dissociation of the reactants [Eq. (III.18)] need be taken into consideration in the assessment of the reactivity sequence (see the penultimate line in Table III.5), which is fairly obvious if the exothermic conversion in Fig. III.15a is considered from right to left. However, energy might also be required by that step which leads to the formation of adsorbed product molecules. Its value is given by Eq. (III.19). In Fig. III.15b this is illustrated by the second step of the endothermic conversion from left to right in the energy diagram.

In this simplified calculation of the energy conditions, however, it is not necessary always to obtain energy steps which, at first glance, appear acceptable. It has been shown above that due to the simplified assumptions of the theory a startingly distorted picture of reality is obtained: the reaction sequence is determined by the step accompanied by the greater energy increase, or smaller energy decrease, in such a way that the reactivity is the higher, the smaller the energy increase, or the greater the energy decrease involved in this step. In the earlier-discussed cases of hydrogenolysis this step was always the surface dissociation of the reactant. On the other hand, the relative rate of bond rearrangements is the higher, the lower the energy input, or the greater the energy liberated in this step. If the step considered here as the rate-determining one were a true elementary reaction step in which an activated complex is formed (see Figs. III. 15), then the energy decrease associated with its exothermic progress could not affect the relative rate of bond rearrangements.

In general, therefore, energy changes accompanying two steps, namely, the surface dissociation of the reactant and the formation of product molecules on the surface, must both be taken into consideration, even if both steps involve a decrease of energy. In this case, too, the statement that the lesser energy decrease is the decisive factor is still valid.

Let us now consider what generally valid correlations can be obtained on the basis of the above arguments. According to Eq. (III.18), E_1 decreases with increasing $(E_{AC} + E_{BC} + E_{CC} + E_{DC})$, while, at the same time, according to Eq. (III.19), E_2 increases. Such a correlation is obtained when the same reaction is investigated on different catalysts, as in this case only $(E_{AC} + E_{BC} + E_{CC} + E_{DC})$ changes, and $(E_{AB} + E_{CD})$ and $(E_{AC} + E_{BC})$ remain unchanged (see, for example, in Fig. III.16, the correlation obtained for the dehydrogenation of alcohols).

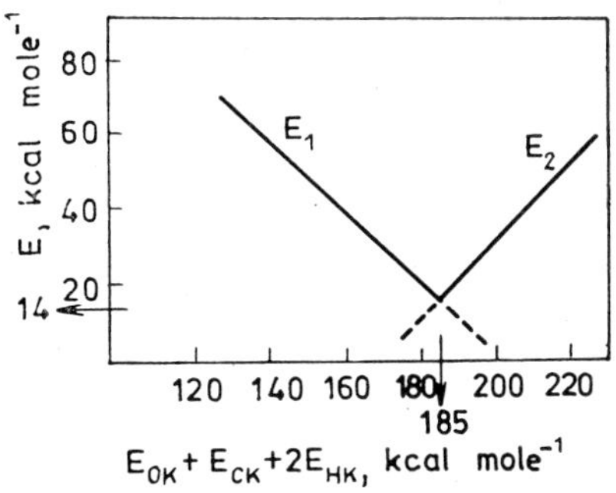

Fig. III.16. Energy barrier E vs. adsorption energy in the dehydrogenation of alcohols (K $\equiv$ C)

It was seen above that, as regards the rates of the part-processes, it is always the higher energy barrier which has to be taken into consideration, i.e. that of E_1 and E_2 which has the higher value. Figure III.16 shows that simplifications (1) — (6) involve a linear decrease, followed by a linear increase, of the energy E, which is decisive when the reaction takes place on the catalyst surface (continuous lines in Fig. III.16). The minimum of E lies at the point of intersection of E_1 and E_2, and the abscissa of this point is obtained from Eqs (III.18) and (III.19) as follows:

$$(E_{AC} + E_{BC} + E_{CC} + E_{DC})_{\mathrm{opt}} = \frac{E_{AB} + E_{CD} + E_{AC} + E_{BD}}{2} \quad \text{(III.20)}$$

At this adsorption energy:

$$E_{\mathrm{min}} = \frac{(E_{AB} + E_{CD}) - (E_{AC} + E_{BD})}{2} = \frac{\Delta E_{\mathrm{r}}}{2} \quad \text{(III.21)}$$

It appears directly from Fig. III.15 that $(E_{AB} + E_{CD}) - (E_{AC} + E_{BD})$ equals ΔE_{r}, and hence E_{min} is half of the heat of reaction. At first glance this may seem strange, since we know that in the course of a homogeneous endothermic reaction the energy input into the reacting system must at least equal the heat of reaction. In this latter case, however, energy is communicated in a single step, in a single activation process, while in catalytic conversions energy input might be divided between two consecutive steps (Fig. III.15), when in the optimum case the energy differences are

$$E_{[AC]^{\ddagger}} - E_{[RC]} = E_{[PC]} - E_{[AC]^{\ddagger}} = \frac{\Delta E_{\mathrm{r}}}{2}$$

Thus, in one of the steps, namely that which controls the reaction rate, and from which the kinetically measurable activation energy is obtained (see Part IV), it seems sufficient to input less energy than required for the entire reaction.

When the conditions are those determined by Eqs III.(20) and (III.21), the "state of energy correspondence" is realized. According to Balandin, such an optimum value can be calculated for every reaction, and can be attained by means of the most suitable catalyst.

However, an important rider must be added to this:

It was pointed out in connection with tetra-atomic rearrangement in the homogeneous phase, that for a reaction to take place it is not necessary to communicate the entire dissociation energy, but sufficient merely to provide for a certain "loosening" of the molecules. To a first approximation the following relationship can be formulated between the energy barrier E and the activation energy $E^{\ddagger}$ as determined from kinetic measurements:

$$E^{\ddagger} = \delta' E \quad \text{(III.22)}$$

Balandin claims [7] that for the types of catalytic reactions discussed earlier $\delta' = 3/4$ can be used in the calculations.

From the Polányi rule for homogeneous reactions [1] it can be expected that

$$E^{\ddagger} = \delta'' E + A$$

would be a more accurate formulation than Eq. (III.22).

The fact that experiment has confirmed the reactivity sequence expected from the energy values required for the total loosening of the bonds is evidence in favour of the approximately identical extent to which the various bonds are loosened.

Hence, from the aspect of catalytic activity, the chemisorption bond energies are of decisive importance. It appears from Fig. III. 16 that if the latter are low, E (i.e. $E^{\ddagger}$) will be high, since chemisorption barely weakens the converted bonds. On the other hand, if the chemisorption bond energies are too high, that is, if the applied catalyst forms a strong bond with one or more atoms, the activation energy will again be high; in other words, in the case of strong adsorption the activity of the catalyst decreases, in complete agreement with experience.

The chemisorption energy values given by $(E_{AC} + E_{BC} + E_{CC} + E_{DC})$ change not only if a catalyst consisting of other chemical elements is chosen, but also when catalyst atoms bound by different energies on the surface form bonds with the corresponding atoms of the substrate. To illustrate this point, it should be mentioned that the bond energy values in the last line of Table III.4 are, in fact, valid only in an idealized extreme case; the bond energies are given by the heat of a reaction in which a molecule consisting of a single catalyst atom and an atom A (to which other atoms might be linked) is dissociated to form A and a free catalyst atom:

$$AC \rightarrow A + C; \quad \text{heat of reaction} = + E_{AC_{\text{atomic}}} \qquad \text{(III.23)}$$

However, in heterogeneous catalysis the catalyst is not in the gas phase, but the reaction proceeds on catalyst atoms which are bound to their neighbours by a certain energy E_{CC}. Using the same approach as before, this might be accounted for by simply subtracting from the energy of the bond $A-C$ the energy of the bond $C-C$:

$$E_{AC_{\text{surface}}} = E_{AC_{\text{atomic}}} - E_{CC_{\text{surface}}} \qquad \text{(III.24)}$$

Since we are dealing here with a catalyst atom on the surface and not within the bulk of the crystalline phase, the energy E_{CC} will probably amount to only a fraction of the heat of evaporation, λ. It is questionable how far such a method of calculation is permissible for a single free catalyst atom or for a catalyst atom forming part of the surface of a crystal lattice.

In addition to the chemical nature of the catalyst, however, the surface bond energies, E_{XC}, are certainly influenced by the immediate neighbourhood of the catalyst atom forming the bond. This immediate environment may vary depending on the method of preparation and possible contaminants. It is often impossible to assign the substrate to a particular catalyst atom, since it is more likely that the substrate forms a bond with certain atomic ensembles on the catalyst surface. In such cases it may occur that $E_{AC_{\text{surface}}}$ is higher than $E_{AC_{\text{atomic}}}$. For this very reason, to obtain reliable adsorption energy values directly, catalytic measurements are indispensable. If we proceed by the method outlined above, that is, if we accept simplifications (1)—(6) together with the respective remarks, the picture obtained of the energy relations of catalytic conversions is again erroneous as regards absolute values, but still relatively correct. This means that E is calculated, as before, from the experimentally determined values of the activation energy with the help of Eq. (III.22), and function $E = f(E_{AC} + E_{BC} + E_{CC} + E_{DC})$ is considered still valid (as shown, for example, in Fig. III.16).

Hence $E_{XC_{\mathrm{surface}}}$ is determined by the following sequence of operations [30]:

The activation energies of various reactions are measured on the same catalyst, and the individual E values are calculated from Eq. (III.22) and substituted into Eq. (III.18). The reactions are chosen in such a way that it is possible to calculate the unknown surface bond energies E_{XC} from the above equations. It has been found that the dehydrogenation of hydrocarbons and the dehydrogenation and dehydration of alcohols are such suitable reaction groups from which the bond energies E_{HC}, E_{CC} and E_{OC} can be determined.

The average bond energies obtained on a number of metals (Fe, Co, Ni, Cu, Ru, Rh, Pd, Ag, Re, Ir, Pt and Au) and metal oxides (oxides of Be, Sc, Ti, U, Cr, Mn, Zn, the rare earth metals, W and Th) are given in Table III.6 [31].

TABLE III.6

Mean values of bond energies (kcal per atom-pair) [31]

	On metals	On oxides
E_{HC}	56±2	49±2
E_{CC}	23±2	18±4
E_{OC}	58±3	63±4

For both types of catalysts the bond energies are very close to each other and their sequence is also the same, verifying the conclusion [31] that in both cases the bonds are formed between the metal ions of the catalyst and the corresponding atoms of the substrate. This is supported by the data of Table III.6, which show that the values are close to the values of the $X—Me$ bonds, thus for example, E_{OO} is only 35 kcal per atom-pair, in contrast to the 63 kcal per atom-pair for E_{OC} in the Table, while the value of E_{HO} is over 100 kcal per atom-pair, a value considerably above the 49 kcal per atom-pair for E_{HC}.

SECTION 2.3. ENERGETICALLY HETEROGENEOUS SURFACE; THE COMPENSATION RULE

All the above results presuppose a homogeneous catalyst surface. It has been proved that the experimentally determined activation energy of a reaction is an unequivocal function of some given adsorption energy. In other words: the reaction takes place only on some parts of the surface characterized by a certain adsorption energy. But what happens if the reaction also takes place somewhere else, and if a series of sites with differing adsorption energies may occur on the catalyst surface? This is obviously the case with the overwhelming majority of catalysts used in practice, where surface catalyst atoms and formations in widely differing energy states might be simultaneously present.

In the following this will be taken into consideration in the interpretation of the catalytic effect.

For the sake of simplicity let us give the symbol E_{AC} to $E_{AC_{\text{surface}}}$ in Eq. (III.24), and let E_{AC} be the strength of the chemisorption bond between the catalyst atom in the ground energy state and the substrate atom A. When A is adsorbed on a catalyst atom more weakly bound on the surface, that is, on one which possesses surplus energy $-\Delta E_{CC}$, then the energy of the chemisorption bond increases by a certain fraction of the liberated bond energy:

$$E_{AC_{\text{active}}} = E_{AC} + \beta \Delta E_{CC} \tag{III.25}$$

where $E_{AC_{\text{active}}}$ is the bond energy of atom A when chemisorbed on an energy-rich catalyst atom, and β is a factor between 0 and 1. In principle, the same method has been used for the formulation of Eq. (III.25) when the energies of the bonds being formed were calculated by *subtracting* the chemisorption bond energies from the initial bond energies: the increase in the energy of the original bond $A-C$ has now been taken into consideration by *adding* an aliquot part of the cohesion energy decrease ΔE_{CC}.

The energies of the bonds eliminated as a result of catalytic conversion are as follows [taking into account Eqs (III.18) and (III.25)]:

$$E_{AB,C} + E_{CD,C} + \ldots = E_{AB} + E_{CD} \ldots - (E_{AC} + E_{BC} + \ldots) + \beta \Delta E_{\text{act}} \tag{III.26}$$

where ΔE_{act} is the surplus energy of the catalyst atoms, that is, of the active surface formation required for catalytic conversion.

However, the disappearing bonds must not necessarily be completely opened in the course of the catalytic conversion process, since, as shown in Eq. (III.22), a certain loosening of the bonds may suffice. If the degree to which the bond is loosened is δ, the value of which presumably lies between 0.5 and 1, the activation energy is obtained from Eq. (III.26):

$$E^{\ddagger} = \delta\{E_{AB} + E_{CD} + \ldots - [(E_{AC} + E_{BC} + \ldots) + \beta \, \Delta E_{\text{act}}]\} \tag{III.27}$$

where β, δ, E_{AB}, E_{CD}, . . . , E_{AC}, E_{BC}, . . . are constants if the same reaction is studied on catalysts of identical chemical compositions. In this case Eq. (III.27) can be written in a simpler form by merging the constants:

$$\Delta E_{\text{act}} = - \text{a} E^{\ddagger} + b \tag{III.28}$$

Equation (III.28) indicates the relationship between the surplus energies of the active formations of the catalyst and the activation energy of the reaction which takes place on them. The question now arises of how the frequency of surface formations possessing this surplus energy ΔE_{act} depends upon this surplus energy, and [by way of Eq. (III.28)] upon the corresponding activation energy. The value of the rate constant is determined partly by $\exp(-E^{\ddagger}/RT)$ and partly by the pre-exponential factor k_0, which is proportional to the number of active formations. (Experience has shown that this statement is valid not only for the true rate constant, but also for the resulting, and even for the apparent, rate constant.)

Different authors have started from different assumptions in determining the distribution functions.

Constable [4] assumed that formations with lattice structures and lattice distances differing from those in the bulk of the catalyst may develop on the surface of the catalyst*. In our approach, all (or only some) of the substrate atoms between which bond rearrangement takes place during the catalytic reaction form bonds with one catalyst atom of the surface, and each of these catalyst atoms has n such neighbours with which it enters into an energy interaction.

The distances of this preferred catalyst atom from these neighbours are $d_1, d_2, \ldots d_n$. In the equilibrium state (in our case this would be the ideal crystalline state) the distances would be $\beta_1, \beta_2, \ldots \beta_n$. For a normal distribution, the probability of a distance d_1 between an atom-pair is:

$$W_{d_1} = \frac{1}{C\sqrt{\pi}} \exp[-(d_1 - \beta_1)^2/C^2] \qquad \text{(III.29)}$$

where C is a constant characteristic of the distribution, while, from Eq. (III.29) the probability of the other atoms being at distances $d_1, d_2, \ldots d_n$ from the atom under consideration is:

$$W_{d_1, d_2 \cdots d_n} = \left(\frac{1}{C\sqrt{\pi}}\right)^n \exp\left[-C^{-2} \sum_{i=1}^{n} (d_i - \beta_i)^2\right] \qquad \text{(III.30)}$$

According to Constable the interaction energy surplus of surface atoms within the ensembles under consideration is proportional to the square of the difference from the equilibrium distance:

$$\eta = \gamma \sum_{i=1}^{n} (d_i - \beta_i)^2 \qquad \text{(III.31)}$$

where γ is a constant.

Let us consider what this means as regards the activation energy.

Let the energy of a catalyst atom in the surface crystal lattice (in the equilibrium state) be $\alpha\lambda$ greater than within the bulk crystal, while the same atom possesses an additional energy surplus due to its random arrangement [given by Eq. (III.31)]:

$$E_{AC_{\text{surface, act}}} = E_{AC_{\text{cryst}}} + \alpha\lambda + \eta \qquad \text{(III.32)}$$

Hence, the term η has to be added to the expression for the adsorption energy:

* In the subsequent discussion it is assumed that the catalyst atoms under consideration are chemically identical elements.

$$(E_{AC_{\text{surface}}} + E_{BC_{\text{surface}}} + \ldots)$$

Since we are within the validity range of Eq. (III.18) in the overwhelming majority of cases (this can be ascertained from $E^{\ddagger}_{\text{exp}}$ and $E^{\ddagger}_{\text{calc}}$ from ideal E_{XC} values):

$$E = \Delta - \nu\eta \tag{III.33}$$

where

$$\Delta \equiv E_{AB} + \ldots - (E_{AC_{\text{cryst}}} + E_{BC_{\text{cryst}}} + \ldots + \nu\alpha\lambda)$$

and ν is the number of catalyst atoms simultaneously participating in the surface reaction. In agreement with Eq. (III.22) the activation energy is

$$E^{\ddagger} = \delta(\Delta - \nu\eta); \quad \delta \approx 3/4 \tag{III.34}$$

It follows from Eqs (III.30) and (III.31) that the probability of an atomic ensemble of activation energy $E^{\ddagger}$ is:

$$W_{E^{\ddagger}} = \left(\frac{1}{C\sqrt{\pi}}\right)^{\nu n} \exp[(E^{\ddagger}/\delta - \Delta)/(\gamma C^2)] \tag{III.35}$$

If the number N of all active formations on unit surface area of the catalyst is multiplied by the probability of Eq. (III.35), the number of formations on which an activation energy $E^{\ddagger}$ is needed for the reaction to take place is obtained:

$$N_{E^{\ddagger}} = N f_1 \exp(h_1 E^{\ddagger}) \tag{III.36}$$

where $N_{E^{\ddagger}}$ is the number of formations characterized by the activation energy $E^{\ddagger}$ on unit surface area of the catalyst, and

$$f_1 \equiv \left(\frac{1}{C\sqrt{\pi}}\right)^{\nu n} \exp[-\Delta/(\gamma C^2)]$$

$$h_1 \equiv \frac{1}{\gamma\delta C^2} > 0$$

Eq. (III.36) was derived by assuming that the various atomic spacings within a single active site on the catalyst surface differ from the values corresponding to the ideal crystalline state. The probability expressed by Eq. (III.30) was accepted as valid for these differences, since this equation defines the most probable, statistical deviations (from the atomic distances in the ideal crystalline state in our case). It may occasionally happen that this non-ordered state formed prior to the development of the crystalline phase persists, but it is also conceivable that the process leading to the ordered state begins at the temperature at which the catalyst is prepared. The rate of this process, that is, the decrease in $N_{E^{\ddagger}}$ may be expressed as follows:

$$-\frac{dN_{E^{\ddagger}}}{dt} = k_s N_{E^{\ddagger}} \nu\eta \tag{III.37}$$

where

$N_{E^{\ddagger}}$ is the number of formations characterized by activation energy $E^{\ddagger}$ on unit surface area of the catalyst at time t;

t is the time of surface arrangement; and

k_s is the rate constant of the process of surface arrangement.

By integrating Eq. (III.37) between the limits $t = 0$, the instant when the active ensembles are formed, and t_f, when the atoms are already fixed as a result of the freezing of the surface transformation (the catalyst has cooled), and by substitution of Eq. (III.34) we obtain:

$$N_{E^\ddagger,t_f} = N_{E^\ddagger,0}\, f_1 \exp(-k_s t_f \Delta) \exp[(h_1 + k_s t_f/\delta)\, E^\ddagger] \qquad \text{(III.38)}$$

$N_{E^\ddagger,0}$ being given by Eq. (III.36) which is valid in this case for $t = 0$.

There is a formal agreement between Eqs (III.38) and (III.36) when the constants are merged. The duration and rate constant of the conversion now occur in the equation and as surface rearrangement is a process requiring activation, its rate constant is temperature-dependent, so that in Eq.(III.38) the temperature at which the catalyst is prepared is also involved.

The weakest, most disputable point of the simplified model and of its derivation is the expression for the interaction energy in Eq. (III.31).

Schwab and Cremer [32], in contrast, admitted only that the catalyst atoms constituting the active formations of the surface must possess some undefined surplus energy ε compared to the other atoms on the catalyst surface. This may mean either the energy term η and/or $\alpha\lambda$ in Eq. (III.32) if they are multiplied by ν.

We are now in a position to draw conclusions concerning the relative quantity of active formations from simple thermodynamic correlations.

If equilibrium is reached at the temperature θ, the ratio of active formations to the total number of formations is given by the following expression, if $N \gg N_{E^\ddagger}$

$$\frac{N_{E^\ddagger}}{N} = \exp(\Delta S_{E^\ddagger}/R - \varepsilon/R\theta) = f_2 \exp(-\varepsilon/R\theta) \qquad \text{(III.39)}$$

where $N_{E^\ddagger}$ is the number of active formations;

N is the number of all formations;

ε is the surplus energy of active formations;

$\Delta S_{E^\ddagger}$ is the entropy change accompanying the development of active formations;

θ is the equilibrium temperature, which can be the same as the temperature at which the catalyst is prepared; and

$f_2 = \exp(\Delta S^\ddagger/R)$ is a constant, characteristic of the active formation under consideration.

From this point the derivation is the same as before. It follows from arguments similar to those leading to Eq. (III.32) that

$$\varepsilon = (E_{AC\text{surface}} + E_{BC\text{surface}} + \ldots)_{\text{act}} - $$
$$- (E_{AC\text{surface}} + E_{BC\text{surface}} + \ldots)_{\text{average}} \qquad \text{(III.40)}$$

It was seen in Section 2.2 that in catalyzed reactions an energy

$$E_{AB,C} + E_{CD,C} + \ldots = E \qquad \text{(III.41)}$$

is needed for the cleavage of the bonds, while in homogeneous reactions the energy requirement is:

$$E_{AB}+E_{CD}+\ldots=E_{\text{hom}} \qquad \text{(III.42)}$$

Thus, in accordance with Eq. (III.18) the following equation is valid for the formations under investigation:

$$E_{\text{hom}}-E=(E_{AC\text{surface}}+E_{BC\text{surface}}+\ldots)_{\text{act}} \qquad \text{(III.43)}$$

From Eq. (III.22):

$$E^{\ddagger}=\delta E \qquad \text{(III.44)}$$

and in the same way:

$$E^{\ddagger}_{\text{hom}}=\delta_{\text{hom}}E_{\text{hom}} \qquad \text{(III.45)}$$

ε from Eq. (III.40) is substituted into Eq. (III.39), and the adsorption energy on the site under investigation is obtained from Eq. (III.43). Using Eqs (III.44) and (III.45), Eq. (III.39) can finally be written in the following form:

$$\frac{N_{E^{\ddagger}}}{N}=f_2\exp[(E_{AC\text{surface}}+E_{BC\text{surface}}+\ldots)/R\theta\,\cdot$$

$$\cdot\exp[-E^{\ddagger}_{\text{hom}}/(\delta_{\text{hom}}R\theta)]\exp[E^{\ddagger}/(\delta R\theta)] \qquad \text{(III.46)}$$

Rearrangement of Eq. (III.46) and reduction of the constants leads to

$$N_{E^{\ddagger}}=N f_3\exp[E^{\ddagger}/(\delta R\theta)] \qquad \text{(III.47)}$$

It is clear from Eqs (III.36), (III.38) and (III.47) that, despite the various models used, a formally identical relationship is obtained:

$$N_{E^{\ddagger}}=A\exp(hE^{\ddagger}) \qquad \text{(III.48)}$$

A and h are constants which, depending upon the initial conditions and the method of derivation, represent ensembles of different constants. There is, as yet, no basis on which these constants can be generally and thoroughly analyzed, since it is not clear which concept is valid for a given case.

It emerges directly from Eq. (III.48) that the number of active sites of the catalyst which may be involved in a given chemical conversion is the greater, the higher the energy available for activation.

Derivation of Eq. (III.48) leads to:

$$\mathrm{d}N_{E^{\ddagger}}=Ah\exp(hE^{\ddagger})\,\mathrm{d}E^{\ddagger} \qquad \text{(III.49)}$$

which defines the change in the number of active formations, $\mathrm{d}N_{E^{\ddagger}}$, when the activation energy changes by $\mathrm{d}E^{\ddagger}$. If the number of active formations characterized by the activation energy $E^{\ddagger}$ changes by $\mathrm{d}N_{E^{\ddagger}}$, the rate constant of the reaction involving activation energy $E^{\ddagger}$ changes by $\mathrm{d}k$. Since the rate constant referred to unit surface area of the catalyst is proportional to the number of those active formations on which the process under investigation takes place, that is

$$k=BN_{E^{\ddagger}}\exp(-E^{\ddagger}/RT)$$

by substituting $N_{E^\ddagger}$ from Eq. (III.48) we may write for the differential change of k:

$$\mathrm{d}k = B\exp(-E^\ddagger/RT)\,\mathrm{d}N_{E^\ddagger} = ABh\exp(-\alpha E^\ddagger)\,\mathrm{d}E^\ddagger \qquad \text{(III.50)}$$

where $\alpha \equiv 1/RT - h$.

The value of $E^\ddagger$ may vary between two extreme limits:

$E^\ddagger_{\text{min}}$ is the lowest activation energy with which the process takes place at the most active sites of the surface:

$E^\ddagger_{\text{max}}$ is the highest activation energy at which the homogeneous conversion does not yet overlap the catalytic reaction; this condition depends on the ratio of the pre-exponential factors of the rate constants of the two types of conversions (catalytic and homogeneous), and also on the ratio of the catalyst to the free space, and on the difference between $E^\ddagger_{\text{hom}}$ and $E^\ddagger_{\text{max}}$.

Thus, the rate constant of the conversion on the entire heterogeneous catalyst surface is:

$$k = \int_{E^\ddagger_{\text{min}}}^{E^\ddagger_{\text{max}}} ABh\exp(-\alpha E^\ddagger)\,\mathrm{d}E^\ddagger = \frac{ABh}{\alpha}[\exp(-\alpha E^\ddagger_{\text{min}}) - \exp(-\alpha E^\ddagger_{\text{max}})] \qquad \text{(III.51)}$$

Since $E^\ddagger_{\text{min}} < E^\ddagger_{\text{max}}$, it follows that, depending on the sign of α, Eq. (III.51) may be written in the simpler forms:

$$k = \frac{ABh}{\alpha}\exp(-\alpha E^\ddagger_{\text{min}}), \qquad \alpha > 0 \qquad \text{(III.52)}$$

$$k = -\frac{ABh}{\alpha}\exp(-\alpha E_{\text{max}}), \qquad \alpha < 0 \qquad \text{(III.53)}$$

However, in the usual manner under no conditions is $E^\ddagger_{\text{min}}$ or $E^\ddagger_{\text{max}}$ determined from the temperature-dependence of the experimentally measurable rate constants, k, but only some average value $E^\ddagger_{\text{av}}$. Conversion proceeds on the entire surface and only on a certain fraction of this surface does the activation energy reach the values of $E^\ddagger_{\text{min}}$ or $E^\ddagger_{\text{max}}$, while on the rest of the surface the values of the activation energy will be between these two extremes.

Hence, the experimentally measurable average activation energy of the reaction on the entire heterogeneous catalyst surface is obtained by weighting the various activation energy values characteristic of the sites of varying activities in relation to the rate $\mathrm{d}j$ of the reaction at the given sites [33]:

$$E^\ddagger_{\text{av}} = \frac{\int_{E^\ddagger_{\text{min}}}^{E^\ddagger_{\text{max}}} E^\ddagger\,\mathrm{d}j}{\int_{E^\ddagger_{\text{min}}}^{E^\ddagger_{\text{max}}} \mathrm{d}j} = \frac{\int_{E^\ddagger_{\text{min}}}^{E^\ddagger_{\text{max}}} E^\ddagger\,\mathrm{d}k}{\int_{E^\ddagger_{\text{min}}}^{E^\ddagger_{\text{max}}} \mathrm{d}k} \qquad \text{(III.54)}$$

since in the generalized form the rate equation of the reaction is

$$j = kf(a_i)$$

where a_i ($i = 1, 2, \ldots$) is the activity of component i of the reacting system.

By substituting Eq. (III.50) into Eq. (III.54) we obtain

$$E^{\ddagger}_{av} = \frac{\int_{E^{\ddagger}_{min}}^{E^{\ddagger}_{max}} E^{\ddagger} ABh \exp(-\alpha E^{\ddagger}) \, dE^{\ddagger}}{\int_{E^{\ddagger}_{min}}^{E^{\ddagger}_{max}} ABh \exp(-\alpha E^{\ddagger}) \, dE^{\ddagger}} =$$

$$= \frac{\exp(-\alpha E^{\ddagger}_{min}) (E^{\ddagger}_{min} + 1/\alpha) - \exp(-\alpha E^{\ddagger}_{max}) (E^{\ddagger}_{max} + 1/\alpha)}{\exp(-\alpha E^{\ddagger}_{min}) - \exp(-\alpha E^{\ddagger}_{max})} \quad \text{(III.55)}$$

It follows from Eq. (III.55) that if

$$\alpha > 0, \; E^{\ddagger}_{av} = E^{\ddagger}_{min} + 1/\alpha \quad \text{(III.56)}$$

$$\text{and if } \alpha < 0, \; E^{\ddagger}_{av} = E^{\ddagger}_{max} + 1/\alpha \quad \text{(III.57)}$$

that is, as expected, at positive values of α an activation energy somewhat higher than $E^{\ddagger}_{min}$ is measured, while at negative values of α the activation energy is somewhat lower than $E^{\ddagger}_{max}$.

Substituting $E^{\ddagger}_{min}$ and $E^{\ddagger}_{max}$ from Eqs (III.56) and (III.57), into Eqs (III.52) and (III.53), we obtain:

$$k = \frac{ABhe}{|\alpha|} \exp(-\alpha E^{\ddagger}_{av}) \quad \text{(III.58)}$$

In the denominator of the coefficient the temperature-dependence of α can be ignored, and thus:

$$k = C \exp(-\alpha E^{\ddagger}_{av}) \quad \text{(III.59)}$$

$$k = C \exp(hE^{\ddagger}_{av}) \exp(-E^{\ddagger}_{av}/RT) \quad \text{(III.60)}$$

$$k_0 = C \exp(hE^{\ddagger}_{av}) \quad \text{(III.61)}$$

That is, because of the increase of the first factor, k_0, the rate of the catalytic reaction decreases only moderately with increasing activation energy, and, for the same reason, the catalytic activity increases only moderately with decreasing activation energy. The fact that, in this sense, the activation energy has an opposite effect on k_0 and on the exponential factor, that is, on the catalytic activity and the value of k, is usually called the *compensation rule*. An alternative expression, the "theta rule", is quite widely used; this follows from the substitution of h by $1/\delta R\theta$ from Eqs (III.47) and (III.48).

Eq. (III.60) was derived from Eq. (III.48) by assuming that $h \neq 0$, that is, the number of active formations on unit surface area of the catalyst is not independent of the activation energy, or, in other words, that the surface is heterogeneous. Hence, the fundamental condition for the experimental proof of Eq. (III.60) is the heterogeneity of the catalyst surface; in addition, h should be constant, while the value of $E^{\ddagger}_{av}$, the experimentally measurable activation energy, changes. In these cases, therefore, $E^{\ddagger}_{min}$ and $E^{\ddagger}_{max}$ have to change in accordance with Eqs (III.56) and (III.57) for $E^{\ddagger}_{av}$.

Changes in the activation energy of heterogeneous catalytic reactions may be studied with the help of the equations derived earlier. It follows from Eqs (III.34) and (III.33) that

$$E^{\ddagger} = \delta[E_{AB} + \ldots - (E_{AC\mathrm{cryst}} + E_{BC\mathrm{cryst}} + \ldots + \nu\alpha\lambda + \nu\eta)] \qquad \text{(III.62)}$$

and from Eqs (III.40), (III.42), (III.43) and (III.44) that

$$E^{\ddagger} = \delta[E_{AB} + \ldots - (E_{AC\mathrm{surface,av}} + E_{BC\mathrm{surface,av}} + \ldots + \varepsilon)] \qquad \text{(III.63)}$$

Hence, if the energies of the cleaving bonds, the strengths of the chemisorption bonds, and/or the energy levels of the active formations, $\nu\eta$ and ε change in the course of some reaction, $E^{\ddagger}$ must also change.

Since $E^{\ddagger}_{min}$ and $E^{\ddagger}_{max}$ determine the experimentally observed value of $E^{\ddagger}_{av}$ at constant h, $E^{\ddagger}_{av}$ depends unequivocally upon the changes of the maximum and minimum energies ($\nu\eta$ and ε) of the active formations, provided the nature of the substrate, the type of bond rearrangement (type of reaction) and the chemical composition of the catalyst remain unchanged, i.e. provided E_{AB}, E_{CD}, ... $E_{AC\mathrm{surface}}$, $E_{BC\mathrm{surface}}$ ... remain unchanged. This is the case when a given reaction of the same substrate is studied on catalysts of the same composition, which differ from each other only in their method of preparation. Thus, in the dehydrogenation of iso-propanol, for instance, it was observed that on zinc oxide catalysts prepared by different methods Eq. (III.61) is valid for the correlation between the activation energy and the pre-exponential factor, k_0 (see Fig. III.17 [34]). From this figure $h = 9.35 \cdot 10^{-4}$ cal^{-1}, and hence α is negative; consequently in accordance with Eq. (III.57) the experimentally determined activation energy values are always lower than the $E^{\ddagger}_{max}$ values, e.g. by 4 kcal at 350 °C.

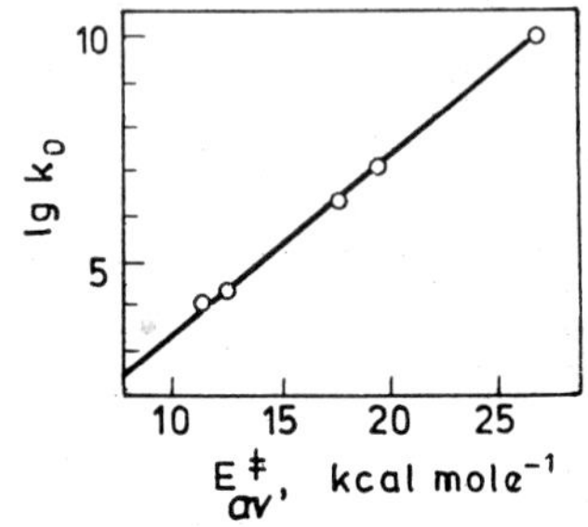

Fig. III.17. Correlation between average activation energy and the logarithm of the pre-exponential factor in the dehydrogenation of *i*-propanol on zinc oxide catalyst prepared by different methods

It follows from Eq. (III.63) that $E^{\ddagger}$ may also change when the geometry of the bond rearrangement changes in a given reaction (see Section 1.1). In this case, however, the structures of the active formations must also differ, so that it seems highly improbable that h will remain the same, or that the energy distributions of active formations of different structures will be the same, which, on the basis of Eq. (III.60), is necessary for the experimental confirmation of the compensation rule.

The dehydrogenation of cyclohexene was investigated on two types of platinum catalysts [35], either pure metal or platinum applied in low concentrations on an inert support. In the case of pure metal h was smaller than $1/RT$ ($\alpha > 0$), while for supported catalysts h was greater than $1/RT$ ($\alpha < 0$). In the derivation of Eqs (III.56) and (III.57) it has already been seen that in the first case $E^{\ddagger}_{av}$ is somewhat higher than $E^{\ddagger}_{min}$, and in the second case somewhat lower than $E^{\ddagger}_{max}$. In fact, the experimentally determined $E^{\ddagger}_{av}$ values are considerably lower for the coherent metal than for supported platinum. From this observation the conclusion was drawn [35] that on pure platinum the reaction takes place on the well-developed octahedral faces, according to the sextet scheme requiring lower activation energy (see Section 1.1), while on the supported platinum particles no octahedral face can develop and dehydrogenation follows the quadruplet scheme, which explains the higher activation energy. For different active formations h is, in fact, not the same.

Eq. (III.63) also means that $E^{\ddagger}$ may change when reactions of the same type take place with different substrates on a given catalyst of heterogeneous surface. In accordance with Eq. (III.61), in the dehydrogenation of various aliphatic amines there is a linear correlation between $\lg k_0$ and $E^{\ddagger}$ (Fig. III.18) [36]. It is surprising, however, that the lines have the same slopes; h is the same for nickel and palladium.

Another phenomenon is presumably involved here. It is known from reaction kinetics that there is a factor $\exp(\Delta S^{\ddagger}/R)$ in the expression of k_0 ($\Delta S^{\ddagger}$ is the entropy change accompanying the formation of the active surface complex). It has generally been proved that in the course of a transformation (e.g. evaporation of the compounds of a homologous series, in reactions of the same type) the entropy difference changes proportionally to the energy difference. If this relationship is extended to $\Delta S^{\ddagger}$ and $E^{\ddagger}$, in catalytic reactions, a result formally agreeing with Eq. (III.61) is obtained: k_0 increases when the activation energy increases. Since this reverse change can be traced back to the change of the substrate, the equation

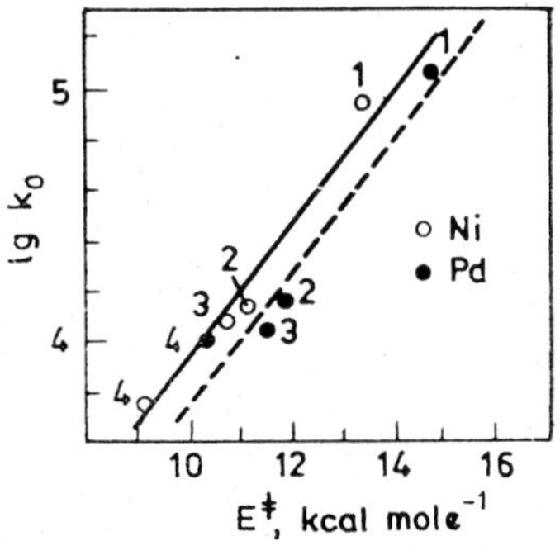

Fig. III.18. Correlation between activation energy and the pre-exponential factor in the dehydrogenation ofd ifferent secondary amines on nickel and palladium catalysts [36]

$$\lg k_0 = h'E^{\ddagger} + B'$$

(B' is a constant for a given catalyst) describes a straight line whose slope h' is the same, even when another catalyst is chosen, providing the substrates are the same [cf. the parallelism of experimentally determined straight lines for the palladium and nickel catalyst (see in Fig. III.18)].

It will be mentioned only for the sake of completeness that a phenomenon in agreement with the compensation rule is also observed when a tunnel effect appears during a chemical conversion [37].

According to the theory of absolute reaction rates the expression [1]

$$k_r = \varkappa \frac{kT}{h} \frac{\Omega^{\ddagger}}{\Omega_A \Omega_B} \exp(-E^{\ddagger}/RT) \tag{III.64}$$

(where k is the Boltzmann constant;
T is the absolute temperature;
h is Planck's constant; and
Ω_A, Ω_B ... $\Omega^{\ddagger}$ are partition functions of the reacting partners and of the active ensemble)

is obtained for the rate constant in which the transmission coefficient is

$$\varkappa = \exp\left\{-\left(\frac{8\pi r}{3h}\right)[2m(E_b - E_p)]^{1/2}\right\} \tag{III.65}$$

where E_b is the height of the energy barrier;
E_p is the surplus energy of the particle (molecular fragment, atom or electron) undergoing rearrangement;
m is the mass of the particle involved; and
r is the width of the barrier.

It follows from Eq. (III.64) that the greater is E_p ($E_p \sim E^{\ddagger}$), the greater is k_0 (since $k_0 \sim \varkappa$), that is, the compensation rule is asserted.

On the other hand, it can be seen from the expression for $\varkappa$ that a significant tunnel effect (high value of $\varkappa$ when the difference between E_b and E_p is great) can appear only on the transition of small particles such as electrons; for example, this phenomenon may occur when charge transfer takes place between ions [37].

As well as the interpretation of the compensation effect, another highly important conclusion can be drawn from Eq. (III.60):

Reactions on catalysts with heterogeneous surfaces can be characterized by a single activation energy just as if the reaction were taking place on a homogeneous catalyst surface. In this way there is a possibility of utilizing and experimentally confirming all the considerations put forward with respect to bond energies in Section 2.2, since catalysts in practice must always be regarded as having heterogeneous surfaces.

Thus, among other things, it is possible, as before, to build on the final results which led, for example, to the relationship illustrated in Fig. III.16, that is, in these cases, too, the activation energy is just as much an unequivocal function of the adsorption energy, as in the case of catalysts having homogeneous surfaces.

SECTION 2.4. INTRODUCTION TO THE ELECTRONIC THEORY OF CATALYSIS

So far nothing has been said about the nature and the deeper physical meaning of the bond energies. The phenomena have essentially been interpreted by phenomenological arguments, and, for this reason, the directions and sizes of bonds formed during the catalytic conversions have been considered separately.

Interaction between the catalyst and the substrate means in fact an overlapping of the electron shells. The theory which interprets the mechanism of catalyst effects on this basis is the electronic theory of catalysis.

Overlapping of the electron shells results in electron migration and can be characterized by the overlapping of the eigenfunctions Ψ, known from quantum mechanics. In other words, a change in the distribution of electron concentrations has to be taken into account, this resulting in the different energy level of the reacting system and different strengths of the rearranged bonds. On the other hand, a suitable approach is necessary for electron transitions to take place, and this can be achieved with appropriate geometric packing. This shows that the ideas described earlier remain valid, and, in this way, gain a definite physical meaning.

Prior to a description of the phenomena, the electron configurations of the catalyst and substrate first have to be known (the electron configurations of solids are discussed in Part I, while bond theory is known from physical chemistry) for the determination of the electron state which develops as a result of interactions. Conclusions on this electron state can also be drawn from ESR and NMR studies (for more details see Chapter 7 of Part VI). At present, we are still far from being able to solve any arbitrarily chosen problem in this way, that is, to forecast a catalytic effect reliably. Among other things, (see p. 337), the exact electron states of the catalyst surface are not known; nor is it known which model will be appropriate, and when the model should be correct, which conversion step (i.e. which electron transition) has primarily to be considered nor, in fact, whethor the calculations performed for the idealized model are permitted, etc. However, despite all these difficulties a correct interpretation of the phenomena has been given in a number of cases, particularly for semiconductor catalysts, which have then been made part of a fairly uniform, concise picture.

Of course, the problem has to be approached in a different manner for metals, semiconductors and insulators.

In the discussion of metals it is necessary to start from a system consisting of a great number of mobile electrons, while in the case of semiconductors only a few mobile electrons or defect electrons (holes) are initially present, and in insulators only localized electrons and defect electrons occur. After interaction with the substrate the bond electrons of all three types of catalysts become localized, that is the corresponding atoms of the substrate are situated on certain lattice points of the catalyst (as if "valences" started from the latter).

CHAPTER 3. METAL CATALYSTS

SECTION 3.1. ADSORPTION DURING CATALYSIS

Before the electronic theory is applied to the metal catalysts it is necessary to deal briefly with adsorption from the aspect with which it primarily occurs in catalytic conversions.

In this respect three idealized cases can be distinguished [38]:

(1) The adsorbate becomes positively charged by donating one electron to the electron system of the metal (cationic adsorption);

(2) the adsorbate takes up one electron from the electron system of the metal and thereby becomes negatively charged (anionic adsorption);

(3) a covalent bond is formed between the metal and the adsorbate.

Let us now examine these cases more closely and illustrate, with some examples taken from practice, what can be expected from these arguments.

PARAGRAPH 3.1.1. CATIONIC ADSORPTION

The formation of positive substrate ions in the surface phase can be divided into the following steps:

$A \rightarrow A^+ + e$ ionization energy $= E_I$

$e + {}^0C \rightarrow C^-$ the electron uptake energy of the catalyst $= -e\Phi'$*

where ${}^0K$ is the vacant electron term in the conduction band of the catalyst in the Fermi level;

$A^+ + C^- \rightarrow A^+C^-$ adsorption energy $= E_a$.

Hence, the ratio of positive ions to non-ionized substrate in the surface phase in the case of equilibrium (by slightly modifying Dowden's equation [38]) is:

$$\frac{N_{A^+}}{N_A} = \frac{\Omega_{A^+}}{\Omega_A} \frac{N_{{}^\circ C}}{\Omega_{{}^\circ C}} \exp[-(E_I - e\Phi' + E_a)/kT] \qquad \text{(III.66)}$$

* Φ' is the electron work function at temperature T

$$\Phi' = \Phi + (\varepsilon_0' - \varepsilon')$$

Φ is the electron work function at $T = 0$ K [see Part I, Eq. (I.45b)]

$$\varepsilon_0' - \varepsilon' = \frac{1}{6}\pi^2 k^2 T^2 \left[\frac{\mathrm{d}\ln n(\varepsilon)}{\mathrm{d}\varepsilon}\right]_{\varepsilon=\varepsilon_0'}$$

where N is the number of some particle on unit surface area of the catalyst;
Ω is the partition function; and
$N_{\circ C}$ is the number of empty electron terms in the *Fermi level*. (The energy change when an electron is captured is $-e\Phi'$ and the negative work function $-\Phi'$ is related to the prevailing ε'.)

It can be seen that the formation of positive ions will be the more extensive,
the lower the electron concentration in the conduction band;
the lower the ionization energy of the substrate;
the higher the work function; and
the more positive the gradient of the density of states in the Fermi level [this latter is of subordinate importance, since even at 1000 °C $(\varepsilon' - \varepsilon_0')$ can only be, depending on the value of ε_0', 0.01 -1% of ε_0'].

Let us now consider the practical conclusions which can be drawn from these statements. N_{A+} can be increased, for instance, by adding substances to which the electrons in the conduction band of the metal can be transferred, whereby $N_{\circ C}$ will be greater. The energy aspects of this process are reflected in the exponent in Eq. (III.66) by $(E_I - e\Phi' + E_a)$, the change of which, in the case of cationic adsorption, is equal to ΔQ, the change in the overall heat of adsorption. At low coverages (when interaction between the adsorbed particles can be ignored) the differences in the heats of adsorption for a given substrate are the same as the differences in the $e\Phi'$ values, multiplied by the number of transmitted electrons (while E_I and E_a remain unchanged, of course):

$$\Delta Q = \frac{n}{2} e \, \Delta\Phi \qquad \text{(III.67)}$$

where n is the number of valence electrons in the newly formed bonds. Boudart found good agreement between this equation and the experimental results (Table III.7 [39]). The greater deviation in the case of the O_2—W pair indicates that the initial formula cannot be used, probably because the condition of cationic adsorption is not encountered here.

TABLE III.7

Heats of adsorption determined experimentally and calculated from Eq. (III. 67) [39] (kcal mole^{-1})

	H_2 on W	H_2 on Ni	O_2 on W	N_2 on W	Cs on W
ΔQ (calculated)	12.5	4	24	15.9	5
ΔQ (observed)	13.5	5	30	15	5.4

PARAGRAPH 3.1.2. ANIONIC ADSORPTION

Reasoning similar to that applied to cationic adsorption permits the derivation of the equilibrium concentration of the negative ions. The following expression is obtained:

$$\frac{N_{B^-}}{N_B} = \frac{\Omega_{B^-}}{\Omega_B} \frac{N_{\bullet C}}{\Omega_{\bullet C}} \exp[-(-E_{\mathrm{I}} + e\Phi' + E_{\mathrm{a}})/kT] \tag{III.68}$$

where $N_{\bullet C}$ is the number of occupied electron terms in the Fermi level, and
E_{I} is the ionization energy of B.

The concentration of negative ions increases

when the electron density in the Fermi level is high;
when the substrate has a high ionization energy;
when the work function decreases; and
when the gradient of the density of states in the Fermi level has an increasingly higher negative value (as has been shown before, this is the least important).

The fact that such ions may occur on the surface during reaction is confirmed by the following observation. On the oxidation of metals [40] it was found [41] that:

$$e\Phi - E^{\ddagger} = \text{const}$$

(where $E^{\ddagger}$ is the activation energy), which means that in the rate-controlling step (i.e. the step the activation energy of which can be measured) electrons must leave the surface; this can happen only if oxygen anions are adsorbed on the metal surface. The same has been confirmed not only for pure metals, but also for Ni—Cu alloys where $e\Phi$ could be varied continuously [38, 42].

PARAGRAPH 3.1.3. COVALENT ADSORPTION

Eqs (III.66) and (III.68) are still valid and permit relevant conclusions when the substrate is no longer fully ionized after adsorption. It is sufficient to take this condition into account by the change in E_{a}. All this indicates that the third extreme case, the covalent bond, can be attained continuously.

Before a detailed discussion of this problem, however, a brief outline of Pauling's metal theory (c.f. Section II.8.1) will be given on the basis of which the prevailing conditions can be interpreted far more easily.

According to their behaviour, Pauling distinguishes three types of electrons in metals: conduction electrons in metal orbitals, electrons in the bonding orbitals ensuring cohesion between the metal atoms, and electrons in atomic orbitals, which are the carriers of the magnetic properties (for details see [43]).

With this in mind, the following two types of covalent adsorption can be distinguished according to Dowden [38]:(1) a covalent bond is formed with the electrons of the bonding orbitals; (2) a covalent bond is formed with the electrons of the atomic orbitals.

Due to the nature of their position, the dsp hybrid orbitals forming bonds on the surface of the metal are asymmetric; they may take up electrons from the substrate and after rehybridization a new, common, substrate—metal dsp bonding-electron system is formed. The higher the percentage d-character of the metal, the stronger the bond, as more electrons from energetically deeper bonding orbitals may participate in the bond.

Unfortunately, to the best of our knowledge, there are no data in the literature which unambiguously refer to such a case.

The strongest covalent bond is formed with the participation of atomic d-orbitals. Dowden claims that the favourable conditions for the formation of such bonds, in order of importance, are as follows:

presence of empty atomic orbitals;

high work function;

high positive value of the gradient $[d \ln n(\varepsilon)/d\varepsilon]$.

Before illustrating this case with appropriate examples, let us point out again the gradual transition between ionic and covalent bonds: for instance, the holes in the atomic d-orbitals may simply be considered as empty electron terms, with the sole additional remark that the developing bond occupies a considerably deeper energy level.

Trapnell [44] gives the following sequence for the adsorptivity of the various metals with respect to the adsorption of oxygen, hydrogen, acetylene, ethylene, carbon monoxide and nitrogen:

$$W, Ta, Mo, Ti, Zr, Fe, Ca, Ba > Ni, Pd, Rh, Pt > Cu, Au, Al > K >$$
$$> Zn, Cd, In, Sn, Pb, Ag$$

This exhibits no relationship whatsoever with the work functions or lattice parameters. On the other hand, the fact that the transition metals are the most active seems to indicate the decisive role of the unsaturation of the d-band (which corresponds to the empty atomic d-orbitals in Pauling's theory). Cu, Au and Al, which have saturated d-bands, adsorb only ethylene and carbon monoxide, and this only because a single $d \rightarrow s$ transition is possible, the process requiring so little activation that it can be even initiated by light. The high activities of calcium and barium can be explained by a certain overlapping of the d-band with the s- and p-bands, so that some electrons may reach the second Brillouin zone, leaving behind them empty sites in the d-band.

Oxygen is adsorbed on all metals, indicating that, here, the unsaturated nature of the d-band is not important. In this connection it has already been seen that O_2^- anions are present in the adsorbed phase, while electron accommodation involves the s- and p-band, and not the more deeply-lying d-band.

Some examples of adsorption will now be given which are of great importance from the aspect of catalyst poisoning (see also Part V, Chapter 2).

The poisonous elements of groups V/b and VI/b of the periodic table (e.g. phosphorus, arsenic and sulphur) are effective only when they possess a free electron pair permitting their "dative" bonding to the metal catalyst:

poisonous   non-poisonous

$$\left[\begin{array}{c} O \\ O\!:\!\ddot{S}\!:\!O \\ \ddot{} \end{array} \right]^{2-} \qquad \left[\begin{array}{c} O \\ O\!:\!\ddot{S}\!:\!O \\ \ddot{O} \end{array} \right]^{2-}$$

$$\begin{array}{c} H \\ H\!:\!\ddot{P}\!:\!H \\ \ddot{} \end{array} \qquad \left[\begin{array}{c} O \\ O\!:\!\ddot{P}\!:\!O \\ \ddot{O} \end{array} \right]^{3-}$$

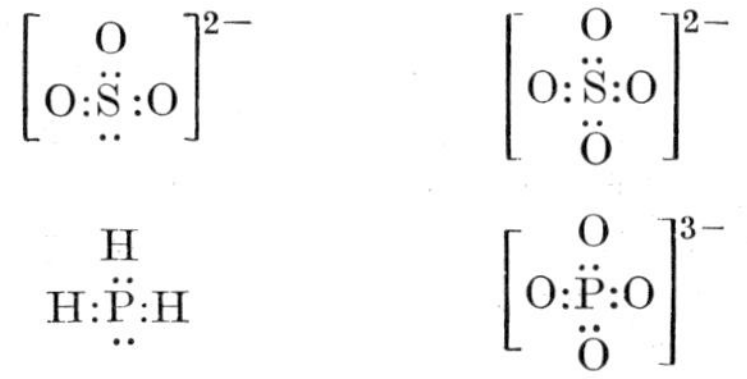

No data are available as to whether polar bonds are formed, or whether a common electron system with the metal develops (that is, overlapping with the bonding-electron orbitals of the metal: first type of covalent bonding; or bonding due to the empty d-band: second type of covalent bonding), or whether we are dealing simply with cationic adsorption. In the adsorption of dimethyl sulphide on palladium it was observed [45] that the magnetic susceptibility of palladium decreased in proportion to the degree of adsorption, indicating that the empty sites of the d-band (in other words, the unsaturated atomic d-orbitals) were gradually filled.

It is not so easy to find a correct interpretation for the poisoning effects of metal ions. Experiments have provided evidence [46] concerning the poisonous nature of ions, all of the d-orbitals of which immediately below the external s- and p-shells contain 1 or 2 electrons, i.e. which have no completely empty d-orbitals. The cause of this phenomenon has not yet been elucidated. Experience has shown that when a bond is formed, the d-orbitals are, in general, hybridized; here too, a special variation of this is probably involved.

SECTION 3.2. INTERPRETATION OF CATALYTIC CONVERSIONS

Because of the very important role played by adsorption in catalysis, it might be expected that catalysis can be studied in accordance with the results of the discussion on adsorption in the preceding parts of this book. Though ideal cases are treated separately according to our classification, they do, however, overlap with each other, and for the sake of an easier survey of the subject the catalytic processes will also be discussed on the basis of a similar classification:

PARAGRAPH 3.2.1. PARTICIPATION OF ADSORBED CATIONS

Some examples of catalytic conversions will first be described in which positive-ion adsorption takes place in the rate-controlling step (the step for which the rate constant and activation energy can be measured).

Schwab [47] investigated the dehydrogenation of formic acid on a series of catalysts. The model reaction was perfectly suitable for the purpose, since it was found to be of zero order; this permitted the calculation of the activation energy directly from the temperature-dependence of the rate constant, since, in such cases, the adsorption coefficient is not included in the rate equations (see Part IV).

The experiments were carried out with various alloys of the metals in group I/b of the periodic system: copper, silver and gold were alloyed with small quantities of the metals in groups II/b—V/b so as not to disturb the α-phase of the initially face-centred cubic lattice structure. In this way the electron concentration of silver could be increased to 1.33 electron per atom by adding elements of period 5/b (Cd, In, Sn and Sb) and to 1.1 electron per atom by alloying with the element in period 6/b (Hg, Tl, Pb, Bi). The activation energy of the decomposition of formic acid was determined for a series of similar alloy combinations, and it was found that

$$E^{\ddagger}_{\text{alloy}} = E^{\ddagger}_{\text{metal}} + Ax(n-1)^2$$

where A is an empirical constant;
x is the atomic fraction of the alloying metal; and
n is the valency of the alloying metal.

In other words, a certain proportionality exists between activation energy and electron concentration. (A formally identical relationship was found between resistance and electron concentrations.) If this empirical relationship is compared to Eq. (III.66), there seems to be a contradiction between the two: electron concentration cannot modify the energy change accompanying the process, since it appears not in the exponent, like the other energy factors, but in the coefficient, and there only to its first power. That this expression is nevertheless correct is probably explained by the change of the work function Φ due to the alloying, experience showing this to be proportional to the square of the electron concentration. The dehydrogenation of formic acid, therefore, proceeds the less readily, the more electrons present in the outer conduction band. This becomes understandable if the quite plausible assumption is made of the splitting-off and adsorption of positive hydrogen ions [47].

The same, in a perhaps even more general form, could be proved for this reaction in experiments with homogeneous and heterogeneous Hume-Rothery alloys [47]. The electron configurations of these alloys can be derived from the geometric structures of their crystal lattices via the relationship between the reciprocal lattice and the Brillouin zone (dealt with in Part I, Section 1.2). Figure III.19a shows the changes in activation

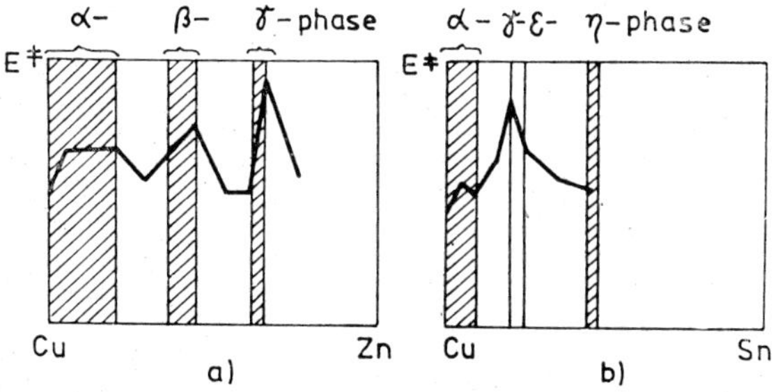

Fig. III.19. Changes in the activation energy of formic acid decomposition on Hume-Rothery alloys based on the experiments of Schwab [47]

energy vs. composition for Cu—Zn alloys. It is seen that the activation energy increases within each homogeneous phase, corresponding to the gradual filling of the available Brillouin zone as the zinc concentration is raised, that is, as the electron concentration in the conduction band increases (see above). When the boundary of a phase is reached, that is, when the available electron terms are quite full, the activation energy has a maximum value. The successive maxima are larger and larger, which appears to indicate simply the decisive role of the electron concentrations pertaining to the different phases (the theoretical value of the electron concentration for complete filling is 1.36 in the α-phase, 1.48 in the β-phase, and 1.54 electron per atom in the γ-phase). When a donor (in the present case, hydrogen) is adsorbed, however, it is not the absolute value of the electron concentration, but the number of empty electron terms which is the essential factor. This latter is the smallest for the γ-phase, while for the further ε- and η-phases it is again larger (in such cases the electrical conductivity undergoes a parallel change), the activation energy again dropping (Fig. III.19b) [47].

Hence, in the heterogeneous regions the activation energy decreases (Fig. III.19a) because the Brillouin zone of the gradually appearing new phase still contains relatively many empty electron terms.

Dowden and Reynolds [48b] determined the catalytic activities of Fe—Ni alloys of various compositions in the hydrogenation of styrene (see curve *A* on the left of Fig. III.20).

Curve *B* shows changes in the constant of the specific heat function measured at low temperature, which is proportional to $g(\varepsilon)$ [48b], the number of electron terms [see Part I, Section 2.1, Eq. (I.50) and the accompanying explanation]. On the addition of iron, *B*, i.e. $g(\varepsilon)$, decreases markedly, and hence the remaining electron terms are increasingly filled with electrons. From the fact that, in this process, *A* also decreases in parallel (at least in the initial stage, on the left of Fig. III.20), it may be concluded that the rate of the reaction depends upon the rate of formation of the chemisorbed positive ions (or radicals) or upon their surface concentration. It should be noted, however, that such a comparison indicates only the trends, for the determination of the values of curve *B* is rather inaccurate, as is the calculation of the function $g(\varepsilon)$ from the latter, while the *A* activity values are not completely reliable either (if for no other reason than the fact that the active surface is not accurately known).

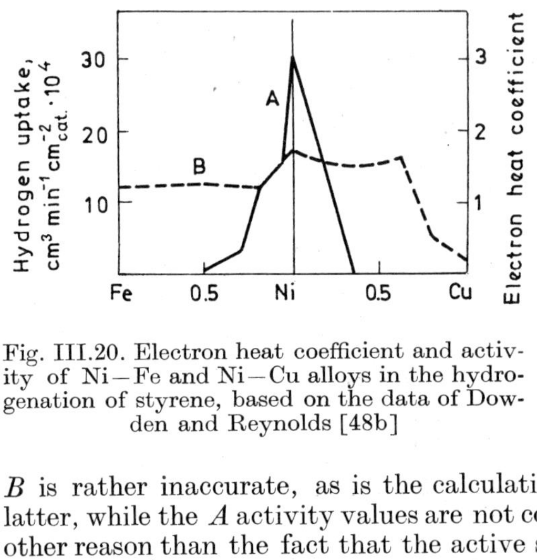

Fig. III.20. Electron heat coefficient and activity of Ni—Fe and Ni—Cu alloys in the hydrogenation of styrene, based on the data of Dowden and Reynolds [48b]

PARAGRAPH 3.2.2. PARTICIPATION OF ADSORBED ANIONS

An example will now be presented for the formation of anions in the course of catalytic conversions.

According to Haber and Weiss [48a], in the first step of homogeneous hydrogen peroxide decomposition catalyzed by Fe^{2+} an electron transfer takes place:

$$H_2O_2 + Fe^{2+} \rightarrow OH^- + Fe^{3+} + OH$$

The same type of electron transition may take place when the peroxide is decomposed on solid catalysts [48b]. This involves the transition of an electron from the metal catalyst to the substrate. It was observed in experiments with various Ni – Cu alloys that the rate of hydrogen peroxide decomposition decreases with increasing quantities of nickel alloyed in the copper [48b]. Curve *B* on the right side of Fig. III.20 shows a marked increase in these cases, indicating a similar rise in the number of electron terms. Thus, when copper is alloyed with nickel an increase in the number of empty electron terms can be expected, which, according to Eq. (III.68), reduces the concentration of negative ions.

PARAGRAPH 3.2.3. PARTICIPATION OF COVALENTLY BOUND SUBSTRATES

In agreement with the two types of "covalent" bonds, we shall treat separately those characteristic examples which indicate either the role of bonding d-orbitals or that of the d-holes.

(1) Role of the percentage d-character.

Films of various metals were prepared by evaporation. It was found that the activation energy of the ammonia-deuterium exchange on the films is the higher, the lower the percentage d-character of the metal (Fig. III.21) [43], presumably meaning that, as expected, the bond with the catalyst is the stronger and, thus, the bonds undergoing rearrangement are the more loosened the more bonding electrons that are available in the d-orbital.

The hydrogenation of ethylene was also studied on metal films [49]. The expected relationship was, in fact, found for various metals: catalytic activity (rate constant) increased as a function of the d-character of the metal (Fig. III.22), but no rule whatsoever could be detected concerning a relationship with any other parameter of the metals. This is reminiscent of the earlier example. (The value for tungsten did not fit, probably because

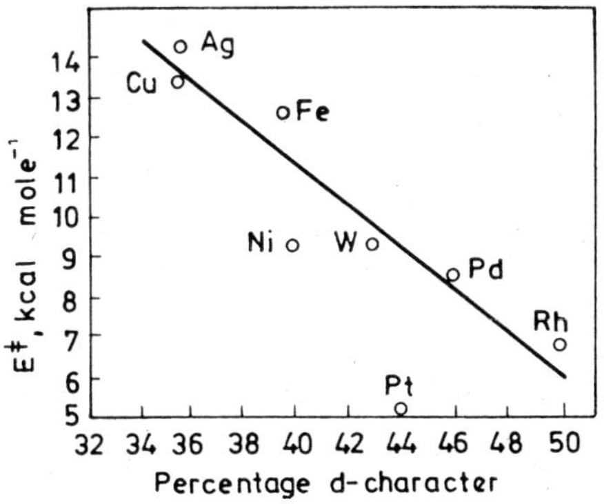

Fig. III.21. Activation energy of ammonia-deuterium exchange vs. the percentage d-character of the catalyst metals [43]

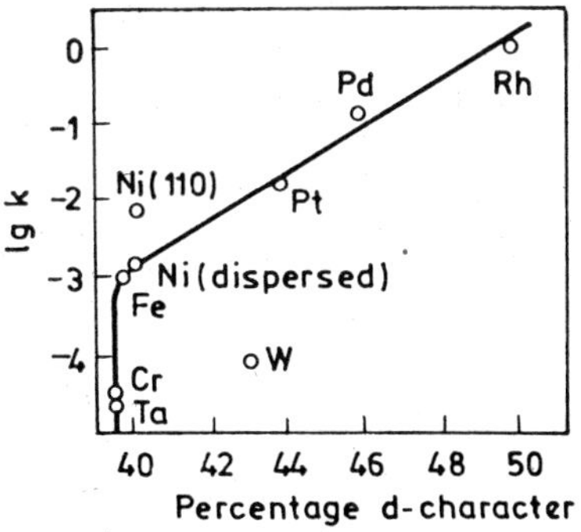

Fig. III.22. Changes in the rate constant of ethylene hydrogenation vs. the percentage d-character of the catalyst [49]

face-centred tungsten was used in the experiments, while the percentage d-character was calculated for the body-centred crystal.) It is, nevertheless, impossible to evaluate the results in the same way, as unfortunately no activation energy values are available for this example.

A more thorough study of the empirical relationships, however, fails to confirm expectation. The heat of adsorption of ethylene decreases as a function of the d-character of the metals (Fig III.23), that is, the bond strength between the metal catalyst and the substrate decreases, which is in contradiction with the conclusion discussed earlier that a higher number of metallic d-orbitals produces stronger bonds.

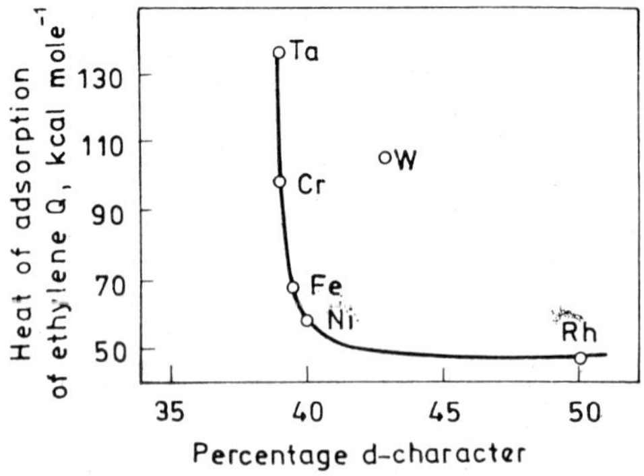

Fig. III.23. Changes in the heat of adsorption of ethylene vs. the percentage d-character of the metals [49]

The result that the stronger the ethylene is bound, the lower is the rate of conversion (cf. Figs III.23 and III.21) is otherwise in agreement with the observation that because of strong chemisorption an excess of ethylene poisons the catalyst during hydrogenation. Beeck [49] claims that on a given metal the activity of the catalyst changes in the opposite direction to the strength of the chemisorption of hydrogen. All this, of course, fits into the general energy pattern of catalysis, as described in Chapter 2, (in Section 2.2 it was shown why strong adsorption, under certain conditions, has an inhibitory effect).

It is an interesting fact that the activity sequence of metals observed in the hydrogenation of ethylene (Fig. III.22) was found inapplicable to the hydrogenation of acetylene; the activity decreased in the following order [50]:

$$Pd > Pt > Ni, Rh > Fe, Cu, Co, Ir > Ru, Os$$

which exhibits no relationship at all with the percentage d-character of these metals. It must be pointed out, however, that such comparative experiments may involve many errors; these will be discussed at the end of this Chapter (in Section 3.3).

It nevertheless seems that for the time being an unequivocal correlation between the percentage d-character and the catalytic activity has not yet been reliably proved, contrary to earlier suggestions.

(2) Role of d-holes.

The next few examples are intended to illustrate the case when catalytic activity is proportional to the unsaturation of the d-band of the metal catalyst.

Couper and Eley [51] studied the o → p hydrogen conversion on Pd—Au alloys. Figure III.24 shows the activation energy and magnetic susceptibility (dashed curve) as functions of the composition. Magnetic susceptibility, of course, is a measure of the unsaturation of the d-band (see Part I,

Section 4.3). The alloy is homogeneous over the entire composition range, the crystal lattice always being face-centred cubic. Over the entire composition range the lattice constant changes by no more than 0.19 Å, this change being linear. Hence, the role of the geometric factors can be ignored.

It was shown in Part I, Section 4.3 that in palladium there are 0.66 unpaired d-electrons (0.66 holes) in the atomic orbitals. These unpaired electrons form pairs with the 1 s electron of gold, whereby the d-band becomes saturated. As long as there are some empty d-sites the activation energy is low. Since this situation practically ceases when the gold content has reached 60 atom%, at this composition the activation energy suddenly rises.

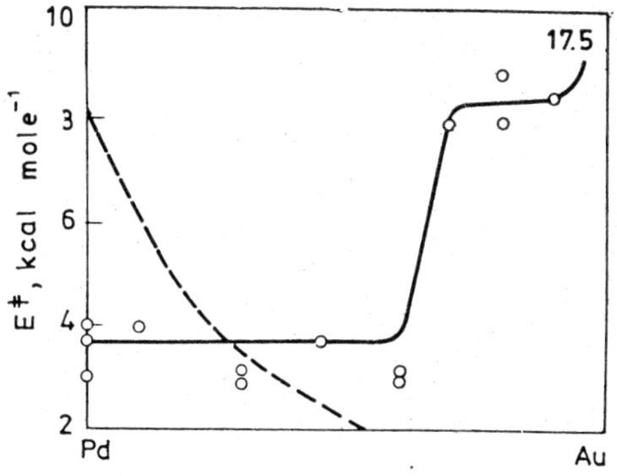

Fig. III.24. Changes in the activation energy of the o → p-hydrogen conversion (———) and the paramagnetic susceptibility (- - - -) vs. the composition of palladium-gold alloys [51]

A sudden change in activation energy in these cases is in full agreement with the kinetic pattern. It is true that the number of active formations, and hence the activity, drops monotonously, but a decisive part of the reaction still takes place on the active formations with unchanged activation energy, and only after these active formations have completely disappeared does the process taking place in some other way with a considerably higher activation energy.

Hydrogen absorbed in palladium has an effect similar to that of gold [51]. The activation energy of the o → p hydrogen conversion rises from 3 to 11 kcal mole^{-1}. It is true that 8 atom% of dissolved hydrogen is sufficient to cause this. If it is borne in mind, however, that this amount of hydrogen penetrates the surface layer only, then with the given specific surface area hydrogen can already be present in a concentration of 50 atom% in the outer 40,000 (100) lattice planes, and thus there is ample to fill the surface d-holes.

All these conclusions and similar relationships have been confirmed by many experiments. Thus, Rienaecker [52] claims that in the o →p hydrogen conversion the activity of Ni—Cu alloys changes suddenly at 70 atom%, that of Pd—Cu alloys at 50 atom%, and that of Pt—Cu alloys at 83 atom% copper, compared to the theoretical 38 atom%. Eley and Luetic [53] observed an increase of the activation energy in the decomposition of formic acid on Pd—Au catalyst at 40 atom% gold, while in the o → p hydrogen conversion such an increase appeared at 60 atom% gold. Dowden and Reynolds [48b] observed an activity change of the Ni-Cu alloys used for the hydrogenation of benzene and styrene at 60 atom% and 30–40 atom% copper, respectively (for the hydrogenation of styrene see curve *A* in the right of Fig. III.20).

One reason for the scattering of the results might be that the alloys were not always completely homogeneous.

"Strong" adsorption may occasionally occur here too, when the great number of d-holes has a marked inhibitory effect. Stowe and Russel hydrogenated carbon monoxide to methane on Fe—Co alloys [54] (the lattice constants and d-characters of the two metals are almost identical). Catalytic activity as a function of composition changed in the opposite direction to that for the magnetic susceptibility, or the unsaturation of the d-band. The number of d-holes reached a maximum at 65 atom% iron, for which the catalytic activity was also the lowest.

PARAGRAPH 3.2.4. ROLE OF DIFFERENTLY ADSORBED SUBSTRATES

For the sake of completeness, mention must be made of complex cases, when, for example, one substrate is adsorbed cationically (as an electron donor) and the other anionically (as an electron acceptor); all this, of course, may occur even within the same molecule.

The results of catalytic investigations of ammonia synthesis, for instance, can be interpreted in this manner [55]. Though it is known that Ni, Pd and Pt are the best hydrogenation catalysts (see Part V, Table V.4), in ammonia synthesis iron was found to be the most suitable. When nitrogen is activated it takes up an electron from the catalyst; in the case of iron, the two free electrons of the first Brillouin zone, are available for this, while the unsaturated d-band is suitable to take up the electron of hydrogen, that is, to activate the latter.

SECTION 3.3. DIFFICULTIES OF THE ELECTRONIC THEORY FOR METALS

By means of the above examples we have tried to demonstrate the applicability of the electronic theory to metal catalysis. Only a qualitative explanation of the phenomenon could be given, as the details of the theory are not yet sufficiently elaborated to permit accurate calculations. Thus, several important physical parameters are unknown, including, for instance, the distribution functions of the electron terms for various alloys containing impurities, the values of the work function, the electron concentrations, etc. Since heterogeneous catalysis is involved, it is highly important that all these parameters should refer to the active surface of the metal catalyst (e.g. to certain crystal elements, or, in the case of heterogeneous alloys, perhaps to the points of contact of the crystallites, etc.). An exact definition of activity and its determination by means of kinetic measurements often cause serious problems (for instance, the surface area to which the conversion values should be referred). The above examples were chosen with the aim of avoiding, as far as possible, the influence of these factors on the conclusions drawn.

To sum up: because our knowledge is incomplete, the electronic theory is only of limited use regarding the interpretation of the phenomena, and in aiding the experimental selection of the suitable catalyst.

CHAPTER 4. SEMICONDUCTOR CATALYSTS

In this Chapter a type of catalyst will be discussed which is highly important from a practical point of view, as it is more often encountered in reality than would be expected. Metal catalysts are usually prepared in a gas atmosphere (e.g. in hydrogen, nitrogen, oxygen, carbon monoxide, etc.), when contamination of the catalyst surface is often unavoidable [56]. This contamination is sometimes only in the form of a monomolecular layer, but it may happen that the surface is converted to a greater depth, so that the metal is, in fact, embedded in a semiconductor coating [57]. In this case the catalytic process takes place on this external coating as the accessible catalyst surface, whereas the metal itself plays hardly any role.

The adsorption phenomena observed on semiconductors and their interpretation (see Part II, Section 8.2) can be used without modification in the study of catalytic conversions, so that their repetition here is superfluous. (The situation differs with respect to metals, since a more detailed analysis of the various types of adsorption is of primary importance in catalysis.) It is necessary only to recall the existence of atoms and molecules bound to the surface by "weak" 1-electron bonds and by "strong" *n*- (acceptor) or *p*- (donor) type bonds. The "strong" bonds can be of a covalent, ionic or transitional nature, depending on the affiliation of the bonding electrons.

SECTION 4.1. MECHANISM OF SURFACE REACTIONS

PARAGRAPH 4.1.1. FUNDAMENTAL PRINCIPLES

It was shown in Part II, Section 8.2 that molecules or atoms adsorbed by "strong bonds" exist for various periods of time in the form of highly active, localized radicals or ion radicals. It seems likely, therefore, that catalytic conversions quite often take place via such radicals. The hydrogenation of ethylene, for instance, can be imagined to proceed in agreement with the scheme of Fig. III.25 as follows:

In the first step, ethylene is adsorbed on a site having an electron defect, thereby forming the radical $^{\bullet}CH_2{-}CH_2{-}$ which is strongly bound to the surface [step (a)]. In the next step, the hydrogen molecule arriving from the gas phase dissociates, one hydrogen atom reacting with the radical $^{\bullet}CH_2{-}$ $-CH_2{-}$, and the other forming a one-electron "weak" bond with the surface [step (b)]. Finally, in step (c) the ethyl radical is desorbed and reacts with the surface hydrogen atom, whereby ethane enters the gas phase and

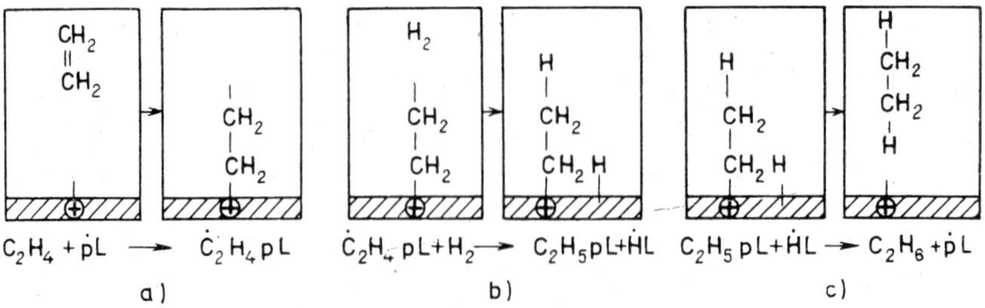

Fig. III.25. Electron scheme of the catalytic hydrogenation of ethylene [58]

the defect remains on the catalyst. (The catalyst may be MoO_3 or $ZnO+$ $+Cr_2O_3$.)

This scheme is only a hypothetical one with which it was intended to illustrate how a process may be divided into elementary steps. The general principle to be borne in mind in such cases is that only a single bond of the molecule can be cleaved at once (see Section 2.1., p. 336) and free radicals, in one form or another, must be present throughout the process. In this respect the catalyst itself can be considered a free radical, since it is an initiator of the reaction in the same way as an introduced free radical, and by its permanent reconversion during the reaction, it plays the same role as the chain carrier in a chain reaction (cf. Introduction of this Part, p. 310). Hence, the greater the number of surface free valences, the higher the rate of the reaction. After "strong" adsorption new free valences may come into being in accordance with the prevailing equilibrium between the surface and the bulk phase, since from the point of view of the surface the bulk of the catalyst is an inexhaustible source of free electrons or defect electrons. All this, however, is limited by the gradually developing high electric charge on the surface (see Part II, Paragraph 8.2.4).

It still remains to be decided how, on the basis of these principles, the expected actual scheme of a reaction can be formulated, that is, how the most probable of all the possible cases can be chosen.

PARAGRAPH 4.1.2. DETERMINATION OF A REACTION SCHEME

A given reaction takes place through molecules with either strong p- or strong n-type bonds (or perhaps through weakly bound molecules). Hence, changes in the surface coverages x^+, x^- and x^0 (see Part II, Paragraph 8.2.4) have unequivocal effects on the reaction rate, from which conclusions can be drawn regarding mechanism being sought.

Let us consider a particular example.

Ethanol may undergo two parallel types of decomposition (e.g. on ZnO samples containing different additives), viz. dehydrogenation and dehydration:

$$C_2H_5OH \begin{cases} \nearrow CH_3CHO + H_2 \\ \searrow C_2H_4 + H_2O \end{cases}$$

Using Wolkenstein's approach the two reaction schemes can be formulated in the following manner [58]:

First of all, we have to start from the fact that in the ethanol molecule the O—H and C—OH bonds are polarized in such a way that the centre of gravity of the electron distribution lies closer to the oxygen in the former, and to the OH group in the latter. Accordingly, after the cleavage of the O—H bond in the field of the catalyst both electrons remain on the oxygen $(C_2H_5O:)^-$, which with these two free electrons, is bound to the surface. (The same formation arises as would arise if one surplus electron, denoted symbolically by *eL*, was present at the lattice point of the catalyst, and one non-bonded electron on the oxygen of the electrically neutral $C_2H_5O\cdot$. In this system of one negative charge the bond is formed by these two free electrons.) This new formation is given the symbol C_2H_5OeL (Part II, Paragraph 8.2.1). The split-off proton forms a bond by using the two electrons of the neighbouring lattice point; in other words, a formation comes into being as if a hydrogen atom were adsorbed on a lattice point of the catalyst which possesses one electron defect, and hence one positive charge. This ensemble possessing one positive charge may be given the symbol *HpL* in the usual manner.

In the case of dehydration, that is, when the C—OH bond is cleaved, the formations C_2H_4pL and $OHeL$ are produced in the manner described above.

Surface conversion (b) now follows this adsorption step (a), and the reaction cycle closes with the desorption of the product, (c). All this has been summarized in an illustrative manner in Fig. III.26 for dehydrogenation, and in Fig. III.27 for dehydration. For the various electron transitions the symbols introduced in the Part on Adsorption have been used.

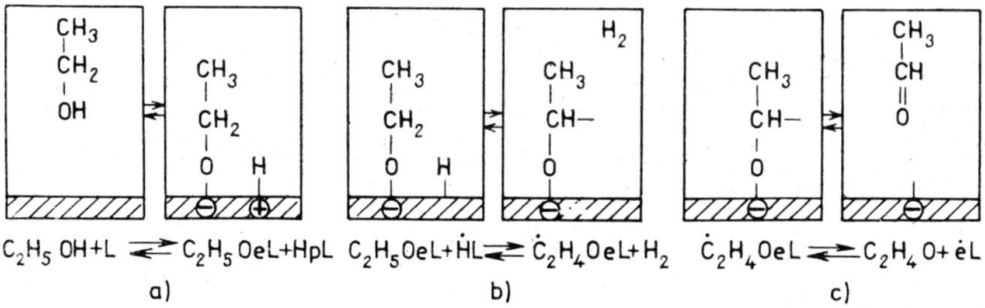

Fig. III.26. Electron scheme of the catalytic dehydrogenation of ethanol [58]

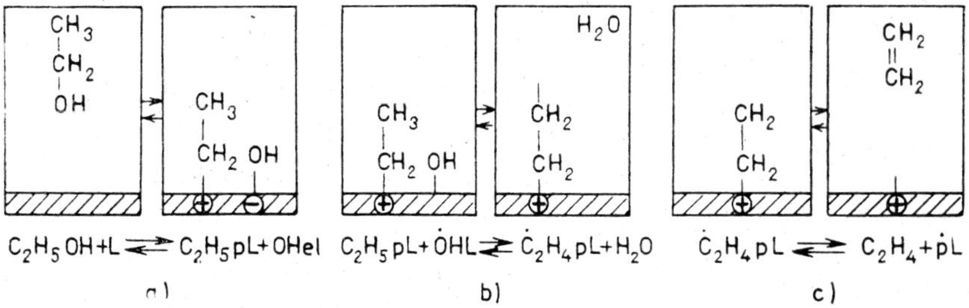

Fig. III.27. Electron scheme of the catalytic dehydration of ethanol [58]

Both transformations have already been mentioned in Section 1.1, during the discussion of the geometric multiplet theory. It was shown that, according to the latter, the processes take place on four catalyst atoms. An analogous conception can also be formulated on the basis of the electron theory of catalysis:

For adsorption (a) two catalyst atoms are needed (one is polarized positively and one negatively). In surface reaction (b) one "weakly" bound H or OH reacts, which may also have been formed by the interaction of eL or pL (both being catalyst lattice points formed in the previous reaction cycle) with $\overset{\text{H}}{\underset{\oplus}{|}}$ or $\overset{\text{OH}}{\underset{\ominus}{|}}$ ($\oplus$ and $\ominus$ each representing a new catalyst atom). Thus, further two catalyst atoms participate in the conversion.

Our current knowledge, however, is still far from sufficient for a detailed comparison of these two concepts (both of which are, to some degree, one-sided) of the mechanism of action of catalysts.

Let us now return to the reaction scheme of Figs III.26 and III.27, and consider in somewhat greater detail what definite conclusions can be drawn from the pertaining kinetic correlations on the basis of the electron theory of semiconductor catalysis.

For dehydrogenation the rates of the three steps of catalytic conversion may be given by the following equations:

Rate of adsorption:

$$J_{(\mathrm{a})} = k_{(\mathrm{a})}p - k_{(-\mathrm{a})}N_{R^-}N_{\mathrm{H}^+} \tag{III.69}$$

Rate of the surface reaction:

$$J_{(\mathrm{b})} = k_{(\mathrm{b})}N_{R^-}N_{\mathrm{H}} - k_{(-\mathrm{b})}N_{\mathrm{A}^-}\,p_{\mathrm{H}_2} \tag{III.70}$$

Rate of desorption:

$$J_{(\mathrm{c})} = k_{(\mathrm{c})}N_{\mathrm{A}^-} - k_{(-\mathrm{c})}N_{\mathrm{e}}p_{\mathrm{A}} \tag{III.71}$$

[where N is the number of some particle on unit surface area of the catalyst;
k is a rate constant;
p is the partial pressure of the alcohol;
p_{H_2} is the partial pressure of hydrogen;
p_{A} is the partial pressure of the aldehyde;
while for the indices,
R^- is the negatively charged $(CH_3CH_2O\!:)^-$ ion;
H^+ is the proton, the ionized hydrogen atom;
H is the hydrogen atom;
A^- is the negatively charged $(CH_3\dot{C}HO\!:)^-$ radical ion;
e is a surplus electron in the conduction band;
(a), (b) and (c) are transitions according to Fig. III.26.]

In step (a) HpL is formed, in step (b) $\dot{\mathrm{H}}$L reacts, and in step (c) $\dot{e}$L is liberated. For this reason, the reaction scheme of Fig. III.26 has to be supplemented by an additional conversion:

$$\text{step (d)}, \; \dot{e}L + \mathrm{H}\,pL \rightleftarrows \dot{\mathrm{H}}\,L$$

to close the cycle of catalytic conversion. The rate of recombination of the strong hydrogen bond is:

$$J_{(d)} = k_{(d)} N_e N_{H^+} - k_{(-d)} N_H \tag{III.72}$$

where N_e is the number of free electrons on unit surface area of the catalyst.

If adsorption is the rate-determining step and equilibrium has been reached in the other three steps, the rate equation of dehydrogenation is

$$J_A = k_{(a)} \left[p - \frac{1}{K_{(a)} K_{(b)} K_{(c)} K_{(d)}} p_A p_{H_2} \right] \tag{III.73}$$

where K is the equilibrium constant of the corresponding step.

If the surface reaction is the rate-determining step the rate equation is

$$J_A = k_{(b)} \frac{N_H}{N_{H^+}} \left[K_{(a)} p - \frac{1}{K_{(b)} K_{(c)} K_{(d)}} p_A p_{H_2} \right] \tag{III.74}$$

and if desorption is rate-determining, then

$$J_A = k_{(c)} \frac{N_H}{N_{H^+}} \left[K_{(a)} K_{(b)} \frac{p}{p_{H_2}} - \frac{1}{K_{(c)} K_{(d)}} p_A \right] \tag{III.75}$$

Step (d) cannot be rate-determining, since if equilibrium has been attained in part-processes (a), (b) and (c) this necessarily involves the equilibrium of step (d), as being a reaction which is not independent of the first three.

In a similar manner, completing the scheme of Fig. III.27 with the reaction

$$\text{(d)} \qquad \dot{p}L + \text{OH}eL \rightleftarrows \dot{\text{O}}\text{H}L$$

the rate equation of ethylene formation by dehydration is obtained:

for adsorptional inhibition the rate equation is

$$J_E = k_{(a)} \left[p - \frac{1}{K_{(a)} K_{(b)} K_{(c)} K_{(d)}} p_E p_{H_2O} \right] \tag{III.76}$$

when the surface reaction is the rate-determining step:

$$J_E = k_{(b)} \frac{N_{OH}}{N_{OH^-}} \left[K_{(a)} p - \frac{1}{K_{(b)} K_{(c)} K_{(d)}} p_E p_{H_2O} \right] \tag{III.77}$$

and, finally, for desorptional inhibition:

$$J_E = k_{(c)} \frac{N_{OH}}{N_{OH^-}} \left[K_{(a)} K_{(b)} \frac{p}{p_{H_2O}} - \frac{1}{K_{(c)} K_{(d)}} p_E \right] \tag{III.78}$$

where p_E is the partial pressure of ethylene.

Since the expressions for conversion rates Eqs (III.73)–(III.78) are known for different cases, the question arises as to how and when the various J values are influenced by the electron state of the catalyst, by which we mean the conditions of electron and defect electron concentrations; in other words, how are the reaction rates influenced when the Fermi level of the catalyst changes?

The equilibrium constants K in the above equations are independent of the Fermi level, and so, naturally, are the partial pressures, but not the fractions N_{H}/N_{H^+} and N_{OH}/N_{OH^-}, which represent the equilibrium ratio of the numbers of weakly bound and strongly bound (ionized) particles. [The steps (d) are always rapid, and thus there will always be an equilibrium between H and H^+, and OH and OH^- on the surface.] The ratio between the ionized and non-ionized donor and acceptor particles can be given with the expressions known from Part I, Chapter 3 and Part II, Section 8.2, i.e.

$$N_{H}/N_{H^+} = \exp[-(E_{H} - E_{F,s})/kT];$$
$$N_{OH}/N_{OH^-} = \exp[(E_{OH} - E_{F,s})/kT] \qquad \text{(III.79)}$$

where E_{H} is the energy level of the donor hydrogen atom in the non-ionized state;
$E_{F,s}$ is the Fermi level on the catalyst surface; and
E_{OH} is the energy level of the acceptor hydroxyl group in the non-ionized state, prior to electron uptake,

that is, N_{H}/N_{H^+} is the greater, the higher the Fermi level, and N_{OH}/N_{OH^-} is the greater, the lower the Fermi level. Hence, the rate of dehydrogenation J_A increases as the Fermi level increases, and the rate of dehydration J_E decreases under the same conditions, provided the adsorption is not the rate-determining step. In this case Eq. (III.73) would be valid for dehydrogenation and Eq. (III.76) for dehydration, and neither of these equations contains a term depending upon E_F. It follows from the reaction schemes in Figs III.26 and III.27 that, in both cases, adsorption takes place on an unperturbed surface, so that the rate of this step is to be independent of the electron concentrations. Experience has shown, however, (see later) that this is not the case, because the rates obtained for the two types of ethanol decomposition depended on the Fermi level in the expected manner (see Fig. III.28), confirming the validity of the reaction schemes in Figs III.26 and III.27.

In agreement with the experimental results, however, the schematic diagram of Fig. III.28 is also obtained when calculating with the adsorption step

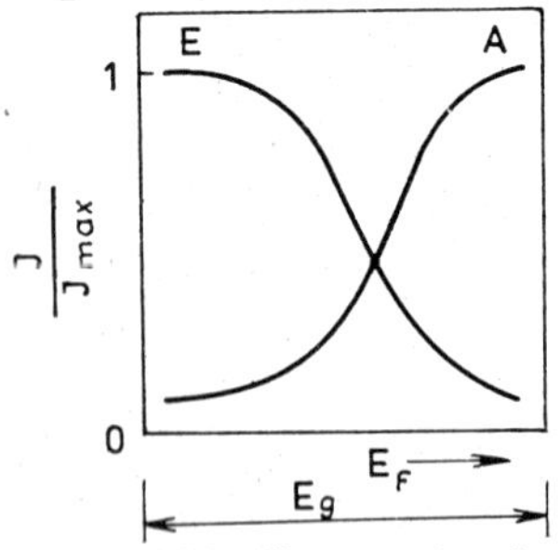

Fig. III.28. Changes in the rates of formation of acetaldehyde (A) and ethylene (E) vs. changes in the Fermi level

$$C_2H_5OH + \dot{e}L = C_2H_5OeL + \dot{H}L$$

in the dehydrogenation of ethanol, and with the adsorption step

$$C_2H_5OH + \dot{p}L = C_2H_5pL + \dot{O}HL$$

in the dehydration of ethanol, followed, according to Figs III.26 and III.27, by steps (b) and (c). In this case, whichever of the three steps is considered the rate-controlling one, the rate of dehydrogenation is

$$J_A \sim N_e \sim \exp[(E_{F,s} - E_g)/kT] \qquad \text{(III.80a)}$$

and the rate of dehydration

$$J_E \sim N_p \sim \exp[-E_{F,s}/kT] \tag{III.80b}$$

The relative change in the rates of the two-directional decomposition of ethanol with the Fermi level can be proved for both the 4-step and the 3-step mechanisms. Thus, only extremely accurate measurements could decide whether Eqs (III.79) or Eqs (III.80) provide a correct description of the conditions.

SECTION 4.2. FACTORS INFLUENCING CATALYTIC ACTIVITY

In the foregoing a relatively simple example was used to indicate the very important role played by the electron concentrations in semiconductor catalysts from the point of view of the electronic theory. We shall now investigate somewhat more thoroughly, with the application of experimental results, how various factors may modify the electron and defect electron concentrations of the catalyst, and the effects which such modifications may have on the catalytic conversion. Based on the results of theoretical and experimental investigations the factors influencing the electron state and thereby the activity of catalysts can be classified into four groups: (1) lattice defects and foreign substances, (2) irradiation, (3) dispersity and (4) electric field.

PARAGRAPH 4.2.1. LATTICE DEFECTS AND FOREIGN SUBSTANCES

Prior to a discussion of these (perhaps the most thoroughly studied) problems of the electronic theory, it seems expedient to briefly summarize certain important results of the physics of semiconductors. In the course of this somewhat sketchy discussion, however, stress will be laid on some factors which are of major importance as regards catalysis.

Let us consider, first of all, a perfect semiconductor crystal in the lattice structure of which there are neither vacancies nor interstitial incorporations nor impurities. As a result of thermal excitation some bonding electrons in the valence energy band will be raised to the conduction band. The resulting electron deficiency is the defect electron. The number of defect electrons, p, is exactly the same as n, the number of electrons in the conduction band. Accordingly, the Fermi level, E_F, will be situated in the middle of the forbidden band: $E_F = 1/2\, E_g$.

If one of the composing elements is lacking at certain points of the crystal lattice, or appears as an excess in the form of an interstitial incorporation in the lattice, it will cause a shift in the initial $n = p$ equilibrium distribution, and thus the Fermi level, expressing this shift, will also change with it. When a negatively charged ion is removed, the negative charge deficiency is relieved by a "captured" electron: this lattice vacancy behaves as an acceptor and reduces the electron concentration in the conduction band. An interstitial lattice component atom, for example, which is transformed into an anion, can also be considered an acceptor, while a deficiency due to neutralization by electron loss corresponds to a donor, increasing the value of n. All other possible cases, such as cation removal, insertion, etc., can be interpreted by a similar approach (see Part I, Section 6.3).

A semiconductor in which the free electrons predominate is an n-type conductor (a negative charge carrier), while one with more defect electrons is a p-type conductor (a positive charge carrier).

Thus, if a semiconductor catalyst, for example, an oxide or sulphide, contains no impurity whatsoever (see below), but nevertheless behaves as an n- or p-type conductor, this is due to the composition deviating somewhat from the stoichiometric one. The definite n-type conducting nature of certain oxides can be attributed to this, e.g.:

$$ZnO, CdO, Al_2O_3, TiO_2, V_2O_5, \text{etc.}$$

while others are p-type conductors, e.g.:

$$CuO_2, NiO, CoO, MnO, Cr_2O_3, \text{etc.}$$

(this has been discussed in greater detail in Part I, Chapter 6).

The nature of individual semiconductors determine whether the metallic or the non-metallic component preponderates during their formation. All this can vary within the limits permitted by homogeneity. In the case of oxide catalysts, for instance, the oxide content of the catalyst can be changed by varying the partial pressure of oxygen during the preparation of the catalyst.

From the aspect of applicability it is essential that the catalyst preserves its initial activity during use, and thus its Fermi level must remain unchanged. It may happen that in the course of the reaction the initially present lattice defects disappear, whereby the Fermi level suffers an undesirable shift. In order to prevent such "burn-out" of the catalyst and to maintain the defect concentration, the substrate has occasionally or steadily to be contaminated. That is why, in hydrocracking on WS_2 or MoS catalysts, a certain definite partial pressure of H_2S has to be maintained (Varga-process).

When the temperature is raised, the lattice atoms may move as a result of thermal excitation and various lattice point vacancy—interstitial incorporation pairs are formed, which tend to change the electron concentration and energy state of the catalyst (see Section 2.3).

Various impurities are usually present in the semiconductor crystal.

There are impurities which dissociate to form ions and are incorporated as such in the crystal lattice, where they replace a cation and an anion. If the ratio of the numbers of cations and anions from the impurity is the same as the numerical ratio of cation/anion in the semiconductor, no change occurs in the electron/defect electron concentration ratio. If the former ratio differs from that of the crystal, however, the n/p ratio is shifted, since a cation or anion deficiency is brought about. As already pointed out, an anion deficiency reduces the number of free electrons, while a cation deficiency causes a drop in the number of defect electrons, both to an extent proportional to their charge number. If Na_2SO_4 is added to CdO, for instance, in agreement with the ideas of our simplified model, 2 Na^+ ions occupy the place of 2 Cd^{2+} ions, while one SO_4^{2-} ion takes the place of a single O^{2-}-ion, the other O^{2-} ion site, fixed by the lattice structure, remaining unoccupied. The lack of 2 negative charges is relieved by the capture of 2 electrons from the conduction band, causing a decrease in the electron concentration of the n-type conductor CdO, that is, a drop of the Fermi level of the latter.

Isopropanol is dehydrogenated and, to a lesser extent, dehydrated on CdO at temperatures between 200 and 310 °C [59]. If the catalyst is prepared with an increasing Na_2SO_4 content, thereby reducing n, the number of electrons in the conduction band, the ratio of the partial pressures of hydrogen and propylene in the end-product decreases (Fig. III.29), that is, the selectivity is improved with respect to dehydration. This is in full agreement with expectation, as it has already been shown that the decrease of the Fermi level (i.e. of n) is accompanied by an increase in the dehydration rate compared to that of dehydrogenation (see the schematic diagram of Fig. III.28).

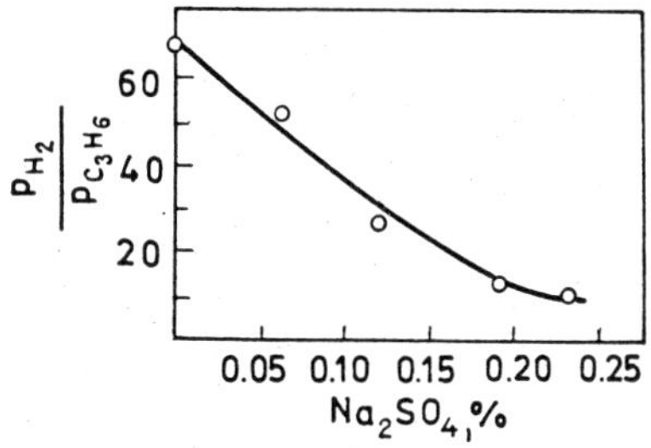

Fig. III.29. Ratio of the dehydrogenation and dehydration of isopropanol on a Na_2SO_4-doped CdO catalyst at 304 °C, based on the data of Krilov, Roginskii and Fokina [59]

The catalysts may contain non-dissociable impurities of course, and, depending upon the conditions of electron affinity, they may accept or donate electrons from or to the electron system of the crystal.

The defect sites and impurities in the catalyst contribute as either donors or acceptors to the overall electron or defect electron concentration and determine the activity conditions. Several factors combine to determine when, and to what extent, this effect of the foreign substance or lattice defect is exerted. A concise summary of this is given by a correlation known from semiconductor physics, the principle of electrical neutrality (according to which the number of positive charges originating from defect electrons and from ionized donor particles must be the same as the number of negative charges originating from electrons and ionized acceptor particles), which may be obtained from the equations derived in Chapter 3 of Part I:

$$2(2\pi m_h^* kT/h^2)^{1/2} \exp(-E_F/kT) + N_{d_1}\{1 + \exp[-(E_{d_1} - E_F)/kT]\}^{-1} + N_{d_2}\{1 + \exp[-(E_{d_2} - E_F)/kT]\}^{-1} + \ldots = 2(2\pi m_e^* kT/h^2)^{1/2} \exp[(E_F - E_g)/kT] + N_{a_1}\{1 + \exp[(E_{a_1} - E_F)/kT]\}^{-1} + N_{a_2}\{1 + \exp[(E_{a_2} - E_F)/kT]\}^{-1} + \ldots \quad \text{(III.81)}$$

where m_h^* is the effective mass of the defect electron;
m_e^* is the effective mass of the electron;
E_F is the Fermi level;
E_g is the width of the forbidden band;
E_d is the energy level of the donor;

E_a is the energy level of the acceptor;
N_d is the total quantity of some donor species; and
N_a is the total quantity of some acceptor species.

The first term on the left hand side in Eq. (III.81) gives the concentration of holes in the valence band, and the first term on the right hand side the electron concentration in the conduction band. The other terms on the left are the positively charged ionized donor concentrations, and on the right hand side the negatively charged ionized acceptor concentrations.

Several conclusions can be drawn drom Eq. (III.81) as to the electronic stucture of the catalyst:

The electron concentration and thus the Fermi level increases with an increase of the number of donor particles N_d, the more so, the higher the ionization energy level of the donor, E_d.

The defect electron concentration increases, and the Fermi level decreases, with the increase of the number of acceptor particles, N_a, the more so, the lower E_a.

In the case of strong n-type conduction the ionization and strong binding of further donor impurities is possible even when the Fermi level is high. The quantitative relationships are greatly influenced by the ratio of E_d and E_F, and if $E_d > E_F$ a significant donor ionization must, by all means, be taken into consideration. The conditions for acceptor particles can obviously be interpreted by similar arguments;

The electron concentration in the conduction band and the concentration of holes in the valence band decrease with the width of the forbidden band, E_g;

As the temperature rises $E_F \rightarrow 1/2\ E_g$, that is, beyond a certain temperature the intrinsic conductivity is so large that additives no longer have any effect.

So far, changes have been considered in the electron—defect electron concentrations in the bulk of the semiconductor crystal. Heterogeneous catalytic processes, on the other hand, always take place on the surface, and the question now arises as to what these conditions on the surface are during catalysis.

The adsorption of the reactant(s) and the desorption of the product(s) are part-processes of the catalytic conversion. However, adsorption occurs not only as an indispensable part of the desired conversion, but often as an undesirable phenomenon, e.g. in a large variety of poisonings and inhibitions.

The adsorbed particle may be considered an impurity incapable of penetrating the inside of the crystal, but remaining on the surface and causing changes in the electron concentration of the surface layer of the crystal. In this way a boundary layer (Randschicht) is formed, in which the conditions are different from those in the bulk of the crystal. As a result merely of the fact that (depending on the crystallographic position of the boundary surface) either a cation surplus or an anion surplus develops in the surface lattice planes of the crystal, the electron concentration on the surface changes in the same way as shown for deviations from the ideal crystal stoichiometry.

For n-type conductors the adsorption of a donor, and, for p-type conductors the adsorption of an acceptor, causes an increase of the concentration of the charge carrier, while a decrease of its concentration is caused in the case of n-type conductors by the adsorption of an acceptor and in the case of p-type conductors by the adsorption of a donor. Schottky suggested the name "Anrecherungsrandschicht" (enrichment boundary layer) for the former, and "Verarmungsrandschicht" (exhaustion boundary layer) for the latter [60]. Hauffe derived exact correlations for the electron – defect electron concentration changes on adsorption, progressing from the surface towards the inside of the semiconductor [61]. The results are illustrated by means of the band model in Fig. III.30. The initial band structure (c) is shifted,

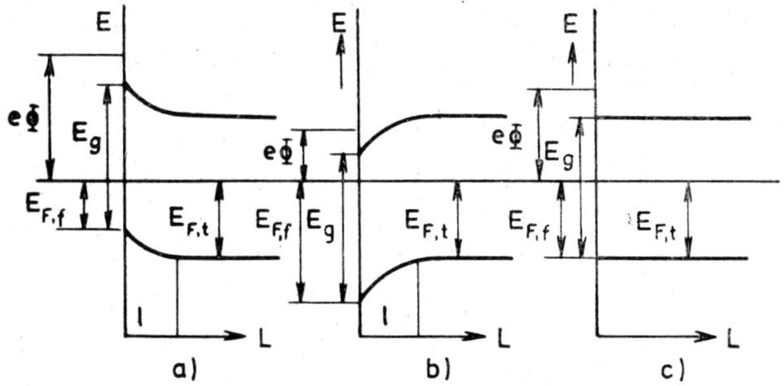

Fig. III.30. Bending of the band structure as a result of adsorption

upwards when an acceptor is adsorbed [case (a)], and downwards when a donor is adsorbed [case (b)], while E_g remains unchanged, so that in (a) the Fermi level approaches the valence band (more correctly, the valence band approaches the Fermi level) on the surface of the crystal, i.e. $E_{F,s} < E_{F,v}$, while in case (b) the situation is reversed. In the first case, the conduction electron concentration decreases, in the second case, increases as the surface is approached. Calculations have shown [61] that, depending on the circumstances, such appreciable shift in concentration may occur down to a depth of 100–1000 Å.

Hence, the surface Fermi level and, with it, the activity of the catalyst can be influenced by means of adsorption too.

Hoover and Rideal investigated the catalytic dehydrogenation and dehydration of ethanol on thorium oxide [62]. The adsorption of water as a donor causes a slowing down of dehydration, while, particularly at low water vapour concentration, the rate of dehydrogenation increases. A rise in the Fermi level must cause an acceleration of dehydrogenation and a deceleration of dehydration (see Fig. III.28). Chloroform has the reverse effect: it greatly accelerates dehydration, while dehydrogenation is almost completely suppressed.

In the case of isopropanol, Wicke [63] found that when the surface of the $ZnO/\gamma Al_2O_3$ catalyst is preliminarily saturated with acetone, which is adsorbed as an electron acceptor (just like the aldehyde in Fig. III.26) dehydrogenation is suppressed as expected and predominantly dehydration takes place.

If only for the sake of completeness, it should be mentioned that a local bending of the band structure can take place not only as a result of adsorption, but also generally in the presence of any inhomogeneously distributed impurity (see Fig. III.31, where f is a position coordinate of arbitrary direction in the bulk crystal or on its surface, while I, II and III are differently contaminated places). As shown earlier in the discussion of Eq. (III.81), when the temperature is raised the effect of impurities is suppressed by the preponderance of intrinsic conductivity, which results, here, in an evening-out of the band structure and homogenization of the catalyst.

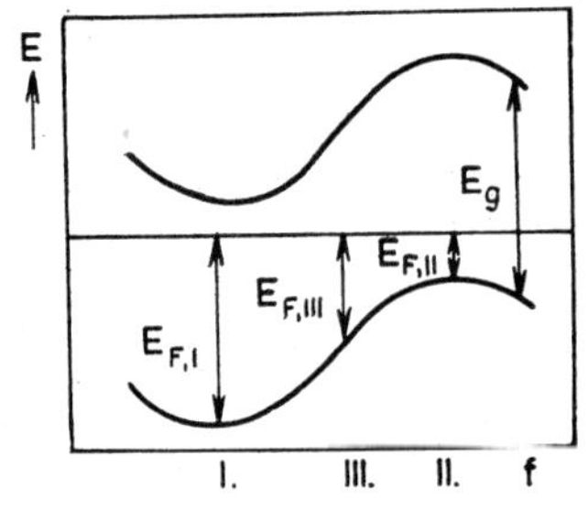

Fig. III.31. Changes in the band structure of semiconductors due to an inhomogeneous concentration distribution of impurities

These examples demonstrate the increasingly higher rate of ethanol dehydrogenation as the number of free electrons in the conduction band increases, for, in the rate-determining step, a low equilibrium concentration of hydrogen adsorbed as a donor, H^+, is advantageous. In dehydration the situation is the reverse: a high number of defect electrons tends to reduce the equilibrium concentration of surface OH^-, thereby accelerating the rate-determining step. Phenomenologically, this appears as if electrons were needed in dehydrogenation, and defect electrons in dehydration. Generally speaking, conversions which are accelerated by increasing electron concentration are of n (negative) type, while those accelerated by the increase of the concentration of defect electrons are p (positive) type reactions.

When the conductivity of the n-type conductor ZnO is reduced by the addition of Li_2O, the rate of the conversion $H_2 + D_2 \rightleftarrows 2\,HD$ decreases, while of the conductivity of the ZnO is increased by the addition of Ga_2O_3 the rate of the same conversion increases [64]. This reaction takes place on the p-type conductor Cr_2O_3 too, the conductivity of which decreases when it is treated with hydrogen and increases on treatment with oxygen. (This is quite general, since in all practically used semiconductor catalysts the donor energy level of hydrogen, $E_{d,H}$, is very high, so that it is bound in great quantities with electron loss. The acceptor level of oxygen, $E_{a,O}$, on the other hand, is very low and almost its entire quantity is always ionized by electron uptake.) Here, the change in conductivity is of opposite direction to the change in the reaction rate [65]. Thus, in both cases, the hydrogen-deuterium exchange can be considered a reaction of n-type.

Methanol decomposition, $CH_3OH \rightarrow CO + 2\,H_2$, is also a reaction of n-type on the n-type conductor ZnO. The addition of excess zinc raises the conductivity, as Zn^{2+} is formed from Zn by the loss of electrons, and simultaneously the rate of methanol decomposition also increases [66].

The dehydrogenation of isopropanol, $i\text{-}C_3H_7OH \rightarrow (CH_3)_2CO + H_2$, on the n-type conductor ZnO is naturally of n-type, because, after treatment with oxygen, both the conductivity and the activity of the catalyst decrease [67].

Hauffe et al. [68] subjected the catalytic decomposition of dinitrogen oxide to thorough study. The rate of the reaction $2\,N_2O \rightarrow 2\,N_2 + O_2$ decreased when In_2O_3 was added to the p-type conductor NiO, and increased considerably when 0.1 mole% Li_2O was added to the same catalyst. Thus, the decomposition is a p-type process, as its rate changes in the same way as the number of defect electrons, which is reduced by the addition of indium oxide or raised by the addition of lithium oxide. On the other hand, when the Li_2O content reaches 0.5 mole% the rate of the reaction begins to drop, and at 3 mole% Li_2O it is considerably lower than the rate on pure NiO. The reaction then appears to be of n-type, because, due to the high defect electron concentration, the rate-determining step lending the p-type character to the conversion has been accelerated to such a degree that another step requiring free electrons has become the rate-determining one [68].

It appears from the above that the same additive may display different effects: it may accelerate one reaction and yet suppress another one (as shown in the relationship between the dehydrogenation and dehydration of alcohols), while in small quantities it may accelerate and in large quantities slow down the same conversion (as demonstrated by the role of lithium oxide in the decomposition of N_2O). It is impossible, therefore, to state, about some foreign substance, simply that it is a poison or a promoter (or accelerator), since this will always depend on the actual conditions (see also Part V).

Let us finally examine somewhat more thoroughly a catalytic conversion in which it is possible to follow the above changes in the properties of a semiconductor catalyst, and also to observe the reaction mechanism and the changes of the latter. This conversion is the catalytic oxidation of CO on semiconductors:

$$CO + 1/2\ O_2 \rightarrow CO_2$$

In accordance with the Langmuir–Hinshelwood mechanism (see Part IV) the overall process can be divided into the following steps: conductivity measurements show that carbon monoxide is adsorbed as a donor [69]*:

$$\text{(a)} \qquad CO \rightleftarrows CO^+ + nL$$

while oxygen is always adsorbed as an acceptor [70] and dissociates:

$$\text{(b)} \qquad 1/2\ O_2 + nL \rightleftarrows O^-$$

The surface reaction takes place between the chemisorbed particles ("ions"):

* The simpler symbols used here can be identified with Wolkenstein's symbols in Section 4.1 as follows:

$$\text{(a)} \qquad CO + L \rightleftarrows COeL = CO^+ + \dot{e}L, \text{ and}$$
$$CO + \dot{p}L \rightleftarrows COpL = CO^+ + L$$

$$\text{(b)} \qquad O + \dot{e}L \rightleftarrows OeL = O^- + L, \text{ and}$$
$$O + L \rightleftarrows OpL = O^- + \dot{p}L$$

$$(c) \quad CO^+ + O^- \rightleftarrows CO_{2\,ads}$$

Finally, the carbon dioxide formed is desorbed:

$$(d) \quad CO_{2\,ads} \rightleftarrows CO_2$$

Depending upon the actual conditions carbon dioxide can be adsorbed both as a donor and as an acceptor [58]; on p-type conductors it loses an electron, while on n-type conductors it takes up electrons. Because of its "amphoteric" nature, nothing definite can be said about this bonding in this brief discussion.

On the basis of the Rideal–Eley mechanism, the following two variations of the reaction scheme are possible:

(1) Conversion begins by the chemisorption of carbon monoxide:

$$(a) \quad CO \rightleftarrows CO^+ + nL$$

Gaseous oxygen reacts with the CO^+ and a defect is formed on the catalyst:

$$(e') \quad CO^+ + 1/2\,O_2 \rightleftarrows CO_{2\,ads} + pL$$

followed by the desorption of CO_2.

(2) The other possibility involves a reaction between chemisorbed oxygen and gaseous carbon monoxide:

$$(b) \quad 1/2\,O_2 + nL \rightleftarrows O^-$$

$$(f) \quad CO + O^- \rightleftarrows CO_{2\,ads} + nL$$

The final step of the reaction cycle [step (d)] is always fast: on semiconductor oxides the adsorption—desorption equilibrium of carbon dioxide is reached instantaneously. For just this reason, in the investigation of the problems of the electron concentration conditions of the catalyst in the mechanism of CO oxidation (d) is simply merged into the preceding step and desorbed CO_2 is considered as the immediate product of the surface reaction.

At 20 °C the adsorption of CO on the p-type Cu_2O proceeds at a considerably higher rate than the adsorption of O_2 [69], the rate of the reaction is dependent on the partial pressure of oxygen, p_{O_2}, but is independent of p_{CO}. Under such conditions the reaction may take place as follows:

either,	(a′)	$CO + pL \rightleftarrows CO^+$	fast	
	(b′)	$1/2\,O_2 \rightleftarrows O^- + pL$	slow, and	(III.82a)
	(c)	$CO^+ + O^- \rightleftarrows CO_2$	fast or slow	
or,	(a′)	$CO + pL \rightleftarrows CO^+$	fast, and	(III.82b)
	(e′)	$CO^+ + 1/2\,O_2 \rightleftarrows CO_2 + pL$	slow	

The second of these possibilities can be expressed with other words that the catalyst surface covered with CO^+ reacts with oxygen and CO_2 is formed.

At temperatures above 250 °C the conditions are quite different. In contact with CO/O_2 mixtures, p-type conductor NiO and CuO catalysts show a slow decrease in conductivity, which is not apparent in the presence of N_2/O_2 mixtures. Hence, rapid oxygen chemisorption is followed by the slow-donor-type adsorption of CO [71–73]. At the same time, the rate of formation of carbon dioxide is independent of p_{O_2} and proportional to p_{CO}. In this case the two variations of the reaction scheme are:

$$
\begin{array}{lll}
(a') & CO + pL \rightleftarrows CO^+ & \text{slow} \\
(b') & 1/2\, O_2 \rightleftarrows O^- + pL & \text{fast, and} \\
(c) & CO^+ + O^- \rightleftarrows CO_2 & \text{slow or fast}
\end{array}
$$

or, (III.83)

$$
\begin{array}{lll}
(b') & 1/2\, O_2 \rightleftarrows O^- + pL & \text{fast, and} \\
(f') & CO + O^- + pL \rightleftarrows CO_2 & \text{slow}
\end{array}
$$

Reversal of the rate ratios of (a′) and (b′), which, in the case of the Rideal–Eley mechanism, means that when the temperature is raised from 20 to 250 °C or above, (b′) will take place instead of (a′), suggests that on p-type conductors (b′), the acceptor-type adsorption of oxygen requires a higher activation energy than (a′), the donor-type bonding of carbon monoxide.

If doping causes a shift of the Fermi level in the bulk of the semiconductor, a symbate displacement will take place on the catalyst surface [74]:

$$dE_{F,s}/dE_{F,v} > 0$$

($E_{F,s}$ and $E_{F,v}$ are the Fermi levels on the surface and in the bulk of the catalyst) and thus the electron concentration conditions also undergo a similar change there.

When the oxide of a metal of higher valency is added to the p-type conductor NiO, the defect electron concentration diminishes. Schwab and Block [73] found that added Cr_2O_3 causes a rise in the activation energy of carbon monoxide oxidation (see Fig. III.32a), as might be expected from the defect electron requirement of the rate-determining steps (a′) and (f′).

For similar reasons, the addition of Li_2O causes a decrease in the activation energy: the slow steps (a′) and (f′) proceed easier. It is interesting, however, that above a certain Li_2O content the activation energy begins to increase and approaches a constant value (Fig. III.32a), while the reaction rate remains proportional to p_{CO}. In his interpretation of the phenomenon, Hauffe [61] started from the inhibition of the formation of chemisorbed O^- when the Li_2O concentration, that is, the number of defect electrons, is raised (ascending stage), until finally, completely displaced from the surface, CO_2 will be formed by an entirely different mechanism: though (a′) remains a slow step, (b′) and (c) are completely suppressed, and gaseous oxygen reacts with chemisorbed CO^+

$$
\begin{array}{lll}
(a') & CO + pL \rightleftarrows CO^+ & \text{slow} \\
(e') & CO^+ + 1/2\, O_2 \rightleftarrows CO_2 + pL & \text{fast}
\end{array}
\qquad \text{(III.84)}
$$

Hence, the reaction rate is proportional to p_{CO} just as before.

The effect of additives on the activation energy was also determined on n-type conductor catalysts at similar temperatures above 250 °C [73]. When Ga_2O_3 is added to the n-type conductor ZnO the activation energy decreases, while the addition of Li_2O to the ZnO causes a rise in the activation energy (Fig. III.32b). The expected reaction scheme is therefore:

$$
\begin{array}{lll}
\text{(a)} & CO \rightleftarrows CO^+ + nL & \text{fast} \\
\text{(b)} & 1/2\, O_2 + nL \rightleftarrows O^- & \text{slow, and} \\
\text{(c)} & CO^+ + O^- \rightleftarrows CO_2 & \text{slow or fast}
\end{array}
$$

or, (III.85)

$$
\begin{array}{lll}
\text{(a)} & CO \rightleftarrows CO^+ + nL & \text{fast} \\
\text{(e)} & CO^+ + 1/2\, O_2 + nL \rightleftarrows CO_2 & \text{slow}
\end{array}
$$

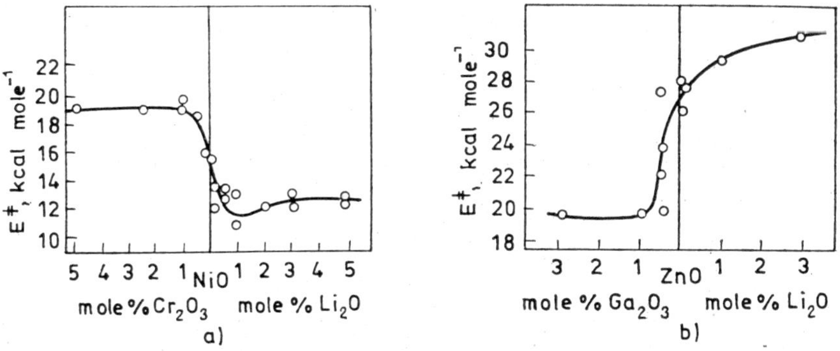

Fig. III.32. Changes in the activation energy of carbon monoxide oxidation on doped (a) NiO, and (b) ZnO catalysts, from the data of Schwab and Block [73]

In the strong n-type conductor state of the catalyst (left side of Figs III.32) the conversion requires the same activation energy, so that the rate pattern of the conversion is presumably the same on two kinds of catalysts.

Comparison of the results obtained on NiO and on ZnO shows the changes in the rate ratios of the part-processes and in the scheme of the conversion as a function of the nature of the conduction of the catalyst. This means not simply the reversal of the situation on changing from a p-type conductor to an n-type conductor, to give the mirror image of the original, but also a modification of the quantitative relations: the "mirror image is distorted". The activation energy on pure NiO different from that on pure ZnO, and after doping with additives the activation energies of the two catalysts change differently, etc. All this is due to the excessively qualitative nature of our conclusions:

(1) Quantitative data, which are difficult to obtain, would be needed, for instance, for the calculation of the concentrations from Eq. (III.81).

(2) The exact correlation between the Fermi level and the activation

energy is not known, though this was required in the earlier arguments.

Essentially, these two problems mean the following.

The catalytic conversion proceeds in several steps. Of these, only one will be a slow, rate-determining step, while the others are so fast that they can be considered to be in equilibrium.

There might be rate-determining steps whose rates are not influenced by the electron concentrations of the catalyst [e.g. in the dehydrogenation and dehydration of alcohols when adsorption is the rate-controlling step and rate equations (III.73) and (III.76) are valid], and there might be others which are greatly modified by changes in the electron concentration. In this latter case two possibilities have to be considered:

(1) The surface concentration of the chemisorbed particles participating in the rate-determining step depends on the Fermi level of the semiconductor catalyst. In principle, these surface concentrations could be determined from Eq. (III.81), provided the Fermi level on the surface and the relevant donor and acceptor levels were known. For want of these, we must be satisfied with the qualitative conclusions obtained from Eq. (III.81) (see the discussion of the examples relating to n- and p-type reactions).

(2) In the rate-determining step, electron transfer takes place between the catalyst and the substrate [see e.g. step (b) in Eq. (III.85) and step (a′) in Eq. (III.83)] when the rate of the involved step, and thus of the entire catalytic conversion, is proportional to the surface electron concentration, c_{eL}, i.e. to the defect electron concentration, c_{pL}, of the catalyst. Since it refers to unit surface area of the catalyst, and not to a single eL or pL, the experimentally determined rate constant is, in fact, a product of k, the true or resulting rate constant (see in Introduction, Section 2) referring to a single electron defect, and c_{eL} or c_{pL}. For the reasons already mentioned, the numerical values of these latter factors are unknown, and only

$$k_{\text{exp}} = (k\, c_{eL}) \quad \text{or}$$

$$k_{\text{exp}} = (k\, c_{pL})$$

is obtained (k_{exp} is the experimentally determined rate constant); by applying Eq. (III.81) the expressions

$$k_{\text{exp}} = (k_{\text{exp}})_0 \exp(-E^{\ddagger}_{\text{exp}}/kT) = k_0 \exp(-E^{\ddagger}/kT)\, A \exp[(E_{\text{F}} - E_{\text{g}})/kT]$$

and

$$k_{\text{exp}} = (k_{\text{exp}})_0 \exp(-E^{\ddagger}_{\text{exp}}/kT) = k_0 \exp(-E^{\ddagger}/kT)\, B \exp(-E_{\text{F}}/kT)$$

are obtained for the temperature-dependence of these from the Arrhenius equation (see Part IV, Paragraph 2.2.1).

The apparent activation energy (Fig. III.32) in one case is $(E^{\ddagger} - E_{\text{F}} + E_{\text{g}})$, and in the other $(E^{\ddagger} + E_{\text{F}})$. Thus, if there is a shift in the Fermi level, E_{F}, a corresponding shift will be observed in the experimentally determined activation energy; for example, the apparent activation energy decreases with increase of the Fermi level when an electron passes from the catalyst to the substrate in the rate-determining step, and increases when the substrate donates an electron to the catalyst.

The types of irradiation which can alter the properties of a solid, for instance, of a catalyst, are: electromagnetic radiation, electron radiation, irradiation with charged heavy particles or neutrons. Their effects can appear in different forms. Without any claim to completeness, we shall mention only a few of these effects, in particular those which can be more or less clearly followed and seem to be suitable for discussion from the aspect of catalytic activity.

(1) Electron excitation.

There is an essential agreement between the effects of electromagnetic (mainly X-ray and gamma-ray) irradiation, and electron irradiation: both primarily cause electron excitation, as a result of which the electron and defect electron concentrations, that is n and p, change.

As long as there is no irradiation the electron and defect electron concentrations can be determined from Eq. (III.81), provided E_F and E_g are known. Illumination causes a rise in both n and p, so that if the terms in Eq. (III.81) were considered formally valid, n should be calculated with a higher, and p with a lower, Fermi energy, that is, a temperature higher than the real one ought to be taken into consideration. The concentrations of ionized donors and acceptors could also be given formally from their expressions in Eq. (III.81). This approach alone suggests the cessation of the validity of the Fermi–Dirac statistics, which was postulated in the derivation of Eq. (III.81) (see Part I, Section 3.3.) Another, so-called kinetic concept is more appropriate for the description of the conditions [58].

In the absence of irradiation, when a particle is adsorbed as a donor on the catalyst and electron equilibrium has been reached between the conduction band and the donor level, as well as between the donor level and the valence band, the partial equilibria are

$$\begin{aligned} \alpha_1 N_0 - \alpha_2 n_0 N_0^+ &= 0 \\ \alpha_3 N_0^+ - \alpha_4 p_0 N_0 &= 0 \end{aligned} \qquad \text{(III.86)}$$

where the α terms are the rate constants of a single electron transition (their ratio is the corresponding equilibrium constant);
1 is the transition from the donor level to the conduction band;
2 is the reverse of 1;
3 is the transition from the valence band to the donor level;
4 is the reverse of 3; and
N_0 and N_0^+ are the numbers of neutral and positively charged particles in the absence of irradiation.

As an effect of irradiation, equilibria [Eq. (III.86)] are upset and stationary state is established, the number of electrons passing from the donor level into the conduction band being the same as that reaching the donor level from the valence band:

$$\alpha_1 N - \alpha_2 n N^+ = \alpha_3 N^+ - \alpha_4 p N \qquad \text{(III.87)}$$

New n and p concentrations develop, with N and N^+ particles pertaining to them under these conditions. From Eqs (III.86) and (III.87) we obtain

$$\frac{\Delta N^+}{N_0} = \frac{\dfrac{\Delta p}{\Delta n} - \alpha \dfrac{p_0}{n_0}}{p_0 \left[\dfrac{1}{\Delta n} + \alpha \left(\dfrac{1}{\Delta n} + \dfrac{1}{n_0} \right) \right]} \tag{III.88}$$

since $N \equiv N_0 = b'p$, as shown by the isotherm in Part II, Paragraph 8.2.4.

In Eq. (III.88) $\Delta N^+ = N^+ - N_0^+$, the change in the number of "strongly" bound donor particles as a result of irradiation, while Δp and Δn are the changes in the concentrations of defect electrons and free electrons, respectively, and $\alpha = \alpha_1 N_0/\alpha_3 N_0^+$ is a constant characteristic of the given adsorption system.

Since Δp and Δn are always positive, from Eq. (III.88), the condition of photoadsorption ($\Delta N^+ > 0$) is that

$$\frac{\Delta p}{\Delta n} > \alpha \frac{p_0}{n_0} \tag{III.89a}$$

irradiation is indifferent ($\Delta N^+ = 0$) when

$$\frac{\Delta p}{\Delta n} = \alpha \frac{p_0}{n_0} \tag{III.89b}$$

and photoadsorption takes place, ($\Delta N^+ < 0$), if

$$\frac{\Delta p}{\Delta n} < \alpha \frac{p_0}{n_0} \tag{III.89c}$$

A similar result is obtained when the calculation is performed for the acceptor particles, but for ΔN^- the reverse conditions are true; if

$$\alpha' = \frac{\alpha_1'}{\alpha_3'} \frac{N_0^-}{N_0}$$

where α_1' is the rate constant of the electron transition from the acceptor level to the conduction band, and α_3' is the rate constant of the electron transition from the valence band to the acceptor level.

For the same catalyst, α always appears as a constant characteristic of the given adsorbate, provided there is no interaction between the contamination levels.

In this way, a theoretical explanation of both radioadsorption and radiodesorption is obtained. Though both have been experimentally observed, the results have so far not been subjected to accurate evaluation.

In the case of semiconductors, changes following the absorption of energy are relatively long-lived, so that is possible to demonstrate their effect. (In metals electron dislocation has an extremely short lifetime, so that no noticeable concentration shift takes place.)

With regard to the dehydrogenation and dehydration of ethanol, Donato studied the changes in the catalytic activity of the n-type conductor ZnO subjected to γ- and neutron-irradiation [75]. As a result of neutron-irradiation β-emitting zinc was formed. As the intensity of the γ-irradiation and the radiation level of the zinc increased, the selectivity of the catalyst changed and dehydrogenation predominated even more markedly (Fig. III.33). In accordance with what has been said in Paragraph 4.1.2. the rate of dehydrogenation increases when N_{H}/N_{H^+} increases, i.e. N_{H^+} decreases and ΔN_{H^+} will be less than zero as a result of irradiation. At the same time the rate of dehydration, which is proportional to N_{OH}/N_{OH^-}, decreases when ΔN_{OH^-} is greater than zero. Case 3 of criteria (III.89c) probably holds:

$$\frac{\Delta p}{\Delta n} < \alpha \frac{p_0}{n_0}$$

which is the condition of photodesorption ($\Delta N^+ < 0$) for donor particles, and of photoadsorption ($\Delta N^- > 0$) for acceptor particles.

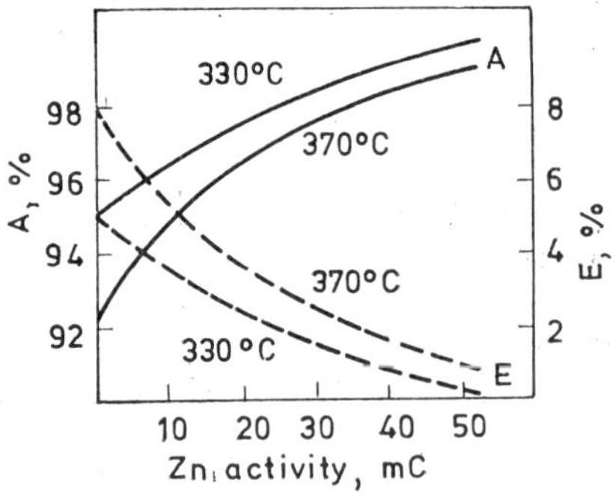

Fig. III.33. Changes in the yield of acetaldehyde (A) and ethylene (E) at a constant contact time vs. radiation level [75]

(2) Dislocation of lattice atoms.

Electron irradiation may result not only in electron excitation, but also, though to only a slight extent, in atomic rearrangement. Incorporations and lattice defects may be formed in much greater numbers, however, when the crystal is irradiated with heavy particles; for instance, the ratio of atoms dislocated by 2 MeV neutron irradiation and by 1.3 MeV gamma-irradiation of the same flux is $10^5 : 1$.

The following correlation is valid between the number of dislocations and the irradiation parameter [76]:

$$N'_{disl} = N \sigma_d \bar{\nu} \varphi t \tag{III.90}$$

where N'_{disl} is the number of atoms dislocated in unit volume;
σ_d is the cross-section of the effect of collision;
$\bar{\nu}$ is the average number of atoms dislocated by the collision of a single particle;
φ is the flux, the number of particles colliding with unit surface area in unit time; and
t is the period of irradiation.

The number of atoms dislocated in unit volume can be calculated from Eq. (III.90). The period for which such a state can be maintained depends greatly upon the temperature. Unfortunately, such an example is available only for copper, where restoration at 20 K is infinitesimally slow and above 450 K instantaneous. Otherwise these irregularities brought about by irradiation differ in no way from the earlier-discussed lattice defects: they modify the Fermi level and thus the catalytic activity in exactly the same manner.

The n-type conductor Al_2O_3 catalyst is a good example of electron excitation and atom dislocation [77]. The activity of this catalyst in the $H_2 - D_2$

exchange reaction increases on exposure to gamma-irradiation. This increase, however, does not last, and some days after irradiation no difference can be observed compared to the initial activity level.

On the other hand, irradiation with α-particles or neutrons again increased the activity in the same reaction, but this did not decline with time.

This observation might be explained by the shorter life-time of the excited electrons, due to their greater mobility, compared to that of the atomic dislocations caused by the heavy particles.

(3) Nuclear conversion in the lattice.

Neutron-irradiation may provoke nuclear reactions resulting in the formation of foreign atoms, mostly in a percentage not exceeding $10^{-4}-10^{-5}$ atom%, in the catalyst lattice points themselves. Because of their special positions, the "micromodifiers" thus formed "*in situ*", may have entirely different effects from those of added impurities. This is illustrated by the following example:

Balandin et al. [78] investigated the dehydration of cyclohexanol on $Na_2SO_4 - MgSO_4$ in which, by substitution with ^{35}S, the specific activity of the catalyst was varied between 1.25 and 105.2 mCi g^{-1}. This resulted in a 1.1–2.8 times increase of conversion at 410 °C. Let us consider what may have caused this increase in catalytic activity.

During irradiation with soft β-rays ^{35}S is converted into ^{35}Cl, but when the catalyst was contaminated with ^{35}Cl no change in its activity could be detected.

Similarly, irradiation of the entire mass of the solid with γ-rays, and thus the excitation of photoelectrons equivalent to the soft β-radiation of the ^{35}S, had no effect on the activity of the catalyst [79].

Since both reproduced, perfectly, ^{35}S radiation and the subsequent introduction of ^{35}Cl was ineffective, the increase in activity was presumably due to the special position of the chlorine atoms formed and incorporated in the crystal.

*

In the above we have tried to give a brief summary of the effect of irradiation treatment. This field has been described for the various types of irradiation, using the fundamental principles discussed in Paragraph 4.2.1 in the discussion of the results. To sum up, it may be said that with an appropriate choice of the type of irradiation it might be possible to deliberately modify the properties of catalysts, a fact which would undoubtedly help to enlarge the scope of catalysis.

PARAGRAPH 4.2.3. ROLE OF THE DEGREE OF DISPERSITY

It has already been seen that adsorption, which is unavoidable in the preparation of catalysts, changes the concentration of electrons in the vicinity of the surface because of equilibrating the charges on the surface. If the electron concentrations are illustrated by means of the band structure, this change can be described by the bending of the latter (see Fig. III.30). Deformation may extend to various depths [61, 80]. Let us consider what happens when the size of the solid is reduced, using the example of a plate which is very large in two dimensions, but very thin. When the thickness is varied three stages can be distinguished (Fig. III.34):

(a) the bending of bands on the opposite sides are separated by straight sections of various length, in other words $L > 2\,l$ (L is the thickness of the plate, and l the depth to which the bending is still significant);

(b) there is a certain overlapping between the bendings of the bands, so that the curvature decreases: $L \approx 2\,l$;

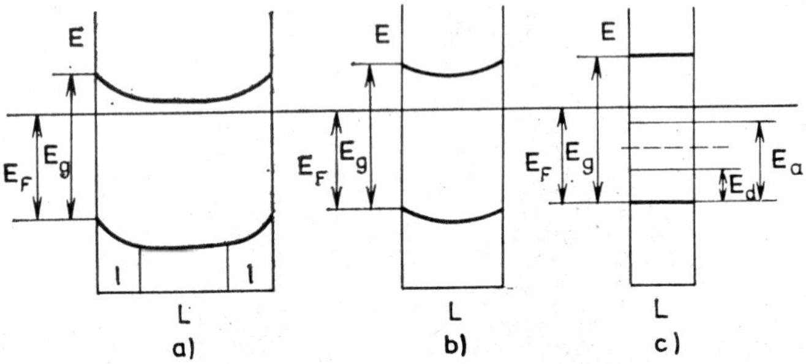

Fig. III.34. Changes in band structure if the thickness (L) of the solid is varied

(c) the interaction is so great that no bending appears, but the straightened band structure is simply shifted: $L \ll l$. In this case the Fermi level is the same in the surface layers and in the bulk of the solid ($E_{F,s} = E_{F,v}$). Let us consider this change as a function of the thickness of the solid plate.

Since the bulk concentrations of n and p, i.e. of the electrons and defect electrons, along L are the same, the conditions of electrical neutrality can be given by means of the following expression

$$e(p - n)L + \sigma_s = 0 \tag{III.91}$$

where e is the charge of a single electron;
p and n are the numbers of defect electrons and electrons in unit volume of the crystal;
L is the thickness of the crystal plate; and
σ_s is the charge per unit surface.

It can be seen from Eq. (III.81) that

$$p = A \exp\,(-E_F/kT) \tag{III.92}$$

$$n = B \exp\,[-(E_g - E_F)/kT] \tag{III.93}$$

whatever irregularities are contained in the semiconductor under discussion.

If there are no impurities or other lattice defects in the crystal, $p = n$, and thus from Eqs (III.92) and (III.93)

$$\exp\,(E_{F,i}/kT) = \sqrt{A/B}\, \exp\,(E_g/2\,kT) \tag{III.94}$$

Eq. (III.94), known from Part I, Chapter 3, in such cases determines $E_{F,i}$, the Fermi level for intrinsic conduction.

In Part II, Paragraph 8.2.4, equations are given from which the following expression can be derived, if the adsorption of only one type of donor and one type of acceptor is considered:

$$\sigma = 2\,\mathrm{e}\,b\,p\,\exp[-(E_{\mathrm{a}} - E_{\mathrm{d}})/kT]\,\sinh[E_{\mathrm{F}} - \alpha)/kT] \qquad \text{(III.95)}$$

where b is a constant;
p is the pressure of the adsorbed component(s); and
α is $(E_{\mathrm{a}} + E_{\mathrm{d}})/2$.

Substitution of Eqs (III.92), (III.93) and (III.95) into Eq. (III.91) and use of Eq. (III.94) leads to:

$$\exp[(E_{\mathrm{F}} - \alpha)/kT] = \left[\frac{\beta p/L + \exp[(E_{\mathrm{F,i}} - \alpha)/kT]}{\beta p/L + \exp[-(E_{\mathrm{F,i}} - \alpha)/kT]}\right]^{1/2} \qquad \text{(III.96)}$$

where $\beta = \dfrac{-b}{\sqrt{AB}} \exp\{[E_{\mathrm{g}} - (E_{\mathrm{a}} - E_{\mathrm{d}})]/kT\}$ is a constant characteristic of the given system.

In a thin crystal the change of E_{F}, the Fermi level, with L the thickness of the crystal, can be obtained from Eq. (III.96) when the pressure, p, is constant (the other parameters, $E_{\mathrm{F,i}}$, E_{a}, E_{d} and E_{g} are, of course, fixed in a given system). When L decreases, the Fermi level will finally be equal to α (see the dashed line in Fig. III.34c), and approaches α on decreasing if $E_{\mathrm{F,i}} > \alpha$, and on increasing if $E_{\mathrm{F,i}} < \alpha$.

When empirical values are substituted it may be expected from Eq. (III.96) that with a specific surface of some 10 $\mathrm{m^2\,g^{-1}}$ a shift in the Fermi level has to be considered, involving a change in the catalytic activity. There are no data in the literature referring directly to such an idealized state. The experimental data may, nevertheless, be evaluated, at least qualitatively, by considering that:

(1) the conclusions are, by and large, valid not only for $L \ll l$, but also for $L \approx l$ [81];

(2) the particle size can be taken as being proportional to the thickness of the large crystal plate used in the model calculation.

All this together might explain the often encountered observation that with increasing specific surface the activity of unit mass of catalyst does not change linearly (see the experiment of Schwab and Rudolph [3] in Section 3 of the Introduction of this Part). Impurities and lattice defects may interfere with these considerations, however, particularly if their concentrations change proportionally to the specific surface.

PARAGRAPH 4.2.4. EFFECT OF THE ELECTRICAL FIELD

A shift in the band structure near the surface can be a result not only of changes in the crystal dimensions, but also of the effect of the electric field.

Conditions can again be studied on a plate crystal on which some substance (e.g. an acceptor) is adsorbed. In Fig. III.35 the forbidden band is bounded by **two** parallel, symmetric curves, and as regards adsorption and catalysis both sides of the plate crystal are entirely equivalent. When this crystal is placed in the electric field so that the vectors of the force are perpendicular to the boundary plane, the band structure no longer possesses a symmetric bending (dashed line in Fig. III.35), and the catalyst is polarized.

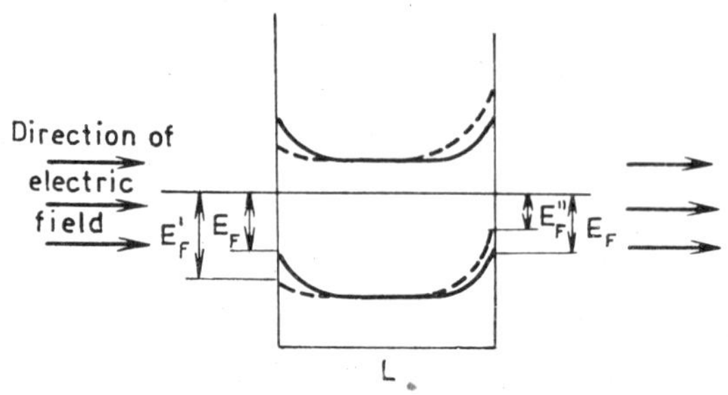

Fig. III.35. Displacement of the band structure under the effect of an electric field

Figure III.35 illustrates the case when the adsorption of an acceptor causes a decrease in the electron concentration of the surface layers and this is modified by the electric field directed from left to right. On the left side the electric field slightly increases, and on the right side it causes a slight decrease in the electron concentration, thereby polarizing the catalyst. Consequently, the two sides of the plate crystal are no longer equivalent. The reverse shift of the Fermi level, however, does not mean that the overall gross measurable values for the two sides of the crystal remain unchanged, or that the compensation is catalytically perfect. This could not happen even if the shift of the Fermi level, though of opposite direction, had the same value, as the adsorption does not depend linearly on E_F, their relation being far more complicated [see Eq. (III.81)].

Adsorption causes a change in Φ, the work function (see Fig. III.30). Thus, if the quantity of adsorbed substance changes under the influence of the electric field, Φ changes too, as confirmed by the experiments of Pratt and Kolm [82]. It should be noted that the change in Φ in these experiments was *greater* than could be ascribed solely to the polarization of the adsorbent.

PARAGRAPH 4.2.5. ADDITIONAL REMARKS

This survey of the electronic theory is incomplete, not only with regard to some important details, as pointed out repeatedly in the discussion, but also in the description of the electron overlappings and the exact potential path in the catalysis, and to the determination of the rates of the elementary steps.

An attempt has been made only to give some points which might be of use regarding the conscious choice of the catalyst and in the modification of its properties.

CHAPTER 5. SUPPORTED CATALYSTS

F. Solymosi

In connection with heterogeneous catalytic processes, it has long been known that the efficiency of certain catalysts can be greatly improved by depositing them on solids with very large surfaces, i.e. on supports. In earlier investigations the effect of the support was interpreted as the stabilization of the state of the active component, the creation of a special distribution, and the increase of the area of the active surface. The experiments of Adadurov et al. [83] can be considered work of fundamental importance in the deeper interpretation of the support effect. These authors demonstrated the chemical interaction between catalyst and support. Depending on the radius and valency of its metal ion the support polarizes the catalyst molecule, and as a result of this polarization the properties of the latter undergo a significant modification. This interaction obviously extends only to that part of the catalyst in direct contact with the support.

Selwood [84] subjected the interaction between support and catalyst to a thorough investigation. From the magnetic and chemical analysis of transition metal oxides deposited in thin layers on various diamagnetic materials (magnesia, alumina and titania), it was found that in the contact zone the ions of the transition metals (Mn, Ni, Fe, etc.) adopt the valency of the metal ion of the support (valency inductivity).

Further progress was made in the interpretation of the support effect by the important discovery in semiconductor physics that there is an electronic interaction between the metal and the semiconductor oxide.

According to the electronic theory of catalysis, the rate and activation energy of the catalytic reaction depend upon the Fermi level of the catalyst, and thus it was expected that electric interaction between metal and oxide (which modifies the position of the Fermi level of the metal catalyst and the support oxide in contact with it) would contribute to the development of the support effect. Experiments supporting this assumption were started in several places at roughly the same time.

SECTION 5.1. EXPERIMENTS WITH SUPPORTED CATALYSTS

PARAGRAPH 5.1.1. DECOMPOSITION OF FORMIC ACID

The decomposition of formic acid was first investigated on a metallic nickel catalyst supported on semiconductor oxides. In accordance with the results described earlier, this reaction is very sensitive to the position of the Fermi level of the catalyst. Of the n-type conductor oxides, the weakly n-type conducting alumina [85, 86], and, later, the far better conductor titania [87]–[89] were used as supports. According to the semiconductor theory, the electron concentrations of n-type oxides are reduced by ions of lower valency, and increased by those of higher valency (for details see Part I, Chapter 6). The activation energy of formic acid decomposition on nickel decreases with the decreasing conductivity of the support, and vice versa (Table III.8).

TABLE III.8

Activation energy of formic acid decomposition (dehydrogenation) on Ni/TiO_2 and Ni/Al_2O_3 catalyst systems

Catalyst	Activation energy, kcal $mole^{-1}$
Ni/TiO_2	22.3
Ni/TiO_2	23.6
Ni/TiO_2	22.6
Ni/TiO_2 + 2% Sb_2O_5	25.0
Ni/TiO_2 + 0.3% BeO	22.0
Ni/TiO_2 + 1% BeO	20.5
Ni/TiO_2 + 5% BeO	18.5
Ni/TiO_2 + 0.5% Al_2O_3	19.4
Ni/TiO_2 + 2% Al_2O_3	17.0
Ni/TiO_2 + 0.3% Cr_2O_3	15.4
Ni/TiO_2 + 1% Cr_2O_3	15.1
Ni/TiO_2 + 2% Cr_2O_3	15.0
Ni/TiO_2 + 1% NiO	17.0
Ni/TiO_2 + 5% NiO	13.0
Ni	26.5
Ni/Al_2O_3	20.5
Ni/Al_2O_3 + 2% BeO	19.0
Ni/Al_2O_3 + 5% NiO	7.0
Ni/Al_2O_3 + 5% TiO_2	24.0
Ni/Al_2O_3 + 2% GeO_2	23.0

Schwab et al. arrived at similar results when using the Co/Al_2O_3 and the Ag/Al_2O_3 catalyst systems [86]. Figure III.36 shows the relationship between the activation energy of formic acid decomposition and the conductivity of the support. It can be expected from these results that for the system formic acid–nickel catalyst–support, p-type oxides will be considerably more powerful supports than n-type conductors. Experiments on Ni/Cr_2O_3, and particularly on Ni/NiO systems, have fulfilled this expectation, since on

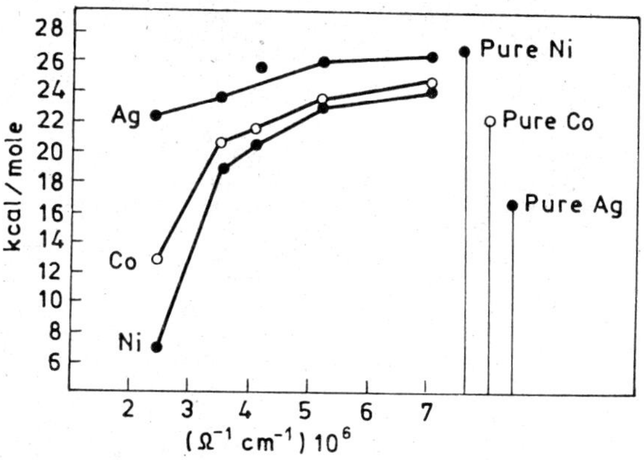

Fig. III.36. Activation energies of formic acid decomposition on Ni, Co and Ag catalysts vs. the conductivity of the alumina support

Ni/Cr_2O_3 and Ni/NiO activation energies of 19.8 [87], [89] and 8 kcal mole^{-1} [90], respectively, were measured. The conductivities and defect electron concentrations of p-type conductor supports also influenced the activation energy of formic acid decomposition. In this case, an increased conductivity reduced the activation energy of the decomposition, and vice versa. Results obtained with defect conductor supports are illustrated in Table III.9 and Fig. III.37.

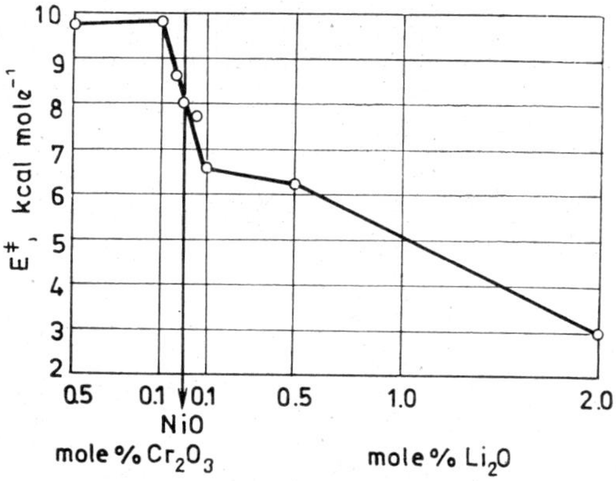

Fig. III.37. Activation energy of formic acid decomposition on a nickel catalyst vs. the doping of the nickel oxide support

TABLE III.9

Activation energy of formic acid decomposition (dehydrogenation) on Ni/Cr_2O_3 catalyst system

Catalyst	Activation energy kcal mole^{-1}	Temperature range, °C
Ni/Cr_2O_3	19.8	200—240
Ni/Cr_2O_3 + 0.5% TiO_2	19.0	200—240
Ni/Cr_2O_3 + 1% TiO_2	21.5	195—235
Ni/Cr_2O_3 + 2.5% TiO_2	23.5	200—245
Ni/Cr_2O_3 + 5% TiO_2	23.6	200—240
Ni/Cr_2O_3 + 0.3% NiO	19.5	200—240
Ni/Cr_2O_3 + 1% NiO	18.0	195—230
Ni/Cr_2O_3 + 3% NiO	17.5	205—250
Ni/Cr_2O_3 + 5% NiO	16.3	195—235

For the evaluation and understanding of the results it is necessary to consider the electric phenomena in the metal—semiconductor oxide contact zone, the electronic structures of the semiconductor oxide support and the metal catalyst, and the mechanism of HCOOH decomposition.

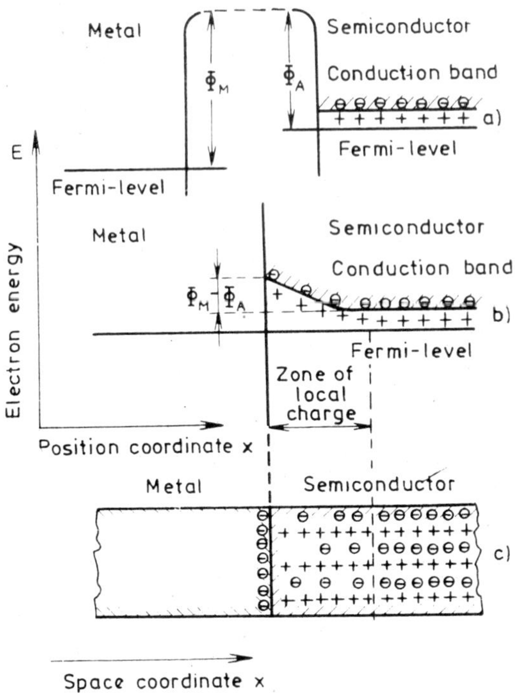

Fig. III.38. Energy diagram of the metal — semiconductor system

(a) prior to contact; Φ_M: work function of the metal; Φ_A: work function of the semiconductor
(b) after contact
(c) the same as in (b) in schematic representation

According to the theory of metal—semiconductor contacts, if a metal comes into contact with a semiconductor then at thermodynamic equilibrium the Fermi level must be the same at the interface. If the work function of the electrons in the metal is higher than the semiconductor, electrons pass from the semiconductor into the metal until equilibrium is reached. If the work function of the metal is lower than that of the semiconductor, electrons flow from the metal into the semiconductor. In the first case, the flow of electrons causes a positive local charge in the semiconductor and an induced negative charge on the surface of the metal, which finally limits the number of transmitted electrons.

Figure III.38 shows the electron energy band of a metal and a semiconductor before, and after, contact. In this example the work function of the metal is higher than that of the semiconductor.

As a rule, the work functions of n-type conductor oxides are lower than those of metals. Hence, electrons flow into the metal; the higher conductivity of the support increases the number of electrons passing into the metal, while a lower electron concentration brings about a decrease of this number.

The decomposition of formic acid is known to belong in the group of donor reactions (see Paragraph 3.2.1), and hence the activation energy of the reaction will be low on catalysts with low Fermi levels. With this knowledge, it is easy to interpret the activation energy values obtained when the electron concentrations of n-type conductor oxides change, because the transition of an electron of formic acid to the metal surface, and the formation of chemisorption bonds and of intermediate states obviously proceed more easily when only few electrons pass from the supporting oxide into the metal. Since the concentration of electrons is very low in defect conductor oxides, electrons pass from the metal into the semiconductor. This process leads to a decrease of the electron concentration and to an increase in the electron affinity of the metal. This explains the considerably higher efficiency of p-type conductor oxides compared to n-type conductor supports. A rise in the defect electron concentration leads to a further decrease in the activation energy, since a higher defect electron concentration of the support means that metal electrons continue to flow to the support. Electronic interaction between catalyst and support in these systems has been confirmed by magnetic measurements.

It is known that if the 0.6 electron hole in the 3d-orbital of nickel is filled up with electrons, the magnetic susceptibility decreases due to the increased number of paired spin moments. This assumed charge-carrier exchange between metal and oxide can be followed approximately by magnetic measurements [86]. Schwab et al. carried out magnetic measurements on a nickel-alumina catalyst system such as that used in the decomposition of formic acid [86]. The saturation magnetization of $Ni-Al_2O_3$ was 49 G but when 5% of titania was incorporated into the alumina this value dropped to 44 G; in contrast, it increased to 56 G on the addition of 5% of nickel oxide. The average μ_B values for one Ni atom were 0.51, 0.46 and 0.59 respectively. Hence, these magnetic measurements indicate that the number of unpaired electrons in nickel decreases as a result of pairing with electrons originating from the semiconductor oxide.

This experiment and its conclusions, however, were criticized by Selwood [91], who pointed out that to obtain the saturation magnetization of nickel, $\mathbf{M}_{Ni}$, the magnetization and not the susceptibility ought to have been extrapolated to an infinite field, while even this procedure could scarcely yield $\mathbf{M}_{Ni}$ values of sufficient accuracy for this purpose.

In the cases discussed so far the supports were homogeneous oxides, perhaps containing small quantities of foreign ions. Supports used in practice generally consist of a mixture of oxides between which reactions, for example the formation of spinels, may take place at higher temperatures.

Change in catalytic activity can usually be attributed to modification of the electronic stucture of the oxide mixture. It is to be expected from what has been said above that a similar correlation exists between the carrier effect of the mixed oxide support and its electrical properties. The importance of electronic interaction has been demonstrated for the supports $NiO-Cr_2O_3$ and $MgO-Cr_2O_3$ [92]. Figure III.39 shows the changes in the resistance and in the activation energies of the conductivity of $NiO-Cr_2O_3$ and of formic acid decomposition (measured on metallic nickel deposited on the oxide mixture) plotted vs. the ignition temperature. It can be seen that at 800 °C, the temperature at which spinel begins to form, both activation energies are at a minimum, indicating the special characteristics of the state which develops during the formation of spinels. Since electrical conductivity measurements have proved the defect conductor nature of both oxides and of the spinel formed from them, the results can be interpreted in

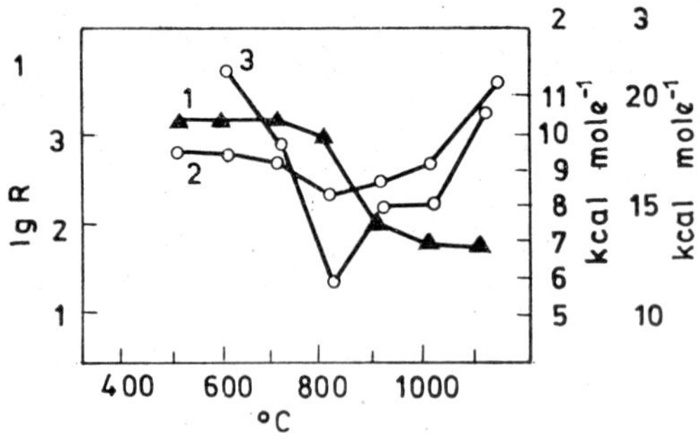

Fig. III.39. Resistivity of $NiO-Cr_2O_3$, and the activation energies of $NiO-Cr_2O_3$ conductivity and of formic acid decomposition vs. the temperature of pretreatment of the mixed oxide support

1. electrical resistivity of $NiO-Cr_2O_3$
2. activation energy of conductivity
3. activation energy of HCOOH decomposition on reduced $NiO-Cr_2O_3$ catalyst systems

a manner similar to that applied to nickel deposited on defect conductor supports.

A very valuable contribution to the solution of this problem has recently been made by Baddour and Deibert [93], who investigated the thermal decomposition of formic acid on nickel deposited on various types of powdered germanium support. The purpose of this study was the quantitative demonstration of the promoting effect of support-catalyst interaction. As germanium is a more thoroughly studied semiconductor than the former refractory oxides, it was possible to predict the extent of electronic interaction at the germanium-nickel interface from the changes in the electronic properties of the support.

Nickel was deposited on germanium by means of the decomposition of purified nickel carbonyl vapour. The kinetic data are listed in Table III.10. The following conclusions were drawn:

TABLE III.10

Catalytic properties of germanium-nickel and pure nickel catalysts in formic acid dehydrogenation at 240 °C and 19.5 torr

Germanium type	Average thickness of Ni layer, atomic layers	Initial activity mole m^{-2} hr^{-1}	Activation energy kcal $mole^{-1}$	Frequency factor mole m^{-2} hr^{-1}
n-type	1.0	none		
Sb-doped	none	2.4×10^{-4}		
	1.0	3.4×10^{-4}		
	3.1	7.8×10^{-4}		
	6.2	14.2×10^{-4}	17.0–18.9	7.1×10^{4}
	18.5	23.7×10^{-4}	10.6–11.0	9.6×10^{1}
p-type	4.4	2.9×10^{-4}	33.6	6.0×10^{10}
Zn-doped	13.1	7.9×10^{-4}	10.0	1.46×10^{1}
n-type	10.7	25.4×10^{-4}	13.8	1.9×10^{3}
As-doped				
pure Ni		1.54 ± 0.06	28.7 ± 1.1	2.2×10^{12}

(1) On all the supports the initial activity of the catalyst increases with increasing thickness of the nickel film.

(2) The value of the activation energy shows an opposite trend and the catalyst consisting of a thin nickel layer has the highest activation energy.

(3) The ratio of the relative activities of catalysts with an average of 10.7 atomic layers of nickel on Zn-, Sb- and As-doped germanium is 1 : 3 : 4. In the light of a charge transfer between nickel and varyingly doped germanium this means that an increase in electron transfer from 7.3×10^{10} to 5.3×10^{11} and then to 3.7×10^{12} electron $\cdot$ cm^{-2}, causes an activity increase as expressed by the above ratio. Considering the number of nickel atoms in a film of 10.7 atomic layers, it appears that an increase of only 2×10^{-5} electron/nickel atom in the interface charge transfer results in a tripling of the catalytic activity of nickel. These results suggest that the rate-controlling step in the decomposition of formic acid is probably the dissociative adsorption of the acid. In this step, electrons are transferred from the catalyst surface to the adsorbed formate group.

If the value of the activation energy is also taken into account, as in the former case, it is difficult to arrive at a conclusion, partly owing to the marked effect of the thickness of the nickel layer and partly because of the very few available activation energy values. Nevertheless, disregarding the value obtained on the very thin layer (4.4 atomic layers), it appears that a change of the type of support from a p-type conductor to an n-type conductor involves a slight increase of the activation energy. If these few activation energy values are interpolated, as carried out by the authors for the initial activity, the result is that on nickel 10.7 atomic layers in thickness the activation energy of decomposition is practically independent of the doping of the support.

PARAGRAPH 5.1.2. SELECTIVE DEHYDROGENATION OF CYCLOHEXANE

The importance of the electronic interaction between the support oxide and the metal catalyst has also been demonstrated for other catalysts and substrates. Langenbeck et al. studied the influence of the semiconductor properties of the support on the selectivity of nickel in the conversion of cyclohexane [94].

Cyclohexane conversion can be described by the following reactions:

$$C_6H_{12} = C_6H_6 + 3\ H_2 \qquad \Delta H^0_r = 49.8\ \text{kcal}$$

$$C_6H_{12} + 6\ H_2 = 6\ CH_4 \qquad \Delta H^0_r = -69.7\ \text{kcal}$$

The electron concentration of the support was modified not by the incorporation of ions of different valency, but by the addition of n-type conductor oxides. Catalytic dehydrogenation of six-membered naphthene rings into aromatic compounds is highly important in gasoline manufacture. Dehydrogenation reactions are usually accompanied by the undesirable destructive hydrogenation of cyclic hydrocarbons, leading to lower members, primarily methane. In practice, platinum deposited on alumina with a low silica content has proved the best dehydrogenation catalyst in this case. On the considerably cheaper Ni/SiO_2, Ni/Al_2O_3 and Ni/MgO systems under the same conditions, mainly destructive hydrogenation takes place. In Langenbeck's experiment the addition of n-type conductor oxides, primarily zinc oxide and cadmium oxide, to the support improved the selectivity and dehydrogenating power of the above systems and made them equivalent to the platinum-alumina catalysts. Application of oxides other than n-type-conductors, e.g. barium oxide or manganese(II) oxide, had no influence whatsoever on the selectivity of nickel (Table III.11).

TABLE III.11

Conversion of cyclohexane on various nickel catalyst systems

Catalyst	Added metal oxide or metal	Reaction tempera-ture, °C	Reaction products		
			benzene	methane	cyclohexane
			%		
Ni/MgO*	250 mg ZnO	500	72	3	25
		400	63	1	36
Ni/Al_2O_3**	250 mg ZnO	500	83	4	13
		400	80	3	17
Ni/SiO_2***	250 mg ZnO	500	67	12	21
		400	37	10	53
Ni/MgO	200 mg Zn	500	74	5	21
		400	70	3	27
Ni/Al_2O_3	200 mg Zn	500	87	5	8
		400	83	3	14
Ni/SiO_2	200 mg Zn	500	35	6	59
		400	68	2	30
Ni/MgO	—	500	—	100	—
	—	400	—	100	—
Ni/Al_2O_3	—	500	—	100	—
		400	12	85	3
Ni/SiO_2	—	500	—	100	—
		400	—	100	—
Ni/MgO	15 mg CdO	400	40	35	25
Ni/Al_2O_3	30 mg CdO	400	62	25	13
Ni/MgO	354 mg TiO_2	500	1	98	1
		400	—	100	—
Ni/Al_2O_3	354 mg TiO_2	500	35	42	23
		400	16	75	9
Ni/SiO_2	354 mg TiO_2	500	6	86	8
		400	3	95	2
Ni/MgO	288 mg ThO_2	500	—	100	—
		400	—	100	—
Ni/Al_2O_3	288 mg ThO_2	500	53	35	12
		400	29	67	4
Ni/Al_2O_3	300 mg BaO	500	—	100	—
Ni/MgO	300 mg MnO	500	—	100	—

* 55.5% Ni — 44.5% MgO
** 62.5% Ni — 37.5% Al_2O_3
*** 17.9% Ni — 82.1% SiO_2

quantity of catalyst: 100 mg

Nehring and Dreyer later modified the electronic properties of the support in the usual manner, by the incorporation of foreign ions [95]. They doped the alumina support with thoria and titania. Though no direct relationship was found between the increased conductivity of the support and the considerable dehydrogenating power of the catalyst system after doping, as the conductivity of the catalyst increased (in the range of low dopants) the dehydrogenation of cyclohexane again came into the foreground. The experimental results are shown in Table III.12.

TABLE III.12

Conversion of cyclohexane on Ni/Al_2O_3 catalyst systems at 400 °C

Doping ion mole%	Addition of ThO_2			Addition of TiO_2		
	benzene	methane	cyclohexane	benzene	methane	cyclohexane
	%			%		
0	9	64	27	9	64	27
0.5	37	34	29	41	25	34
1	37	30	33	44	27	29
2	37	26	37	38	26	36
5	45	18	37	34	12	54
10	60	13	27	32	12	56
20	31	8	61	—	—	—
50	6	3	91	13	11	76
90	—	—	—	7	9	84
100	0	0	100	5	10	85

In the interpretation of the results the following two mechanisms were considered:

(1) The electrons of the hydrogen are transferred to the nickel catalyst with the formation of nickel hydride and activated hydrogen. The destructive hydrogenation of cyclohexane actually takes place on this catalyst. As a result of the electrons of the support occupying the 3d-orbital of the nickel, the electrons of the hydrogen will not be bound, or to only a small extent, on the nickel, and consequently the hydrogen will not be sufficiently active in the destructive hydrogenation reaction. It is difficult, however, to interpret the dehydrogenation reaction with the help of this model.

(2) In the other explanation it is assumed that electrons of the cyclohexane are transferred to the metallic nickel catalyst; this involves the deformation of the chemisorbed molecule and leads to the breaking of the C—C bonds. In this case, hydrogenation is a process following the decomposition. It appears from the experimental results that an increase in the electron concentration of the oxide (incorporation of ions of higher valency) favours the dehydrogenation reaction. Since a similar promoter effect resulted when n-type conductor oxides were mechanically mixed into the support, it is justifiable to assume that, in this case too, the n-type conductor nature of the support was enhanced. Supposing the earlier discussed charge-carrier exchange between the metal catalyst and the n-type conductor oxide, the favourable effect of the high electron concentration of the support can be interpreted as the effect of the partial filling of the empty d-band of the

nickel by the electrons of the support; this causes a decrease in the electronic interaction between the cyclohexane and the metal and thereby a decrease in the extent of deformation. Instead of the destruction the dehydrogenation reaction becomes preponderant. These two mechanisms probably represent extreme cases of true activation and as long as at least the charge-carrier exchanges in the chemisorption of cyclohexane are not fully known, it is extremely difficult to decide between the two mechanisms.

The electronic interaction between the support oxide and the metal catalyst is also manifested when the same model reaction is performed on a platinum catalyst. The data of Table III.13 show that the percentage benzene content

TABLE III.13

Conversion of cyclohexane on various platinum catalyst systems at 500 °C

Catalyst	Benzene %	Non-condensed part, %	Non-converted cyclohexane, %
Pt/ZnO	—	—	100
Pt/TiO_2	76.1	3.2	20.7
Pt/Al_2O_3 (a) pH 6	63.1	2.5	34.4
Pt/Al_2O_3 (b) pH 8	59.8	2.8	37.4
Pt/MgO	32.3	22.0	45.7
Pt/SiO_2	23.1	20.1	56.8
Pt/activated carbon	55.0	14.0	31.0

of the product decreases on changing from a titania support to a silica support [96]. Taking into consideration the semiconductor properties of the oxides, this change coincides with the decrease of the n-type conductor nature of the oxides. It is interesting to note, however, that on zinc oxide, the electron concentration of which is several orders higher than that of the best conductor of the group, titania, no dehydrogenation reaction at all takes place. This seems to indicate that with platinum catalysts the oxides of lowest conductivity are the most efficient supports. Interpretation of the results obtained for platinum catalysts is the same as for nickel, involving the transition of the electrons of the support to the 5d-band of the platinum.

PARAGRAPH 5.1.3. SELECTIVE HYDROGENATION OF CYCLOPROPANE AND PROPYLENE

Langenbeck et al. also performed experiments on the hydrogenation of propylene and the destructive hydrogenation of cyclopropane [97]. The nickel catalyst was deposited on alumina, magnesia, silica and nickel(II) oxide. The electron concentration of the support was again modified by the addition of n-type conductor oxides. In the experiments without additives only destructive hydrogenation proceeded, and the product consisted of a mixture of propane, ethane and methane. Addition of zinc(II) oxide and cadmium(II) oxide to the support reduced this reaction, which ceased completely in the presence of higher quantities of the additives (Table III.14).

TABLE III.14

Effect of the zinc oxide or cadmium oxide content of the support on the conversion of cyclopropane on nickel catalysts

Catalysts	Reaction products			
	cyclopropane	propane	ethane	methane
	%			
Ni/Al_2O_3	0.0	61.0	26.6	12.4
Ni/Al_2O_3 + 10 mg ZnO	3.4	46.6	29.1	20.9
Ni/Al_2O_3 + 50 mg ZnO	38.1	29.6	23.2	9.1
Ni/Al_2O_3 + 150 mg ZnO	74.8	16.3	7.4	1.5
Ni/Al_2O_3 + 5 mg CdO	55.2	21.3	14.7	8.8
Ni/Al_2O_3 + 50 mg CdO	100.0	0.0	0.0	0.0
Ni/Al_2O_3 + 10 mg HgO	58.0	24.2	8.3	9.5
Ni/Al_2O_3 + 30 mg HgO	100.0	0.0	0.0	0.0
Ni/MgO + 30 mg ZnO	30.7	36.0	17.3	16.0
Ni/MgO + 100 mg ZnO	76.1	16.2	4.1	3.6
Ni/MgO + 5 mg CdO	79.8	13.2	2.8	4.2
Ni/MgO + 10 mg CdO	100.0	0.0	0.0	0.0
Ni/SiO_2 + 10 mg ZnO	6.7	59.1	26.3	7.9
Ni/SiO_2 + 200 mg ZnO	83.0	9.3	5.3	2.4
Ni/SiO_2 + 5 mg CdO	63.6	19.8	12.7	3.9
Ni/SiO_2 + 50 mg CdO	100.0	0.0	0.0	0.0

The catalyst systems generally contained 100 mg of nickel, corresponding to about 48.3% of Ni.

This behaviour indicates that the transition of the electrons of n-type conductor oxides to the 3d-orbital of metallic nickel prevents the transition of the electrons of the substrate to the metallic catalyst.

The mechanism of the destructive hydrogenation of cyclopropane is not yet clear. One of the possible mechanisms would again involve the activation of hydrogen or of cyclopropane on the surface of the catalyst and this would be the electron-donating step. The other mechanism assumes the chemisorption of cyclopropane on one or two sites. When chemisorption is on one site, the ring can be opened only in one place and the radical formed in this manner is able to form a propane molecule by only taking up hydrogen. The fact that lower homologues, such as ethane and methane, are also formed in the course of hydrogenation, seems to support chemisorption on two sites. It appears probable that these mechanisms again represent extreme cases of activation and that the true process takes place via some intermediate state. Changes in selectivity on the addition of n-type conductor components suggest that the decisive process leading to destructive hydrogenation requires catalyst systems poor in electrons, which is a further indication of the importance of the activation of hydrogen. For a clarification of the true mechanism it would again be desirable to measure the charge-carrier exchanges caused by the chemisorption of propane.

Since the hydrogenation of propylene is not affected by the presence of n-type conductor oxides under these conditions, in the mixture of cyclopropane and propylene the latter can be selectively hydrogenated on a catalyst system of this type.

PARAGRAPH 5.1.4. DECOMPOSITION OF HYDRAZINE

The electron concentration of the support affects the decomposition of hydrazine catalyzed by metallic nickel [98]. If ions of higher valency (e.g. gallium) was added to the magnesia support, the activation energy of the hydrazine decomposition decreased, whereas the addition of lithium caused it to increase. Changes in the activity of the catalyst showed the same trend. The experimental results are shown in Fig. III.40. Despite the

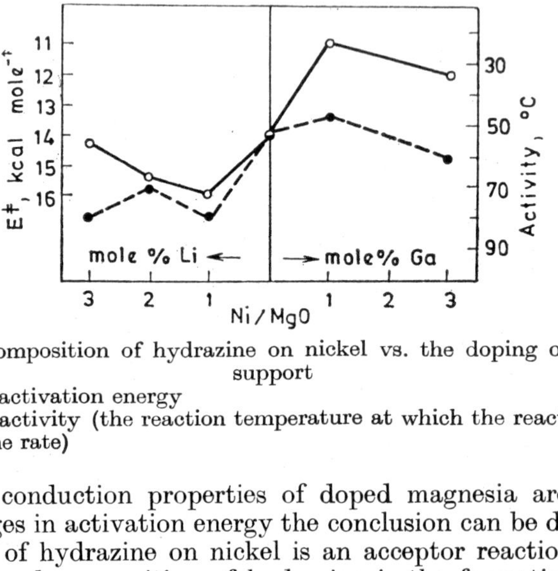

Fig. III.40. Decomposition of hydrazine on nickel vs. the doping of the magnesia support

——— activation energy

– – – – activity (the reaction temperature at which the reaction proceeds at the same rate)

fact that the conduction properties of doped magnesia are not known, from the changes in activation energy the conclusion can be drawn that the decomposition of hydrazine on nickel is an acceptor reaction. Hence, the first step in the decomposition of hydrazine is the formation of the NH_2 radical, as the N—N bond is much weaker than the N—H bond. To attain the noble gas configuration the NH_2 radical needs one electron. The higher electron concentration of the metal obviously favours the bonding of the radical, i.e. the decomposition of hydrazine.

PARAGRAPH 5.1.5. INVESTIGATION OF CATALYST SYSTEMS USED IN AMMONIA SYNTHESIS

The favourable effects of an alumina support and a potassium oxide activator on the iron catalyst used in ammonia synthesis have been known for a long time. In an effort to interpret the effect of alumina, it was assumed that this consists mainly in preventing the sintering of small iron crystals. However, very little is known about the effect of the potassium oxide additive. The questions arise as to whether the role of the alumina is in fact restricted to the prevention of sintering, or whether it also has some chemical influence on the catalytic behaviour of iron, and as to whether the role of the potassium oxide may be responsible for a change in the electron concentration of alumina. In an attempt to find answers to these questions, the simple and well-known model reaction, the decomposition of formic acid, was first investigated [99]. On a completely reduced iron catalyst Schwab et al. measured an activation energy of 19.5 kcal $mole^{-1}$. When reduction was not complete and the catalyst contained appreciable amounts of Fe_3O_4 the activation energy rose to 30–40 kcal $mole^{-1}$. On the system Fe/Al_2O_3 the activation energy dropped to 17–18 kcal $mole^{-1}$.

The activation energy fell more markedly when the alumina support also contained potassium oxide. When germanium(IV) oxide was incorporated into the alumina, however, the activation energy rose to 22 kcal $mole^{-1}$ (Table III.15).

TABLE III.15

Activation energy of the decomposition of formic acid on iron catalyst systems

Catalyst	Activation energy, kcal $mole^{-1}$
Fe/Fe_3O_4	30–40
Fe	19.5
Fe/Al_2O_3	17–18
$Fe/Al_2O_3 + 1\%\ K_2O$	13–13.5
$Fe/Al_2O_3 + 2\%\ K_2O$	7
$Fe/Al_2O_3 + 1\%\ TiO_2$	13
$Fe/Al_2O_3 + 2\%\ GeO_2$	22

Results with alumina containing various additives can be interpreted quite similarly to the phenomena observed on the nickel catalyst system in the decomposition of formic acid. It seems probable that in this case, too, electrons flow from the semiconductor towards the metal.

A few words must be said about the behaviour of the Fe/Fe_3O_4 system. Magnetite is known as an extremely good n-type conductor, its conductivity even at room temperature being about 10^{10} times higher than that of alumina. Because of its high Fermi level, therefore, Fe_3O_4 raises the Fermi level of the metal electrons far more than does Al_2O_3. These results provide convincing evidence that the effects of support and activator on industrially important catalyst systems can be attributed to the electronic interaction between catalyst and support, and to the changes in the electrical properties of the support.

The question remains as to whether this effect also plays a role in reactions of other electronic character, such as the industrially highly important synthesis of ammonia. This is an acceptor reaction in which the rate-controlling step is the chemisorption of nitrogen, and thus it might be expected that the reaction takes place with the lowest activation energy on catalyst systems rich in electrons. The results of Schwab and Putzar partly confirm this expectation, since the addition of lanthanum oxide resulted in a slight decrease of the activation energy, while the presence of potassium oxide caused an increase [100]. As regards the activation energy, therefore, the addition of potassium is detrimental.

The situation was somewhat different on sintered catalysts, the lowest activation energy being measured when the system contained potassium oxide. Experiments showed that catalysts containing potassium oxide maintain their activity even after heat-treatment at high temperatures, while the other catalysts listed in Table III.16 are extremely sensitive to

TABLE III.16

Activation energy of ammonia synthesis on iron catalyst systems

Catalyst	Activation energy kcal mole^{-1}	
	non-sintered	sintered
Fe, pure	13.5	—
Fe/Al_2O_3 ($+ La_2O_3$)	10–12	17–18
$Fe/Al_2O_3 + K_2O$	13–15	13–15
$Fe/Al_2O_3 + GeO_2$	16	22

heat. The high free-energy sites of the system, produced by the alumina support, probably disappear on heat-treatment. The observation that this process does not take place in the presence of potassium oxide can be explained by a change in the defect structure of the alumina, the incorporation of potassium inhibiting the diffusion leading to sintering. Alumina being an n-type conductor, the incorporation of potassium oxide causes a decrease in the electron concentration:

$$K_2O + O_2 + 4\,e' = 2\,K\,|Al\,|'' + Al_2O_3$$

From the laws of defect equilibrium and mass action:

$$2\,|\,Al\,|''' + Al_2O_3 = 6\,e' + 3/2\,O_2$$

$$[e']^3/[\,|\,Al\,|'''] = K/p_{O_2}^{3/4}$$

which means that there is a decrease in the number of empty aluminium sites. According to the theory of diffusion in solids, the lowering of the degree of disorder is accompanied by a decrease in the diffusion rate. Since germanium is an ion of higher valency than aluminium, its incorporation raises the concentration of electrons and thus exerts an opposite effect on the reactions.

The hydrogenation of ethylene has been studied on Ni/ZnO catalyst systems [101], and the support was also doped with lithium oxide or gallium oxide. Figure III.41 shows changes in the rate constant of the reaction as a function of the Ni : ZnO ratio.

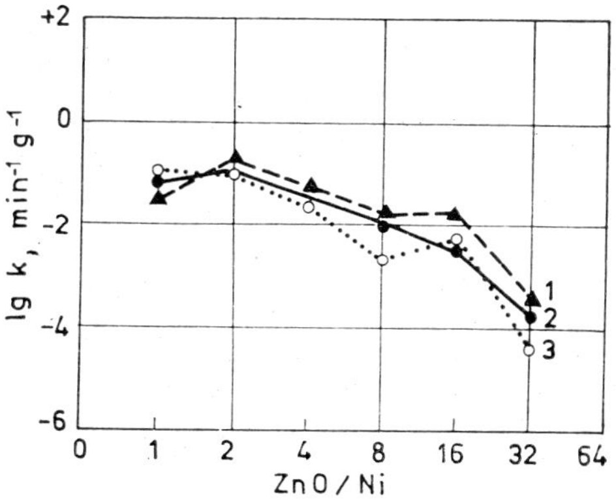

Fig. III.41. The rate constant of ethylene hydrogenation vs. the ratio between the ZnO support and the nickel catalyst

1. ZnO + 0.1 mole% Li_2O
2. ZnO
3. ZnO + 0.1 mole% Ga_2O_3

Despite the slight difference, it can be stated that the catalyst doped with lithium oxide possesses the highest activity, and the catalyst contaminated with gallium oxide the lowest activity. Since zinc oxide is an n-type oxide, these results confirm the donor nature of the hydrogenation, which means that in the activation process electrons of ethylene or hydrogen are transferred to the nickel metal. A detailed kinetic study of the effect of the temperature and pressure of the reaction components disclosed the finer mechanism of the reaction. The energy data are listed in Table III.17. The temperature optimum results from the heats of adsorption and desorption of ethylene. On the lithium-containing catalyst, as with the usual nickel catalyst, the apparent activation energy (last column in Table III.17) is negative at high temperatures, and thus the reaction rate exhibits a maximum at a certain temperature. On catalyst systems doped with gallium a positive apparent activation energy was obtained. Accordingly, a decrease in the electron concentration of the support led to a higher adsorption energy and a lower activation energy. An increased electron concentration had the opposite effect.

All these results indicate that the activity of the catalyst can be very efficiently influenced by means of a metal oxide in some contact with it.

TABLE III.17

Energy parameters of the hydrogenation of ethylene on Ni/ZnO catalysts

Catalyst, Ni/ZnO	$(E^{\ddagger} - Q_{a,H})$	$Q_{a,E}$	$(E - Q_{a,H} - Q_{a,E})$
0.2 atom% of Li^+	13.4±1.5	13.8±2	−0.4
0.2 atom% of Ga^{3+}	16.5±1	11.6±1.5	+4.9

E = "true" activation energy
$Q_{a,H}$ = heat of adsorption of hydrogen
$Q_{a,E}$ = heat of adsorption of ethylene

This fact drew attention to the importance of the experimental conditions under which the catalyst is prepared. The metal catalysts used in both practical and theoretical work are, in most cases, prepared by reduction. In earlier investigations several authors observed the sensitivity of, for instance, nickel catalysts to the conditions of preparation of the nickel (temperature and duration of reduction). The activation energy values obtained by different authors for the same catalyst and the same model reaction exhibit a wide scatter. Important roles are undoubtedly played in this by the dispersity and preliminary heat-treatment of the catalyst, but in our opinion the extent to which the nickel oxide was reduced, that is, the presence of unreduced nickel oxide, is of decisive importance here. Results obtained with the Ni/NiO catalyst system offer a striking illustration of the effect of the presence of nickel oxide on the catalytic properties of metallic nickel (a 15–20 kcal mole^{-1} decrease in activation energy [90]). This effect is particularly apparent in the preparation of nickel catalysts when the supporting oxide is suspended in a solution of a nickel salt, and the nickel oxide layer obtained by thermal decomposition is reduced with hydrogen. The oxides used as supports are a great hindrance to the reduction of nickel oxide with which they are in contact, and the conditions found suitable for the preparation of metallic nickel from pure nickel oxides cannot be used without modification. This is shown by the results obtained on the Ni/TiO_2 and Ni/Cr_2O_3 catalyst systems [87], [88], [89], when the activation energy of formic acid decomposition increased with increasing temperature and duration of the reduction. Analysis of nickel catalyst systems confirmed that even after a reduction of 10 hours at 400 °C the system still contained unchanged nickel oxide. The fact that Ni/NiO is highly effective in the decomposition of formic acid, explains the lower activation energy values obtained with Ni/TiO_2 and Ni/Cr_2O_3 systems containing unreduced nickel oxide.

A certain amount of caution is necessary in the interpretation of results obtained with supports doped with foreign ions. In our view it is necessary to consider the presence of effective Ni/NiO in the Ni/Al_2O_3 + 5% NiO catalyst investigated by Schwab et al. [86], who found a 7 kcal mole^{-1} activation energy with this catalyst (in the decomposition of formic acid) compared to the 20.5 kcal mole^{-1} measured on Ni/Al_2O_3 and the 19 kcal mole^{-1} on Ni/Al_2O_3 + 2% BeO. Though these authors attributed this marked decrease in activation energy to the decrease in the conductivity of alumina, this is far from being confirmed by the conductivity values of Al_2O_3, Al_2O_3 + 2% BeO and Al_2O_3 + 5% NiO mixtures.

It seems more probable that, because of the limited solubility of nickel oxide in alumina, part of the nickel oxide is not incorporated in the alumina lattice, so that when metallic nickel is deposited on the support by evaporation, the highly efficient Ni/NiO catalyst is formed.

PARAGRAPH 5.1.7. INVERSE CATALYSTS

In the examples mentioned so far it was always the metal which acted as catalyst, while the semiconductor was the support. It follows from the theory of metal/oxide interface described in the introduction of this

Chapter that electrical interaction between the two conductors affects the catalytic action not only of the metal but of the oxide too. Schwab and Siegert [102] provided experimental evidence of this effect in an investigation of the catalytic oxidation of carbon monoxide. Nickel oxide was used as the catalyst and metallic silver as the support. It was found that the oxidation of carbon monoxide is of the first order with respect to carbon monoxide and of zero order with respect to oxygen. Similarly as for the reaction with unsupported nickel oxide, the chemisorption of carbon monoxide was assumed to be the rate-determining step (c.f. Paragraph 4.2.1):

$$CO + pL = CO^{+}_{(chem)}$$

The activation energy of the reaction depended on the thickness of the nickel oxide layer (Fig. III.42). On nickel oxide layers of 760–15.200 Å activation energies of between 15 and 16 kcal mole^{-1} were measured, in agreement with the activation energy obtained on pure nickel oxide. With decreasing thickness of the layer there was a sharp rise in activation energy between 380 and 124 Å, which was interpreted as the transition of electrons from metallic silver into the nickel oxide catalyst. This reduced the defect electron concentration of the oxide, a condition unfavourable from the aspect of the activation and chemisorption of carbon monoxide. To explain the decrease of the activation energy with catalysts containing 76 Å thick nickel oxide it was assumed that, at this point, metallic silver, too, begins to participate in the catalysis of the reaction. The activation energy of the oxidation of carbon monoxide on silver is 14 kcal mole^{-1}. When the electron concentration of the silver support was raised by alloying it with antimony, a marked decrease in the catalytic activity of the nickel was observed.

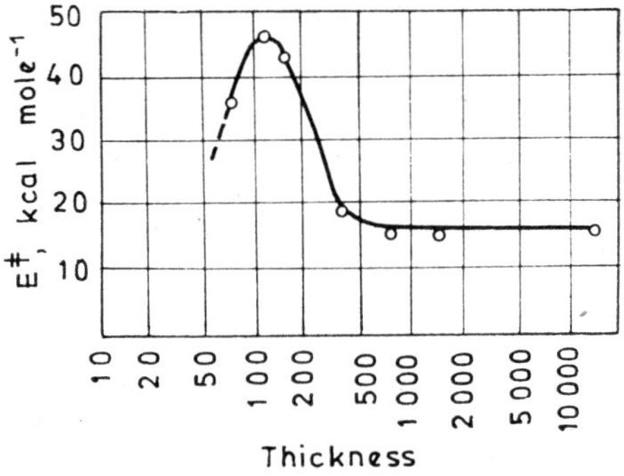

Fig. III.42. Activation energy of carbon monoxide oxidation vs. the logarithm of the thickness of the nickel oxide layer

Similar results were obtained in the investigation of the oxidation of sulphur dioxide on an iron oxide catalyst deposited on a silver support [103]. The electron concentration of silver was modified by alloying it with palladium or mercury. The former alloying element decreased, and the latter raised, the electron concentration of the silver. The rate-controlling step of the oxidation of sulphur dioxide on iron(III) oxide is the acceptor-type chemisorption of oxygen. The activation energy of the oxidation on pure iron oxide is 31 kcal mole^{-1}. By applying a silver support this value drops to 13 kcal mole^{-1}. An increase in the electron concentration of the support brought about a further decrease in the activation energy, while a lower electron concentration resulted in a higher activation energy. The kinetic data are given in Table III.18.

TABLE III.18

Oxidation of sulphur dioxide on Fe_2O_3/Ag catalyst [103]

Support	Activation energy kcal $mole^{-1}$	Temperature range °C	Electron concentration of support
—	31	480–600	—
Ag + Pd	24	250–340	0.8
Ag	13	200–360	1.0
Ag + Hg	7	100–300	1.05

The effect of electron transition between metal and oxide is also manifested in the catalytic oxidation of methanol [104]. Figure **III.43** shows the temperature-dependence of the partial pressure of water formed in the course of the oxidation on the various catalysts. The Ag/ZnO system is seen to have the highest catalytic activity. The higher activity of the mixed powders of Ag and ZnO compared to the activities of the components was explained by an electron transition from the metallic silver into the zinc oxide, whereby the chemisorption of oxygen is promoted.

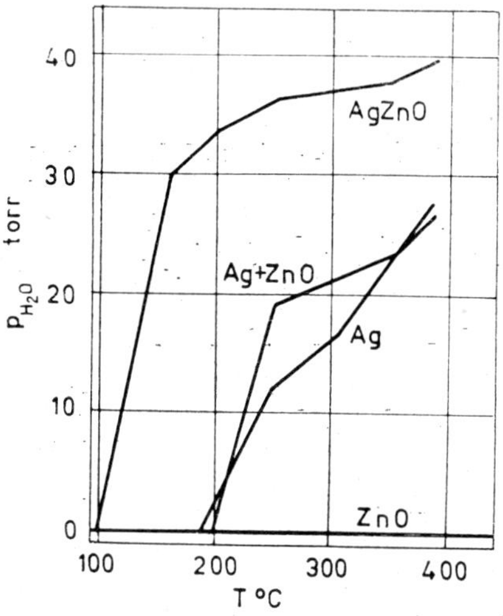

Fig. III.43. Oxidation of methanol on silver, on zinc oxide, and on a mixture of the two

Ag · ZnO: homogenized mixture of Ag and ZnO

Ag + ZnO: the two substances in separate boats

Steinbach's [105] investigations of the photocatalytic properties of various metal—oxide catalyst systems also confirms the importance of electron migration at the interface of the two solids.

PARAGRAPH 5.1.8. MISCELLANEOUS OBSERVATIONS

As well as observations resulting from catalytic investigations, certain other observations have been made which draw attention to the importance of the electrical properties of the support. Comparison of the infrared spectra of carbon monoxide chemisorbed on Pt, Pt/SiO_2 and Pt/Al_2O_3 shows that for the platinum catalyst deposited on alumina the band due to the linearly situated carbon monoxide is shifted towards lower frequencies and the quantity of bridged carbon monoxide is higher [106]. It was suggested that the n-type conductor alumina influences the electron configuration of the platinum with which it is in contact, and thereby promotes the chemisorption of carbon monoxide.

Almost all of the catalytic reactions described above involved the activation of hydrogen, and it was repeatedly assumed in the interpretation of the results that the filling of the empty d-bands of the catalyst with the electrons of the support inhibits the chemisorption of hydrogen. This has recently been proved experimentally by Lewis who found a decrease in the chemisorption of hydrogen on a nickel catalyst supported by alumina [107]. Independently of this observation, it is worth noting that the X-ray absorption edges of pure and alumina-supported nickel were found to be the same. This naturally does not mean the absence of an electronic interaction between the support and the catalyst, but merely indicates that, if there is such an interaction, it is not extensive enough to appear in the X-ray absorption spectrum.

Electron transition between metal and oxide is further supported by the latest experiments of Schwab and Kritikos [108]. Electrical conductivity measurements showed that metallic silver added to semiconductor oxides has a marked effect on the conductivity of the oxide. The conductivity of n-type zinc oxide is higher in the presence of silver, the increase being considerably greater than might be expected from the conductivity of silver. On the addition of silver the conductivity of p-type nickel oxide decreases in spite of the higher conductivity of silver. This behaviour can be understood only if electrons pass from the metal into the semiconductor, that is, if there is an electronic interaction between the metal and the semiconductor oxide. The existence of an electronic interaction is confirmed by the change in the conductivities of silver and chromia mixtures. The conduction type of chromia can be modified by changes in the pretreatment and in the gas atmosphere. In accordance with the above the conductivity of n-type chromia increased in the presence of silver, whereas the conductivity of p-type chromia decreased (Fig. III.44).

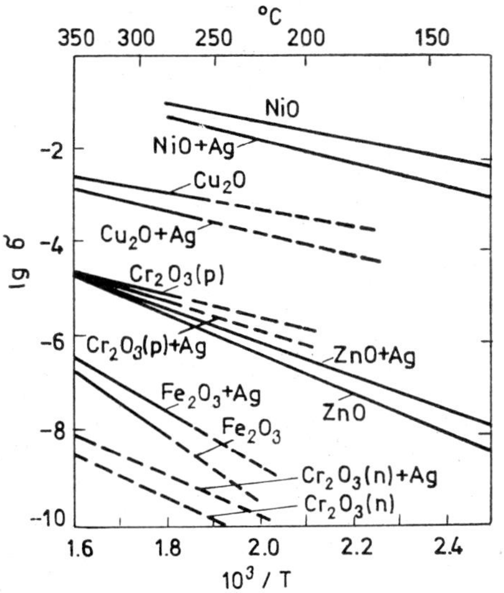

Fig. III.44. Conductivities of various semiconductor oxides in the presence of 10% of silver

Despite the apparent proof of an electronic interaction between an oxide support and a metal catalyst furnished by catalytic investigations, which might open new ways in the study and interpretation of the effects of supports, there are still many doubts and objections concerning electronic interaction. The greatest problem is the extent of the interaction, or, in other words, the question of whether it is sufficiently pronounced to be manifested in the catalytic behaviour of the metal. According to semiconductor theory, the interaction extends only to crystals in contact with one another, and thus it is easy to understand that the influence of the electron concentration of the support is displayed in the activity of the catalyst only if the metal is deposited in low concentration on the support, so that it forms the thinnest possible layer on the surface of the latter. In some of the works cited above, this limiting factor has, in fact, been taken into consideration and the quantity of the catalyst was kept at the lowest possible level. It is difficult to imagine the assertion of an interaction between metal and oxide, however, when the concentration of the catalyst reaches 20–50% of the weight of the system. In these cases, metal crystals, whose electron configurations are influenced by the semiconductor oxide, comprise a substantially lower fraction of the metal on which the catalytic process takes place. The catalyst content of industrially employed systems lie roughly between these limits, and substances of similar compositions were also used in the experiments of Langenbeck et al. described above. However, the fact that the effect of the electronic properties of the support was still observed either indicates a more marked interaction than could be expected from the theory of metal—semiconductor catalyst systems, or else there is some other way by which the electron concentration of the support is capable of asserting itself. By the extension of the theory of active sites it is conceivable that, although the support activates only a fraction of the metal catalyst, the substrate molecules migrating on the surface are capable of finding these active sites, the reaction primarily taking place on these. Hence, in accordance with the degree and extent of the electron interaction, an extremely wide scale of active sites would develop, the distribution of which would depend on the concentration of the metal, the nature and extent of the contact between metal and oxide, the surface of the support, the electrical conductivity, etc. It can also be assumed that the effect of the support alters not the electron concentration of the metal, but the surface potential of the metal crystal. However, there is no evidence confirming this.

It has also been suggested that as a result of the interaction between the metal and the support oxide the initially inactive support is activated, and the support takes over certain of the functions of the catalyst. In other words, molecules adsorbed on the metal would migrate from the metal to the support where they would react according to the chemical properties of the support. In our opinion, though some activation of the support cannot be completely excluded, particularly since the electron interaction between metal and oxide involves both substances, the problem is here more or less the same as in the previous case, i.e. the extent of the interaction between catalyst and support. It seems the most probable solution to assume the boundary lines between the metal and the oxide as the sites where the catalytic reaction actually takes place, since, due to electron interaction, the activity on these lines is far higher than anywhere else in the catalyst.

CHAPTER 6. INSULATOR CATALYSTS

Similar irregularities of the electron configuration occur in catalytically active insulators as in semiconductors. However, in the latter, for example by thermal excitation, electron—defect electron pairs may be brought about (resulting in intrinsic conductivity), because of the very high excitation energies this is a practical impossibility in insulators. Electron defect sites due to interstitial incorporations and lattice defects cannot develop in the crystal lattice itself, that is, electron defects are not a result of non-stoichiometric composition which, essentially, is a consequence of the variable valency of the metal ions being generally constituents of the semiconductor lattice. In insulator catalysts, defect electron sites are caused by the incorporation of ions, mainly cations, whose valency differs from that of the dominant ions in the lattice. This type of incorporation is similar to the doping of semiconductors. While electrons in the conduction band or defects (holes) in the valence band move more or less freely in an electric field in elemental semiconductors, in oxide and sulphide semiconductors this may require a certain activation, and in insulators it follows as a matter of course that their displacement is greatly hindered and that one cannot speak about a collective electron system for this type of solid. Hence, the defect and surplus electrons are localized, their free movement being limited to the immediate environment of a single lattice atom or lattice ion.

SECTION 6.1. STRUCTURES OF INSULATOR CATALYSTS

In Part I only solids possessing collective electron systems were discussed, and nothing was said about the structure of isolated formations which, as catalytically active ensembles, are primarily typical of insulator catalysts. Starting from the analogy with doped semiconductors a picture of the active sites can be obtained. Although, in principle, the type of doping is irrelevant, because of its paramount importance we shall discuss only the case when a cation of lower valency is incorporated into an unperturbed lattice, the substitution of Si^{4+} by Al^{3+} in the SiO_2 lattice. The results of the very comprehensive experimental work on silica-alumina catalysts and the conclusions drawn can be reasonably applied to other systems too, such as silica-magnesia and boria-titania, while in boria-alumina catalysts it is the different coordination number of aluminium which is the source of the positive charge deficiency [109].

In the SiO_2 lattice Al^{3+} will have a tetrahedral coordination just as Si^{4+}, so that in the neighbourhood of the aluminium an electron deficiency will arise (Fig. III.45). However, a very high activation energy would be needed for the displacement of this hole in the conduction band.

A localized hole, as a "free valency", is suitable for the binding of atoms with electron affinities lower than that of the defect site. In the ionized state all metals can be bound on these sites (see right hand side of Fig III.45). In principle, this would lead to formations with compensated

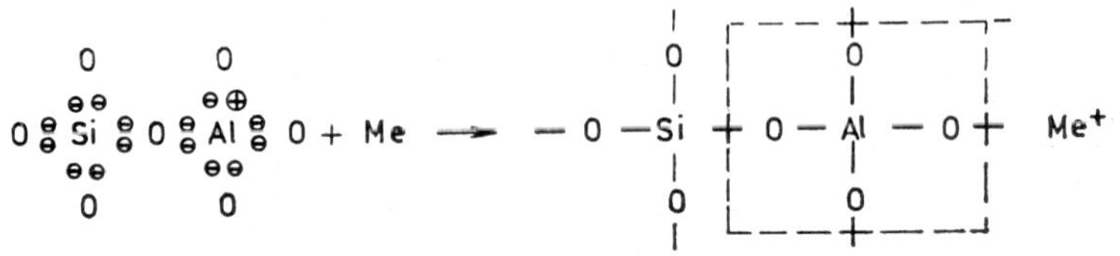

Fig. III.45

electron systems, the ground form of insulator catalysts. This route is only a theoretical one, however, because, to the best of our knowledge, the formation with electron deficiency used as the starting substance cannot be prepared. In the preparation of catalysts this formation is obtained directly, since the presence of foreign components is unavoidable during the process of preparation and these will be incorporated into the catalyst in their ionized form, so that no ensemble with a deficient electron system can be formed.

To illustrate the above, Fig. III.46 gives the scheme of the formation of the silicate lattice: the tetrahedral three-dimensional lattice is formed from

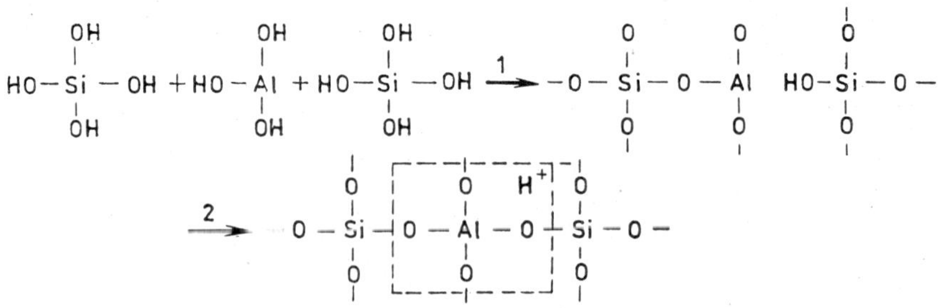

Fig. III.46

the hydroxides by the loss of water. However, the hydrogen in hydroxides can generally be replaced by alkali metals, introduced in one of the initial compounds, e.g. sodium silicate.

The metal ions are bound electrostatically to the

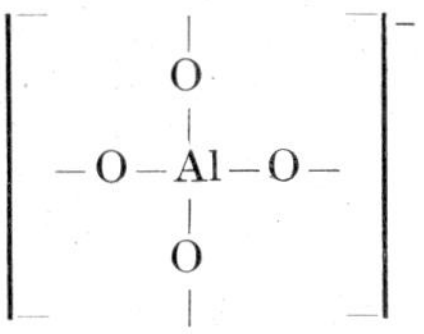

tetrahedron. On binding, multivalent cations neutralize the negative charge of several aluminium-centred tetrahedra.

The possibility exists of substituting various cations by each other by means of simple ion-exchange. When hydrolyzed the salt-type ion-exchangers will bind different amounts of H^+ in accordance with the equilibrium of the conversion:

$$xZ^-Me^{x+} + xH_2O \rightleftharpoons xZ^-xH^+ + Me^{x+} + xOH^-$$

Cations can be more perfectly exchanged for H^+ by means of acid treatment. Since this may occasionally be accompanied by the destruction of the lattice, NH_4^+ ion-exchange is more favourable, followed by the thermal removal of NH_3, leaving behind hydrogen ion centres [110–115]. The H-form obtained in this way contains very loosely bound, easily dissociable protons. The substance is therefore acidic. Its catalytic properties are fundamentally determined by this acidic nature.

Hence, ion-exchange solids which can be prepared in their protonic form may be ordered in the group of insulator catalysts. This includes the acidic ion-exchange resins, and primarily those containing strongly acidic, e.g. $-SO_3H$, groups [116]. In these, however, the acidity is no longer the consequence of crystal structure, but can be attributed to the functional group of the monomer molecular unit. From the point of view of solid-state structure these can no longer be derived from the semiconductors and, in many respects, can be considered as being related to the insulator catalysts only because of the similarity of their active formations. At this point there is an obvious link with the liquid-phase acid catalysts.

PARAGRAPH 6.1.2. CATALYTICALLY ACTIVE FORMATIONS

Insulator catalysts thus give rise to proton-donor, acid formations. The proton is also bound electrostatically to the silica-alumina lattice. Since the lattice is made up of SiO_4^{4-} and AlO_4^{5-} tetrahedra linked to each other by their O^{2-} apices, the acid formation can be illustrated as follows:

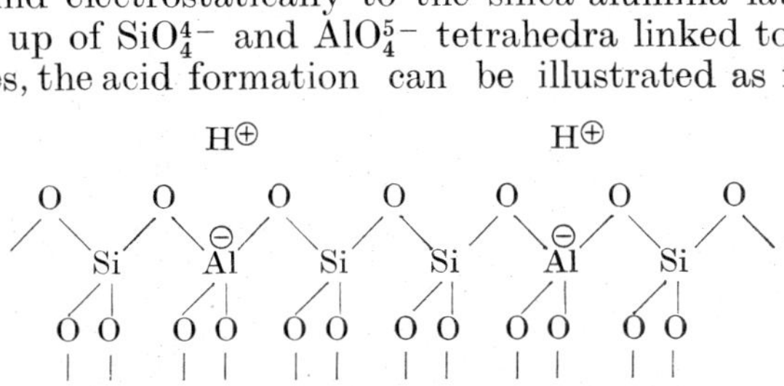

(*A*)

When the temperature is increased, a bond rearrangement takes place in the Y-type molecular sieve (see later in Paragraph 6.1.2) at temperatures between 200 and 500 °C, with the formation of the "acidic structural" SiOH group; this has been identified by Ward [117]:

(A) 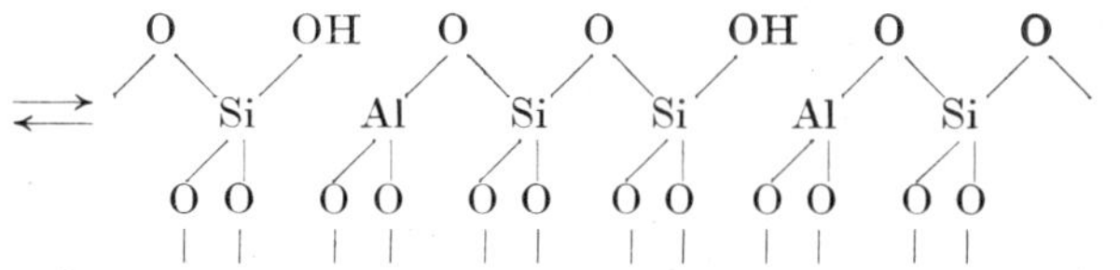 (B)

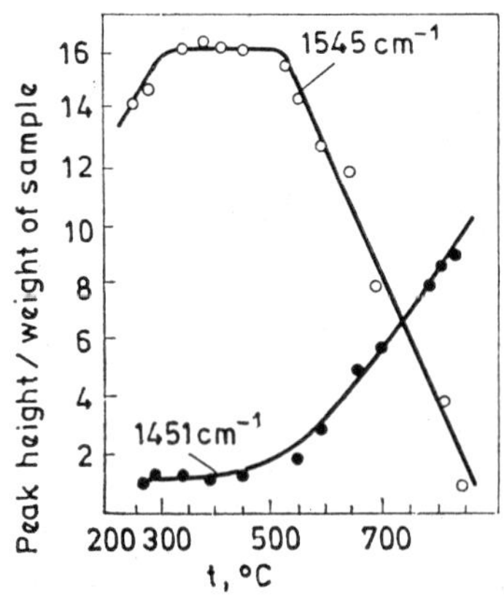

Fig. III.47. Changes in Brönsted (○) and Lewis (●) acidities on HY-type molecular sieves vs. the temperature (according to Ward [117]). The former is obtained from the height of the absorption peak at 1545 cm^{-1}, which characterizes the adsorption of the pyridinium ion, and the latter from the height of the absorption peak at 1451 cm^{-1}, characteristic of coordinatively bound pyridine

(The crystal structure does not collapse as a result of a bond opening of this type, since the ensembles are still fixed to the lattice by other bonds.) The Si—OH group in (*B*) is not identical with the "amorphous structural" terminal Si—OH group of the lattice [117]. The wavenumber of the IR absorption band of the former —OH group, e.g. in the Y-type molecular sieve, is 3540 and 3640 cm^{-1}, depending on the position of the OH group, while that of the latter is greater, 3740 cm^{-1}, in agreement with a stronger O—H bond. The "structural" terminal Al—OH group of the lattice appears in a similar manner at 3688 cm^{-1} [117].

In the case of oxides, at lower temperatures the charge of the surface oxide ions is similarly compensated by protons, and thus their hydroxyl coverage is extremely high after their precipitation, so that occasionally, depending mainly on the lattice cations, they may possess a considerable acidity (see later). Consequently, catalytic conversions with the same mechanisms as on insulators may take place on semiconductors. (At higher temperatures the surface becomes dehydroxylated by the loss of water, and a structure with /O\ bridges is created.)

At temperatures above about 550 °C formation (*B*) is transformed by loss of water to give an ensemble of one AlO_4^{5-}, one SiO_3^{2-} and one AlO_3^{3-} [112], [118]:

(B) 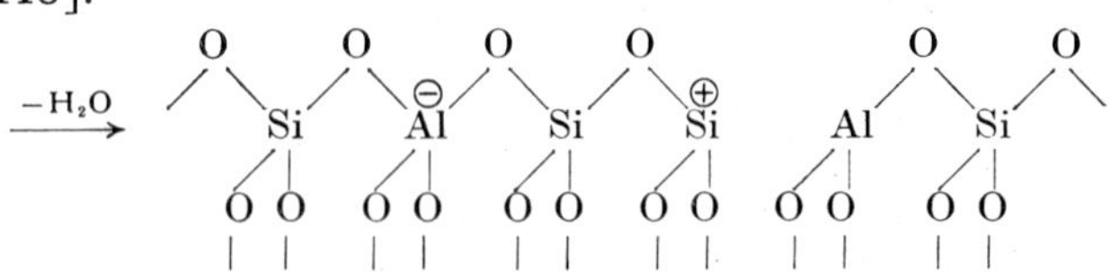 (C)

Dehydration is no longer completely reversible because of lattice deformations. (*B*), the Brönsted acid form is converted by the loss of water into

(C), the Lewis acid form, as shown by the antibate change of their concentrations in Fig. III.47 [117]. The two acid formations can easily be distinguished spectroscopically, because pyridine and ammonia are adsorbed as pyridinium and ammonium ions on the Brönsted acid centre, producing characteristic absorption bands, while with the Lewis centre of high electron affinity each base forms a dative bond, which is characterized by another absorption band.

The formation of both acid centres is attached to the aluminium in the lattice. Their acidity depends on the environment, and, thus, on the distance between the aluminium ions: the more closely they are placed, that is, the higher the aluminium concentration, the weaker the acidity of the formations [119].

If the protons in (A) are substituted by alkali metals, their catalytic activity practically ceases, whereas upon the introduction of bivalent alkaline earth metal, or trivalent earth metal, cations active catalysts are obtained. Exactly what sort of structure should be attributed to the active formations on the basis of the observations has been the subject of long discussion, and has still not been fully resolved.

We shall next consider the most probable structural forms after exchange with Mg^{2+} ions. The hydrated cation is bound asymmetrically to two tetrahedra with aluminium centres [120]

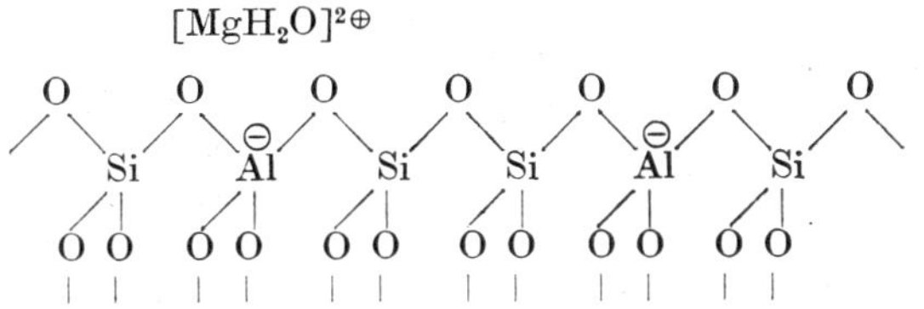

(D)

At temperatures between 300 and 500 °C, depending on the structure of the crystal and on the cation, the water, bound to the cation in the lattice, dissociates to a proton and a hydroxyl ion under the influence of the electrostatic field [121]. The proton attacks the oxide ion of the lattice and the group identified in (B) as an "acidic structural" Si—OH group is formed, while the hydroxyl is bound to the metal ion and can be detected as "cationic MeOH" [115, 122].

(D) ⟶ 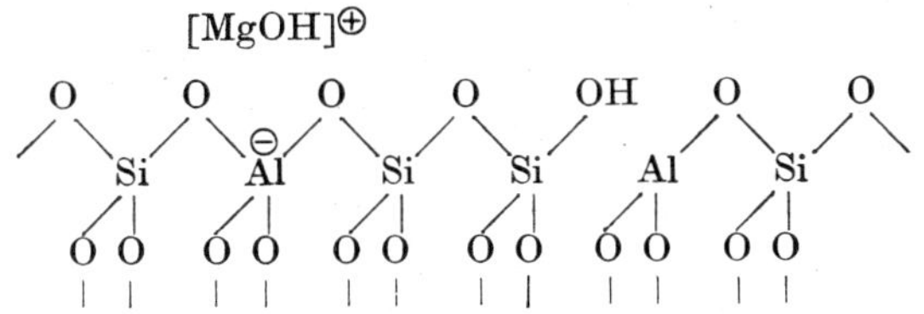 (E)

The characteristic wavenumber of the MeOH depends upon the cation (in the Y-type molecular sieve it is 3610 cm^{-1} for Mg and 3585 cm^{-1} for Ca) [115, 121]. The extent to which these acidic ensembles are formed and the acidity of the acid centres resulting from conversion into (E) through the action of cations, depend, in the same manner, on the cation, or, more accurately, on its electrostatic potential, ne/r (where ne is the charge and r the radius of the cation).

A process similar to the formation of (*E*) can be observed when the metal ions in the surface lattice plane of semiconductor oxides are hydrated:

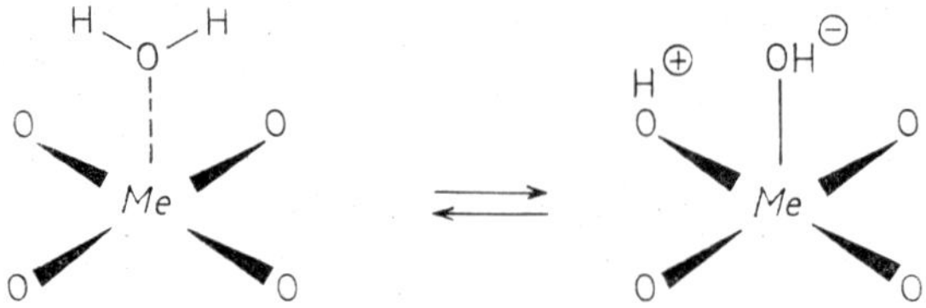

This is related to the fact that the aquo-complexes are more acidic than water, so that the acidic dissociation of water takes place more easily in the complexes, which contributes to the formation of (*E*) [123]. Hence, coordinatively unsaturated surface cations can be regarded as Lewis acid centres and in a complementary way, the oxide ions may be regarded as Lewis base centres.

At temperatures above 500 °C dehydration may lead to the same type of formations as in (*C*) [117, 124–126]:

(*E*) ⟶ 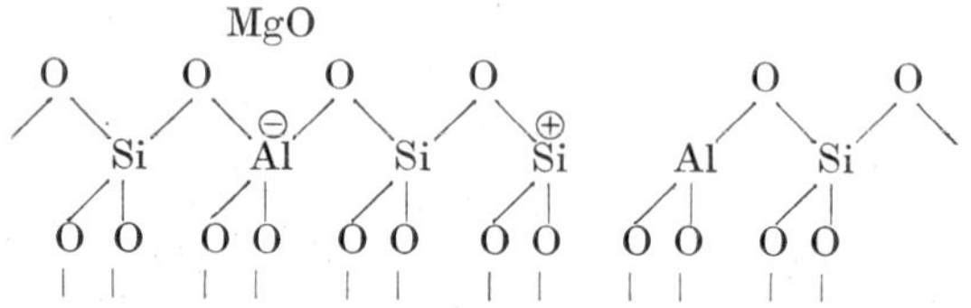 (*F*)

The conversions are the same when trivalent earth metal cations are introduced, and merely the formulae have to be changed accordingly; in this case three (*F*) formations are finally created in the neighbourhood of two cations. Since the electrostatic field of these cations is considerably stronger, the formation of type (*E*) is achieved far more easily and its catalytic activity is also considerably higher [120, 122, 125, 127, 128].

The structural identity of formations (*C*) and (*F*), both of high electron affinity, is supported by their cation-independent catalytic activity in the hydrodemethylation of toluene [129].

Occasionally, however, the cations act as electron acceptors, as, for example, in the adsorption of pyridine [130]. The behaviour of these cations resembles that of cations in semiconductors, as already shown above. It is not surprising, therefore, that cations introduced by ion-exchange, and primarily ions of transition metals, exert a specific modifying effect on the catalytic activity [131]; this effect cannot be correlated with their valencies or their electrostatic potentials, i.e. with properties which play a role in the development of formations (*E*) and (*F*). The isomerization of 1-butene into cis- and trans-2-butene is an ionic reaction on Mg, Ca and CeX molecular sieves, but proceeds by a partly radical mechanism on ZnX and as an almost entirely radical reaction on NiX, this is also supported by the increase of the reaction rate and in the change of the selectivity [132].

The earlier extensively used amorphous silica-alumina catalysts contain irregularly arranged SiO_4^{4-} and AlO_4^{5-} tetrahedra. The extreme variety of packing results in a wide range of active formations leading to far less favourable results of specific activity and selectivity than would be expected for crystalline systems.

The preponderant majority of crystalline silica-alumina catalysts are zeolites in which the tetrahedra, whose apices are in contact, form a regular channel system which the substrate is capable of penetrating. In these catalysts the ratio (Si + Al) : O is always 1 : 2, but the ratio Al : Si may vary between 1 : 1 and 1 : 6, and the equivalent quantity of cations is the same as the number of aluminium atoms. At lower temperatures water may be present, adsorbed in the pores and in the hydrate shells of the cations. In such a lattice the cations can be moved by an electrostatic field, so that these typically insulator catalysts may have a secondary conductivity [133].

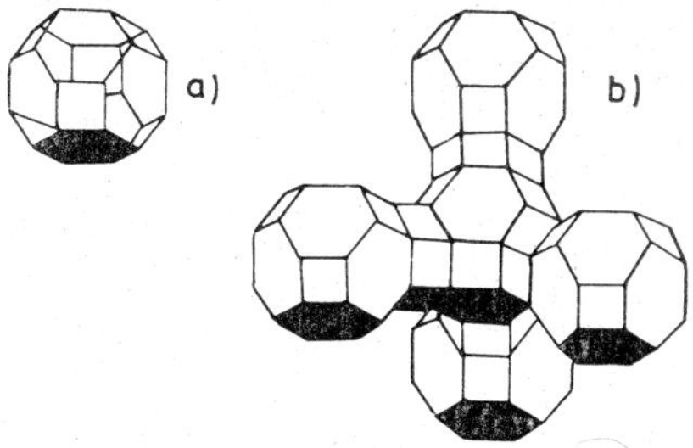

Fig. III.48. [135]
(a) sodalite unit
(b) packing of sodalite units in secondary tetrahedron structure

Many structures are possible for these crystal lattices built up from tetrahedra, and most of them are known [134]. The structure of those most used, the sodalite-zeolites, will be described.

As the primary structural elements, the SiO_4^{4-} and AlO_4^{5-} tetrahedra are arranged in the form of a truncated octahedron as shown schematically in Fig. III.48a. The central Si^{4+} and Al^{3+} of the tetrahedra are at the apices of the octahedron, while the O^{2-}'s are in the middle of the edges, 3 O^{2-}'s of each tetrahedron being in this secondary structural element, while the fourth O^{2-} is situated out from the apices, forming the linkage between the truncated octahedra. The arrangements of how these sodalite units are linked to each other afford the different variants of sodalite type zeolites. In faujasite, and the X and Y-type molecular sieves, the sodalite units are arranged in a tetrahedron (Fig. III.48b), so that the apices of tetrahedrally positioned hexagonal sides are bound through O^{2-}'s forming hexagonal prisms. Four other sodalite units are linked in this way to the four oppositely situated hexagonal faces of the sodalite units, thereby leading to the secondary tetrahedral structure. In this way, the sodalite units enclose a regular pore system, the "supercage". The elementary cell of the crystal consists of 8 sodalite units, each unit comprising 24 tetrahedra. Thus, the elementary cell is made up from (192-x)SiO_2 and $x AlO_2^-$ with xMe^+ to compensate the negative charges. The composition of a NaY molecular sieve with respect to its elementary cell is: $Na_{56}(AlO_2)_{56}(SiO_2)_{136}264H_2O$ [135]. The sites of the components within the crystal structure can be accurately defined. It is sufficient to consider a single sodalite unit with the attached hexagonal prisms (Fig. III.49).

Si^{4+} and Al^{3+} are situated at the apices of the polyhedron. The oxygen ions are at the middle of the edges (their different positions are designated by Arabic numerals: O1, O2, . . .). The various cation sites are given by Roman figures: I is in the centre of the hexagonal prism, II lies opposite the centre of the hexagonal face of the sodalite unit in the supercage. In the "sodalite cage" within the sodalite unit, proceeding towards its centre from

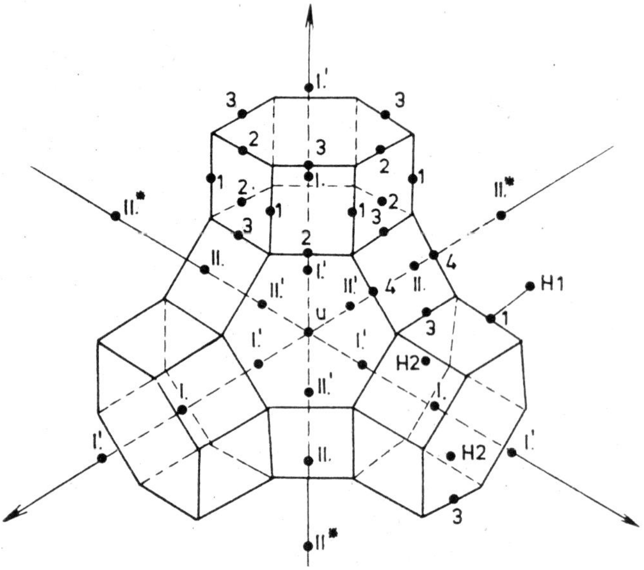

Fig. III.49. Sodalite units with hexagonal prisms (according to Smith [cited in 143]) I, I′, II, II′, II*: cation positions; 1, 2, 3, 4: oxide ion positions (in the text as O1, O2, O3, O4);H1, H2: hydrogen positions in the —OH group; U: centre of the sodalite unit

position I, I′ is reached; and from position II, position II′; in a radial direction from the latter position II* is situated deep in the supercage.

The coordination states of the metal ions are different at the various sites, so that they are not statistically arranged. Position I is generally the most stable, while position II is less stable, and the distribution between the other positions always depends on the number and valency of the cations [136] and on their hydration [137].

The hydrogen of the "acidic structural" hydroxyl group in position H1 is linked to O1 towards the "supercage", while H2 is linked to O3 and situated inside the hexagonal prism [138]. The former are more easily accessible to larger molecules; their bond vibrations give rise to an absorption band at 3650 cm^{-1}, while the absorption band for the latter is at 3550 cm^{-1}, indicating, in agreement with experience, a weaker hydrogen bond and a higher acidity [121, 139]. At temperatures above 300 °C, however, the proton is capable of migrating from one of its positions to the other [131, 138].

The catalytic activity is influenced by the arrangement of the protons, as well as by the positions of the various cations. It should be remembered, however, that these settlements may change as a result of migration either during pretreatment, or in the course of the reaction.

SECTION 6.2. CONVERSIONS ON ACIDIC INSULATOR CATALYSTS

Since the acidic centres determine the catalytic properties of insulators, in the majority of practical cases it seems advisable to consider the methods by which their acidic character can be determined. These will, at the same time, provide information concerning the adsorption process taking place on the active sites.

PARAGRAPH 6.2.1. BASE ADSORPTION ON ACID CATALYSTS

There are, essentially, two methods of investigation, namely, liquid-phase and gas-phase adsorption measurements; the former, in effect, comprise various titration methods.

Since acid has to be determined, simple alkaline titration in an aqueous medium appeared the most obvious procedure. However, under these conditions the catalyst undergoes a change: as a result of hydrolysis its structure damages and the dissolution of Al^{3+} can be observed. Ion-exchange between the catalyst protons and the alkali metal ions in neutral aqueous solutions of alkali halides or acetates, appeared to be more advantageous, when the acidity of the catalyst can be calculated from the pH of the solution. During the time-consuming process of ion-exchange, however, the catalyst network reacts with water and new acidic formations are created [140]; in addition, ion-exchange is a process always leading to equilibrium, the final value of which, i.e. the equilibrium partition, is unknown and is, in fact, the parameter which we are trying to determine.

All these disadvantages can be avoided by working, not in an aqueous medium, but in a neutral non-polar solvent. Titration in benzene was found to be the most adequate, using n-butylamine as the titrating agent in the presence of an adsorption indicator [141]. The indicators are bases adsorbed on the acidic surface centres of the catalyst where they are converted into their conjugated acidic form. It must be possible to distinguish the two forms by simple visual observation or spectroscopically.

The acidity at which the conjugated forms of the indicator are present in the same concentrations is characterized by the H_0 value, after a suggestion of Hammett [142]. H_0 is the negative logarithm of the acidic dissociation of the indicator:

$$H_0 \equiv \mathrm{p}K_\mathrm{a}$$

where K_a is the equilibrium constant of the conversion:

$$B\mathrm{H}^+ \rightleftarrows B + \mathrm{H}^+$$

that is

$$K_\mathrm{a} = [\mathrm{H}^+][B]/[B\mathrm{H}^+] \cdot (\gamma_{\mathrm{H}^+}\gamma_\mathrm{B})/\gamma_{\mathrm{BH}^+} \tag{III.97}$$

(γ is the activity coefficient).

At the equivalence point ($[\mathrm{B}]/[B\mathrm{H}^+] = 1$)

$$H_0 = \mathrm{p}K_\mathrm{a} = -\lg a_{\mathrm{H}^+} + \lg(\gamma_{B\mathrm{H}^+}/\gamma_B) \tag{III.98}$$

Since it is found that the ratio γ_{BH^+}/γ_B is the same for all Hammett indicators (see e.g. [143]), the H_0-scale of the indicators is consistent, that is to say, it indicates a value which is always proportional to the proton activity. Hence, the value of H_0 is independent of the medium, and is, therefore, the same on the liquid-solid interface as, for example, in a dilute aqueous solution (where $\gamma_{H^+} = \gamma_{BH^+} = \gamma_B = 1$) when H_0 will be equal to the pH of the solution.

The scale of the Hammett indicators extends from $H_0 = 2.77$ (p-aminoazonbenzene) to $H_0 = -11.38$ (nitrobenzene) [143] or more; values of -0.3, -3.0 and -8.27 correspond to the proton activities (acidities) of 10%, 48% and 90% sulphuric acid solutions, respectively.

When catalysts are titrated stepwise with n-butylamine, using indicators with increasingly higher H_0 values, the distribution of the number of acid centres according to the H_0 scale is obtained [141]. The number of all acid centres is generally about 10^{14} cm^{-2}, while that of the very strong ones is two orders lower.

In the presence of Hammett indicators the Brönsted and Lewis acid centres can be determined simultaneously by titration.

There is a possibility of the specific detection of the Brönsted centres by performing the titration in the presence of triphenylmethanol derivatives [144]. When these compounds react with protons, coloured carbonium ions are obtained:

$$R - OH + H^+ \rightleftarrows R^+ + H_2O$$

The equilibrium constant of the reverse conversion, K_R, defines acidity, i.e. proton activity, in the same way as K_a:

$$H_R \equiv pK_R = -\lg a_{H^+} + \lg(\gamma_{R^+}/\gamma_{ROH}) + \lg a_{H_2O} \qquad \text{(III.99}$$

The same acidity appears with a different numerical value on the H_R-scale as on the H_0-scale, because of the term $\lg a_{H_2O}$ in Eq. (III.99). In the majority of cases, H_R, characteristic of the Brönsted acidity, correlates better with the catalytic activity than H_0, which is characteristic of the overall acidity.

It is also possible to detect the Lewis-type acid formations directly from the adsorption of the leuco-bases of various dyes of the triphenylmethane type; from their colour, conclusions can be drawn regarding the concentration and strength of the acidic centres [145].

Beside the liquid-phase adsorption of various organic bases it is possible to draw conclusions regarding the strength and type of the acid formations from the changes in certain $-$OH bond frequencies, and from the appearance of frequencies indicating the formation of new bonds [146].

The conclusions which can be drawn from measurements in the liquid phase, however, are limited considerably by the experimental conditions, which differ from those of the catalytic reaction; thus, first of all, the temperature is much lower than in the catalytic pretreatment or during the catalytic reaction, whereas it is commonly known that the temperature has a marked effect on catalysts. Measurements in the gaseous phase help to eliminate errors of this type.

The main demands which the adsorbate has to meet are an adequate thermal stability, and dimensions which will not hinder its penetration into the relatively small pores.

Adsorption measurements in the gaseous phase have already been encountered in the preceding paragraphs dealing with the structures of catalysts. On Brönsted centres ammonia is bound in the form of the ammonium ion, and pyridine in the form of the pyridinium ion. The dative bonds formed on the Lewis centres have a modifying effect on the initial bond frequencies [118, 147]. As the molecular size of the base increases, the acidic —OH in position H2 of the Y-type molecular sieve is no longer accessible, indicating its internal position (Fig. III.49) [117].

The acidic centres can be determined quantitatively by spectroscopic measurements or by traditional gravimetric-volumetric adsorption measurements. The joint application of these methods is the most practicable procedure, as the quantitative evaluation of spectroscopic data requires special calibration, while the latter do not allow the qualitative distinction of simultaneous adsorption processes. Certain conclusions concerning the number and strength of the acid centres can also be drawn from investigating desorption, for example, by gravimetric methods, as functions of temperature and time [148].

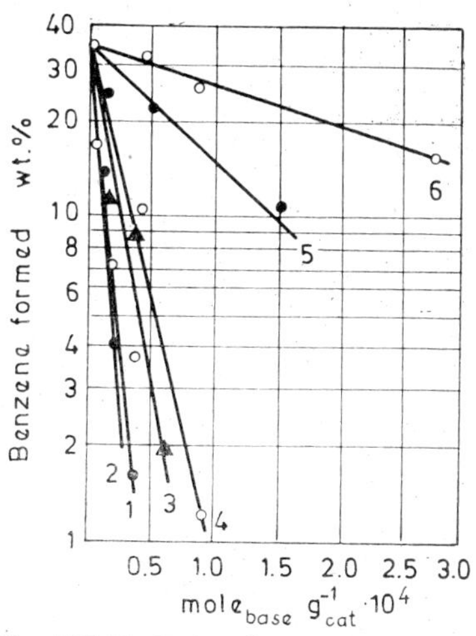

Fig. III.50. Poisoning of silica-alumina catalyst with various nitrogen bases. Reaction under investigation: dealkylation of cumene at identical temperatures and identical contact times [148].

1. quinoline
2. quinaldine
3. pyrrole
4. piperidine
5. decylamine
6. aniline

PARAGRAPH 6.2.2. CATALYTIC CONVERSIONS

Though catalytic effects of insulators can be observed in a number of reactions [149], mainly in relation to their acidic nature, for an easier survey of the problem and because of its importance, only C—C bond conversions will be discussed.

The role of acidic centres in these reactions has been turned out in the reduction of the activity in proportion to the strength of the bases adsorbed on the catalyst. This can be observed on amorphous silica-alumina catalysts, for example in the cracking of cumene (Fig. III.50 [148]) when competitive adsorption takes place between the nitrogen bases and the "basic" substrate.

Various interactions are possible between the substrate and the acid formations, depending on the type of the latter.

The hydrogen ion of Brönsted centres forms a bond with unsaturated systems, so that addition takes place on the $>C=C<$ group by σ-bonding or π-bonding. A σ-bond is formed when the empty s-shell of H^+ overlaps at the negatively polarized end of the bond with the filled sp^3-orbital, while a π-bond results from the combination of the empty s-shell of H^+ with the full π-orbital of the double bond:

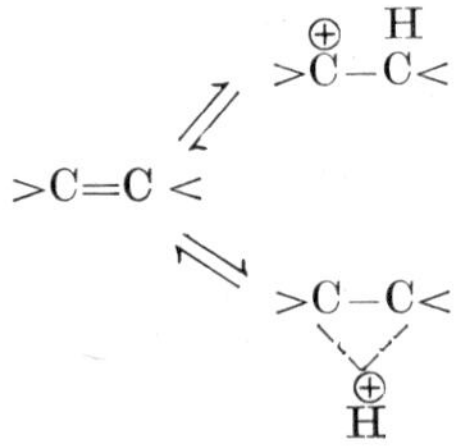

$$>C=C< \qquad \overset{\oplus}{>}\overset{H}{C}-C< \quad \sigma\text{-complex}$$

$$>\underset{\underset{\oplus}{H}}{\overset{}{C}}-C< \quad \pi\text{-complex}$$

The stability of the σ complex, indicated as the carbonium ion, depends upon the position of the charge-carrier carbon atom [150]:

$$CH_2=CH_2+H^\oplus \rightleftarrows CH_3-\overset{\oplus}{C}H_2 \qquad \text{(primary)} \quad -152 \text{ kcal mole}^{-1}$$

$$CH_2=CH-CH_3+H^\oplus \rightleftarrows CH_3-\overset{\oplus}{C}H-CH_3 \quad \text{(secondary)} \quad -175.5 \text{ kcal mole}^{-1}$$

$$\overset{\oplus}{C}H_2-CH_2-CH_3 \quad \text{(primary)} \quad -168.5 \text{ kcal mole}^{-1}$$

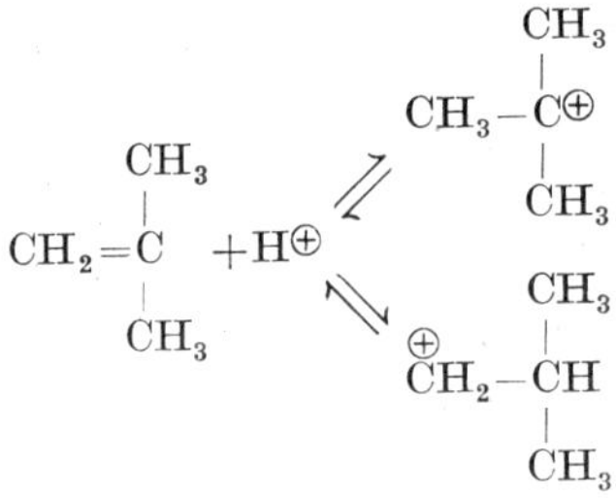

$$CH_2=\underset{CH_3}{\overset{CH_3}{C}} \;+H^\oplus$$

$$CH_3-\overset{CH_3}{\underset{CH_3}{\overset{|}{C}\oplus}} \qquad \text{(tertiary)} \quad -189 \text{ kcal mole}^{-1}$$

$$\overset{\oplus}{C}H_2-\underset{CH_3}{\overset{CH_3}{\overset{|}{C}H}} \qquad \text{(primary)} \quad -168 \text{ kcal mole}^{-1}$$

The carbonium ion is most stable when the charge is on a tertiary, less stable when on a secondary, and the least stable when on a primary carbon atom. Consequently, in desorption the proton most easily leaves a tertiary, less easily a secondary, and most reluctantly a primary carbon atom of the carbonium ion. Thus, in agreement with the Markownikoff rule, the proton is added to the C-atom of the $>C=C<$ group richer in hydrogen, and leaves the C-atom poorer in hydrogen. According to Whitmore's theory, catalytic conversions can be interpreted on the basis of the formation and decomposition of carbonium ions [151].

In accordance with the above, the rate of displacement of the double bond is found to be as follows [152]:

$$\underset{H}{\overset{H}{\overset{|}{C}}}=\underset{}{\overset{CH_3}{\overset{|}{C}}}-\underset{H}{\overset{H}{\overset{|}{C}}}-\underset{H}{\overset{H}{\overset{|}{C}}}-\underset{H}{\overset{H}{\overset{|}{C}}}-H + H^\oplus \rightleftarrows H-\underset{H}{\overset{H}{\overset{|}{C}}}-\underset{}{\overset{CH_3}{\overset{|}{C}\oplus}}-\underset{H}{\overset{H}{\overset{|}{C}}}-\underset{H}{\overset{H}{\overset{|}{C}}}-\underset{H}{\overset{H}{\overset{|}{C}}}-H$$

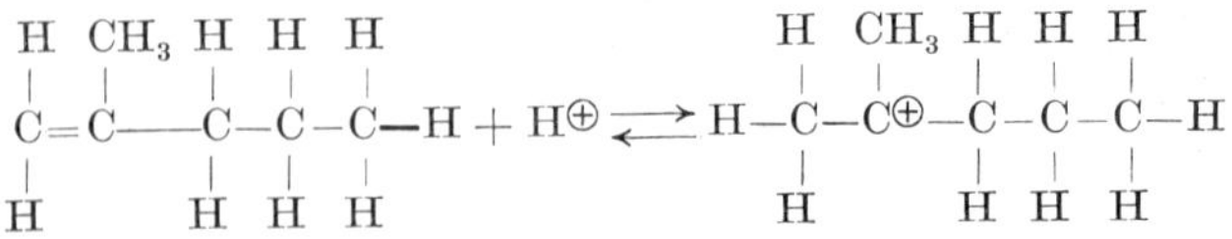

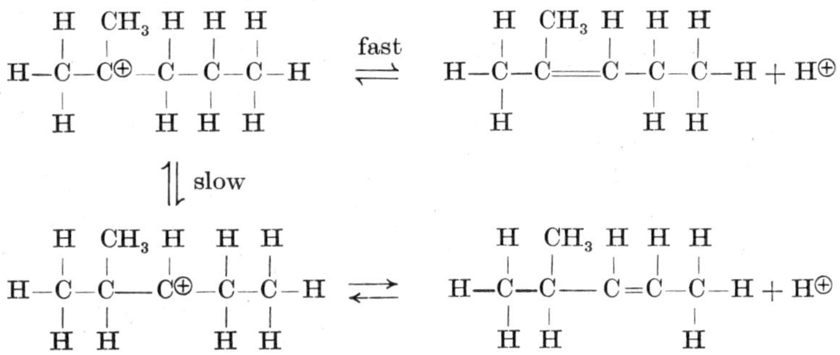
fast
slow

$$\begin{array}{c}
\text{H CH}_3\text{ H H H} \\
\overset{|}{\underset{|}{\text{H}}}-\overset{|}{\underset{|}{\text{C}}}-\overset{|}{\underset{|}{\text{C}}}-\overset{}{\text{C}^{\oplus}}-\overset{|}{\underset{|}{\text{C}}}-\overset{|}{\underset{|}{\text{C}}}-\text{H} \\
\text{H H H H}
\end{array}
\rightleftharpoons
\begin{array}{c}
\text{H CH}_3\text{ H H H} \\
\overset{|}{\underset{|}{\text{H}}}-\overset{|}{\underset{|}{\text{C}}}-\overset{|}{\underset{|}{\text{C}}}-\text{C}=\overset{|}{\underset{}{\text{C}}}-\overset{|}{\underset{|}{\text{C}}}-\text{H}+\text{H}^{\oplus} \\
\text{H H H}
\end{array}$$

2-Methyl-1-pentene is rapidly converted into 2-methyl-2-pentene, but its conversion into 2-methyl-3-pentene is a slow process, since this step requires the conversion of the more stable tertiary carbonium ion into the less stable secondary carbonium ion on the surface. This conversion means formally that the ⊕-charge and the hydride ion change place within the chain:

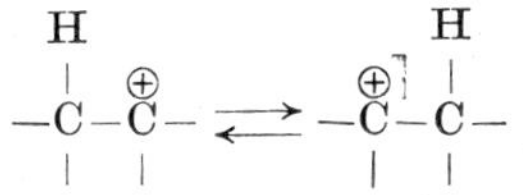

and the probability of this process is higher when it leads to a more stable carbonium ion. For the same reason, the alkyl group might be displaced along the chain:

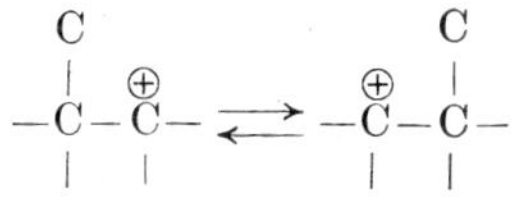

leading to skeletal isomerization. The relative rates of the following conversions can be interpreted similarly (using simplified formulae) [152]:

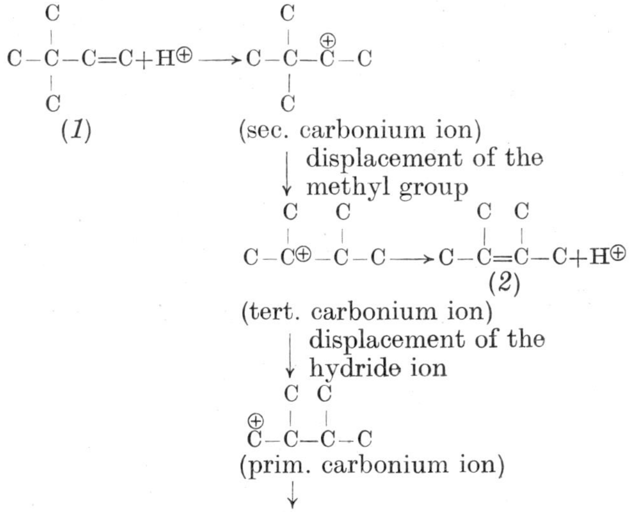
(1)
(sec. carbonium ion)
displacement of the methyl group
(2)
(tert. carbonium ion)
displacement of the hydride ion
(prim. carbonium ion)

displacement of the
methyl group

$$
\begin{array}{ccc}
\text{C} & \text{C} & \text{C} \\
| & \overset{\oplus}{|} & | \\
\text{C}-\text{C}-\text{C}-\text{C} & \longrightarrow & \text{C}-\text{C}-\text{C}=\text{C}-\text{C}+\text{H}^{\oplus}
\end{array}
$$

$$(3)$$

(sec. carbonium ion)

displacement of the
methyl group

$$
\begin{array}{cc}
\text{C} & \text{C} \\
| & | \\
\text{C}-\text{C}-\text{C}-\overset{\oplus}{\text{C}}-\text{C} \longrightarrow \text{C}-\text{C}-\text{C}=\text{C}-\text{C}+\text{H}^{\oplus}
\end{array}
$$

$$(4)$$

(sec. carbonium ion)

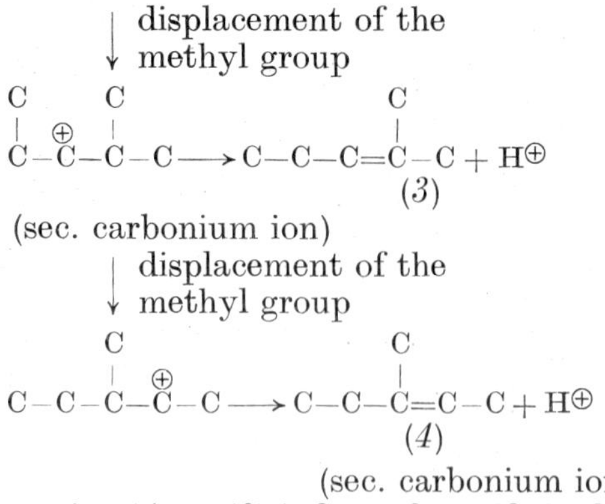

Consequently, conversion $(1) \to (2)$ is fast, $(2) \to (3)$ is slow and $(3) \to (4)$ relatively fast.

The isomerization of alkanes can be interpreted in a similar manner, assuming the formation of a small amount of olefin as the result of thermal, catalytic, or oxidative dehydrogenation and an intermolecular hydride ion transfer between the carbonium ion formed from the olefin and the gas-phase paraffin:

$$
-\overset{|}{\underset{|}{\text{C}}}{}^{\oplus} + \text{H}-\overset{|}{\underset{|}{\text{C}}}- \; \rightleftharpoons \; -\overset{|}{\underset{|}{\text{C}}}-\text{H} + {}^{\oplus}\overset{|}{\underset{|}{\text{C}}}-
$$

which ensures the continuity of the reactions. Isomerization conversion then takes place via the carbonium ions [153].

The polymerization of olefine can also be interpreted with the carbonium ion theory. One of the carbon atoms of the double bond in the attacking molecule forms a bond with the charge-carrier C-atom. In the case of the dimerization of i-butene the reaction might be as follows:

$$
\begin{array}{cc}
\text{C} & \text{C} \\
| & | \\
\text{C}-\overset{\oplus}{\text{C}} + \;\; \text{C}-\text{C} \\
| & | \\
\text{C} & \text{C}
\end{array}
$$

$$
\begin{array}{ccc}
\text{C} & & \text{C} \\
| & & | \\
\text{C}-\text{C}-\text{C}-\overset{\oplus}{\text{C}}-\text{C} & & \text{(tert. carbonium ion)} \\
| & & \\
\text{C} & &
\end{array}
$$

$$
\begin{array}{cc}
\text{C} & \text{C} \\
| & | \\
\text{C}-\text{C}-\text{C}-\text{C}^{\oplus} & \text{(prim. carbonium ion)} \\
| & | \\
\text{C} & \text{C}
\end{array}
$$

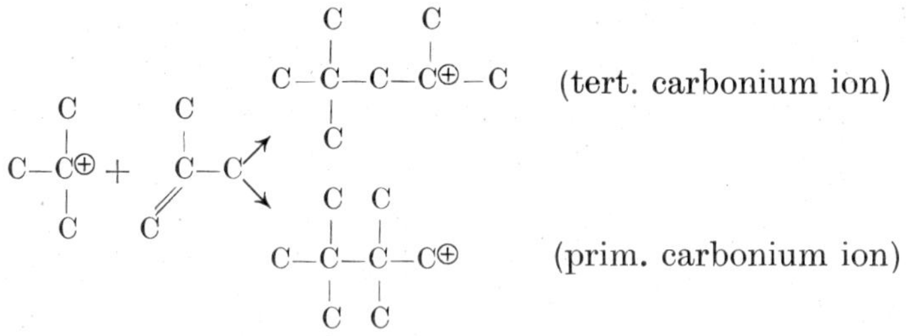

It is evident that experimentally only 2,2,4-trimethylpentene has been detected [154]. The length of the polymer chain is determined within the limits of thermodynamic equilibrium by the relative rates of chain growth (see above) and proton loss (desorption). Besides the primary products, various isomers may be formed as a result of bond rearrangements within the carbonium ions. In addition to the above described isomerization, self-alkylation (ring closure) may also occur. The faster these internal rearrangements of the carbonium ion compared to chain growth by addition, the more complicated the structure of the product [152].

In the alkylation of paraffins with olefins on acid catalysts it is assumed that the first step of the reaction is proton addition to the olefin [155]:

$$CH_2{=}CH_2 + H^{\oplus} \rightleftarrows CH_3{-}\overset{\oplus}{C}H_2$$

followed by transfer of the hydride ion:

$$CH_3{-}\overset{\oplus}{C}H_2 + H{-}\underset{\underset{C}{|}}{\overset{\overset{C}{|}}{C}}{-}C \longrightarrow CH_3{-}CH_3 + {\oplus}\underset{\underset{C}{|}}{\overset{\overset{C}{|}}{C}}{-}C$$

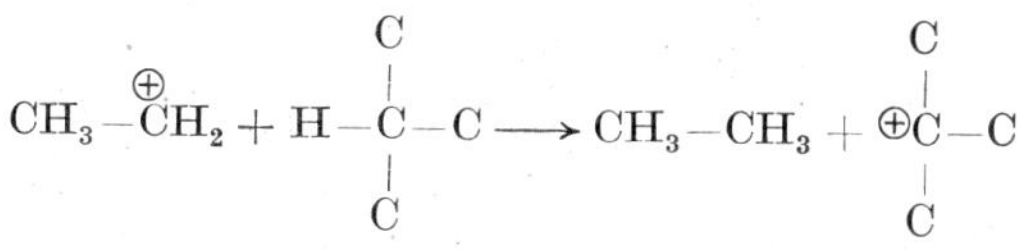

and the tertiary carbonium ion reacts with the olefin

$$\underset{\underset{C}{|}}{\overset{}{C}}{-}\overset{}{C}{\oplus} + CH_2{=}CH_2 \longrightarrow \underset{\underset{C}{|}}{\overset{}{C}}{-}\overset{}{C}{-}C{-}C{\oplus} \longrightarrow \underset{\underset{C}{|}}{\overset{}{C}}{-}\overset{}{C}{-}\overset{\oplus}{C}{-}C$$

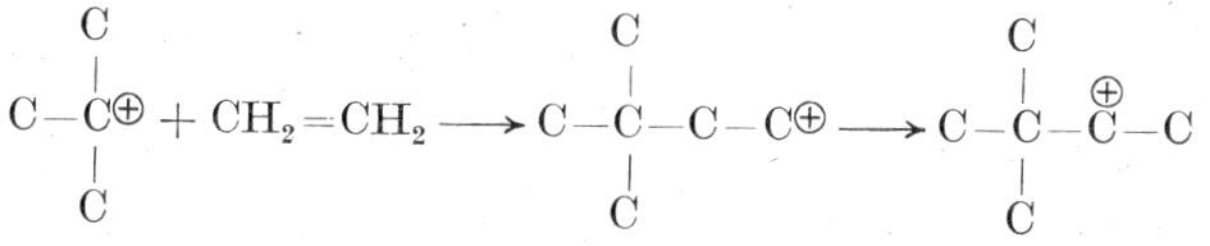

The propagation of the reaction takes place with the cycle being repeated, another paraffin molecule transmitting a hydride ion to the alkylate-carbonium ion with the formation of a tertiary-butyl carbonium ion and an alkylate molecule.

It should be added, however, that it is difficult to conceive the mechanism of alkylation as described above: the second step, involving the transfer of the hydride ion, would require that the two partners change place on the active centre of the catalyst, while, in the third step, the insertion would require a displacement of the positive charge, which is not a very probable event (see p. 426–8), and the "repetition" of the cycle after the third step would again demand that two partners, now greater in volume, change place.

Cracking of paraffins is the reverse process of alkylation. Proton addition to the olefine present in small quantities is again assumed followed by hydride ion transfer. Bond rearrangement may take place within the carbonium ion formed from the paraffin molecule, accompanied by the splitting of the C—C bond in the β-position to the charge-carrier carbon atom. Though it has thus proved possible to find a satisfactory explanation for the distribution of the products [156], the initial step of the reaction, the protonization of an olefin is rather improbable, as more accurate studies have shown that olefins have no influence on the initiation of cracking of paraffins [157]. A direct interaction between surface protons and paraffin hydrocarbons must probably be taken into consideration.

The conversion of aromatic hydrocarbons is also initiated by proton addition. The proton is bound to different carbon atoms of the aromatic ring. When mesitylene is protonated, the resonance forms of the 1,3,5-trimethylbenzonium ion are obtained [158]:

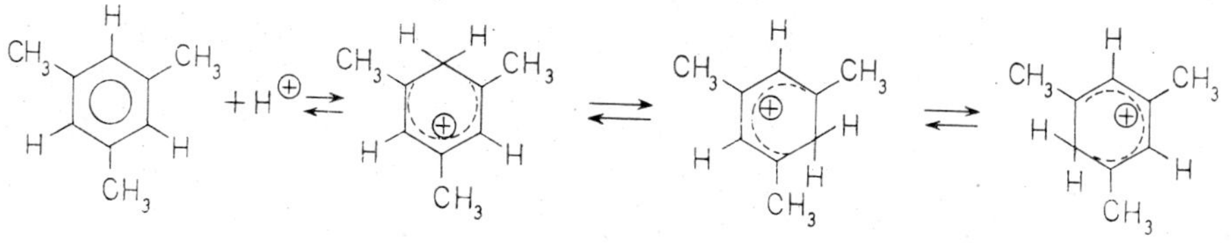

The various reactions, the isomerization, formation and cracking (splitting-off the side-chains) of alkyl-aromatics, proceed through carbonium ions of this type [152].

Compared to the above, reasonably convincing evidence on the role of Lewis centres in the conversion of hydrocarbons has been found in only very few cases. Although the electron affinities of these centres are very high, as exhibited in their strong radical-stabilizing powers [126], their catalytic activities have been demonstrated in only a few alkyl-aromatic reactions, e.g. the alkylation of benzene with propylene [159] or the hydrodemethylation of toluene [129]. In these cases the Lewis centres probably cause a deformation of the π-electron system of the benzene ring and thereby weaken the bonds on the ring.

Although the strengths of acid formations are of decisive importance as regards catalytic activity, this should not be the only basis for the interpretation of the observed phenomena: specific effects of solid catalysts have to be considered since conversions on solid catalysts and in homogeneous acidic media of the same acidity may be different. Secondary effects attributable to the environment of the acid formations can also be involved. These effects may be purely steric [160], while others might be ascribed to interaction with other sites of the crystal lattice of the catalyst.

REFERENCES TO PART III

1. e.g. Moelwyn-Hughes, E. A.: Physical Chemistry. Pergamon Press, New York–London–Paris, 1957
2a. Gauger, A. W. and Taylor, H. S.: J. Am. Chem. Soc. **45**, 920 (1923)
2b. Armstrong, E. F. and Hilditch, T. P.: Trans. Faraday Soc. **17**, 669 (1921)
2c. Pease, R. N.: J. Am. Chem. Soc. **45**, 1196, 2235, 2297 (1923)
3. Schwab, G.-M. and Rudolph, L.: Z. phys. Chem. **B 12**, 427 (1931)
4. Constable, F. H.: Proc. Roy. Soc. **A 108**, 355 (1925)
5. Burk, R. E.: J. Phys. Chem. **30**, 1134 (1926)
6. Balandin, A. A.: Z. phys. Chem. (Leipzig) **B 2**, 289 (1929)
7. Balandin, A. A.: Voprosi khimitseskoi kinetiki kataliza i reakcionnoi sposobnostii. Izdatielstvo Akademia Nauk SSSR, Moscow, 1955
8a. Horiuchi, H. and Polányi, M.: Trans. Faraday Soc. **30**, 1164 (1934)
8b. Twigg, G. H.: Trans. Faraday Soc. **35**, 934 (1939)
8c. Griffith, R. H.: Adv. Cat. **1**, 91 (1948)
9. Crawley, B. and Griffith, R. H.: J. Chem. Soc. 717, 720, 2034 (1938)
10a. Burwell, R. L.: Chem. Rev. **57**, 895 (1957)
10b. Polányi, M.: Z. Elektrochem. angew. physik. Chem. **27**, 143 (1921)
11. Rice, F. O. and Teller, E.: J. Chem. Phys. **6**, 489 (1938); Balandin, A. A.: Adv. Cat. **10**, 93 (1958); Uts. Zap. Mosc. Gos. Univ. **175**, 97 (1956)
12. Rubinstein, A. M. and Pribitkova, N. A.: Izv. Akad. Nauk SSSR, Otdiel, Khim. Nauk **5**, 509 (1945)
13. Tétényi, P., Babernics, L. and Thompson, S. J.: Acta Chim. Acad. Sci. Hung. **34**, 335 (1962)
14. Trapnell, B. M. W.: Adv. Cat. **3**, 13 (1951)
15. Rubinstein, A. M., Suikin, N. I. and Minatsev, H. M.: Dokl. Akad. Nauk. SSSR **67**, 287 (1948)
16. Balandin, A. A. and Isaguliants, G. V.: Dokl. Akad. Nauk SSSR **64**, 207 (1949)
17. Rubinstein, A. M. and Pribitkova, N. A.: Dokl. Akad. Nauk SSSR **61**, 285 (1948)
18. Balandin, A. A. and Klabunovskii, E. I.: Dokl. Akad. Nauk SSSR **110**, 571 (1956)
19a. Lebedev, W. P.: Chem. Techn. **10**, 267 (1958)
19b. Kobozev, N. I. *et al.*: J. physik. Chem. **13**, 1, 28 (1939)
19c. Kobozev, N. I. *et al.*: J. physik. Chem. **15**, 882 (1941); *loc. cit.* **19**, 71 (1945); *loc. cit.* **20**, 145 (1946); *loc. cit.* **23**, 388 (1949)
19d. Shekhobolova, V. I., Krylova, I. V. and Kobozev, N. I.: J. physik. Chem. **26**, 703, 1666 (1952)
19e. Kobozev, N. I.: J. physik. Chem. **14**, 663 (1940)
20. Tschernitz, J., Borstein, S., Beckmann, R. B. and Hougen, O. A.: Trans. Am. Inst. Chem. Engrs **42**, 883 (1946)
21. Poltorak, O. M.: Zhur. Fiz. Khim. **29**, 1650 (1955)
22. Griasnov, V. M., Shiamanov, Yu. M., Usova, L. K. and Frost, A. V.: Dokl. Akad. Nauk SSSR **65**, 867 (1949)
23. Sherman, A. and Eyring, H. J.: J. Am. Chem. Soc. **54**, 2661 (1932)
24. Cunningham, R. E. and Gwathmey, A. T.: Adv. Cat. **9**, 25 (1957); Gwathmey, A. T. and Cunningham, R. E.: Adv. Cat. **10**, 57 (1958)
25. Szabó, Z.: Kémiai Tudományok Osztályának Közleményei **19**, 291 (1963)
26. Bérces, T. and Szabó, Z.: Kémiai Tudományok Osztályának Közleményei **19**, 303 (1963)

27. Cottrell, T. L.: The Strength of Chemical Bonds. Butterworth Sci. Publ. London, 1958
28. Levi, G. J. and Balandin, A. A.: Izvest. Akad. Nauk SSSR Otdiel Khim. Nauk No. 2, 157 (1960)
29. Balandin, A. A. and Titova, A. N.: Uts. Zap. Mosc. Gos. Univ. **2**, 229 (1934)
30. Balandin, A. A. and Tolstopiatova, A. A.: Dokl. Akad. Nauk SSSR **94**, 49 (1954)
31. Tétényi, P.: Kémiai Közlemények **36**, 59 (1971)
32. Schwab, G.-M. and Cremer, E.: Z. phys. Chem. **A 144**, 243; **B 5**, 406 (1929)
33. Constable, F. H.: Active Centres from the Point of View of Kinetics. (Schwab, G. M.: Handbuch der Katalyse Vol. 5, Springer Verlag, Wien), 1957
34. Agronomov, A. E.: Vestnik Moskva Univ. Ser. Khim. **2**, 109; **11**, 41 (1951)
35. Balandin, A. A.: Dokl. Akad. Nauk SSSR **93**, 475 (1953)
36. Balandin, A. A. and Vasiunina, N. A.: Zhur. Obshts. Khim. **18**, 398 (1948); Dokl. Akad. Nauk SSSR **105**, 981 (1955)
37. Cremer, E.: Adv. Cat. **7**, 75 (1955)
38. Dowden, D. A.: J. Chem. Soc. 242 (1952)
39. Boudart, M.: J. Am. Chem. Soc. **74**, 3556 (1952)
40. Ward, A. F. H. and Bharucha, N. R.: Rec. trav. chim. **72**, 735 (1953)
41. Rideal, E. K. and Wansbrough-Jones, O. H.: Proc. Roy. Soc. **A 123**, 202 (1929)
42. Pilling, N. B. and Bedworth, R. E.: Ind. Eng. Chem. **17**, 372 (1925)
43. Baker, M. M. and Jenkins, G. I.: Adv. Cat. **7**, 1 (1955)
44. Trapnell, B. M. W.: Proc. Roy. Soc. **A 218**, 566 (1953)
45. Dilke, M. H., Eley, D. D. and Maxted, E. B.: Nature **161**, 804 (1948)
46. Maxted, E. B.: J. Chem. Soc. 1987 (1949)
47. Schwab, G. M.: Trans. Faraday Soc. **42**, 689 (1946); Schwab, G. M. and Psematjoglou, S. J.: Phys. and Coll. Chem. **52**, 1046 (1948); Schwab, G. M.: Disc. Faraday Soc. **No. 8**, 166 (1950)
48a. Haber, F. and Weiss, J.: Naturwissenschaften **20**, 948 (1932); Proc. Roy. Soc. **A 147**, 332 (1934)
48b. Dowden, D. A. and Reynolds, P. W.: Disc. Faraday Soc. **No. 8**, 184 (1950); Reynolds, P. W.: J. Chem. Soc. 265 (1950)
49. Beeck, O.: Revs. Mod. Phys. **17**, 61 (1945); Disc. Faraday Soc. **No. 8**, 117 (1950)
50. Sheridan, J. and Reid, W. D.: J. Chem. Soc. 2962 (1952)
51. Couper, A. and Eley, D. D.: Disc. Faraday Soc. **No. 8**, 172 (1950)
52. Rienaecker, G.: Abh. d. Dtsch. Akad. d. Wiss. Berlin Kl. Chemie, Geol. Biol. **A 1956**, No. 3, 33
53. Eley, D. D. and Luetic, P.: Trans. Faraday Soc. **53**, 1483 (1957)
54. Stowe, R. A. and Russel, W. W.: J. Am. Chem. Soc. **76**, 319 (1954)
55. Szabó, Z. and Solymosi, F.: Kémiai Tudományok Osztályának Közleményei **13**, 81 (1960)
56. Tolpin, J. G., John, G. S. and Field, E.: Adv. Cat. **5**, 234 (1953)
57. Roginskii, S. Z.: Izvest. Akad. Nauk SSSR Ser. Fiz. **22**, 163 (1957)
58. Wolkenstein, F. F.: Elektronnaia tieoria kataliza na poluprovidniikah. Gos. Izd. Fiz. Mat. Literaturi, Moscow, 1960; Adv. Cat. **12**, 189 (1960)
59. Krilov, O. V., Roginskii, S. Z. and Fokina, E. A.: Izvest. Akad. Nauk SSSR Otdel. Khim. 421 (1957)
60. Schottky, W.: Naturwissenschaften **26**, 843 (1938); Z. Physik **113**, 367 (1939); *loc. cit.* **118**, 539 (1942)
61. Hauffe, K.: Angew. Chem. **67**, 189 (1955);
Hauffe, K.: Reaktionen in und an festen Stoffen. Springer Verlag, Berlin, Heidelberg, New York, 1966
62. Hoover, G. I. and Rideal, E. K.: J. Am. Chem. Soc. **49**, 104, 116 (1927)
63. Wicke, E.: Z. Elektrochemie **53**, 279 (1949)
64. Molinari, E. and Parravano, G.: J. Am. Chem. Soc. **75**, 5233 (1953)
65. Wellner, S. W. and Voltz, S. E.: J. Am. Chem. Soc. **75**, 5227; Z. phys. Chem. N. F. **5**, 100 (1955)
66. Mateiev, K. I. and Boreskov, G. K.: Problemi kinetiki i kataliza **8**, 165 (1955)
67. Miasnikov, I. A. and Pshezshecki, S. I.: Problemi kinetiki i kataliza **8**, 175 (1955)
68. Hauffe, K., Glang, R. and Engell, H. J.: Z. phys. Chem. **201**, 223 (1952); Engell, H. J. and Hauffe, K.: Elektrochem. **57**, 776 (1953)

69. Stone, F. S. and Tiley, P. F.: Disc. Faraday Soc. **8**, 256 (1950); Nature (London) **167**, 654 (1951)
70. Gray, T. J.: Proc. Roy. Soc. (London) **A 197**, 314 (1949);
Garner, W. E., Gray, T. J. and Stone, F. S.: Disc. Faraday Soc. **8**, 246 (1950);
Garner, W. E., Stone, F. S. and Tiley, P. F.: Proc. Roy. Soc. **A 211**, 472 (1952)
71. Roginskii, S. Z. and Tselinskaya, T. S.: Zhur. Fiz. khim. **22**, 1350 (1948)
72. Parravano, G.: J. Am. Chem. Soc. **75**, 1448 (1952)
73. Schwab, G. M. and Block, J.: Z. Elektrochemie **58**, 756 (1954)
74. Wolkenstein, F. F. and Kogan, S. M.: Z. phys. Chem. (Leipzig) **211**, 282 (1959)
75. Donato, N.: J. Phys. Chem. **67**, 773 (1963)
76. Guczi, L. and Tétényi, P.: Kémiai Tudományok Osztályának Közleményei **18**, 585 (1962);
Guczi, L.: Besugárzás hatása szilárd testek adszorpciós és katalitikus tulajdonságaira (Effect of irradiation on the catalytic and adsorption properties of solids) private communication, 1963
77. Kohn, H. W. and Taylor, E. H.: J. Am. Chem. Soc. **79**, 252 (1957); J. Phys. Chem. **63**, 500, 966 (1955); Acta du IIe Congrès Int. de Catalyse, Paris, 1960 p. 1461 (1961)
78. Balandin, A. A., Spitzin, V. I., Dobrozelskaya, N. P. and Mihailenko, L. E.: Dokl. Akad. Nauk SSSR **121**, 495 (1958);
Spitzin, V. I.: Izvest. Akad. Nauk SSSR Otdiel. Khim. Nauk 1296 (1958)
79. Krohn, N. A. and Smith, M. A.: J. Phys. Chem. **65**, 1919 (1961)
80. Hauffe, K.: Adv. Cat. **7**, 213 (1955)
81. Kogan, S. M.: Problemi kinetiiki i kataliza **10**, 52 (1960)
82. Pratt, J. W. and Kolm, H. H.: Semiconductor Surface Physics. University of Pennsylvania Press Philadelphia, 1957
83. Adadurov, M. E.: Zhurn. fiz. khim. SSSR **12**, 445 (1938); ibid. **5**, 136, 1139 (1934)
84. Selwood, P. W.: Adv. Cat. **3**, 28 (1951)
85. Schwab, G. M., Block, J., Müller, W. and Schultze, D.: Naturwissenschaften **44**, 582 (1957)
86. Schwab, G. M., Block, J. and Schultze, D.: Angew. Chem. **71**, 101 (1958)
87. Szabó, Z. G. and Solymosi, F.: Acta du IIe Congrès Int. de Catalyse. Paris, 1960 p. 1627 (1961)
88. Solymosi, F. and Szabó, Z.: Magyar Kémiai Folyóirat **66**, 289 (1960)
89. Szabó, Z., Solymosi, F. and Egri, L.: Kémiai Tudományok Osztályának Közleményei, **18**, 447 (1962)
90. Szabó, Z. G. and Solymosi, F.: Z. phys. Chem. N. F. **17**, 128 (1958);
Szabó, Z., Solymosi, F. and Batta, I.: Kémiai Tudományok Osztályának Közleményei **11**,147 (1959);
Szabó, Z. G., Solymosi, F., and Batta, I.: Z. phys. Chem. N. F. **23**, 56 (1960)
91. Selwood, P. W.: Adsorption and Collective Paramagnetism. Academic Press, New York, 1962 pp. 78, 79
92. Batta, I., Solymosi, F. and Szabó, Z.: Magyar Kémiai Folyóirat **69**, 261 (1963)
93. Baddour, R. F. and Deibert, M. C.: J. Phys. Chem. **70**, 2173 (1966)
94. Langenbeck, W., Nehring, D. and Dreyer, H.: Z. anorg. Chem. **304**, 37 (1960)
95. Nehring, D. and Dreyer, H.: Z. anorg. Chem. **315**, 27 (1962)
96. Nehring, D. and Dreyer, H.: Chem. Techn. **12**, 343 (1960)
97. Langenbeck, W., Nehring, D., Dreyer, H. and Furmann, H.: Z. anorg. Chem. **314**, 167, 179 (1962)
98. Völter, J.: Z. Chem. **3**, 323 (1963)
99. Schwab, G. M. and Putzar, R.: Chem. Ber. **92**, 2132 (1959)
100. Schwab, G. M. and Putzar, R.: Z. phys. Chem. N. F. **31**, 341 (1962)
101. Schwab, G. M. and Mutzbauer, G.: Z. phys. Chem. **32**, 367 (1962)
102. Schwab, G. M. and Siegert, R. Z.: Z. phys. Chem. N. F. **50**, 191 (1966)
103. Schwab, G. M. and Derloth, H.: Z. phys. Chem. N. F. **53**, 1 (1967)
104. Schwab, G. M. and Kollar, K.: J. Am. Chem. Soc. **90**, 3078 (1968)
105. Steinbach, F.: Z. phys. Chem. N. F. **61**, 235 (1968);
Steinbach, F. and Krieger, K. A.: Z. phys. Chem. N. F. **58**, 290 (1968);
Steinbach, F.: Z. phys. Chem. N. F. **71**, 14 (1970)

106. Eischens, R. P. and Pliskin, W. A.: Adv. Cat. **10,** 1 (1958)
107. Lewis, P. H.: Private Communication (see Gatos, H. C.: The Surface Chemistry of Metals and Semiconductors, John Wiley Inc. Publ. New York, 1960 p. 434)
108. Schwab, G. M. and Kritikos, A.: Naturwissenschaften, **55,** 228 (1968); Schwab, G. M. and Kritikos, A.: Helvetia Phys. Acta **41,** 1166 (1968)
109. Thomas, C. L.: Ind. Eng. Chem. **41,** 2564 (1949)
110. Frilette, V. J., Weisz, P. B. and Golden, R. L.: J. Catalysis **1,** 301 (1962); Weisz, P. B., Frilette, V. J. and Mower, E. B.: ibid. 307
111. Stamires, D. N. and Turkevich, J.: J. Am. Chem. Soc. **86,** 749 (1964)
112. Uytterhoeven, J. B., Christner, L. C. and Hall, W. K.: J. Phys. Chem. **69,** 2117 (1965)
113. Kerr, G. T.: J. Phys. Chem. **71,** 4155 (1967)
114. McDaniel, C. V. and Maher, P. K.: Molecular Sieves Soc. Chem. Ind., London, 1968 p. 186
115. Uytterhoeven, J. B., Schoonheydt, R., Liengme, B. V. and Hall, W. K.: J. Catalysis **13,** 425 (1969)
116. Manassen, J. and Khalif, Sh.: J. Catalysis **7,** 110 (1967); Kalló, D., Preszler, I. and Schay, G.: Acta Chim. Acad. Sci. Hung. **64,** 211 (1970)
117. Ward, W. J.: J. Catalysis **9,** 225, 396 (1967); ibid. **10,** 34 (1968); ibid. **11,** 238, 251, 259 (1968)
118. Hughes, T. R. and White, H. M.: J. Phys. Chem. **71,** 2192 (1967)
119. Ward, J. W.: J. Catalysis **17,** 355 (1970)
120. Rabo, J. A., Pickert, P. E., Stamires, D. N. and Boyle, D. N.: Acta du IIe Congrès Int. de Catalyse, Paris, 1960 Vol. 2, p. 2055
121. Ward, J. W.: J. Phys. Chem. **72,** 4211 (1968)
122. Christner, L. G., Liengme, B. V. and Hall, W. K.: Trans. Faraday Soc. **64,** 1679 (1968)
123. Steward, R., O'Donnel, J. P., Cram, D. J. and Rickborn, B.: Tetrahedron Lett. **18,** 317 (1962)
124. Venuto, P. B.: J. Catalysis **5,** 484 (1966)
125. Rabo, J. A., Angell, G. L., Kassai, P. H. and Schomaker, V.: Disc. Faraday Soc. **41,** 328 (1966)
126. Papp, J., Miheikin, I. D. and Kazanskii, V. B.: Kin. i. Kataliz **11,** 812 (1970)
127. Ward, J. W.: J. Catalysis **13,** 321 (1969)
128. Eberly, P. E. jr. and Kimberlin, C. N. jr.: in Advances in Chemistry Series, Am. Chem. Soc. Washington, 1971 Vol. 102, p. 374
129. Papp, J., Kalló, D. and Schay, G.: in Advances in Chemistry Series. Am. Chem. Soc. Washington. 1971 Vol. 102, p. 362
130. Ben Taarit, Y., Mathieu, M. V. and Naccache, C.: in Advances in Chemistry Series, Am. Chem. Soc. Washington, 1971 Vol. 102, p. 362
131. Ward, W. J.: J. Catalysis **14,** 365 (1969)
132. Cross, N. E., Kemball, C. and Leach, H. F.: in Advances in Chemistry Series, Am. Chem. Soc. Washington, 1971 Vol. 102, p. 389
133. Schoonheydt, R. A., Uytterhoeven, J. B.: in Advances in Chemistry Series, Am. Chem. Soc. Washington, 1971 Vol. 101. p. 456 Lohse, U., Stach, H. and Schirmer, W.: Z. Phys. Chem. (Leipzig) **246,** 91 (1971)
134. Meier, W. M. and Olson, D. H.: in Advances in Chemistry Series, Am. Chem. Soc. Washington, 1971, Vol. 101, p. 155
135. Breck, D. W.: J. Chem. Education **41,** 678 (1964)
136. Dempsey, E.: Molecular Sieves. Soc. Chem. Ind. London, 1968, p. 293; J. Phys. Chem. **73,** 3660 (1969)
137. Moscou, L. and Lakemann, M.: J. Catalysis **16,** 173 (1970)
138. Olson, D. H. and Dempsey, E. J.: J. Catalysis **13,** 221 (1969)
139. Sherry, N. S.: J. Phys. Chem. **70,** 1158 (1966)
140. Danforth, J. D.: Acta du IIe Congrès Int. de Catalyse, Paris, 1960 p. 61
141. Benesi, H. A.: J. Phys. Chem. **61,** 970 (1957)
142. Hammett, L. P. and Deyrup, A. J.: J. Am. Chem. Soc. **54,** 2721 (1932); Hammett, L. P. and Paul, M. A.: ibid. **56,** 327 (1934)
143. Paul, M. A. and Long, F. A.: Chem. Rev. **1957,** 1
144. Hirschler, A. E.: J. Catalysis **2,** 428 (1963)
145. Pines, H. and Haag, W. O.: J. Am. Chem. Soc. **82,** 2471 (1960)

146. Borisova, M. S., Dzisko, V. A., Ignatieva, L. A. and Timoriba, L. N.: Kin. i Kataliz **IV/3,** 461 (1963)
147. Geodokian, K. T., Kiselev, A. V. and Lygin, V. J.: Zhur. Fiz. Khim. **43,** 106 (1969)
148. Oblad, A. G., Milliken, T. H. and Mills, G. A.; Adv. Cat. **3,** 199 (1951)
149. Venuto, P. B.: 2nd Int. Conf. on Molecular Sieves, Worcester, Mass., USA, 1970 p. 186
150. Evans, A. G. und Polányi, M.: J. Chem. Soc. **1947,** 252
151. Whitmore, F. C.: Chem. Eng. News **26,** 668 (1947)
152. Germain, J. E.: Catalytic Conversion of Hydrocarbons. Academic Press, London–New York, 1969
153. Condon, F. E.: in "Catalysis" (Editor P. H. Emmett) Reinhold Publ., New York, 1958 Vol. 6, p. 83
154. Oblad, A. G., Mills, G. A. and Heinemann, H.: in "Catalysis" (Editor P. H. Emmett) Reinhold Publ., New York, 1958 Vol. 6, p. 341
155. Ipatieff, V. N. and Schmerling, L.: Adv. Cat. **1,** 27 (1948)
156. Greensfelder, B. S., Voge, H. H. and Good, G. M.: Ind. Eng. Chem. **41,** 2573 (1949)
157. Beyer, H.: Acta Chim. Acad. Sci. Hung. **84,** 25 (1975)
158. McLean, C. and Mackor, E. L.: J. Chem. Phys. **34,** 2207 (1961)
159. Pickert, P. E., Bolton, A. P. and Lanewala, M. A.: A. I. Ch. E. Meeting, 59th, Columbus, Ohio, 1966
160. Weisz, P. B. and Frilette, V. J.: J. Phys. Chem. **64,** 382 (1960);
Chen, N. Y. and Weisz, P. B.: Chem. Eng. Progr. Sym. Ser. **63,** 86 (1967)

HANDBOOKS

1. Advances in Catalysis, Vols. 1, 2, etc. Academic Press Inc., New York, from 1949
2. Emmett, P. H.: Catalysis, Vols. 1., 2, etc. Reinhold Publ. Co., New York, from 1952
3. Schwab, G. M.: Handbuch der Katalyse. Vols. 1, 2, etc. Springer Verlag, Wien, from **1941**

PART IV

KINETICS OF HETEROGENEOUS CATALYTIC REACTIONS

F. Nagy

INTRODUCTION

Catalysts are applied to control chemical processes in order to increase the rate of the reaction producing the required substance. Thus, essential characteristics of all heterogeneous catalytic reactions are the magnitude of the reaction rate and the relationship (the rate equation) expressing the effects of various parameters upon the rate. Catalytic activity, which provides a measure of the effectiveness of catalysts, is also related to the rate. The relation between the activities of two catalysts (which are different or were prepared in different ways) is usually expressed by the ratio of rates observed for the respective samples, with the catalyst amounts, the component concentrations and the temperature identical. The mathematical form of the rate equation and the parameters it involves are usually specific to the mechanism of the catalytic process. Thus, a knowledge of rate equations is important from both a practical and a theoretical point of view.

It follows from the general definition of catalysis that heterogeneous catalytic reactions can be regarded as heterogeneous consecutive reactions whose components include surface elements, i.e. active centres of the catalyst. Thus, the problems arising with regard to the rates of heterogeneous catalytic processes are somewhat similar to those of heterogeneous chemical reactions. Accordingly, the kinetic laws of heterogeneous reactions will also be dealt with in the forthcoming discussion.

In this Part the general problems connected with the rates of chemical processes are considered, and relationships are sought between the kinetics of heterogeneous reactions and of heterogeneous catalytic reactions. In conformity with the character of this book, only the kinetics of catalytic systems consisting of gaseous (vapour phase) reactants and solid catalysts are discussed. The construction and working principle of laboratory reactors used in rate determination experiments, and the experimental verification of theoretically derived rate equations are also considered.

(The notations used in this Part are listed in the Table entitled "Notations and Dimensions", p. 538.)

CHAPTER 1. RELATIONSHIPS FOR THE RATES OF HETEROGENEOUS CATALYTIC REACTIONS

SECTION 1.1. THE CONCEPT, MEASURE AND DIMENSIONS OF THE REACTION RATE

Chemical reactions are time-dependent processes in the course of which chemically distinct species (molecules, atoms, ions, radicals, etc.) are converted into new species. A distinction is made between complex and simple chemical reactions by the criterion of whether a chemical species which is generated only temporarily and disappears during the reaction (or enters a reaction only partially) is formed in the process or not. (For the sake of simplicity the term chemical reaction refers, hereafter, to simple chemical reactions.)

Chemical reactions are, in principle, reversible, and consequently each chemical reaction is a result of two opposite processes. Naturally, under conditions far from equilibrium one of these processes predominates so that only this one need be considered. It is convenient to make a distinction, even in the notation, between the, by definition, one-directional processes, i.e. elementary chemical reactions and processes which are reversible in principle and which can be regarded as resultants of elementary chemical reactions of opposite directions.

The definition of the reaction rate is based upon the stoichiometric equation and the extent of the reaction (ξ). If the reactants and products of an elementary chemical reaction are denoted by A_i and B_j, and the stoichiometric coefficients by ν_{A_i} and ν_{B_j}, the stoichiometric equation of the reaction is:

$$0 = \sum_i \nu_{A_i} A_i + \sum_j \nu_{B_j} B_j \tag{IV.1}$$

By definition, the value of ν_{A_i} is negative and that of ν_{B_j} positive. In the case of elementary reactions ν can have only integral values. The equation defining the extent of the reaction is:

$$\mathrm{d}\xi = \nu_{A_i}^{-1}\, \mathrm{d}n_{A_i} = \nu_{B_j}^{-1}\, \mathrm{d}n_{B_j} \tag{IV.2}$$

where n is the amount of a component. The reaction rate (J), which is essentially the rate of increase of the extent of the reaction ($\dot{\xi}$), is:

$$J \equiv \dot{\xi} \equiv \frac{\mathrm{d}\xi}{\mathrm{d}t} = \nu_{A_i}^{-1} \frac{\mathrm{d}n_{A_i}}{\mathrm{d}t} = \nu_{B_j}^{-1} \frac{\mathrm{d}n_{B_j}}{\mathrm{d}t} \tag{IV.3}$$

If several elementary reactions take place within the system, then the rate of the k-th elementary reaction (J_k) is

$$J_k = (\nu_{A_i})_k^{-1} \frac{\mathrm{d}_k n_{A_i}}{\mathrm{d}t} = (\nu_{B_j})_k^{-1} \frac{\mathrm{d}_k n_{B_j}}{\mathrm{d}t} \tag{IV.4}$$

Elementary chemical reactions are, by definition, one-directional processes, and thus the rate definition is unambiguous and the numerical value of the rate is always positive.

In the case of simple equilibrium reactions the sign of the rate is unequivocal only together with the stoichiometric equation, and prior to the formulation of the rate expression, therefore, the stoichiometric equation must be written, too. It is practical, therefore, to use a notation providing an unambiguous distinction between starting (reactants) and resulting components (products). Such notations are, for example, oxidation, hydrogenation, dehydrogenation, chlorination, etc. The rates of simple chemical reactions may assume positive or negative values. The sign is positive if the reaction proceeds in the direction defined by the stoichiometric equation (the amounts of the reactants decrease, and those of the products increase). According to the definition the rate of a simple reaction can be written as the resultant of two opposed elementary reactions, i.e.

$$J = \vec{J} - \overleftarrow{J} \tag{IV.5}$$

In Eq. (IV.5) the arrow drawn above the rate symbol indicates that the direction of the elementary reaction is identical with ($\vec{J}$) or opposite to ($\overleftarrow{J}$), that of the simple chemical reaction involved. Another common notation is the sign + or − in the subscript of the rate symbol. In this case, the sign + stands for the identical, and − for the opposite, direction.

The rate of a simple chemical reaction can be written in a similar manner to Eq. (IV.4), for an elementary reaction, i.e.

$$J_k = \vec{J}_k - \overleftarrow{J}_k = (\nu_{A_i})_k^{-1} \frac{d_k n_{A_i}}{dt} = (\nu_{B_j})_k^{-1} \frac{d_k n_{B_j}}{dt} \tag{IV.6}$$

In the case of homogeneous reactions the reaction volume (the volume in which the reaction proceeds) is identical with the macroscopic phase in which concentration changes of the components can be observed. In this case, therefore, the volumetric reaction rate, dc/dt is also used, where c is the concentration of the component referred to.

In heterogeneous reactions the reaction volume constitutes only a fraction of the macroscopic phase, and the concentration changes cannot be directly measured in it. Moreover, the magnitude of the reaction volume is independent of the volume of the macroscopic phase (e.g. of the gas phase) in which the concentration changes (caused by the reaction) of the components can be measured. Consequently, in this case the volumetric reaction rate has no meaning.

The rate of a heterogeneous catalytic reaction depends upon the activity and the number of active centres of the catalyst. The direct measurement of these quantities is not feasible. Thus, at best, for a given sort of catalyst the rate related to unit surface or unit mass can be chosen as a specific quantity. Neither of these can be treated as an absolute quantity, even when chemically identical reactions and catalysts are studied, since the ageing of catalysts and the effect of the preparation upon the activity and number of active centres render the specific rates ambiguous: these latter change with time. Since catalyst amounts are easy to measure, in practice (bearing the above in mind) the rate referred to unit mass of catalyst is used as the specific quantity ($\bar{j}$):

$$\bar{j} = \frac{J}{m_s} \quad . \tag{IV.7}$$

where m_s is the catalyst mass.

In agreement with Eqs (IV.6) and (IV.7) the dimensions of the reaction rates $\bar{j}$ and j are:

$$j : nt^{-1}; \qquad \bar{j} : nm^{-1}t^{-1}$$

where n, t, m are the dimensions of the component amount, the mole, the time and the mass, respectively.

The choice of units corresponding to the above dimensions is arbitrary in principle and depends upon the nature of the problem involved.

SECTION 1.2. PARTIAL PROCESSES OF HETEROGENEOUS CATALYTIC REACTIONS

The preliminary condition of elementary chemical reactions in which several molecules take part is the encounter of these molecules, i.e. their collision according to the kinetic theory. This is essentially a process of a physica nature, the velocity of which depends upon the number of phases containing the components. For elementary chemical reactions taking place in a homogeneous phase the rate expression involves the rate of the collision process. This is because, according to the kinetic theory, the rate of an elementary chemical reaction is proportional to the number of collisions in which the participants possess at least the activation energy and, additionally, the collision takes place in a configuration favourable for the reaction. In the case of heterogeneous reactions, also including catalytic heterogeneous reactions, the individual components also appear in those phases where the reaction does not proceed. In these processes the conditions necessary for the encounter of components, and for the occurrence of the elementary chemical reactions, involve not only collision of the component molecules within the reaction phase. The reactants must be transported from the various phases into the reaction phase and the products must leave the latter phase. In other words, the mass transport of various components must proceed simultaneously with the elementary chemical reaction.

For gas-phase heterogeneous catalytic reactions the following phases can be distinguished from the point of mass transport: external phase outside the system bounds (e), free gas phase (g); pore phase, i.e. the gas phase located in the catalyst pores (p) and adsorption phase (a). (The *system* comprises those parts of the catalytic reactor between which transport of components occurs.)

As regards the various components, the following elementary transport processes can be mentioned: transport from the external phase into the gas phase; transport from the gas phase into the pore phase; transport from the pore phase into the adsorption phase, and the corresponding reverse processes.

The rates of the above transport processes are denoted as follows:

$$(\vec{J}_{A_i})_g,\ (\overleftarrow{J}_{A_i})_g,\ (\vec{J}_{A_i})_p,\ (\overleftarrow{J}_{A_i})_p,\ (\vec{J}_{A_i})_a \text{ and } (\overleftarrow{J}_{A_i})_a,$$

and

$$(\vec{J}_{B_j})_g,\ (\overleftarrow{J}_{B_j})_g,\ (\vec{J}_{B_j})_p,\ (\overleftarrow{J}_{B_j})_p,\ (\vec{J}_{B_j})_a \text{ and } (\overleftarrow{J}_{B_j})_a$$

Consider a chemical reaction proceeding in the adsorption phase with reactants A_i and products B_j. Each phase in the system is treated as a homogeneous one. Considering the principle of conservation of matter and the directions of the above transport processes, the time-dependences of the component amounts in the various phases can be expressed as follows:

$$\left(\frac{dn_{A_i}}{dt}\right)_e = -\,[(\vec{J}_{A_i})_g - (\overleftarrow{J}_{A_i})_g]$$

$$\left(\frac{dn_{A_i}}{dt}\right)_g = [(\vec{J}_{A_i})_g - (\overleftarrow{J}_{A_i})_g] - [(\vec{J}_{A_i})_p - (\overleftarrow{J}_{A_i})_p]$$

$$\left(\frac{dn_{A_i}}{dt}\right)_p = [(\vec{J}_{A_i})_p - (\overleftarrow{J}_{A_i})_p] - [(\vec{J}_{A_i})_a - (\overleftarrow{J}_{A_i})_a]$$

$$\left(\frac{dn_{A_i}}{dt}\right)_a = [(\vec{J}_{A_i})_a - (\overleftarrow{J}_{A_i})_a] + \nu_{A_i} J$$

$$i = 1, \ldots, n \qquad \text{(IV.8)}$$

$$\left(\frac{dn_{B_j}}{dt}\right)_e = -\,[(\vec{J}_{B_j})_g - (\overleftarrow{J}_{B_j})_g]$$

$$\left(\frac{dn_{B_j}}{dt}\right)_g = [(\vec{J}_{B_j})_g - (\overleftarrow{J}_{B_j})_g] - [(\vec{J}_{B_j})_p - (\overleftarrow{J}_{B_j})_p]$$

$$\left(\frac{dn_{B_j}}{dt}\right)_p = [(\vec{J}_{B_j})_p - (\overleftarrow{J}_{B_j})_p] - [(\vec{J}_{B_j})_a - (\overleftarrow{J}_{B_j})_a]$$

$$\left(\frac{dn_{B_j}}{dt}\right)_a = [(\vec{J}_{B_j})_a - (\overleftarrow{J}_{B_j})_a] + \nu_{B_j} J$$

$$j = 1, \ldots, l \qquad \text{(IV.8)}$$

From Eq. (IV.8) the following expression is easily obtained:

$$\left(\frac{dn_{A_i}}{dt}\right)_e + \left(\frac{dn_{A_i}}{dt}\right)_g + \left(\frac{dn_{A_i}}{dt}\right)_p + \left(\frac{dn_{A_i}}{dt}\right)_a = \nu_{A_i} J\,; \quad i = 1, \ldots, n$$

$$\text{(IV.9)}$$

$$\left(\frac{dn_{B_j}}{dt}\right)_e + \left(\frac{dn_{B_j}}{dt}\right)_g + \left(\frac{dn_{B_j}}{dt}\right)_p + \left(\frac{dn_{B_j}}{dt}\right)_a = \nu_{B_t} J\,; \quad j = 1, \ldots, l$$

If there are stationary conditions within the pore and adsorption phases, or the amounts of components present in these phases are negligible compared with those in the external and gas phases, i.e. a quasi-steady state can be assumed within the pore and gas phases (see the Bodenstein principle for homogeneous reactions), then:

$$\left(\frac{dn_{A_i}}{dt}\right)_p = 0, \quad \left(\frac{dn_{Ai}}{dt}\right)_a = 0, \qquad i = 1, \ldots, n$$

$$\left(\frac{dn_{B_j}}{dt}\right)_p = 0, \quad \left(\frac{dn_{B_j}}{dt}\right)_a = 0, \qquad j = 1, \ldots, l$$

$$\left(\frac{dn_{A_i}}{dt}\right)_p \approx 0, \quad \left(\frac{dn_{A_i}}{dt}\right)_a \approx 0, \qquad i = 1, \ldots, n$$

$$\left(\frac{dn_{B_j}}{dt}\right)_p \approx 0, \quad \left(\frac{dn_{B_j}}{dt}\right)_p \approx 0, \qquad j = 1, \ldots, l \tag{IV.10}$$

Using the approximations given in Eqs (IV.10), Eq. (IV.9) can be rewritten:

$$\left(\frac{dn_{A_i}}{dt}\right)_e + \left(\frac{dn_{A_i}}{dt}\right)_g = \nu_{A_i} J; \quad \left(\frac{dn_{B_j}}{dt}\right)_e + \left(\frac{dn_{B_j}}{dt}\right)_g = \nu_{B_j} J$$

$$i = 1, \ldots, n \qquad\qquad j = 1, \ldots, l$$

$$\left(\frac{dn_{A_i}}{dt}\right)_e + \left(\frac{dn_{A_j}}{dt}\right)_g \approx \nu_{A_i} J; \quad \left(\frac{dn_{B_j}}{dt}\right)_e + \left(\frac{dn_{B_j}}{dt}\right)_g \approx \nu_{B_j} J$$

$$i = 1, \ldots, n \qquad\qquad j = 1, \ldots, l \tag{IV.11}$$

Since $(n_{A_i})_e$, $(n_{A_i})_g$ and $(n_{B_j})_e$, $(n_{B_j})_g$ are measurable quantities, Eq. (IV.11) provides a theoretical basis for the calculation of the rates of heterogeneous catalytic reactions from experimental data [the validity of Eq. (IV.10) is assumed].

SECTION 1.3. RATE-"CONTROLLING" ROLE OF PARTIAL PROCESSES

In the proceding Section the partial processes of heterogeneous catalytic reactions were dealt with. It was established that these processes are coupled, i.e. the occurrence of one of them is the precondition for the occurrence of the other. In this respect, similarly as for all coupled processes, the question arises of the role of the rate-controlling partial process. It is worth paying some attention to this question since erroneous opinions are frequently encountered with regard to the rate-controlling step, as "the slowest partial process". The definition "slowest partial process" is particularly incorrect, since the coupled processes are consecutive, and consequently their rates are identical in the steady or quasi-stationary state, and thus there is no point in talking of the "slowest partial process".

Let us first consider a simple case, e.g. current conduction through resistors connected in series. If the conductivities of the resistors are denoted by $\sigma_1 \ldots, \sigma_i \ldots, \sigma_n$, the corresponding voltage drops by $(v_0 - v_1) \ldots, (v_{i-1} - v_i) \ldots, (v_{n-1} - v_n)$ and the currents flowing through them by $I_1 \ldots, I_i \ldots, I_n$, then from the Ohm and Kirchhoff laws we have:

$$I_i = \sigma_i(v_{i-1} - v_i)$$

$$I = I_1 = I_2 = \ldots = I_i = \ldots = I_n \qquad \text{(IV.12)}$$

$$I = \frac{1}{\sum_i \frac{1}{\sigma_i}}(v_0 - v_n); \; v_{i-1} - v_i = \frac{\frac{1}{\sigma_i}}{\sum_i \frac{1}{\sigma_i}}(v_0 - v_n)$$

Eq. (IV.12) shows that the velocities of the partial processes $(I_1 \ldots, I_i \ldots, I_n)$ are naturally equal, and consequently there is no "slowest step". However, the role of the rate-controlling step can be elucidated. If the conductivity of one resistor is much less than those of the others (i.e. its resistance is much larger than the overall resistances of the others), the following relationship can be obtained from Eq. (IV.12):

$$\sigma_k \ll \sum_i \sigma_i ; \quad i \neq k$$

$$I \approx \sigma_k(v_0 - v_n) ; \; (v_{k-1} - v_k) \approx (v_0 - v_n) \qquad \text{(IV.13)}$$

It is seen that the entire chain behaves as if it consisted of the resistor of lowest conductivity only. The current passing through the resistor series is equal to the conductivity of the above resistor multiplied by the voltage drop over the total chain.

In the case of consecutive chemical reactions the role of the rate-controlling step is not so obvious. Consider the following consecutive homogeneous reaction system:

$$A \overset{1}{\rightleftarrows} B \overset{2}{\rightleftarrows} C \qquad \text{(IV.14)}$$

where B is an unstable intermediate. The rate equations for the partial processes (i.e. the mathematical expressions giving functional relationships between the rate and the component concentrations) are as follows:

$$J_1/V = \vec{k}_1 c_A - \overleftarrow{k}_1 c_B$$

$$J_2/V = \vec{k}_2 c_B - \overleftarrow{k}_2 c_C$$

(J is the rate of the homogeneous reaction,
V is the volume of the system,
k is the rate constant, and
c is the concentration).

Introducing the equilibrium constants of reactions 1 and 2 (K_1 and K_2):

$$K_1 \equiv \frac{\vec{k}_1}{\overleftarrow{k}_1} ; \; K_2 \equiv \frac{\vec{k}_2}{\overleftarrow{k}_2} \qquad \text{(IV.15)}$$

the above rate expression can be rewritten as

$$J_1/V = \vec{k}_1 \left(c_A - \frac{1}{K_1} c_B\right) = \overleftarrow{k}_1(K_1 c_A - c_B)$$
$$J_2/V = \vec{k}_2 \left(c_B - \frac{1}{K_2} c_C\right) = \overleftarrow{k}_2(K_2 c_B - c_C) \tag{IV.16}$$

For closed systems the time-dependences of the component concentrations are:

$$-\frac{dc_A}{dt} = J_1/V; \frac{dc_B}{dt} = J_1/V - J_2/V; \frac{dc_C}{dt} = J_2/V \tag{IV.17}$$

If B is an intermediate for which the Bodenstein principle is valid, i.e. a quasi-steady state can be reached

$$\frac{dc_B}{dt} \approx 0$$

then, under these conditions, on the basis of Eqs (IV.16) and (IV.17) we have

$$J/V \approx -\frac{dc_A}{dt} \approx \frac{dc_C}{dt} \approx J_1/V \approx J_2/V \tag{IV.18}$$

Consequently the rates of the consecutive reactions 1 and 2 are equal. Now let us define the following quantities:

$$K_1 \equiv \frac{c_B}{c_A^*} = \frac{c_B^*}{c_A}; \; K_2 \equiv \frac{c_C}{c_B^{**}} = \frac{c_C^*}{c_B} \tag{IV.19}$$

In accordance with the above defining equations c_A^*, c_B^*, c_B^{**} and c_C^* are the equilibrium concentrations relating to c_B, c_A, c_C and c_B, respectively.

Using Eqs (IV.15) and (IV.19) the rate equations (IV.16) for the partial processes can be transformed:

(a) $$J/V \approx \vec{k}_1 c_A \left(1 - \frac{c_A^*}{c_A}\right) \approx \overleftarrow{k}_2 c_C \left(\frac{c_C^*}{c_C} - 1\right)$$

(b) $$J/V \approx \vec{k}_1 c_B \left(\frac{c_B^*}{c_B} - 1\right) \approx \overleftarrow{k}_2 c_B \left(1 - \frac{c_B^*}{c_B}\right) \tag{IV.20}$$

It is seen from Eq. (IV.20) that in the case of consecutive chemical reactions leading to equilibrium, as for the current conduction, the rates of the partial processes are obtained as the products of two terms, the second of which can be regarded as the driving force of the chemical process.

Eq. (IV.20) is suitable for the elucidation of the role of the rate-controlling step.

(a) The rate-controlling step is process 1:

$$\vec{k}_1 c_A \ll \overleftarrow{k}_2 c_C \tag{IV.21}$$

From Eqs (IV.19), (IV.20) and (IV.21) we have

$$c_C^* \approx c_C \approx k_2 c_B; \; c_A^* \approx \frac{1}{k_1 k_2} c_C \tag{IV.22}$$

$$J/V \approx \vec{k}_1 \left(c_A - \frac{1}{k_1 k_2} c_C \right); \; \vec{k}_1 c_A \ll \overleftarrow{k}_2 c_C \tag{IV.23}$$

(b) The rate-controlling step is process 2:

$$\vec{k}_2 \ll \overleftarrow{k}_1 \tag{IV.24}$$

From Eqs (IV.19), (IV.20) and (IV.24) we obtain:

$$c_B^* \approx c_B \approx K_1 c_A; \; c_B^{**} \approx \frac{1}{k_2} c_C \tag{IV.25}$$

$$J/V \approx K_1 \vec{k}_2 \left(c_A - \frac{1}{k_1 k_2} c_C \right); \; \vec{k}_2 \ll \overleftarrow{k}_1 \tag{IV.26}$$

The above example illustrates the method of treating the case when, in a consecutive reaction system, the limiting rate of one partial process is much smaller than those of the others. The overall rate consists of the product of two terms. The first of these involves not only the rate constant of the rate-controlling step, but equilibrium constants of partial processes preceding that step, too. The other factor, which is a difference of intensive quantities, is a function of equilibrium constants and component concentrations. The role of the rate-controlling step is primarily manifested in the fact that it leads to the other partial processes establishing a near-equilibrium state.

In the case of consecutive chemical reactions it may be that the order of the reaction depends upon which is the rate-controlling step. Let us consider the following homogeneous reaction system:

$$A + B \overset{1}{\rightleftarrows} AB \tag{IV.27}$$

$$AB + C \overset{2}{\longrightarrow} D$$

where AB is an unstable intermediate.

Analogously to Eqs (IV.25) and (IV.26) we write

$$J/V \approx \vec{k}_1 c_A c_B; \; \vec{k}_1 \ll \overleftarrow{k}_2 c_C$$

$$J/V \approx K_1 \vec{k}_2 c_A c_B c_C; \; \vec{k}_2 c_C \ll \overleftarrow{k}_1 \tag{IV.28}$$

In the above case we see that if reaction 1 is rate-controlling then the chemical reaction is of the second order and its rate is independent of the concentration of C. However, if the second reaction is the rate-determining step then the chemical reaction will be of the third order. In the former case the apparent rate constant is equal to that of the elementary chemical reaction ($\vec{k}_1$) of the first process, while in the second case it includes the rate constant of the elementary chemical reaction of the second process ($\vec{k}_2$) and the equilibrium constant (K_1) of the first step.

It can be clearly seen from the above examples that information concerning the partial processes and their kinetic relationships is of primary importance. The most important partial processes of heterogeneous catalytic reactions, and mainly the transport processes, were discussed in the previous Section. The kinetic relationships for mass transport through the pore phase, i.e. pore diffusion, will be considered in Section 1.4. A detailed discussion of the kinetic relationships of adsorption and desorption can be found in Chapters II.6 and II.7 and, therefore, it is only referred to in this Section. The rates of surface reactions give rise to the most difficult problem, because surface reactions themselves are composite processes, particularly in the case of complex reactions. According to the conclusions drawn in Chapter III (in which the mechanisms of catalytic reactions are discussed), the recognition of these partial processes requires special study for each reaction. Only a small number of catalytic reactions are known for which the processes occurring on the catalyst surface have been elucidated. Information concerning the kinetics of surface processes is particularly incomplete. In the following, the treatment is confined to the kinetic study of the simplest reactions, with the assumptions that the surface reaction proceeds in one step (cf. Section III.2.1) and that the active centres of the catalyst are energetically homogeneous. This latter is equivalent to stating that the same rate constant can be employed, irrespectively of which active centre contributes to the reaction [experience indicates that this latter assumption is justified (cf. the conclusions at the end of Section III.2.3)]. The types of reactions whose kinetic relationships will be discussed are

$$\begin{aligned} &\text{I} \quad A \rightleftarrows B \\ &\text{II} \quad A \rightleftarrows B_1 + B_2 \end{aligned} \qquad \text{(IV.29)}$$

Type I represents monomolecular conversion, e.g. isomerization, whereas type II corresponds in the forward direction to monomolecular decomposition, e.g. dehydrogenation, and in the reverse direction to bimolecular addition, e.g. hydrogenation (II.a and II.b).

For the given conditions, the kinetic equations for the above reactions can be formulated as follows:

$$\begin{aligned} &\text{I} \quad J = A_s(\vec{k}[A]_a - \overleftarrow{k}[B]_a) \\ &\text{II} \quad J = A_s(\vec{k}[A]_a - \overleftarrow{k}[B_1]_a[B_2]_a) \end{aligned} \qquad \text{(IV.30)}$$

where $[\;]_a$ denotes the surface concentrations of the individual components in the adsorption phase, and A_s is the catalyst surface in contact with the gas phase of homogeneous composition and temperature.

Before considering the question of the kinetic relationships for these types of reactions when both adsorption and desorption may be the rate-controlling step, the rate-limiting role of mass transport in the pores must be discussed.

SECTION 1.4. MASS TRANSPORT WITHIN THE CATALYST GRAIN

D. Kalló

Three consecutive partial processes of heterogeneous catalytic reactions have so far been considered: adsorption, surface reaction and desorption. All three processes take place on the active catalyst surface and so are confined to it.

The catalytic conversion becomes observable in the gas phase surrounding the catalyst grain in that the reactants are steadily transported to the large, inner surfaces of the porous catalyst grain and the products continuously leave into the gas phase. If the functioning of a single catalyst grain is studied macroscopically, then the intragranular transport processes must be considered as connected in series with the multi-step catalytic conversion. In this sense, one can speak of their rate-controlling role, which can affect or even determine the catalyst activity. In the case of a packed reactor, transport processes must proceed in the intergranular space as well. These processes, however, are responsible for concentration inhomogeneities, and can be correlated with the problems arising in the study of reactors operating under non-ideal conditions (see "Component transport" in Paragraph VII.3.2.3).

Porous catalysts are widely used in practice. The reason for this is that an active catalyst surface of several hundred m^2 can be formed in 1 cm^3 of reactor volume, thereby increasing the capacity of the reactor significantly. Owing to the large specific surface, smaller amounts of catalysts made of expensive materials are sufficient, when used either in pure form or dispersed over a catalyst support.

The high increase of the specific activity, however, may result in the rate of transport of the reactants along the porous channels being too low.

PARAGRAPH 1.4.1. MASS TRANSPORT PROCESSES IN POROUS GRAINS

In order that a stationary interaction is established between a porous solid and a fluid, some sort of mass transport must proceed within the pores. This is also true for heterogeneous catalytic processes.

The reactants enter the active catalyst grain and the products leave. If the reaction results in a pressure change a convection flow occurs in the pores. It can also happen that the product concentrations are greater inside the grain than in its environment, the concentrations of the starting materials dropping progressively towards the grain centre. A concentration gradient develops between the centre and the outer surface of the catalyst grain, and gives rise to diffusional mass transport. This can be gas diffusion or Knudsen diffusion depending upon the conditions, the pore dimensions, the pressure, the temperature and the characteristics of the given substance

In some cases, the catalytic transformation can be accompanied by physical adsorption; the concentration difference in the gaseous phase leads to a corresponding difference in the adsorbed phase. This gives the possibility of surface migration, and hence mass transport.

(1) Convective mass flow and diffusion.

The starting point in the study of the mass transport within a catalyst grain is that the driving force is either the total pressure difference or simply the concentration difference both due to the reaction. The former appears mainly in reactions in which the number of moles is altered, with a resultant mass velocity G g s^{-1}. The latter is characteristic of any process and can be termed diffusion. In the present case this relates to normal gas and liquid diffusion, Knudsen diffusion and surface migration.

The convective mass flow is proportional to the pressure gradient in the flow direction and inversely proportional to the flow resistance, R_f:

$$G = -\frac{1}{R_f}\varrho\frac{dP}{dl_p} \tag{IV.31}$$

where ϱ is the density of the fluid, g cm^{-3}, and
l_p is the pore length, cm.

In the case of Poiseuille flow we have

$$\frac{1}{R_f} = \frac{\pi r_p^4}{8\eta}$$

where r_p is the pore radius, cm, and
η is the viscosity coefficient, g cm^{-1} s^{-1}.

As pointed out earlier, the study is confined to gas phase reactions; convective mass transport will be defined in moles of the i-th component transported through unit area of pore cross-section in unit time:

$$\dot{N}_{c,i} = \frac{G}{\varrho r_p^2\pi}c_i = -\frac{1}{R_f r_p^2\pi}\frac{dP}{dl_p}c_i \tag{IV.32}$$

where $\dot{N}_{c,i}$ is the convective mass transport of the i-th components, mole cm^{-2} s^{-1}, and
c_i is the volume concentration of the i-th component, mole cm^{-3}.

If the ideal gas law is valid

$$\dot{N}_{c,i} = -\frac{RT}{\varkappa}\frac{dc_g}{dl_p}c_i \tag{IV.33}$$

where $\varkappa = R_f r_p^2\pi$, and
c_g is the sum of the volume concentrations, mole cm^{-3}.

The convective mass transport is brought about by the difference of the total concentrations, c_g, which can appear in processes characterized by change of the mole number, e.g. the burning of coke or wood charcoal. In the process of lime kilning the CO_2 evolved can produce a pressure difference between the inner and outer parts of the limestone grain sufficient to cause its mechanical disintegration.

Naturally, such transport can also occur in catalytic reactions, when the total concentration is changed either because of the stoichiometry or, in the case of the imperfect thermostating of the catalyst grain because of the heat evolved in the reaction. The latter has been mentioned for the sake of completeness only and will hereafter be neglected.

Diffusional mass transport proceeds simultaneously with the convective mass transport. For the i-th component it is proportional to its concentration gradient:

$$\dot{N}_{\mathrm{d},i} = -D_i \frac{\mathrm{d}c_i}{\mathrm{d}l_\mathrm{p}} \tag{IV.34}$$

where $\dot{N}_{\mathrm{d},i}$ is the diffusional mass transport of the i-th component, mole $\mathrm{cm}^{-2}\,\mathrm{s}^{-1}$, and

D_i is the total diffusion coefficient of the i-th component, $\mathrm{cm}^2\,\mathrm{s}^{-1}$ [for a detailed discussion see Point (2)].

The resultant mass transport of the i-th component, $\dot{N}_i$, is given by the sum of Eqs (IV.33) and (IV.34):

$$-D_i \frac{\mathrm{d}c_i}{\mathrm{d}l_\mathrm{p}} + \dot{N}_{\mathrm{c},i} = \dot{N}_i \tag{IV.35}$$

If β is the convective portion of the total mass transport, then its value for the i-th component is

$$\beta_i = \frac{\dot{N}_{\mathrm{c},i}}{\dot{N}_i} \tag{IV.36}$$

Substituting Eq. (IV.36) into Eq. (IV.35) and rearranging

$$\dot{N}_i = -\frac{D_i}{1-\beta_i} \frac{\mathrm{d}c_i}{\mathrm{d}l_\mathrm{p}} \tag{IV.37}$$

This expression is formally analogous with Eq. (IV.34). When simultaneous convection and diffusion occur, the mass transport can be described by a simple diffusion equation if, as shown above, the diffusion coefficient includes the convection as well.

In view of the subsequent discussion of simultaneous mass transport and chemical reaction, the assumption that $D_i/(1-\beta_i)$ is constant, i.e. independent of the concentration, would constitute a significant simplification. The phenomenological constant

$$D'_i = \frac{D_i}{1-\beta_i}$$

can be regarded invariant only within certain limits, even if the concentration dependence of D_i is neglected and only variations in β_i are taken into account (see later).

Dividing Eq. (IV.35) by D_i, and summing over i, $\sum_i \frac{\mathrm{d}c_i}{\mathrm{d}l_\mathrm{p}} = \frac{\mathrm{d}c_\mathrm{g}}{\mathrm{d}l_\mathrm{p}}$, which is substituted from Eq. (IV.33):

$$\frac{\varkappa}{RT}\frac{\dot{N}_{c,i}}{c_i} + \sum_i \frac{\dot{N}_{c,i}}{D_i} = \sum_i \frac{\dot{N}_i}{D_i}$$

Since $\frac{\dot{N}_{c,i}}{\dot{N}_c} = \frac{c_i}{c_g}$ and $p_i = c_i RT$, it follows that

$$\dot{N}_{c,i} = \frac{\sum_i \frac{\dot{N}_i}{D_i}}{\varkappa + \sum_i \frac{p_i}{D_i}} p_i \qquad \text{(IV.38)}$$

The chemical transformation is expressed by the following overall equation

$$\nu_{A_1}A_1 + \nu_{A_2}A_2 = \nu_{B_1}B_1 + \nu_{B_2}B_2$$

where ν is the stoichiometric coefficient, and the subscripts refer to the reaction components. Thus, at a site, l_p, of the pore structure the following relationship is valid between the mass transport, $\dot{N}_i$ and the stoichiometric coefficients of the components, ν_i, (the sign of ν is positive for the products and negative for the reactants):

$$\frac{\dot{N}_{A_1}}{\nu_{A_1}} = \ldots = \frac{\dot{N}_{B_1}}{\nu_{B_1}} = \ldots = \frac{\dot{N}_i}{\nu_i}$$

Hence, from Eq. (IV.33) (extending the numerator by ν_i)

$$\dot{N}_{c,i} = \frac{\sum_i \frac{\nu_i}{D_i}}{\varkappa + \sum_i \frac{p_i}{D_i}} p_i \frac{\dot{N}_i}{\nu_i}$$

i.e.

$$\beta_i \equiv \frac{\dot{N}_{c,i}}{\dot{N}_i} = \frac{\sum_i \frac{\nu_i}{D_i}}{\varkappa + \sum_i \frac{p_i}{D_i}} \frac{p_i}{\nu_i} \qquad \text{(IV.39)}$$

D_i' remains constant if the value of β_i is negligibly small compared with 1, or its variation is small.

β_i practically vanishes i.e. no convection need be considered, if

(a) the number of moles remains unchanged in the reaction, $\sum \nu_i = 0$, since the values of D_i are regarded as being roughly equal, or

(b) the value of $\varkappa$ is large; since

$$\varkappa = \frac{8\,\eta}{r_p^2}$$

its value is large when r_p is very small, but the Poiseuille flow is still maintained. If the value of r_p becomes smaller than the free path of the molecules, only Knudsen diffusion can be taken into account and there is no convective flow.

A phenomenological constant, D'_i, with a constant numerical value can be defined instead of $D_i/(1-\beta_i)$ even when β_i is practically constant. According to Eq. (IV.39) the above can be expected when the variation of p_i or $\sum_i p_i$ within a catalyst grain is low, i.e. in the region of low conversions. Otherwise, the value of β_i, and consequently that of D_i, varies from site to site along the pores.

(2) Various types of diffusion.

It was mentioned in the introduction to this Section that various types of diffusion are encountered within a catalyst grain during gas-phase heterogeneous catalytic reactions, e.g. gas diffusion, Knudsen diffusion and surface diffusion (i.e. migration). The common characteristic of these processes is that the rates of all diffusional mass transport are proportional to the concentration gradient; thus D_i in Eq. (IV.34) should be interpreted as $\sum_j m_j D_{i,j}$. In this sum the subscript j refers to the type of diffusion contributing to the mass transport of the i-th component, while m_j is a weighting factor used to conform to the real situation.

In agreement with experimental observation the various types of diffusion can be characterized and distinguished in the following way.

Gas diffusion takes place in pores the diameters of which are large compared with the mean free path of the molecules; the value of the latter is about 1000 Å at atmospheric pressure. Pores with diameters exceeding 1000 Å can usually be found in the catalyst grains formed by pressing or sintering powders (cf. Paragraph V.1.3.6). The gas diffusion is independent of both the pore diameter and the adsorption conditions, but the gas diffusion coefficient of the i-th component, $D_{i,g}$, may depend upon the number of components and the concentration conditions.

If there is only one component (e.g. in isomerization or in the exchange of labelled atoms between otherwise identical molecules), the self-diffusion coefficient of the given substance can be estimated from the following relationship:

$$D_{1,g} = \frac{1}{3}\bar{c}_1\lambda_1 \qquad \text{(IV.40)}$$

where $\bar{c}_1 = \sqrt{\dfrac{8\,RT}{\pi M_1}}$ is the mean velocity of the molecule, cm s^{-1},

M_1 is the molecular weight,

$\lambda_1 = 3.065\ 10^{-23}\ \dfrac{T}{p_1 d_1^2}$ is the mean free path, cm,

p_1 is the gas pressure, atm, and

d_1 is the effective molecular diameter, cm.

Consequently, the diffusion coefficient increases in direct proportion to $T^{1.5}$; however, it is inversely proportional to the pressure. Its value for benzene is 0.103 cm^2 s^{-1} at 200 °C and 1 atm; under the same conditions $\lambda_1 = 863$ Å.

If there are two components present then the following relationship, a rigorous consequence of Eq. (IV.34), is valid in the entire composition range:

$$D_{1,g} = D_{2,g} = D_{1-2,g} \tag{IV.41}$$

$D_{1-2,g}$ depends upon the mixing ratio; it can still be calculated exactly using a relationship which is more complex than Eq. (IV.40). In the case of gas mixtures where both $\bar{c}_1$ and $\bar{c}_2$, and λ_1 and λ_2 are different, only approximate values can be obtained using Eq. (IV.40).

If the number of components exceeds two the conditions are usually highly complicated, since the diffusion of each component is controlled by its own diffusion coefficient. In addition, the diffusion coefficients depend upon the mixing ratio, and consequently upon the depth, l_p, since the concentration varies from site to site along the pores.

Naturally, the diffusion coefficients can also be determined experimentally in the gas mixtures. However, if the obtained results, i.e. the values of $D_{i,g}$, are to be used to calculate the diffusional mass transport within the porous grain the following facts must be taken into account: the available free area of the pore cross-section is only a fraction ϑ of the external surface of the grain, while, in addition, the diffusion path is lengthened owing to the zigzag nature of the pores, i. e.:

$$D_{i,g}^{+} = D_{i,g}\vartheta\chi \tag{IV.42}$$

where $D_{i,g}^{+}$ is the pore diffusion coefficient referring to the porous grain, $cm^2\ s^{-1}$,

$\vartheta \approx$ the specific pore volume, $cm^3\ cm^{-3}$, and

$\chi \approx \frac{1}{\sqrt{2}}$ is the labyrinth factor.

Because of the uncertainties in the values of ϑ and χ it is reasonable to determine $D_{i,g}^{+}$ directly, i.e. to measure the diffusion within the grain itself. In this case the concentration gradient in expression (IV.34) for the diffusional mass transport should be taken not in the actual direction, l_p, of the pore but uniformly, in an arbitrary direction x.

Knudsen diffusion occurs in pores with diameters smaller than the mean free path:

$$2\,r_p < \lambda \approx 1000\ \text{Å}$$

Since intermolecular collision in the gas is negligible, the diffusion coefficients of the components are independent of the pressure, the gas mixture composition and the pressure of other components:

$$D_{i,K} = \frac{2}{3}\bar{c}\bar{r}_{ip} \tag{IV.43}$$

where $\bar{c}_i$ is the mean velocity of the molecule (see above), and

$\bar{r}_p$ is the mean pore radius, cm.

It can be seen from a comparison of Eqs (IV.40) and (IV.43) that, in spite of the criterion $2\,r_p < \lambda$,

$$D_g > D_K$$

This is very important, since it makes clear that the presence of macropores in the large-surface microporous catalyst grain is necessary for the inner surfaces to be accessible. The value of D_K for benzene is 0.006 cm^2 s^{-1} at 200 °C in pores 50 Å in diameter, whereas it is 0.103 cm^2 s^{-1} when the diameter is 1000 Å. D_K can only be estimated on the basis of Eq. (IV.43). For its more exact calculation, a knowledge of r_p is necessary. The pore sizes of porous materials determined by adsorption or other methods are only approximately equal to r_p, since the pore dimensions are far from uniform. Direct experimental determination is the most reliable method in this case too; the theoretical formulae provide only a guide.

Surface diffusion is a type of mass transport occurring independently of, and simultaneously with, the above two types of diffusion. Its magnitude depends exclusively on the concentration gradient in the adsorbed phase. This can be substituted by the concentration gradient for the gas phase only if there is a linear relationship between the surface concentration and the pressure (e.g. if the adsorption isotherm is linear in the case of an established adsorption equilibrium).

When the adsorption equilibrium is attained, the surface concentration decreases as $\exp(|Q|/RT)$ with the temperature, where $|Q|$ is the adsorption heat; the rate constant of the surface diffusion increases as $\exp(-E_m^\ddagger/RT)$. The activation energy of migration is $E_m^\ddagger \approx 1/2\ Q$. The rate of mass transport due to the surface diffusion decreases exponentially with increasing temperature as $\exp(|Q|/2\ RT)$.

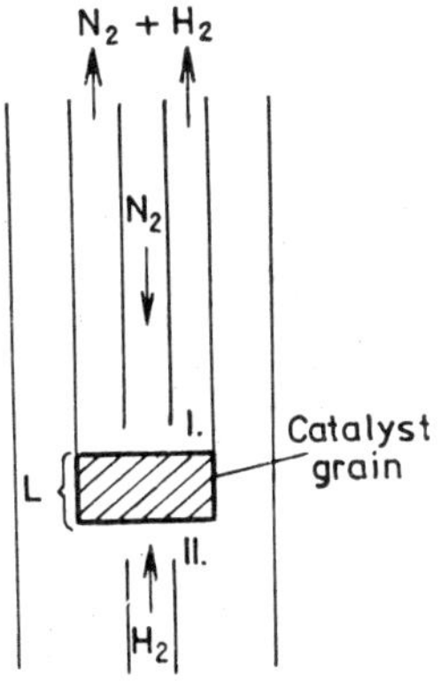

Fig. IV.1. Experimental set-up for measurement of the effective diffusion coefficient [1]

Experiment itself seems to be the most reliable method for the determination of the diffusional mass transport. The result is the value of D_i or, by generalization of Eq. (IV.42), that of D_i^+ which is characteristic of the porous grain. This latter is generally referred to as the effective diffusion coefficient, D_{eff}.

A cylindrical or prism-shaped specimen of cross-section F and length L is formed from the porous solid to be studied, and its superficies are coated. Constant gas concentrations c_I and c_{II} are adjusted in the volumes in front of the specimen faces I and II (see Fig. IV.1). Under these conditions $P_I - P_{II} = 0$, and consequently there is no convective mass transport. Thus

$$\dot{N}_i^+ = \dot{N}_{d,i}^+ = -D_i^+ \frac{c_{I,i} - c_{II,i}}{L} \tag{IV.44}$$

(the sign + refers to the total face area).

If only Knudsen diffusion occurs, then $D_{i,K}$ can be calculated for an arbitrary gas and temperature from the experimental data for N_2 and H_2, for instance, using Eq. (IV.43). If gas diffusion takes place the recalculation can be performed in some cases [1]. However, not even informative results can be obtained for the surface diffusion by this method. It is generally

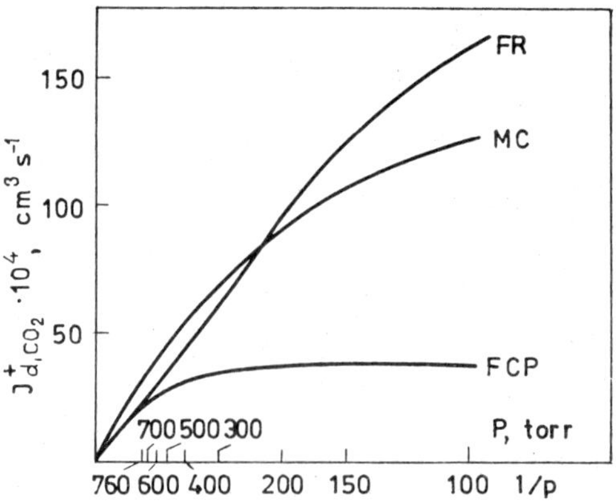

Fig. IV.2. Dependence of the total diffusion on the pressure at 0 °C [1]

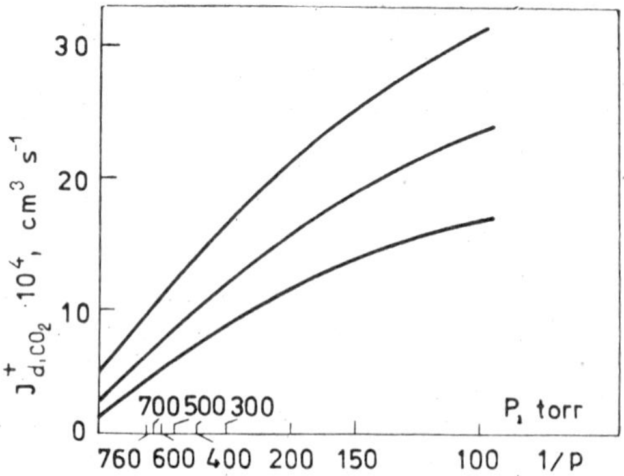

Fig. IV.3. Dependence of the total diffusion on the pressure at 0 °C for some technically active carbons [1]

reasonable to perform the measurements upon the studied catalyst using the reacting substrates themselves if possible (the reaction can be prevented by poisoning the active centres, or by altering the concentrations and the temperature appropriately).

What has been said above will now be illustrated with experimental results.

The dependence of $J^+_{d,i}$ ($\sim \dot{N}^+_{d,i}$) upon $1/P$ is shown in Fig. IV.2 [1]. The studied gas was CO_2, and the porous materials frit (FR), medicinal carbon (MC) and fired clay plate (FCP). During the measurements, the concentration difference for the CO_2 ($c_{I,CO_2} - c_{II,CO_2}$) was constant and equivalent to a pressure difference of 100 torr. On fired clay plate J^+_{d,CO_2} was independent of pressure up to about 300 torr. This indicates a Knudsen diffusional mass transport (i.e. the pores are small). On frit J^+_{d,CO_2} was proportional to $1/P$, i.e. gas diffusion takes place (the pores are large). The effect of the Knudsen diffusion can also be observed in this case below 150 torr: the straight line curves downwards. The medicinal carbon constitutes a transition between the two extreme cases, i.e. the resultant mass transport is affected by both types of diffusion.

In other measurements the J^+_{d,CO_2} vs. $1/P$ plot does not start from the origin (see Fig. IV.3, [1]), i.e. diffusional mass transport occurs even in the limiting case $P \to \infty$ ($1/P \to 0$). It was shown in the previous examples that gas diffusion replaces the Knudsen diffusion as the pressure increases, the former being inversely proportional to the pressure and vanishing when $P \to \infty$. The remaining diffusional mass transport can thus be attributed only to the surface diffusion.

PARAGRAPH 1.4.2. KINETIC CONDITIONS WITHIN THE CATALYST GRAIN

(1) Chemical reaction and mass transport

Within a working catalyst grain positive and negative source effects appear because of the reaction, simultaneously with the corresponding mass transport. Our study will be confined to a one-dimensional catalyst grain. The model is a catalyst cylinder or prism with finite cross-section, whose superficies are impermeable to the reactants. This can be achieved by inserting the catalyst grain into a socket which is open at both ends, or a pellet is formed the thickness of which is negligible compared with the face dimensions. Under stationary conditions the variation due to the reaction in the amount of each component in a volume element $q\mathrm{d}x$ of this grain (where q is the area of the surface perpendicular to the mass transport, cm^2, and x is the distance from the outer surface, cm, ranging from 0 to L, i.e. to half the grain thickness), is equal to the mass transport change for the given component in the x direction:

$$j_i q\,\mathrm{d}x = -q\frac{\mathrm{d}\dot{N}_i^+}{\mathrm{d}x}\,\mathrm{d}x \tag{IV.45}$$

where j_i is the rate of conversion for the i-th component per unit volume of the catalyst grain, mole cm^{-3} s^{-1}, and $\dot{N}_i^+$ is the mass transport for the i-th component in the x direction per unit surface area of the catalyst cross-section, mole cm^{-2} s^{-1}.

By generalization of Eq. (IV.42) and substituting $D_i^+(\equiv D_{\mathrm{eff},i})$ into Eq. (IV.37) we obtain

$$\dot{N}_i^+ = -\frac{D_i^+}{1-\beta_i}\frac{\mathrm{d}c_i}{\mathrm{d}x} \tag{IV.46}$$

where

$$\frac{D_i^+}{1-\beta_i} \equiv D'_{\mathrm{eff},i} \tag{IV.47}$$

Substituting Eq. (IV.47) into (IV.46):

$$\frac{\mathrm{d}\dot{N}_i^+}{\mathrm{d}x} = -D'_{\mathrm{eff},i}\frac{\mathrm{d}^2c_i}{\mathrm{d}x^2} \tag{IV.48}$$

Eqs (IV.45) and (IV.48) give rise to the following differential equation

$$j_i q\,\mathrm{d}x = D'_{\mathrm{eff},i}\,q\frac{\mathrm{d}^2c_i}{\mathrm{d}x^2}\,\mathrm{d}x \tag{IV.49}$$

The latter expresses a general relationship between the reaction rate and the mass transport for a one-dimensional catalyst grain [Eqs (IV.48) and (IV.49) are valid only if $D'_{\mathrm{eff},i}$ is constant. This can be decided either from the condition following Eq. (IV.39), or by making use of the information given in Point 1.4.1. (2)].

Substitution of an exact rate equation reflecting the reaction mechanism (see Section 1.5) into Eq. (IV.49) in place of j_i is not necessary, since the nature of this phenomenon means that an empirical equation $j_i = f(c_1, c_2, \ldots, c_i)$ is sufficient, and only changes of the component amounts due to the reaction need be expressed.

(2) Concentration distribution within the catalyst grain

For the sake of simplicity let us consider a first-order, irreversible reaction described by the following rate equation:

$$j_i = kc_1 \qquad \text{(IV.50)}$$

(k is the rate constant, s^{-1}; since j_i has units of mole cm^{-3} s^{-1}, it is related to the porous catalyst volume just like the concentration c_i, mole cm^{-3}). Other simplifications are: the reaction proceeds at constant volume; thus $A \rightarrow B$ and $\beta_1 = 0$, i.e.

$$D'_{\text{eff},1} = D_{\text{eff},1} = D_1^+$$

and the total diffusion coefficient for A and B remains constant independently of the concentration relations

$$D_1^+ = D_2^+ = D_{\text{eff}}$$

(The subscript 1 refers to A, and 2 to B.) Substituting the latter expression and Eq. (IV.50) into Eq. (IV.49) and simplifying, we obtain

$$D_{\text{eff}} \frac{d^2 c_1}{dx^2} = kc_1 \qquad \text{(IV.51)}$$

i.e. an ordinary second-order differential equation for the concentration distribution within the catalyst grain. If the length of the one-dimensional grain is $2L$, then, at the position L, i.e. at the centre of the grain,

$$\left(\frac{dc_1}{dx}\right)_L = 0$$

At the face of the grain, i.e. where $x = 0$, $c_1 = c_{1,0}$ ($c_{1,0}$ is the concentration of component A in the outer surface layer of the grain; at the same site the concentration in the external gas atmosphere is

$$c_{1,g} = \frac{c_{1,0}}{\vartheta}$$

where ϑ is the specific pore volume, cm^3 cm^{-3}). With these boundary conditions the solution of Eq. (IV.51) is

$$\frac{c_1}{c_{1,0}} = \frac{\cosh\left[\varphi\left(1 - \frac{x}{L}\right)\right]}{\cosh \varphi}; \quad \varphi = L\sqrt{\frac{k}{D_{\text{eff}}}} \qquad \text{(IV.52)}$$

The variable φ, in this case, is dimensionless and is usually referred to as the Thiele modulus. It can be seen from Fig. IV.4 that the higher the value of φ, the lower the concentration of the reactant progressing towards the grain centre, and the "poorer" the catalyst with respect to the reactant; if

$\varphi \geqq 3$, the concentration is practically zero at the catalyst centre. Thus, if the expression for φ given by Eq. (IV.52) is considered, then we see that the concentration within the catalyst grain increasingly diminishes, the greater the thickness of the grain, L, the larger the value of k, and the smaller the diffusion coefficient, D_{eff}.

(3) Effectiveness of the catalyst grain

The concentration distribution within the catalyst grain neither provides directly exact values for the observable rate of conversion within a single grain, nor indicates how this rate can be related to the maximum rate which could be achieved in the case of "infinitely rapid" mass transport.

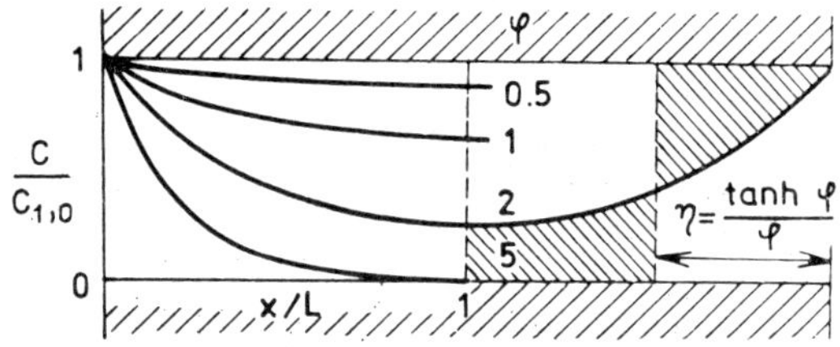

Fig. IV.4. Concentration distributions of the reacting component within a one-dimensional catalyst grain for different values of the modulus φ

Eq. (IV.51) gives the reaction rate, and the change in the diffusional mass transport at the site x of the grain. Half of the value for the total grain can be obtained by integrating Eq. (IV.51) between 0 and L:

$$\int_0^L D_{\text{eff}} \left(\frac{\mathrm{d}^2 c_1}{\mathrm{d}x^2} \right) \mathrm{d}x = \int_0^L k c_1 \, \mathrm{d}x$$

i.e.

$$- D_{\text{eff}} \left(\frac{\mathrm{d}c_1}{\mathrm{d}x} \right)_{x=0} = k c_{1,0} \int_0^L \frac{c_1}{c_{1,0}} \mathrm{d}x = J_1 \qquad \text{(IV.53)}$$

$$\left[\text{since} \left(\frac{\mathrm{d}c_1}{\mathrm{d}x} \right)_{x=L} = 0 \right]$$

(J_1 is half of the conversion rate related to unit cross-sectional area observed within a one-dimensional catalyst grain of length $2L$, mole cm^{-2} s^{-1}.)

The meaning of the equation obtained for J_1 is that the amount of component transported by diffusion at the face is equal to the rate of conversion taking place in the whole grain. This is a consequence of the stationary criterion in the derivation of the starting equation (IV.45).

Eq. (IV.53) offers two different ways of calculating the conversion rate, J_1:

(a) the concentration distribution according to Eq. (IV.52) is substituted into the rate equation of the reaction, and the expression obtained is integrated between 0 and L;

(b) the value of (dc_1/dx) is calculated for the site $x = 0$ and is multiplied by D_{eff}.

If D_{eff} is very large then the intragranular diffusional mass transport can supply the mass demands of the reaction even at vanishingly small values of $(dc_1/dx)_{x=0}$. With regard to Fig. IV.4, this state is equivalent to the condition that $c_1 = c_{1,0}$ throughout the entire grain thickness. In this case, i.e. when the process is not diffusion-inhibited, the highest rate is obtained for the grain from Eq. (IV.53):

$$J_{1,\max} = kc_{1,0}L \qquad \text{(IV.54)}$$

The ratio of the actual reaction rate, J_1 and the maximum reaction rate $J_{1,\max}$, is known as the effectiveness of the catalyst grain and is denoted by η:

$$\eta \equiv \frac{J_1}{J_{1,\max}} \qquad \text{(IV.55)}$$

Substituting Eq. (IV.52) into Eq. (IV.53), dividing by Eq. (IV.54) and integrating:

$$\eta = \frac{1}{L}\int_0^L \frac{c_1}{c_{1,0}}\,dx = \frac{\tanh \varphi}{\varphi} \qquad \text{(IV.56)}$$

[φ is the modulus, defined in Eq. (IV.52)].

In Fig. IV.4 the value of η is shown for $\varphi = 2$; η can be visualized as follows: from the definition of η, Eq. (IV.55) and Eq. (IV.54) we have

$$J_1 = \eta J_{1,\max} = \eta kc_{1,0}L \qquad \text{(IV.57)}$$

In Eq. (IV.57) the term ηL is regarded as L_{eff}, the effective length of the one-dimensional catalyst. In other words, the catalyst is hypothetically divided into two parts; in the first part, the outer crust, the diffusional mass transport supplies the mass demands of the reaction completely; in the inner part, no reaction can be observed and concentration conditions corresponding to the thermodynamically highest possible conversion(s) are established as a result of the infinite residence time of the substrate. This explains why η is referred to as "Eindringtiefe" in the German, and "fraction of surface available" in the English literature.

In spite of this descriptiveness the product ηk is used in practice, since, experimentally, the rate constant, k_{eff}, is determined (the effective mass of the catalyst is also involved in this term).

From a similar consideration the formulae corresponding to Eqs (IV.52) and (IV.56) can be obtained for a reversible reaction, $A \rightleftarrows B$; in this case, c_1 must be replaced by the concentration value

$$c_i^* = c_i - c_{i,\infty}; \qquad i = 1, 2$$

($c_{i,\infty}$ is the equilbrium concentration).

If consecutive and competitive, reversible or irreversible, reactions occur, and all of them can be described by a first-order rate equation, then the corresponding reaction rate, J_i, is written for each component and the system of differential equations constructed similarly to (IV.49) is solved. The formulae corresponding to Eqs (IV.52) and (IV.56) for the scheme

$$\begin{array}{ccc} & A & \\ \nearrow\swarrow & & \nwarrow\searrow \\ C & \rightleftarrows & B \end{array}$$

and for all the simpler systems derived from it can be found in the relevant literature [2].

The relationships were derived for a second-order reaction with no change in the mole number: $2A \rightarrow B + C$ [3], when

$$\varphi = L\sqrt{\frac{k^+ c_{1,g}}{D_{eff}}}$$

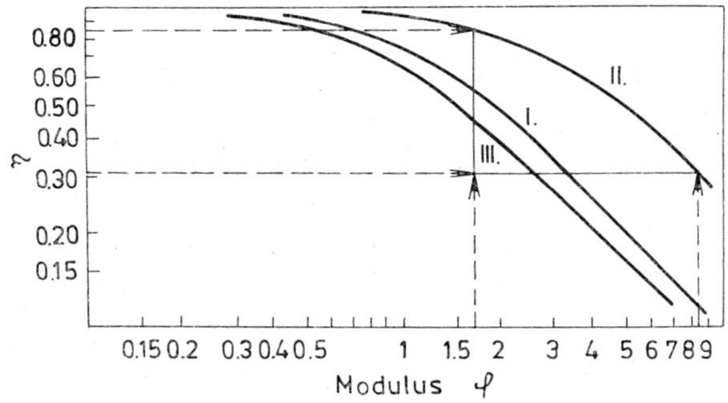

Fig. IV.5. Relationships between η and φ

I. first-order reaction — one-dimensional catalyst grain
II. first-order reaction — spherical catalyst grain
III. second-order reaction — one-dimensional catalyst grain

or, generally, in the case of an n-th order reaction with no change in the mole number

$$\varphi = L\sqrt{\frac{k^+ c_{1,g}^{n-1}}{D_{eff}}}$$

The same derivation was performed for the first-order reaction, $A \rightarrow B + C$, in which the mole number changes [4].

Figure IV.5 depicts $\eta = f(\varphi)$ functions similar to Eq. (IV.56). It is striking that the shape of the curves is only slightly affected by the reaction order. From similar calculated curves the conclusion can be drawn that in the case of spherical catalyst grain for a zero-order reaction at $\varphi \lesssim 1.5$, for a first-order reaction at $\varphi \lesssim 1$, and for a second-order reaction at $\varphi \lesssim 0.5$, $\eta = 1$, i.e. there is no diffusional inhibition. On the other hand, the relationship $\eta \approx \frac{1}{\varphi}$ is generally valid if $\varphi \gtrsim 3$.

(4) Determination of the rate constant, the effectiveness and the effective diffusion coefficient

The rate constant in Eq. (IV.50) can be determined directly experimentally if the process is not diffusion-inhibited, i.e.

$$k_{\text{eff}} = k \quad \text{and} \quad \eta = 1$$

As shown above, this condition is satisfied for small values of φ. At given values of other reaction parameters (substrate, catalyst, pressure, temperature, i.e. k and D_{eff} are fixed), it can be seen from Eq. (IV.52) that φ can be decreased only if the catalyst size, L, is decreased. Since the value of k_{eff} remains constant beyond a certain grain size, independently of further crushing, the full effectiveness is reached, i.e. the value of k is obtained. The question may next arise of why catalysts of the above grain size are not always used. This is actually the aim in kinetic studies since, as will be seen in the following, the diffusional inhibition may not only affect the measurable rate but may also distort the entire kinetic picture. However, the application of catalysts of high effectiveness but of small size in industrial reactors is limited by the high flow resistance of the catalyst bed. In these cases one must be satisfied with lower effectiveness in the interest of lower bed resistance. In any such consideration a knowledge of η is necessary.

The value of η can be determined experimentally on the basis of definition (IV.55). The reaction rate (or the rate constant) is measured on catalysts for which, owing to their sizes, the value of η is equal to 1, and the same is performed for the studied catalyst grain under analogous conditions. The ratio of the two measured quantities gives η.

The experimental methods outlined above can be applied to determine k and η in any case provided that the conversion is described by a single rate equation (i.e. with the exception of complex reactions).

The value of η can be calculated by partly theoretical methods, too. Various cases can be encountered:

(a) k_{eff} is known from a single kinetic measurement, D_{eff} is calculated as shown in Point 1.4.1. (2), and thus the value of η and k are obtained.

(b) k_{eff} is determined for two catalyst grains of different sizes, and from these values η, k and φ are calculated; using the latter data D_{eff} is obtained.

Let us consider an example for the first case. From Eq. (IV.56), after substituting φ according to Eq. (IV.52), we obtain:

$$k_{\text{eff}} = \eta k = \frac{1}{L}\sqrt{kD_{\text{eff}}}\tanh\left(L\sqrt{\frac{k}{D_{\text{eff}}}}\right) \tag{IV.58}$$

Example. In the cracking of cetane at 500 °C the first-order decomposition constant, k_{eff}, was found to be 0.167 s^{-1} on a catalyst with $2L = 0.30$ cm [3]. The mean pore radius and the porosity of the silica-alumina cracking catalyst were 30 Å and $\vartheta = 0.50$, respectively. Since only Knudsen diffusional mass transport can occur

$$D_{\text{eff}} \approx D_{i,\text{K}}\,\vartheta\chi$$

The value of $D_{C_{16}H_{34},\text{K}}$ can be estimated from Eq. (IV.43) given the temperature, the molecular weight and the pore radius:

$$D_{C_{16}H_{34},\text{K}} = 0.0072 \ \text{cm}^2\ \text{s}^{-1}$$

Hence, knowing k_{eff}, L and D_{eff}, the approximate value of k can be obtained using Eq. (IV.58).

Finally, if k 0.214 s^{-1}, k_{eff} 0.167 s^{-1}, and thus

$$\eta = \frac{0.167}{0.214} = 0.78$$

In the second case, k_{eff,L_1} and k_{eff,L_2}, i.e. the rate constants determined on catalyst grains of sizes L_1 and L_2, respectively, are given. Here the procedure is as follows:

Eq. (IV.56), or another expression of the function $\eta = f(\varphi)$ valid in the given case, is plotted in the following form

$$\lg \eta = f'(\lg \varphi)$$

The value of the ratio

$$\lg \frac{\varphi_{L_1}}{\varphi_{L_2}} = \lg \frac{L_1}{L_2}$$

is known and corresponds to a definite interval on the abscissa. Similarly, the distance on the ordinate

$$\lg \frac{\eta_{L_1}}{\eta_{L_2}} = \lg \frac{k_{eff,L_1}}{k} - \lg \frac{k_{eff,L_2}}{k} = \lg \frac{k_{eff,L_1}}{k_{eff,L_2}}$$

can be calculated from the experimental results. These mutually perpendicular intervals are fitted parallel with the axes to the corresponding curve (as shown in Fig. IV.5) so that one of the end-point of each coincide, while the other ends are located on the curves. Subsequently, the values of η and φ can be read off. Given the values of k, L_1, L_2 and φ, D_{eff} can easily be calculated.

There is another, more direct method of determining the effective diffusion coefficient during the reaction from the experimental data.

In the case of full diffusional inhibition we have:

$$\varphi \geqq 3 \quad \text{and} \quad \eta \approx \frac{1}{\varphi}$$

If only gas diffusion takes place

$$\varphi \sim D_g^{+\,-1/2} \sim P^{1/2}, \text{ i.e.}$$

$$\eta \sim P^{-1/2}$$

Example. The o → p-hydrogen transformation, kinetically a first-order process, was studied on nickel turnings 0.1–0.5 mm thick [4]. The turnings were placed in a tube 10 mm in diameter, closed at one end and connected with the hydrogen gas atmosphere at the other, to give a one-dimensional catalyst model. In the large, empty interspaces of this model only gas diffusion could take place. The value of k could be determined for a packing length of 5 mm. The conversion, which was strongly inhibited by the diffusional mass transport, was measured for a 90 mm packing length, k_{eff} was determined at various pressures between 160 and 760 torr. It is seen in Fig.

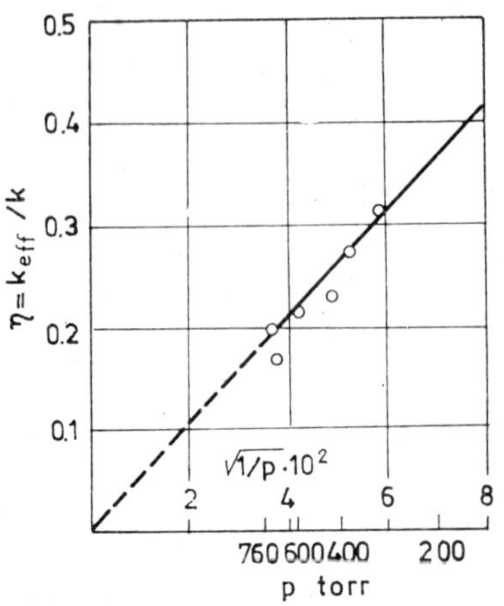

Fig. IV.6. The effectivity, η, in the o → p-hydrogen conversion [4]

IV.6 that η is proportional to $\sqrt{1/P}$, i.e. the gas diffusion retards the reaction. According to the experiments

$$k = 0.28 \text{ s}^{-1}, \text{ and at 760 torr}$$

$$\eta = 0.20$$

Since $\tanh \varphi = 1$, then from Eq. (IV.58) we have

$$D_{\text{eff,H}_2,\text{g}} = \eta^2 L^2 k = 0.86 \text{ cm}^2 \text{ s}^{-1}$$

From Eq. (IV.40) the self-diffusion coefficient of hydrogen at 404 °C and 760 torr is

$$D_{\text{H}_2,\text{g}} = 3.5 \text{ cm}^2 \text{ s}^{-1}$$

i.e. according to Eq. (IV.42)

$$\frac{D_{\text{eff,H}_2,\text{g}}}{D_{\text{H}_2,\text{g}}} = \vartheta\chi = 0.25$$

Consequently $\vartheta\chi$ can significantly modify the value of D_g determined by calculation.

PARAGRAPH 1.4.3. CHANGE OF THE ACTIVATION ENERGY, THE REACTION ORDER, AND THE SELECTIVITY IN THE CASE OF DIFFUSIONAL INHIBITION

The investigation is limited to the simplest cases, since the study of complicated reactions would lead too deeply into questions of special details.

For strong diffusional inhibition $\varphi \geq 3$, i.e. $\tanh \varphi \approx 1$; hence, from Eq. (IV.58) we have

$$k_{\text{eff}} = \frac{1}{L}\sqrt{kD_{\text{eff}}} \qquad \text{(IV.59)}$$

The temperature-dependence of the rate constant is given by the Arrhenius equation:

$$k = A\text{e}^{-\frac{E^\ddagger}{RT}} \qquad \text{(IV.60)}$$

where $E^\ddagger$ is derived from the Arrhenius plot of the single temperature-dependent constant of the rate equation for the catalytic conversion. If the temperature-dependence is correctly reflected by the $E^\ddagger$ value, then it is unimportant whether this is the true, resulting or apparent activation energy (for definitions see Section 2 of the Introduction to Part III); just as it makes no difference in the calculation of the diffusional mass transport whether the conversion rate is expressed by an equation faithfully reflecting the mechanism or by an empirical formula. Substituting Eq. (IV.60) into

Eq. (IV.59) the temperature-dependence of the rate constant measured at full diffusional inhibition is given by the following expression

$$k_{\text{eff}} \sim \exp\left(-E^{\ddagger}/2\,RT\right) \qquad \text{(IV.61)}$$

since the temperature-dependence of D_{eff} is usually negligible compared with that of k. Consequently, under these conditions only half the activation energy is measured.

In accordance with Eq. (IV.52), φ also increases with temperature. If the measurements are performed over a wide temperature range it may be that below a certain temperature $\varphi < 1$, i.e. $k_{\text{eff}} = k$, and the activation energy is measured, whereas at higher temperatures $\varphi > 3$, and in agreement with Eq. (IV.51) an activation energy only half as large is observed. In a log k_{eff} vs. $1/T$ plot the straight line exhibits a break point, its slope in the higher-temperature region being half that at lower temperatures.

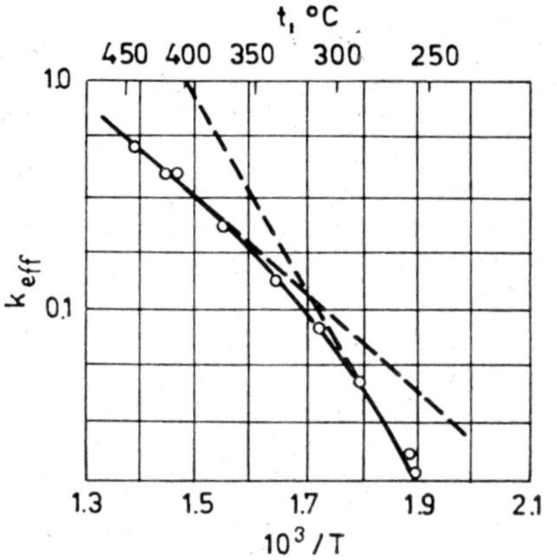

Fig. IV.7. Dependence of the effective rate constant on temperature for the decomposition of methanol on a ZnO catalyst [4]

A further increase in temperature may result in the diffusional mass transport between the outer surface of the catalyst grain and the bulk gas phase becoming the rate-controlling process. Near zero "activation energy" is obtained from the temperature-dependence of this process according to the representation given by Eq. (IV.60), i.e. the second linear interval may be followed by a third one after another break point.

The above is illustrated by the relationship obtained for the decomposition of methanol (Fig. IV.7) [4]. The decomposition is first order on a zinc oxide catalyst. Below 280 °C the catalyst is fully active and the activation energy is 19 kcal mole^{-1}; its value is 9.4. kcal mole^{-1} after the transitional bonding, above 350 °C.

The reaction order determined from the empirical rate equation can change, too, under the effect of the diffusional inhibition. Taking into account Eq. (IV.57), the rate equation is written in the following form:

$$J_i = \eta_n k c_{i,\text{g}}^{n} L \qquad \text{(IV.62)}$$

(n is the reaction order, and η_n the effectiveness of the n-th order reaction).

Thiele [5] has shown that for zero, first and second-order reactions, with strong diffusional inhibition:

$$\eta_n = \frac{2}{(n+1)\varphi_n}, \quad \text{where} \quad \varphi_n = L\sqrt{\frac{k c_{i,\text{g}}^{n-1}}{D_{\text{eff}}}}$$

If this is substituted into Eq. (IV.62)

$$J_i \sim \sqrt{k D_{\text{eff}} c_{i,\text{g}}^{(n+1)}} \qquad \text{(IV.63)}$$

Consequently, n is the kinetic order of the reaction if there is no diffusional inhibition; if there is diffusional inhibition the observed reaction order is $(n + 1)/2$. In this case, the value of the order is naturally affected by any concentration-dependence of D_{eff} [from Eq. (IV.40) e.g. $D_{1,g} \sim 1/p_1 \sim 1/c_1$!].

In the event of several simultaneous reactions the diffusional inhibition may affect the rates of the various conversion processes to different extents. This leads to the change in the selectivity, S. In this respect three principal types of complex reaction systems will be considered, all steps being assumed kinetically first order.

(a) Two independent, simultaneous reactions:

$$A \xrightarrow{k_1} B$$

$$X \xrightarrow{k_2} Y$$

The selectivity is the ratio of the two reaction rates; if $c_{A,g} = c_{X,g}$ and there is no diffusional inhibition, then from Eq. (IV.54)

$$S = \frac{k_1}{k_2}$$

With strong diffusional inhibition the selectivity, in accordance with Eqs (IV.52), (IV.56) and (IV.57), is

$$S_{\text{eff}} = \sqrt{\frac{k_1 D_{A,\text{eff}}}{k_2 D_{X,\text{eff}}}} \approx \sqrt{S}$$

(b) Two competitive conversions:

$$A \begin{smallmatrix} \nearrow B \\ \searrow C \end{smallmatrix}$$

η is identical for both reactions, i.e.

$$S_{\text{eff}} = S$$

(c) Consecutive reactions:

$$A \xrightarrow{k_1} B \xrightarrow{k_2} C$$

It is readily seen that the selectivity is influenced considerably by diffusional inhibition, since the intermediate, B, formed in the pores stays longer in the catalyst grain, i.e. the higher its conversion, the stronger the diffusional inhibition hindering its escape. Without going into the details of the expressions derived for this scheme (for a detailed discussion

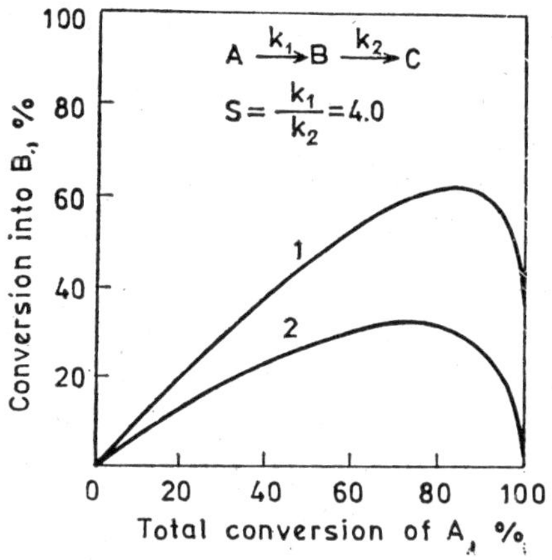

Fig. IV.8. Effect of the pore dimensions (length and size) on the yield of intermediate B in the case of selectivity of type (c)

1: wide pores, or catalyst of small grain size ($\varphi \ll 1$)

2: narrow pores, or catalyst of large grain size ($\varphi > 3$)

see [2]), the theoretical (calculated) product distribution is shown for $\frac{\dot{k}_1}{k_2} = 4$ both without diffusional inhibition ($\varphi < 1$), and when there is strong diffusional inhibition ($\varphi > 3$) (Fig. IV.8). It can be seen that in the inhibited process the yield into B decreases considerably.

During the dehydrogenation of cyclohexane on a chromia-alumina catalyst benzene is formed via the intermediate cyclohexene [6]. In the non-inhibited case the ratio of the rate constants is $S = 1/14$, which decreases to 1/230 when the catalyst grain size is increased.

SECTION 1.5. KINETIC RELATIONSHIPS FOR SOME TYPES OF CATALYTIC REACTIONS, CONSIDERING BOTH ADSORPTION AND DESORPTION RATES

The kinetic relationships for various types of reactions can be derived in an almost uniform manner; thus, a detailed derivation is presented only for a reaction of type I [Eq. (IV.29)], and for the others merely the results and their interpretation will be given.

The rates of adsorption and desorption are defined by the following relationship:

Consider the adsorption as a heterogeneous reaction between gas molecules and adsorption sites of a catalyst surface. Let us denote the free adsorption sites by S, and their surface concentration by $[S]$.

The adsorption then corresponds to the following "reaction"

$$A_i + S \rightleftarrows A_iS \equiv (A_i)_a; \qquad i = 1, \ldots, n$$
$$B_j + S \rightleftarrows B_jS \equiv (B_j)_a; \qquad j = 1, \ldots, l \tag{IV.64}$$

where $A_iS \equiv (A_i)_a$ and $B_jS \equiv (B_j)_a$ represent the adsorbed molecules of components A_i and B_j.

From Eq. (IV.64) the adsorption rates are:

$$J_{A_i} = A_s(\vec{k}_{A_i}\, p_{A_i}[S] - \overleftarrow{k}_{A_i}[A_i]_a); \qquad i = 1, \ldots n,$$
$$J_{B_j} = A_s(\vec{k}_{B_j}\, p_{B_j}[S] - \overleftarrow{k}_{B_j}[B_j]_a); \qquad j = 1, \ldots l$$
$$[S]_0 = [S] + \sum_i [A_i] + \sum_j [B_j]_a \tag{IV.65}$$

where: $\vec{k}_{A_i}$, $\vec{k}_{B_j}$ and $\overleftarrow{k}_{A_i}$, $\overleftarrow{k}_{B_j}$ are the rate constants of the adsorption and desorption for components A_i and B_j, respectively.

$[S]_0$ is the initial surface concentration of the adsorption sites.

In the case of reactions of type I, if pore diffusion is neglected the catalytic reaction consists of the following partial processes:

$$\begin{aligned} &1. \qquad A + S \rightleftarrows (A)_a \\ &2. \qquad (A)_a \rightleftarrows (B)_a \\ &3. \qquad (B)_a \rightleftarrows B + S \end{aligned} \tag{IV.66}$$

From Eqs (IV.30) and (IV.65) the rates of the individual partial processes are:

$$J_1 = A_s(\vec{k}_A p_A[S] - \overleftarrow{k}_A[A]_a) \equiv J_A$$
$$J_2 = A_s(\vec{k}[A]_a - \overleftarrow{k}[B]_a) \equiv J \qquad \text{(IV.67)}$$
$$J_3 = A_s(\overleftarrow{k}_B[B]_a - \vec{k}_B p_B[S]) \equiv -J_B$$

On the basis of the law of the conservation of matter the following expressions, analogous to Eq. (IV.8), can be formulated for the adsorption phase:

$$\left(\frac{dn_A}{dt}\right)_a = J_A - J, \quad \left(\frac{dn_B}{dt}\right)_a = J_B + J$$
$$[S]_0 = [S] + [A]_a + [B]_a \qquad \text{(IV.68)}$$

In steady, or quasi-steady, states, see Eq. (IV.10)], i.e. if

$$\left(\frac{dn_A}{dt}\right)_a = 0, \quad \left(\frac{dn_B}{dt}\right)_a = 0, \quad \text{or} \quad \left(\frac{dn_A}{dt}\right)_a \approx 0, \quad \left(\frac{dn_B}{dt}\right)_a \approx 0$$

then, from Eq. (IV.68) we have

$$J = J_A = -J_B \qquad \text{(IV.69)}$$

(In the discussion below, only the expressions valid for the steady state will be considered. For a quasi-steady state the sign = should be replaced by ≈.)

Let us define the following "equilibrium constants":

$$K_A \equiv \frac{\vec{k}_A}{\overleftarrow{k}_A}; \quad K_B \equiv \frac{\vec{k}_B}{\overleftarrow{k}_B}; \quad K \equiv \frac{\vec{k}}{\overleftarrow{k}} \qquad \text{(IV.70)}$$

where K_A and K_B are the "adsorption equilibrium constants" for components A and B, and
K is the equilibrium constant of the surface reaction.

Using Eq. (IV.70), the expressions in Eq. (IV.67) can be transformed:

$$J = A_s\vec{k}_A\left(p_A[S] - \frac{1}{K_A}[A]_a\right) = A_s\vec{k}\left([A]_a - \frac{1}{K}[B]_a\right) =$$
$$= A_s k_B([B]_a - K_B p_B[S])$$
$$[S]_0 = [S] + [A]_a + [B]_a \qquad \text{(IV.71)}$$

Analogous with Eq. (IV.19) we write:

$$K_A = \frac{[A]_a^*}{p_A[S]} = \frac{[A]_a}{p_A^*[S]}; \quad K = \frac{[B]_a^{**}}{[A]_a} = \frac{[B]_a}{[A]_a^{**}}; \quad K_B = \frac{[B]_a^*}{p_B[S]} = \frac{[B]_a}{p_B^*[S]}$$

Using these terms, three different expressions can be formulated for Eq. (IV.71):

$$\text{(a)}\quad \vec{k}_A p_A[S]\left(1-\frac{p_A^{**}}{p_A}\right)=\overleftarrow{k}[B]_a\left(\frac{[B]_a^{**}}{[B]_a}-1\right)=\overleftarrow{k}_B p_B[S]\left(\frac{p_B^*}{p_B}-1\right)=J/A_s$$

$$\text{(b)}\quad \vec{k}[A]_a\left(1-\frac{[A]_a^{**}}{[A]_a}\right)=\overleftarrow{k}_A[A]\left(\frac{[A]_a^*}{[A]_a}-1\right)=\overleftarrow{k}_B p_B[S]\left(\frac{p_B^*}{p_B}-1\right)=J/A_s$$

$$\text{(c)}\quad \overleftarrow{k}_B[B]_a\left(1-\frac{[B]_a^*}{[B]_a}\right)=\overleftarrow{k}_A[A]_a\left(\frac{[A]_a^*}{[A]_a}-1\right)=\overleftarrow{k}[B]_a\left(\frac{[B]_a^{**}}{[B]_a}\right)=J/A_s$$

(IV.72)

Depending upon the rate-controlling partial process various extreme cases can be distinguished:

(a)
$$\vec{k}_A p_A \ll \vec{k}\overleftarrow{k}_B p_B,\ \overleftarrow{k}_B p_B \qquad \text{(IV.73)}$$

$$p_B^* \approx p_B;\ [B]_a \approx K_B p_B[S];$$

$$[B]_a^{**} \approx [B]_a \equiv k[A]_a;\ p_A^* = \frac{1}{K}\frac{K_B}{K_A}p_B$$

The steady-state rate expression [Eq. (IV.71)] is as follows:

$$J \approx A_s[S]_0\vec{k}_A\frac{p_A-\dfrac{1}{K}\dfrac{K_B}{K_A}p_B}{1+\left(1+\dfrac{1}{K}\right)K_B p_B};\quad \vec{k}_A p_A \ll \vec{k}K_B p_B,\ \overleftarrow{k}_B p_B \qquad \text{(IV.74)}$$

It is seen from Eq. (IV.74) that the rate of the process depends upon the adsorption rate for component A only. Consequently, in this extreme case, the process is controlled by the adsorption of this component. From similar considerations the rate expressions can be obtained for the other extreme cases.

$$\text{(b)}\qquad J \approx A_s[S]_0\vec{k}K_A\frac{p_A-\dfrac{1}{K}\dfrac{K_A}{K_B}p_B}{1+K_A p_A+K_B p_B};\ \vec{k} \ll \overleftarrow{k}_A,\ \overleftarrow{k}_B\frac{p_B}{K_A p_A}$$

(IV.75)

$$\text{(c)}\qquad J \approx A_s[S]_0\overleftarrow{k}_B K_A K\frac{p_A-\dfrac{1}{K}\dfrac{K_B}{K_A}p_B}{1+(1+K)K_A p_A};\ \overleftarrow{k}_B \ll \overleftarrow{k}_A\frac{1}{K},\ \overleftarrow{k}$$

Of these equations the former represents the rate-controlling role of the surface reaction.

It can be seen from Eqs (IV.74) and (IV.75) that compared with the rate-controlling step, the other partial processes can be regarded as being equilibrium ones.

Substituting $K \gg 1$ into Eqs (IV.74) and (IV.75), the one-directional reaction can be treated as a particular extreme case. The partial processes are:

$$(1) \qquad A + S \rightleftarrows (A)_a; \; J_1 = J_A = A_s\vec{k}_A \left(p_A[S] - \frac{1}{K_A}[A]_a\right)$$

$$(2) \qquad A_s \rightarrow (B)_a; \; J_2 = J = A_s\vec{k}[A]_a \qquad \text{(IV.76)}$$

$$(3) \qquad (B)_a \rightleftarrows B + S; \; J_3 = -J_B = \overset{k}{A_s}k_B([B]_a - K_B p_B[S])$$

$$[S]_0 = [S] + [A]_a + [B]_a$$

The rate expressions derived from Eq. (IV.76) are:

$$J = A_s[S]_0\vec{k}_A \frac{p_A}{1 + K_A p_A}; \; \vec{k}_A p_A \ll \vec{k}, \; \overleftarrow{k}_B p_B$$

$$J = A_s[S]_0\vec{k}K_A \frac{p_A}{1 + K_A p_A + K_B p_B}; \; \vec{k} \ll \overleftarrow{k}_A, \frac{\overleftarrow{k}_B p_B}{K_A p_A} \qquad \text{(IV.77)}$$

$$J = A_s[S]_0\overleftarrow{k}_B; \; \vec{k} \ll \overleftarrow{k}_A; \; \overleftarrow{k}_B \ll \frac{\vec{k}K_A p_A}{1 + K_A p_A + K_B p_B}$$

Rate equations for the two reaction types are collected in Tables IV.1 and IV.2, assuming a number of probable reaction mechanism. (Of course, the set of mechanisms listed is not exhaustive.)

The assumptions as to the nature of the active sites and the adsorbed molecules (energetically homogeneous active sites, localized adsorption) used in this Chapter, with the surface reaction as the rate-controlling step, are termed the Langmuir–Hinshelwood mechanism in the literature concerning the kinetics of heterogeneous catalytic reactions. The extreme cases discussed for reactions of type II, II/a and II/b, in which the adsorption of every component is not necessary for a reaction to occur, are collectively known as the Rideal–Eley mechanism.

CHAPTER 2. KINETIC STUDY OF HETEROGENEOUS CATALYTIC REACTIONS

Section 2.1. LABORATORY REACTORS DESIGNED FOR RATE DETERMINATIONS

A chemical reactor is the equipment in which a chemical process can be performed.

A possible classification of chemical reactors is presented below.

- Static
  - Batch
- Dynamic
  - Flow
    - Integ.
    - Diff.
  - Pulse
  - Gradientless
    - Recirculation
    - Stirred
      - Fluidized
    - Flow

In this Section, the working principles and the construction of most common laboratory reactors used in the kinetic study of heterogeneous catalytic reactions are described.

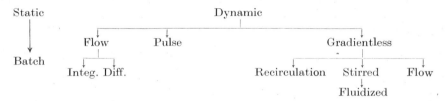

A reactor designed for kinetic investigations should satisfy the requirement that the process occurring in it be unambiguous and reproducible. This means that all the parameters affecting the rate of the process should be maintained at an *a priori* given reproducible level. Of these parameters, it is primarily the temperature upon which the rate of a heterogeneous catalytic reaction depends. Thus, the reactor should be suitable for maintaining a known, or a constant, temperature. Under isothermal conditions the kinetic relationships are much simpler than under adiabatic conditions. Accordingly, the working regime of laboratory reactors is usually isothermal; this is normally provided by keeping a constant temperature in the reaction zone of the reactor. This part of the reactor is, therefore, thermostated by means of auxiliary equipment. No matter how perfect the thermostating, a homogeneous temperature field cannot be expected, because of the finite heat conductivity in the system. Consequently, it is reasonable to decrease the amount of heat to be transferred. Another important factor is the geometric shape of the catalyst-supporting part of the reactor, which should provide the most favourable conditions for heat conduction. This is achieved with the largest possible heat-transfer area and the shortest possible path of heat conduction. The amount of heat to be transferred can be effectively decreased if the temperature of the reaction mixture at the moment it enters the pore phase is equal to the required reaction temperature; i.e. in a flow system, the reaction mixture should be appropriately preheated or cooled (naturally, homogeneous reactions cannot occur). Since chemical reactions are usually accompanied by heat effects, i.e. heat is lost or gained, the smaller the amount of reactants entering the reaction the less the amount of heat to be transferred. According to Eq. (IV.11), the rate can be calculated on the basis of the changes in the composition of the reaction mixture. In order that these changes may be determined with due accuracy, appropriate amounts of reactants must be used and the changes should be as great as possible. This requirement is the opposite of that for the isothermal condition (where the amounts of reactants must be as low as possible). Laboratory reactors must also be suitable for measuring the partial pressures and amounts of the components in the pore phase at an arbitrary moment, while the composition and amount of the reaction mixture in the bulk phase have to be determined as well. In order to perform this latter task, auxiliary equipment is usually employed which is not a standard part of the reactor. The determination of the amount and composition of the gas mixture in the pore phase is a very difficult problem. Since we are dealing with a gaseous reaction mixture, the simplest solution is the evaluation of these variables from data relating to the volume, the pressure and the temperature of the bulk phase, making use of the gas law. This requires information concerning the variation of these parameters along the catalyst layer. It is reasonable to simplify the problem by ensuring a constant pressure along the catalyst bed. To achieve this under conditions of permanent flow of the reaction mixture across the catalyst bed, the flow resistance and the flow rate should be as small as possible. A decrease in the thickness of the catalyst bed in the flow direction means a correspondingly lower flow resistance. (Flow occurs along the catalyst layer not only in flow reactors but also in static ones, if the reactions involve a change in the number of moles.) However, efforts directed towards obtaining the largest concentration change per unit time impose a limit on the thickness of the catalyst layer. A thin layer can be formed from a greater amount of catalyst by increasing the cross-section of the reactor; but this contradicts a condition necessary for the realization of isothermal conditions, namely that the cross-sectional area must be chosen to be as small as possible.

The determination of the gas composition in the pore phase gives rise to similar problems. The sampling may, in principle, be performed at an arbitrary site on the catalyst bed. The sample is then analyzed using auxiliary equipment. This sampling encounters technical problems, and it is therefore reasonable to analyze (with the aid of a suitable apparatus) the gas mixture in contact with the catalyst bed, as well as the gases at the inlet and the outlet of the reactor. From the data obtained in this analysis the concentration along the catalyst bed can be estimated. This procedure is the simplest if the concentration change in the pore phase is negligibly small. To achieve this, either a very thin layer must be formed or the gas must be intensively mixed. The drawbacks of the thin layer were mentioned above, while the mixing raises practical problems.

The equipment used for the analysis of the reaction mixture is an important, though not standard, element of reactors. The accuracy of the determination is reflected in that of the rate equation obtained. The accuracy of the concentration measurement depends mainly on the accuracy of the method involved, and also on the number of samples taken for the analysis, i.e. the number of repetitions. From this aspect it is reasonable to use a reactor functioning in the steady state; in this case the concentration can be measured using an arbitrary number of simultaneous samples. This disadvantage of this type of reactor is that the measurement requires a longer period. If the activity of the catalyst changes with time, and the period necessary for the steady state to be established is significant from the point of view of ageing, the steady state necessary cannot be achieved. Thus, in the above case there is no sense in using such a reactor.

From what has been said it can be seen that heterogeneous catalytic reactors must satisfy several different, partly contradictory, requirements. The laboratory reactors commonly used can be distinguished by the criterion of which of the above-mentioned properties is preferred in their construction. Obviously, these reactors differ from the aspect of the technical level of realization. The conclusion can be drawn that the choice of reactor is influenced primarily by the nature of the task to be solved and by the available technical possibilities.

PARAGRAPH 2.1.1. STATIC (BATCH) REACTORS

Static reactors constitute closed systems, i.e. the overall mass of the reaction components located in the different parts of the reactor connected by mass transport is constant. These types are used mainly in the study of reactions with mole changes. (Naturally they can be used if the amounts of the components remain constant. However, in these cases, the observation of changes in the pressure or the volume is not suitable to determine the variation of the gas composition with time.) The static reactor consists of two basic parts (Fig. IV.9a). These are: the catalyst vessel with free gas volume v_r and with the temperature adjusted to be equal to the reaction temperature T_r. It contains catalyst of mass m_s. This vessel is connected to measuring equipment of volume v_m and temperature T_m. It includes instruments for the determination of changes in pressure and volume. When the amounts of the components are calculated, the connecting tubes of volume v_d and temperature T_d must also be taken into account.

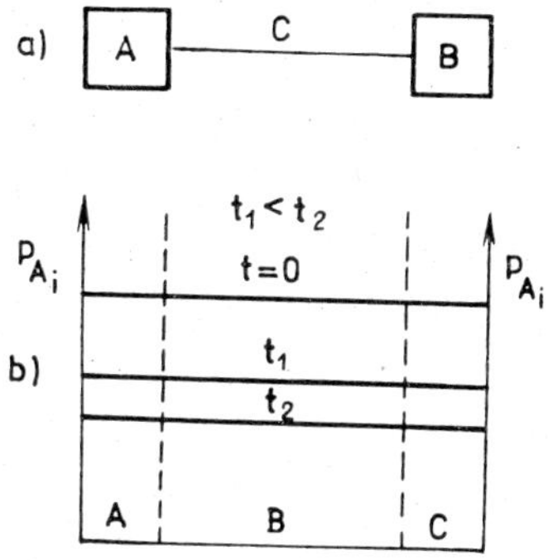

Fig. IV.9.

(a) Scheme of batch reactor A: catalyst vessel; B: auxiliary equipment for the determination of changes in pressure and volume; C: connecting tube
(b) distribution of partial pressure of the starting component along the individual parts of the batch reactor at different moments of time

The rate of the catalytic reaction is calculated according to Eq. (IV.11) with the necessary modifications for the particular closed system involved:

$$\left(\frac{dn_{A_i}}{dt}\right)_e = 0; \left(\frac{dn_{B_j}}{dt}\right)_e = 0$$

$$J = \nu_{A_i}^{-1}\left(\frac{\mathrm{d}n_{A_i}}{\mathrm{d}t}\right)_{\mathrm{g}} = \nu_{B_j}^{-1}\left(\frac{\mathrm{d}n_{B_j}}{\mathrm{d}t}\right)_{\mathrm{g}} \qquad \text{(IV.78)}$$

$$i = 0, \ldots n; \; j = 0, \ldots l$$

(This expression is valid if only one reaction must be considered. In the following this is always assumed.)

The total amount of gas in the system $(n)_{\mathrm{g}}$ is:

$$(n)_{\mathrm{g}} = \sum_i n_{A_i} + \sum_j n_{B_j} \qquad \text{(IV.79)}$$

From Eqs (IV.78) and (IV.79):

$$J = \frac{1}{\sum_i \nu_{A_i} + \sum_j \nu_{B_j}}\left(\frac{\mathrm{d}n}{\mathrm{d}t}\right)_{\mathrm{g}}; \qquad \text{(IV.80)}$$

$$\sum_i \nu_{A_i} + \sum_j \nu_{B_j} \neq 0$$

There are three parts in the reactor with different temperatures. Accordingly, we have:

$$(n)_{\mathrm{g}} = (n)_{\mathrm{p}} + (n)_{\mathrm{m}} + (n)_{\mathrm{d}}$$

If there is no pressure difference between the different parts of the reactor, and the ideal gas law holds, the above expression can be rewritten:

$$(n)_{\mathrm{g}} = \frac{p}{R}\left(\frac{v_{\mathrm{p}}}{T_{\mathrm{p}}} + \frac{v_{\mathrm{m}}}{T_{\mathrm{m}}} + \frac{v_{\mathrm{d}}}{T_{\mathrm{d}}}\right) \qquad \text{(IV.81)}$$

where R is the universal gas constant, and
p is the pressure of the reaction mixture.
$T_{\mathrm{p}} = T_{\mathrm{r}}$.

On the basis of Eqs (IV.80) and (IV.81), if v_{p}, v_{m} and v_{e} are constant the reaction rate can be expressed as follows:

$$J = \frac{1}{\sum_i \nu_{A_i} + \sum_j \nu_{B_j}} \frac{1}{R}\left(\frac{v_{\mathrm{p}}}{T_{\mathrm{p}}} + \frac{v_{\mathrm{m}}}{T_{\mathrm{m}}} + \frac{v_{\mathrm{d}}}{T_{\mathrm{d}}}\right)\frac{\mathrm{d}p}{\mathrm{d}t} \qquad \text{(IV.82)}$$

$$\sum_i \nu_{A_i} + \sum_j \nu_{B_j} \neq 0$$

Eq. (IV.82), derived for the determination of the reaction rate, is valid only if the overall pressures are identical in the various parts of the reactor. If connecting tubes of suitable diameter are used, the above condition is easily satisfied. The objective of the rate determination involves the elucidation of the form of the rate equation, i.e. of the function $J(\ldots p_{A_i} \ldots p_{B_j})$. This task requires data on partial pressures $(p_{A_i})_{\mathrm{r}}$ and $(p_{B_j})_{\mathrm{r}}$ in the pore phase. The values of these partial pressures can be calculated from the total pressure if the partial pressure of the given component is identical in each part of the reactor. In this case Eq. (IV.81) can be written for any of the components:

$$(n_{A_i})_g = \frac{1}{R}\left(\frac{v_p}{T_p} + \frac{v_m}{T_m} + \frac{v_d}{T_d}\right) p_{A_i}; \quad i = 1, \ldots n$$

$$(n_{B_j})_g = \frac{1}{R}\left(\frac{v_p}{T_p} + \frac{v_m}{T_m} + \frac{v_d}{T_d}\right) p_{B_j}; \quad j = 1, \ldots l \qquad \text{(IV.83)}$$

From Eqs (IV.78), (IV.82) and (IV.83) we have:

$$\Delta p_{A_i} = \frac{\nu_{A_i}}{\sum_j \nu_{B_j} + \sum_i \nu_{A_i}} \Delta p$$

$$\Delta p_{B_j} = \frac{\nu_{B_j}}{\sum_j \nu_{B_j} + \sum_i \nu_{A_i}} \Delta p$$

In accordance with Eq. (IV.82), if the volumes (v_p, v_m, v_d) and temperatures (T_p, T_m, T_d) of the different parts, and the stoichiometric numbers (ν_{A_i}, ν_{B_j}) are known, the rate of the heterogeneous catalytic reaction (J) can be obtained by graphical differentiation of the experimentally determined time-dependence of the gas pressure $p(t)$.

The partial pressure differences between the various parts are compensated by diffusion. Thus, the diffusion must be assumed to be fast enough to compensate the partial pressure differences that arise due to the reaction. However, this is not probable, especially if fast reactions are involved, and this type of reactor cannot be used to study rapid reactions. A further disadvantage is that the compensation of the concentration differences along the catalyst layer is accomplished only by diffusion. This is another reason why fast reactions cannot be studied in these reactors if a powdered catalyst is used. The variation of the partial pressure of the starting component (p_{A_i}) along the individual parts of the reactor is shown in Fig. IV.9b for different moments of time (t).

PARAGRAPH 2.1.2. DYNAMIC REACTORS

(1a) Gradientless recirculation reactor

The above-mentioned disadvantages of static reactors (i.e. that the partial pressure differences are compensated by diffusion only) are eliminated in recirculation reactors, which also constitute a closed system (the overall mass of the reaction components is constant) (Fig. 10). If the rate of gas circulation is sufficiently large compared with the reaction rate, i.e. only negligible composition changes occur while the reaction mixture passes over the catalyst layer, the reactor can be specified as gradientless. This condition can be realized by decreasing the amount of catalyst or increasing the rate of circulation. When the amount of catalyst is reduced the reaction time (or, if the pressure is measured, the time necessary for a finite change of pressure) increases. Consequently, the reduction of the amount of catalyst is limited by the magnitude of the reaction rate. On the other hand, the circulation rate cannot be increased indefinitely either, since this would complicate the heating or cooling of the gas mixture. In this case, the temperature of the reaction mixture could be adjusted by using a heat-exchanger of appropriate length. However, if the temperature in the various reactor parts are very different, there is a considerable gas volume in which the temperature is poorly defined. Since Eq. (IV.82), from which the rate

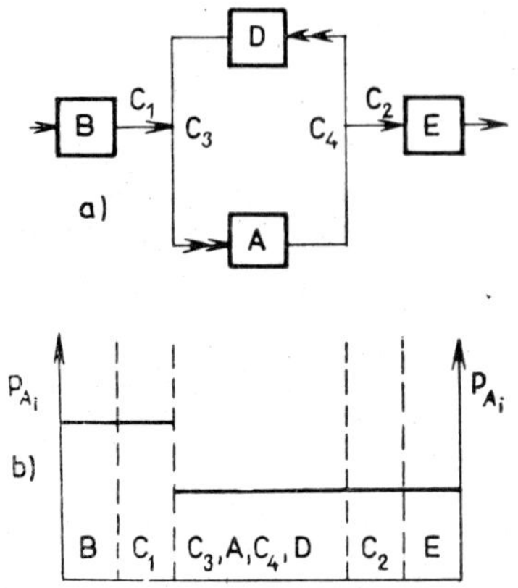

Fig. IV.10. (a) Scheme of gradientless recirculation reactor. *A*: catalyst vessel; *B*: auxiliary part for determination of changes in pressure and volume; *C*: connecting tubes (b) distribution of partial pressure of the starting component along the individual parts of the gradientless recirculation reactor at different moments of time

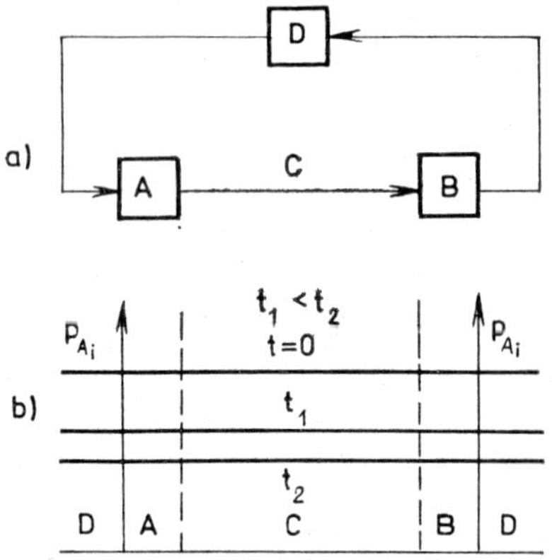

Fig. IV.11. (a) Scheme of gradientless stirred flow reactor. *A*: catalyst vessel; *B*: apparatus providing the feed of the reaction mixture with constant composition and mass flow; *D*: recycling equipment controlling the mass flow of the outgoing reaction mixture; C_1, C_2, C_3 and C_4: connecting tubes (b) Change of partial pressure of the starting component in the various parts of the gradientless stirred flow reactor

is determined, includes the temperature and the volume of the connecting parts, the uncertainty in the value of T_d (with v_d being commensurable to v_m and v_r) causes an appreciable uncertainty in J.

The gradientless recirculation reactor constitutes a closed system. Thus, if the moles of the reaction components are changed during the reaction, Eq. (IV.82) can still be used to determine the reaction rate. However, if the total amount of the components remains constant then the partial pressure of any of the components can be measured by a suitable analytical method (e.g. by gas chromatography). In this case, from Eqs (IV.78) and (IV.83) the reaction rate is:

$$J = \nu_{A_i}^{-1} \frac{1}{R} \left(\frac{v_\mathrm{p}}{T_\mathrm{p}} + \frac{v_\mathrm{m}}{T_\mathrm{m}} + \frac{v_\mathrm{d}}{T_\mathrm{d}} \right) \frac{\mathrm{d}n_{B_j}}{\mathrm{d}t} = \nu_B^{-1} \frac{1}{R} \left(\frac{v_\mathrm{p}}{T_\mathrm{p}} + \frac{v_\mathrm{m}}{T_\mathrm{m}} + \frac{v_\mathrm{d}}{T_\mathrm{d}} \right) \frac{\mathrm{d}n_{B_j}}{\mathrm{d}t} \tag{IV.84}$$

$$i = 1, \ldots n; \; j = 1, \ldots l$$

(1b) Gradientless stirred flow reactors

A gradientless regime can also be established in flow reactors, either by appropriate stirring of the catalyst (fluidized catalyst bed) or by intense mixing of the gas mixture. The fluidization brings about the powdering of the catalyst, and it is difficult to establish conditions that are really gradientless. This type is therefore not used for kinetic purposes. The Tyomkin–Kiperman reactor [7] is of the gas circulation type; it includes two gas circuits (Fig. IV.11a).

The inner circuit is essentially a recirculation reactor, while the outer one ensures continuous flow. The partial pressure change of the starting component (p_{A_i}) in the steady state along the different reactor parts is shown in Fig. IV.11b.

The steady state can be reached by stirred flow gradientless reactors provided that the inlet and outlet rates and the inlet composition remain constant, i.e.:

$$\left(\frac{\mathrm{d}n_{A_i}}{\mathrm{d}t}\right)_{\mathrm{g}} = 0, \quad \left(\frac{\mathrm{d}n_{B_j}}{\mathrm{d}t}\right)_{\mathrm{g}} = 0$$

$$i = 1, \ldots n; \quad j = 1, \ldots l$$

According to Eq. (IV.11), the stationary reaction rate can be expressed as follows:

$$J = \nu_{A_i}^{-1}\left(\frac{\mathrm{d}n_{A_i}}{\mathrm{d}t}\right)_{\mathrm{e}} = \nu_{B_j}^{-1}\left(\frac{\mathrm{d}n_{B_j}}{\mathrm{d}t}\right)_{\mathrm{e}} \qquad \text{(IV.85)}$$

$$i = 1, \ldots n; \quad j = 1, \ldots l$$

Eq. (IV.85) can be rewritten in a form suitable for calculations by expressing the constant inlet rate as mass flow (F), and relating the component concentrations (y) to unit mass of the gas mixture. In this case, according to the law of conservation of matter we have:

$$\left(\frac{\mathrm{d}n_{A_i}}{\mathrm{d}t}\right)_{\mathrm{e}} = F[(y_{A_i})_1 - (y_{A_i})_0]; \quad i = 1, \ldots n$$

$$\left(\frac{\mathrm{d}n_{B_j}}{\mathrm{d}t}\right)_{\mathrm{e}} = F[(y_{B_j})_1 - (y_{Bj})_0]; \quad j = 1, \ldots l \qquad \text{(IV.86)}$$

where the subscripts 0 and 1 refer to mixtures entering and leaving the reactor, respectively.

From Eqs (IV.85) and (IV.86):

$$J = \nu_{A}^{-1} F[(y_{A_i})_1 - (y_{A_i})_0] = \nu_{B_j}^{-1} F[(y_{B_j})_1 - (y_{B_j})_0] \qquad \text{(IV.87)}$$

It is clear from Eq. (IV.87) that direct calculation of the reaction rate is possible from data relating to the mass velocity and the compositions of the incoming and outgoing gases, and no graphical differentiation is necessary. Taking into account the errors inherent in graphical differentiation, the direct determination of the reaction rate from the experimental data is a significant advantage of this type. In accordance with the conditions of the gradientless regime, the concentration of the gas mixture is uniform along the catalyst layer and thus the reaction rate obtained from Eq. (IV.87) relates to concentrations $(y_{A_i})_1$ and $(y_{B_j})_1$.

(2a) Dynamic integral reactor

In this type of reactor (Fig. IV.12a) there is a concentration gradient along the catalyst bed, in contrast to the gradientless dynamic reactors. Thus back-mixing can be neglected. In the steady state, with the inlet rate and composition constant, the infinitely small element of the reactor (r) chosen in the flow direction can be treated as a gradientless reactor to which Eq. (IV.87) can be applied.

$$\mathrm{d}J_r = \nu_{A_i}^{-1} F \, \mathrm{d}y_{A_i} = \nu_B^{-1} F \, \mathrm{d}y_{B_j} \qquad \text{(IV.88)}$$

$$i = 1, \ldots n; \; j = 1, \ldots l$$

If there is an amount of catalyst $\mathrm{d}m_\mathrm{s}$ in this reactor element r, and the reaction rate related to unit mass of catalyst is $\bar{\jmath}$ the following expression holds:

$$\mathrm{d}J_\mathrm{r} = \bar{\jmath}\mathrm{d}m_\mathrm{s} \qquad \text{(IV.89)}$$

From Eqs (IV.88) and (IV.89):

$$\bar{\jmath} = \nu_{A_i}^{-1} F \frac{\mathrm{d}y_{A_i}}{\mathrm{d}m_\mathrm{s}} = \nu_{B_j}^{-1} F \frac{\mathrm{d}y_{B_j}}{\mathrm{d}m_\mathrm{s}} \qquad \text{(IV.90)}$$

$$i = 1, \ldots n; \quad j = 1, \ldots l$$

The specific reaction rate, $\bar{\jmath}$, depends, in principle, on the component concentrations y_{A_i} and y_{B_i}. If only one reaction proceeds in the pore phase, then, in the steady state, there is an unequivocal relationship between the concentration changes of the components in agreement with Eq. (IV.88), i.e.:

$$\nu_{A_i}^{-1}[(y_{A_i})_1 - (y_{A_i})_0] = \nu_{B_j}^{-1}[(y_{B_j})_1 - (y_{B_j})_0]$$

Thus, if the composition of the gas mixture at the inlet $((y_{A_i})_0, (y_{B_j})_0)$ is given, and the concentration of a starting component is determined at an arbitrary site of the reactor, the concentrations of the rest of the components can be calculated using the stoichiometric numbers. Consequently, Eq. (IV.90) can be integrated for the two ends of the catalyst bed (in the flow direction):

$$\frac{1}{S} = \nu_{A_i}^{-1} \int_{(y_{A_i})_0}^{(y_{A_i})_1} \frac{\mathrm{d}y_{A_i}}{\bar{\jmath}} = \nu_{B_j}^{-1} \int_{(y_{B_j})_0}^{(y_{B_j})_1} \frac{\mathrm{d}y_{B_j}}{\bar{\jmath}} \qquad \text{(IV.91)}$$

where S is the space velocity:

$$S = \frac{F}{m_\mathrm{s}}$$

According to Eq. (IV.91), if the inlet concentration has been adjusted suitably, then the composition of the gas mixture leaving the catalyst bed is a function of the specific rate ($\bar{\jmath}$) and the space velocity (S). Thus, from Eq. (IV.91):

$$\bar{j}_1 = \nu_{A_i}^{-1} \frac{\mathrm{d}(y_{A_i})_1}{\mathrm{d}\left(\frac{1}{S}\right)} = \nu_{B_j}^{-1} \frac{\mathrm{d}(y_{B_j})}{\mathrm{d}\left(\frac{1}{S}\right)}; \qquad \begin{array}{l} (y_{A_i})_0 = \text{const.} \\ (y_{B_j})_0 = \text{const.} \end{array} \qquad \text{(IV.92)}$$

By means of Eq. (IV.92) the rate expression $\bar{j}(\ldots (y_{A_i})_1 \ldots (y_{B_j})_1 \ldots)$ can, in principle, be determined in the following way: the value of S is changed by either the catalyst amount, m_s, or the feed rate, F, and the magnitudes of $(y_{A_i})_1$ and $(y_{B_j})_1$ can be determined at constant feed composition. The values of $\bar{j}$ can be obtained as a function of the composition of the gas mixture leaving the catalyst bed by graphical differentiation of the plot of $(y_{A_i})_1$ versus $1/S$.

This type of reactor is commonly used, primarily because of its similarity to industrial tube reactors. However, from the kinetic aspect it has many disadvantages owing to the difficulties encountered in the calculation of the rate.

The variation of the partial pressure of the starting component along the different parts of the reactor is shown in Fig. IV.12b.

(2b) Dynamic differential reactor

The case when the concentration change along the catalyst bed is approximately linear can be treated as a separate type of reactor. This condition is usually satisfied if the extent of conversion is small. Reactors functioning according to the above principle are therefore called dynamic differential reactors (Fig. IV.13a). If the concentration change is assumed to be linear, Eq. (IV.91) can be transformed:

$$\frac{m_s}{F} = \frac{1}{S} = \nu_{A_i}^{-1} \frac{(y_{A_i})_1 - (y_{A_i})_0}{\bar{j}^*} = \nu_{B_j}^{-1} \frac{(y_{B_j})_1 - (y_{B_j})_0}{\bar{j}^*} \qquad \text{(IV.93)}$$

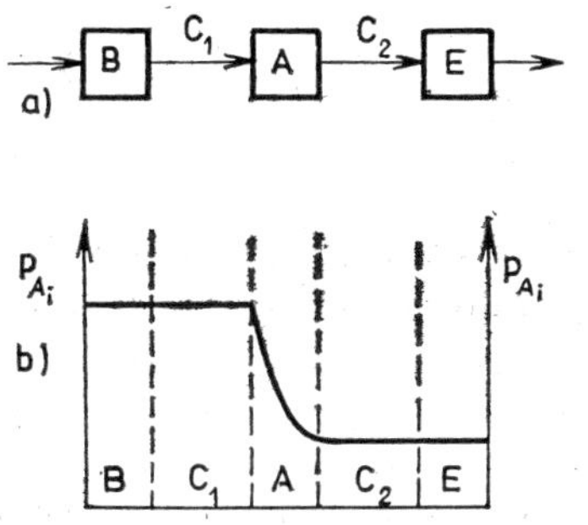

Fig. IV.12. (a) Scheme of integral dynamic reactor. A: catalyst vessel; B: apparatus providing the feed of the reaction mixture with constant composition and mass flow; E: equipment controlling the mass flow of the outgoing reaction mixture; C_1, C_2, C_3 and C_4: connecting tubes (b) Change of partial pressure of the starting component in the various parts of the integral dynamic reactor

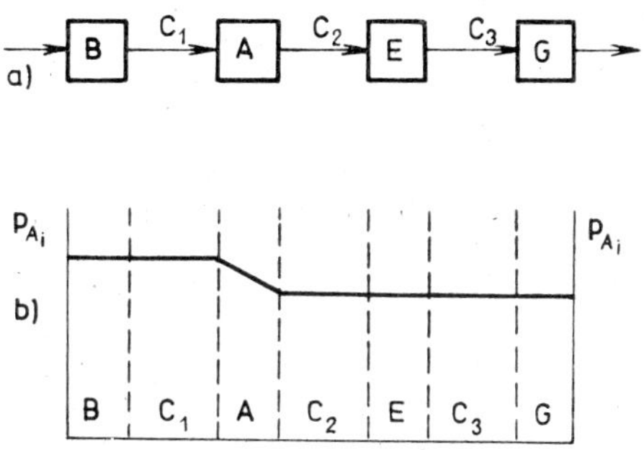

Fig. IV.13. (a) Scheme of differential dynamic reactor. A: catalyst vessel; B: apparatus providing the feed of the reaction mixture with constant composition and mass flow; E: equipment controlling the mass flow of the outgoing reaction mixture; G: equipment for identification of one of the reactants; C_1, C_2, C_3: connecting tubes (b) Change of partial pressure of the starting component in the various parts of the differential dynamic reactor

$$i = 1, \ldots n; \; j = 1, \ldots l$$

where $\bar{j}^*$ is the specific reaction rate referring to the average concentrations (y^*_{Ai}) and (y^*_{Bj}) in the pore phase. These can be calculated for linear concentration change as follows:

$$y^*_{A_i} = \frac{(y_{A_i})_0 + (y_{A_i})_1}{2} \; ; \; y^*_{A_j} = \frac{(y_{B_j})_0 + (y_{B_j})_1}{2}$$

If the differential reactor functions in the steady state for a period Δt, and during this period amounts $(n_{A_i})_1$ and $(n_{B_j})_1$ were fed into the catalyst bed and amounts $(n_{A_i})_1$ and $(n_{B_j})_1$ left, then according to the law of conservation of matter:

$$(n_{A_i})_1 - (n_{A_i})_0 = F\Delta t[(y_{A_i})_1 - (y_{A_i})_0] \quad i = 1, \ldots n \qquad \text{(IV.94)}$$

$$(n_{B_j})_1 - (n_{B_j})_0 = F\Delta t[(y_{B_j})_1 - (y_{B_j})_0] \quad j = 1, \ldots l$$

From Eqs (IV.93) and (IV.94) we have:

$$\bar{j}^* m_s = \nu_{A_i}^{-1} \frac{(n_{A_i})_1 - (n_{A_i})_0}{\Delta t} = \nu_{B_j}^{-1} \frac{(n_{B_j})_1 - (n_{B_j})_0}{\Delta t}$$

$$i = 1, \ldots n; \qquad j = 1, \ldots l \qquad \text{(IV.95)}$$

In principle, $\bar{j}^*$ can be calculated from Eq. (IV.95).

As Eq. (IV.95) shows, it is practical to use the difference $(n_{B_j})_1 - (n_{B_j})_0$ to calculate the rate. Accordingly, this type of reactor provides suitable accuracy primarily when $(n_{B_j})_0 = 0$. Naturally, in the elucidation of the dependence of the rate, the partial pressure of component B_j must be abandoned.

The variation of the partial pressure of the starting component along the different parts of the reactor is shown schematically in Fig. IV.13b.

SECTION 2.2. PRINCIPLES OF THE DETERMINATION OF KINETIC AND RATE EQUATIONS

The preceding Section dealt with the various types of laboratory reactors and the principles of the determination of the rate. However, a sharp distinction must be made between kinetic and rate equations. The rate equation is the mathematical relationship with which the experimentally determined rate values can be described within a certain approximation. The kinetic equation is the rate expression derived from a knowledge of the mechanism and the elementary partial processes. Naturally, the experimentally determined rate values can be described by this latter equation, too. In some favourable cases the kinetic and rate equations are identical. However, this equality is not necessary, since there may be several mathematical expressions (rate equations) which would satisfy the experimental data within a certain approximation. It is obvious that each of these cannot, at the same time, be the kinetic equation. Experimentally determined kinetic or rate equations must therefore be considered with some care. This is particularly so with regard to the physical meanings of the parameters in these equations. Several kinetic equations will be shown below, the physical meanings of which are different, but which nevertheless appear in the same mathematical form. At any event, rate equations are of great use in the field of reactor design, giving a functional description of experimental data. Special care should be exercised if rate equations including fractional powers of concentrations are used. No physical meaning can be attributed to the parameters in these equations. They can be applied in reactor design only for that range of partial pressures in which the determination was accomplished. The same refers to the temperature.

The rate equation is also useful for the determination of the kinetic equation.

The rate equation is a functional relationship between the partial pressures of the components (amount or surface of the catalyst) and the temperature. Since laboratory reactors are usually operated under isothermal conditions, the relationship between the rate and the component concentrations must primarily be elucidated. The temperature-dependence of the parameters in this function, determined in a series of experiments at various temperatures, provides the temperature-dependence of the kinetic or rate equation.

Of the types of reactors mentioned above, rate values can be determined directly in the gradientless, stirred flow, and dynamic differential reactors. In the others the so-called integral curve can be obtained, i.e. the time-dependence of either the total pressure or the partial pressure (or amount) of any of the components. The rate is found by graphical differentiation of these integral curves. This procedure gives only approximate values owing to the errors of graphical differentiation. A practical method of determining the rate equation is the following: the initial partial pressures of the components are varied and, using the graphical representation of the experimental results, the rates corresponding to various partial pressures are found. Those components are then selected whose partial pressures affect the magnitude of the rate. To this end, rate values are chosen corresponding to the variation of the partial pressure of only one of the components. In this way it can be decided whether the rate depends upon the partial pressure of the component concerned. Further, using the method of trial and error a suitable transformation can be found which reveals the mathematical form of the rate equation. This procedure is tedious but the only practical one in a number of cases. Another method of elucidating the kinetic equation is to examine the validity of several, a priori chosen, kinetic equations. This method is suitable primarily for simple reactions. In the case of gradientless recirculation and dynamic integral reactors the adequacy of kinetic equations can be confirmed by examining the integral forms of these equations. The simplest way of testing the validity of a mathematical formula is as follows: a suitable transformation is found (if any) which linearizes this formula. The experimental data are plotted according to this transformation. If a straight line is obtained in this representation then the above formula adequately describes the phenomenon involved: the parameters in this formula, or their functions, can be calculated from the slope and the intercept of the straight line. If there is no suitable transformation for the expected relationship, however, then the following method can be employed: the experimental data are substituted into the variables of the chosen relationship. This leads to a system of algebraic equations the number of which is equal to the number of experimental points. The solution of this system give the parameter values. If these latter are constant the adequacy of the formula is proved. This procedure is naturally very long and tedious, and it is frequently inaccurate owing to the experimental errors. An effort should therefore be made to linearize the formula involved, or another one must be chosen that can be linearized. The latter is advised even if the new relationship does not refer to directly measured data but, for example, to slopes obtained by graphical differentiation of the experimentally recorded integral curve.

The kinetic relationships for two simple reaction types valid under extreme conditions were presented in Section 1.5 (Tables IV.1 and IV.2). These kinetic equations are quite complicated even for these simple cases, and their integration often raises serious problems. The integrated forms cannot usually be linearized. The adequacy of the kinetic equations will therefore be examined further, since in the above extreme cases they can always be linearized. Naturally, it can occur that the integrated form is also amenable to linearization. It is then reasonable to confirm the applicability of this form, too. For the kinetic equations derived in Section 1.5, the conditions of validity were also presented. Accordingly, cases can be encountered to which none of the kinetic equations collected in Tables IV.1 and IV.2 can be applied, or the reaction proceeds by a mechanism different from those considered during the derivation of these equations.

PARAGRAPH 2.2.1. DETERMINATION OF THE KINETIC EQUATION FOR CATALYTIC REACTIONS $A \rightarrow B$

For reactions $A \rightarrow B$ three extreme cases can be distinguished (Table IV.1), according to which of the following steps determines the reaction rate: adsorption of component A (I/a–1), surface reaction (I/a–2), or desorption of the product (I/a–3). There are also several extreme cases as regards the mathematical form of the kinetic equation. These are collected in Table IV.3, which contains the mathematical form of the rate equation; the kinetic equation in accordance with Table IV.1 (where possible, the parameters are collected into one symbol); the kinetic type number (see Table IV.1); the validity region of the kinetic equation; the expression for the collective constants of the kinetic equation; and the relationship between the constants in the rate and kinetic equations. It can be seen from this Table that, if the kinetic equations appearing in Table IV.1 are applicable, three extreme classes can be distinguished from the point of the mathematical form of the rate equation:

TABLE IV.1

Kinetic equations for the catalytic reactions $A \rightleftarrows B$ and $A \rightarrow B$

p = gas pressure; A_s = surface area of catalyst; k = rate constant; $[S]$ = surface concentration of active sites; K = equilibrium constant; J = reaction rate

Type	Partial processes and their rate equations	Rate-controlling elementary partial process	Kinetic equation for the catalytic reaction	Type No.
I	$A + S \rightleftarrows (A)_a$ (1) $J_1 = A_s(\overrightarrow{k}_A p_A[S] - \overleftarrow{k}_A[A]_a)$ $(A)_a \rightleftarrows (B)_a$ (2)	(1) $A + S \rightarrow (A)_a$ $\overrightarrow{k}_A p_A \ll \overleftarrow{k} K_B p_B,\ \overrightarrow{k}_B p_B$	$J \approx A_s[S]_0 \overrightarrow{k}_A \dfrac{p_A - \dfrac{1}{K}\dfrac{K_B}{K_A} p_B}{1 + \left(1 + \dfrac{1}{K}\right) K_B p_B}$	I — 1
	$J_2 = A_s(\overrightarrow{k}[A]_a - \overleftarrow{k}[B]_a)$ $(B)_a \rightleftarrows B + S$ (3)	(2) $(A)_a \rightarrow (B)_a$ $\overrightarrow{k} \ll \overleftarrow{k}_A \dfrac{\overrightarrow{k}_B p_B}{K_A p_A}$	$J \approx A_s[S]_0 K_A \overrightarrow{k} \dfrac{p_A - \dfrac{1}{K}\dfrac{K_B}{K_A} p_B}{1 + K_A p_A + K_B p_B}$	I — 2
	$J_3 = A_s(\overleftarrow{k}_B[B]_a - \overrightarrow{k}_B p_B[S])$ $K_A = \dfrac{\overrightarrow{k}_A}{\overleftarrow{k}_A}$; $K_B = \dfrac{\overrightarrow{k}_B}{\overleftarrow{k}_B}$; $K \equiv \dfrac{\overrightarrow{k}}{\overleftarrow{k}}$ $[S]_0 = [S] + [A]_a + [B]_a$	(3) $(B)_a \rightarrow B + S$ $\overleftarrow{k}_B \ll \dfrac{\overleftarrow{k}_A}{K},\ \overleftarrow{k}$	$J \approx A_s[S]_0 K_A K \overleftarrow{k}_B \dfrac{p_A - \dfrac{1}{K}\dfrac{K_B}{K_A} p_B}{1 + (1 + K) K_A p_A}$	I — 3
I/a	$A + S \rightleftarrows (A)_a$ (1) $J_1 = A_s(\overrightarrow{k}_A p_A[S] - \overleftarrow{k}_A[A]_a)$ $(A_a) \rightarrow (B)_a$ (2)	(1) $A + S \rightarrow (A)_a$ $k_A p_A \ll \overrightarrow{k},\ k_B p_B$	$J \approx A_s[S]_0 \overrightarrow{k}_A \dfrac{p_A}{1 + K_B p_B}$	I/a — 1
	$J_2 = A_s \overrightarrow{k}[A]_a$ $(B)_a \rightleftarrows B + S$ (3)	(2) $(A)_a \rightarrow (B)_a$ $\overrightarrow{k} \ll \overleftarrow{k}_A \dfrac{\overrightarrow{k}_B p_B}{K_A p_A}$	$J \approx A_s[S]_0 K_A \overrightarrow{k} \dfrac{p_A}{1 + K_A p_A + K_B p_B}$	I/a — 2
	$J_3 = A_s[\overleftarrow{k}_B[B]_a - \overrightarrow{k}_B p_B[S])$ $K_A \equiv \dfrac{\overrightarrow{k}_A}{\overleftarrow{k}_A}$, $K_B \equiv \dfrac{\overrightarrow{k}_B}{\overleftarrow{k}_B}$ $[S]_0 = [S] + [A]_a + [B]_a$	(3) $(B)_a \rightarrow B + S$ $\overrightarrow{k} \ll \overleftarrow{k}_A,\ \overleftarrow{k}_B \ll \dfrac{\overrightarrow{k} K_A p_A}{1 + K_A p_A + K_B p_B}$	$J \approx A_s[S]_0 \overleftarrow{k}_B$	I/a — 3

TABLE

Kinetic equations for the catalytic reaction $A \rightleftarrows B_1 + B_2$, p = gas pressure; A_s = sites; K = equilibrium con-

Type	Partial processes and their rate equations	Rate-controlling elementary partial process
II/1	$A + S \rightleftarrows (A)_a$ (1) $J = A_s(\overrightarrow{k}_A p_A[S] - \overleftarrow{k}_A[A]_a)$ $(A_a) \rightleftarrows (B_1)_a + B_2$ (2) $J = A_s(\overrightarrow{k}[A]_a - \overleftarrow{k}[B]_a p_{B_2})$ $(B_1)_a \rightleftarrows B_1 + S$ (3) $J = A_s(\overleftarrow{k}_{B_1}[B_1]_a - \overrightarrow{k}_{B_1} p_{B_1}[S])$ $K_A \equiv \frac{\overrightarrow{k}_A}{\overleftarrow{k}_A}$; $K_{B_1} \equiv \frac{\overrightarrow{k}_{B_1}}{\overleftarrow{k}_{B_1}}$; $K = \frac{\overrightarrow{k}}{\overleftarrow{k}}$ $[S]_0 = [S] + [A]_a + [B_1]_a$	(1) $A + S \rightarrow (A)_a$ $\overrightarrow{k}_A p_A \ll \overleftarrow{k} K_{B_1} p_{B_1} p_{B_2}, \overrightarrow{k}_{B_1} p_{B_2}$ (2) $(A)_a \rightarrow (B_1)_a + B_2$ $\overleftarrow{k} \ll \overleftarrow{k}_A, \frac{\overrightarrow{k}_{B_1} p_{B_1}}{K_A p_A}$ (3) $(B_1)_a \rightarrow B_1 + (S)$ $\overleftarrow{k}_{B_1} \ll \frac{\overleftarrow{k}_A p_{B_2}}{K}, \overleftarrow{k} p_{B_2}$
II/2	$A + S \rightleftarrows (A)_a$ (1) $J_1 = A_s(\overrightarrow{k}_A p_A[S] - \overleftarrow{k}_A[A])_a$ $(A)_a + S \rightleftarrows (B_1)_a + (B_2)_a$ (2) $J_2 = A_s(\overrightarrow{k}[A]_a[S] - \overleftarrow{k}[B_1]_a[B_2]_a)$ $(B_2)_a \rightleftarrows B_2 + S$ (3) $J_3 = A_s(\overleftarrow{k}_{B_2}[B_2]_a - \overrightarrow{k}_{B_2} p_{B_2}[S])$ $(B_1)_a \rightleftarrows B_1 + S$ (4) $J_4 = A_s[\overleftarrow{k}_{B_1}(B_1]_a - \overrightarrow{k}_{B_1} p_{B_1}[S])$ $K_A \equiv \frac{\overrightarrow{k}_A}{\overleftarrow{k}_A}$; $K_{B_1} \equiv \frac{\overrightarrow{k}_{B_1}}{\overleftarrow{k}_{B_1}}$; $K_{B_2} \equiv \frac{\overrightarrow{k}_{B_2}}{\overleftarrow{k}_{B_2}}$; $K \equiv \frac{\overrightarrow{k}}{\overleftarrow{k}}$ $[S]_0 = [S] + [A]_a + [B_1]_a + [B_2]_a$	(1) $A + S \rightarrow (A)_a$ $\overrightarrow{k}_A p_A \ll \overleftarrow{k} K_{B_1} K_{B_2} p_{B_1} p_{B_2}[S], \overrightarrow{k}_{B_2} p_{B_2}, \overleftarrow{k}_{B_1} p_{B_1}$ $[S]/[S]_0 \approx 1 + \left(\frac{1}{K} K_{B_2} p_{B_2}\right) K_{B_1} p_{B_1} + K_{B_2} p_{B_2}$ (2) $(A)_a + S \rightarrow (B_1)_a + (B_2)_a$ $\overrightarrow{k}[S] \ll \overleftarrow{k}_A \frac{\overrightarrow{k}_{B_2} p_{B_2}}{k_A p_A} \frac{\overrightarrow{k}_{B_1} p_{B_1}}{K_A p_A}$ $[S]_0/[S] \approx 1 + K_A p_A + K_{B_1} p_{B_1} + K_{B_2} p_{B_2}$ (3) $(B_2)_a \rightarrow B_2 + S$ $\overleftarrow{k}_{B_2} \ll \frac{\overleftarrow{k}_A K_{B_1} p_{B_1}}{K} \overleftarrow{k} K_{B_1} p_{B_1}[S], \overrightarrow{k}_{B_1} p_{B_1} \frac{K_{B_1} p_{B_1}}{K K_A p_A}$ $[S]_0/[S] \approx 1 + \left(1 + \frac{K}{K_{B_1} p_{B_1}}\right) K_A p_A + K_{B_1} p_{B_1}$ (4) $(B_1)_a \rightarrow B_1 + S$ $\overleftarrow{k}_{B_1} \ll \frac{\overleftarrow{k}_A K_{B_2} p_{B_2}}{K}, \overleftarrow{k} K_{B_2} p_{B_2}[S], \overrightarrow{k}_{B_2} p_{B_2} \frac{K_{B_2} p_{B_2}}{K K_A p_A}$ $[S]_0/[S] \approx 1 + \left(1 + \frac{K}{K_{B_2} p_{B_2}}\right) K_A p_A + K_{B_2} p_{B_2}$
II/3	$A + S \rightleftarrows (A)_a$ (1) $J_1 = A_s(\overrightarrow{k}_A p_A[S] - \overleftarrow{k}_A[A]_a)$ $(A)_a + S \rightleftarrows 2({}^1/_2B_1)_a + B_2$ (2) $J_2 = A_s(\overrightarrow{k}[A]_a[S] - \overleftarrow{k}[{}^1/_2B_1]_a^2 p_{B_2})$ $2({}^1/_2B_1)_a \rightleftarrows B_1 + 2[S]$ (3) $J_3 = A_s(\overleftarrow{k}_{B_1}[{}^1/_2B_1]_a^2 - \overrightarrow{k}_{B_1} p_{B_1}[S]^2)$ $K_A \equiv \frac{\overrightarrow{k}_A}{\overleftarrow{k}_A}$; $K_{B_1} \equiv \frac{\overrightarrow{k}_{B_1}}{\overleftarrow{k}_{B_1}}$; $K = \frac{\overrightarrow{k}}{\overleftarrow{k}}$ $[S]_0 = [S] + [A]_a + [{}^1/_2B_1]_a$	(1) $A + S \rightarrow (A)_a$ $\overrightarrow{k}_A p_A \ll \overleftarrow{k} \frac{K_{B_1} p_{B_1}}{p_A} [S]_2 \overrightarrow{k}_{B_1} p_1[S]$ $[S]_0/[S] \approx 1 + \left(1 + \frac{1}{k} p_{B_2} \sqrt{K_{B_1} p_{B_1}}\right) \sqrt{K_{B_1} p_{B_1}}$ (2) $(A)_a + S \rightarrow 2({}^1/_2B_1)_a + B_2$ $\overrightarrow{k}[S] \gg \overleftarrow{k}_A, \frac{\overrightarrow{k}_{B_1} p_{B_1}}{K_A p_A} [S]$ $[S]_0/[S] \approx 1 + K_A p_A + \sqrt{K_{B_1} p_{B_1}}$ (3) $2({}^1/_2B_1)_a \rightarrow B_1 + 2S$ $\overleftarrow{k}_{B_1} \gg \overleftarrow{k}_A \frac{p_{B_2}}{K[S]}, \overleftarrow{k} p_{B_2}$ $[S]_0/[S] \approx 1 + K_A p_A + \sqrt{K \frac{K_A p_A}{p_{B_2}}}$

surface area of catalyst; k = rate constant; $[S]$ = surface concentration of active stant; J = reaction rate

Kinetic equation for the catalytic reaction	Type number
$J \approx A_s[S]_0 \vec{k}_A \dfrac{p_A - \dfrac{1}{k}\dfrac{K_{B_1}}{K_A} p_{B_1} p_{B_2}}{1+ \left(1+\dfrac{1}{K} p_{B_2}\right) K_{B_1} p_{B_1}}$	II/ — 1
$J \approx A_s [S]_0 K_A \vec{k} \dfrac{p_A - \dfrac{1}{K}\dfrac{K_{B_1}}{K_A} p_{B_1} p_{B_2}}{1 + K_A p_A + K_{B_1} p_{B_1}}$	II/1 — 2
$J \approx A_s[S]_0 K_A K \overleftarrow{k} \dfrac{p_A - \dfrac{1}{K}\dfrac{K_{B_1}}{K_A} p_{B_1} p_{B_2}}{p_{B_2}\left[1 + \left(1+\dfrac{K}{p_{B_2}}\right) K_A p_A\right]}$	II/1 — 3
$J \approx A_s[S]_0 \vec{k}_A \dfrac{p_A - \dfrac{1}{K}\dfrac{K_{B_1} K_{B_2}}{K_A} p_{B_1} p_{B_2}}{1 + \left(1 + \dfrac{1}{K} K_{B_2} p_{B_2}\right) K_{B_1} p_{B_1} + K_{B_2} p_{B_2}}$	II/ — 1
$J \approx A_s[S]_0^2 K_A \vec{k} \dfrac{p_A - \dfrac{1}{K}\dfrac{K_{B_1} K_{B_2}}{K_A} p_{B_1} p_{B_2}}{(1 + K_A p_A + K_{B_1} p_{B_1} + K_{B_2} p_{B_2})^2}$	II/2 — 2
$J \approx A_s[S]_0 K_A K \overleftarrow{k}_{B_2} \dfrac{p_A - \dfrac{1}{K}\dfrac{K_{B_1} K_{B_2}}{K_A} p_{B_1} p_{B_2}}{K_{B_1} p_{B_1}\left[1 + \left(1 + \dfrac{K}{K_{B_1} p_{B_1}}\right) K_A p_A + K_{B_1} p_{B_1}\right]}$	II/2 — 3
$J \approx A_s[S]_0 K_A K \overleftarrow{k}_{B_1} \dfrac{p_A - \dfrac{1}{K}\dfrac{K_{B_1} K_{B_2}}{K_A} p_{B_1} p_{B_2}}{K_{B_2} p_{B_2}\left[1 + \left(1 + \dfrac{K}{K_{B_2} p_{B_2}}\right) K_A p_A + K_{B_2} p_{B_2}\right]}$	II/2 — 4
$J \approx A_s[S]_0 \vec{k}_A \dfrac{p_A - \dfrac{1}{K}\dfrac{K_{B_1}}{K_A} p_{B_1} p_{B_2}}{1 + \sqrt{K_{B_1} p_{B_1}} + \dfrac{1}{K} K_{B_1} p_{B_1} p_{B_2}}$	II/3 — 1
$J \approx A_s[S]_0^2 K_A \vec{k} \dfrac{p_A - \dfrac{1}{K}\dfrac{K_{B_1}}{K_A} p_{B_1} p_{B_2}}{(1 + K_A p_A + \sqrt{K_{B_1} p_{B_1}})^2}$	II/3 — 2
$J \approx A_s[S]_0^2 K_A K \overleftarrow{k}_{B_1} \dfrac{p - \dfrac{1}{K}\dfrac{K_{B_1}}{K_A} p_{B_1} p_{B_2}}{p_{B_2}\left(1 + K_A p_A + \sqrt{K \dfrac{K_A p_A}{p_{B_2}}}\right)^2}$	II/3 — 3

TABLE

Kinetic equations for the catalytic reaction $A \to B_1 + B_2$, p = gas pressure; A_s = sites; K = equilibrium

Type	Partial processes and their rate equations	Rate-controlling elementary partial process
II/a_1	$A + S \rightleftarrows (A)_a$ (1) $J_1 = A_s([\vec{k}_A p_A[S] - \overleftarrow{k}_A[A]_a)$ $(A)_a \to (B_1)_a + B_2$ (2) $J_2 = A_s \vec{k}[A]_a$ $(B_1)_a \rightleftarrows B_1 + S$ (3) $J_3 = A_s(\overleftarrow{k}_{B_1}[B_1]_a - \vec{k}_{B_1} p_{B_1}[S])$ $K_A \equiv \frac{\vec{k}_A}{\overleftarrow{k}_A}$; $K_{B_1} \equiv \frac{\vec{k}_{B_1}}{\overleftarrow{k}_{B_1}}$ $[S]_0 = [S] + [A]_a + [B_1]_a$	(1) $A + S \to (A)_a$ $\vec{k}_A p_A \ll \vec{k}_1 \vec{k}_{B_1} p_{B_1}$ (2) $(A)_a \to (B_1)_a + B_2$ $\vec{k} \ll \overleftarrow{k}_A$, $\frac{\vec{k}_{B_1} p_{B_1}}{K_A p_A}$ (3) $(B_1)_a \to B_1 + S$ $\vec{k} \ll \overleftarrow{k}_A$; $\overleftarrow{k}_{B_1} \ll \frac{\vec{k} K_A p_A}{1 + K_A p_A + K_{B_1} p_{B_1}}$
II/a_2	$A + S \; (A)_a$ (1) $J_1 = A_s(\vec{k}_A p_A[S] - \overleftarrow{k}_A[A_a])$ $(A)_a + S \to (B_1)_a + (B_2)_a$ (2) $J_2 = A_s \vec{k}[A]_a[S]$ $(B_2)_a \rightleftarrows B_2 + \mathrm{S}$ (3) $J_3 = A_s(\overleftarrow{k}_{B_2}[B_2]_a - \vec{k}_{B_2} p_{B_2}[S]$ $(B_1)_a \rightleftarrows B_1 + S$ (4) $J_4 = A_s(\overleftarrow{k}_{B_1}[B_1]_a - \vec{k}_{B_1} p_{B_1}[S])$ $K_A \equiv \frac{\vec{k}_A}{\overleftarrow{k}_A}$ $K_{B_1} \equiv \frac{\vec{k}_{B_1}}{\overleftarrow{k}_{B_1}}$; $K_{B_2} \equiv \frac{\vec{k}_{B_2}}{\overleftarrow{k}_{B_2}}$ $[S] = [S] + [A]_a + [B_1]_a + [B_2]_a$	(1) $A + S \to (A)_a$ $\vec{k}_A p_A \ll \vec{k}[S]$, $\vec{k}_{B_1} p_{B_1}$, $\vec{k}_{B_2} p_{B_2}$ $[S]_0/[S] \approx 1 + K_{B_1} p_{B_1} + K_{B_2} p_{B_2}$ (2) $(A)_a + S \to (B_1)_a + (B_2)_a$ $\vec{k}[S] \ll \overleftarrow{k}_A$, $\frac{\vec{k}_{B_1} p_{B_1}}{K_A p_A}$, $\frac{\vec{k}_{B_2} p_{B_2}}{K_A p_{B_2}}$ $[S]_0/[S] \approx 1 + K_A p_A + K_{B_1} p_{B_1} + K_{B_2} p_{B_2}$ (3) $(B_2)_a \to B_2 + S$ $\vec{k}[S] \ll \overleftarrow{k}_A, \frac{\vec{k}_{B_1} p_{B_1}}{K_A p_A}; \overleftarrow{k}_{B_2} \ll \frac{\vec{k}[S] K_A p_A}{1 + K_A p_A + K_{B_1} p_{B_1} + K_{B_2} p_{B_2}}$ $[S]_0/[S]^2 \approx \frac{\vec{k} K_A p_A}{\vec{k} B_2}$ (4) $(B_1)_a \to B_1 + S$ $\vec{k}[S] \ll \overleftarrow{k}_A, \frac{\vec{k}_{B_2} p_{B_2}}{K_A p_A}; k_{B_1} \ll \frac{\vec{k}[S] K_A p_A}{1 + K_A p_A + K_{B_1} p_1 + K_{B_2} p_B}$ $[S]_0/[S]^2 \approx \frac{\overleftarrow{k} K_A p_A}{\vec{k}_{B_1}}$
II/a_3	$A + S \rightleftarrows (A)_a$ (1) $J_1 = A_s(\vec{k}_A p_A[S] - \overleftarrow{k}_A[A]_a)$ $(A)_a + S \to 2(^1/_2 B_1) + B_2$ (2) $J = A_s \vec{k}[A]_a[S]$ $2(^1/_2 B_1)_a \rightleftarrows B_1 + 2\,S$ (3) $J = A_s(\overleftarrow{k}_{B_1}[^1/_2 B_1]^2_a - \vec{k}_{B_1} p_{B_1}[S]^2)$ $K_a \equiv \frac{\vec{k}_A}{\overleftarrow{k}_A}$; $K_{B_1} \equiv \frac{\vec{k}_{B_1}}{\overleftarrow{k}_{B_1}}$ $[S]_0 = [S] + [A]_a + [^1/_2 B_1]_a$	(1) $A + S \to (A)_a$ $\vec{k}_A p_A \ll \vec{k}[S]$, $\vec{k}_{B_1} p_{B_1}[S]$ $[S]l_0/[S] \approx 1 + \sqrt{K_{B_1} p_{B_1}}$ (2) $(A)_a + S \to 2\,(^1/_2 B_1)_a + B_2$ $\vec{k}[S] \ll \overleftarrow{k}_A$, $\frac{\vec{k}_{B_1} p_{B_1}}{K_A p_A}[S]$ $[S]_0/[S] \approx 1 + K_A p_A + \sqrt{K_{B_1} p_{B_1}}$ (3) $2\,(^1/_2 B_1)_a \to B_1 + 2\,S$ $\vec{k}[S] \ll \overleftarrow{k}_A$; $\overleftarrow{k}_{B_1} \ll \frac{\vec{k} K_A p_A}{(1 + K_A p_A)^2}$ $[S]_0/[S] \approx \sqrt{\frac{\vec{k} K_A p_A}{\overleftarrow{k}_{B_1}}}$

surface area of catalyst; k = rate constant; $[S]$ = surface concentration of active constant; J = reaction rate

Kinetic equation for the catalytic reaction	Type number
$J \approx A_s[S]_0 \vec{k}_A \frac{p_A}{1 + K_{B_1} p_{B_1}}$	$II/a_1 - 1$
$J \approx A_s[S]_0 K_A \vec{k} \frac{p_A}{1 + K_A p_A + K_{B_1} p_{B_1}}$	$II/a_1 - 2$
$J \approx A_s[S]_0 k_{B_1}$	$II/a_1 - 3$
$J \approx A_s[S]_0 \vec{k} \frac{p_A}{1 + K_{B_1} p_{B_1} + K_{B_2} p_{B_2}}$	$II/a_2 - 1$
$J \approx A_s[S]_0^2 K_A \vec{k} \frac{p_A}{(1 + K_A p_A + K_{B_1} p_{B_1} + K_{B_2} p_{B_2})^2}$	$II/a_2 - 2$
$J \approx A_s[S]_0 \overleftarrow{k}_{B_2}$	$II/a_2 - 3$
$J \approx A_s[S]_0 \vec{k}_{B_1}$	$II/a_2 - 4$
$J \approx A_s[S]_0 \vec{k}_A \frac{p_A}{1 + \sqrt{K_{B_1} p_{B_1}}}$	$II/a_3 - 1$
$J \approx A_s[S]_0^2 K_A \vec{k} \frac{p_A}{(1 + K_A p_A + \sqrt{K_{B_1} p_{B_1}})^2}$	$II/a_3 - 2$
$J \approx A[S]_0^2 \overleftarrow{k}_{B_1}$	$II/a_3 - 3$

TABLE

Kinetic equations for the catalytic reaction $A_1 + A_2 \to B$, p = gas pressure; A_s = sites; K = equilibriu m

Type	Partial processes and their rate equations	Rate-controlling elementary partial process
II/b_1	$A_1 + S \rightleftarrows (A_1)_a$ (1) $J_1 = A_s(\vec{k}_{A_1} p_{A_1}[S] - \overleftarrow{k}_{A_1}[A_1]_a)$ $(A_1)_a + A_2 \to (B)_a$ (2) $J_2 = A_s \vec{k} [A_1]_a p_{A_2}$ $(B)_a \rightleftarrows B + S$ (3) $J_3 = A_s(\overleftarrow{k}_B [B]_a - \vec{k}_B p_B[S])$ $K_{A_1} \equiv \frac{\vec{k}_{A_1}}{\overleftarrow{k}_{A_1}}; K_B \equiv \frac{\vec{k}_B}{\overleftarrow{k}_B}$ $[S_0] = [S] + [A_1]_a + [B]_a$	(1) $A_1 + S \to (A_1)_a$ $\vec{k}_{A_1} p_{A_1} \ll \vec{k} p_{A_2}, \vec{k}_B p_B$ (2) $(A_1)_a + A_2 \to (B)_a$ $\vec{k} p_{A_2} \ll \overleftarrow{k}_{A_1}; \frac{\vec{k}_B p_B}{K_{A_1} p_{A_1}}$ (3) $(B)_a \to B + S$ $\vec{k} p_{A_2} \ll \overleftarrow{k}_{A_1}; \overleftarrow{k}_B \ll \frac{\vec{k} p_{A_2} K_{A_1} p_{A_1}}{1 + K_{A_1} p_{A_1} + K_B p_B}$
II/b_2	$A_1 + S \rightleftarrows (A_1)_a$ (1) $J_1 = A_s(\vec{k}_{A_1} p_{A_1}[S] - \overleftarrow{k}_{A_1}[A_1]_a)$ $A_2 + S \rightleftarrows (A_2)_a$ (2) $J_2 = A_s(\vec{k}_{A_2} p_{A_2}[S] - \overleftarrow{k}_{A_2}[A_2]_a)$ $(A_1)_a + (A_2)_a \to (B)_a + S$ (3) $J_3 = A_s \vec{k}[A_1]_a[A_2]_a$ $(B)_a \rightleftarrows B + S$ (4) $J_4 = A_s(\overleftarrow{k}_A[B]_a - \vec{k}_B p_B[S])$ $K_{A_1} \equiv \frac{\vec{k}_{A_1}}{\overleftarrow{k}_{A_1}}; K_{A_2} \equiv \frac{\vec{k}_{A_2}}{\overleftarrow{k}_{A_2}}; K_B \equiv \frac{\vec{k}_B}{\overleftarrow{k}_B}$ $[S]_0 = [S] + [A_1] + [A_2]_a + [B]_a$	(1) $A_1 + S \to (A_1)_a$ $\vec{k}_{A_1} p_{A_1} \ll \vec{k}_{A_2} p_{A_2}, \vec{k}[S] K_{A_2} p_{A_2}; \vec{k}_B p_B$ $[S]_0/[S] \approx 1 + K_{A_2} p_{A_2} + K_B p_B$ (2) $A_2 + S \to (A_2)_a$ $\vec{k}_{A_2} p_{A_2} \ll \vec{k}_{A_1} p_{A_1}, \vec{k}[S] K_{A_1} p_{A_1}, \vec{k}_B p_B$ $[S]_0/[S] \approx 1 + K_{A_1} p_{A_1} + K_B p_B$ (3) $(A_1)_a + (A_2)_a \to (B)_a + S$ $\vec{k}[S] \ll \frac{\overleftarrow{k}_{A_1}}{K_{A_2} p_{A_2}}, \frac{\overleftarrow{k}_{A_2}}{K_{A_1} p_{A_1}}, \frac{\vec{k}_B p_B}{K_{A_1} K_{A_2} p_{A_1} p_{A_2}}$ $[S]_0/[S] \approx 1 + K_{A_1} p_{A_1} + K_{A_2} p_{A_2} + K_B p_B$ (4) $(B)_a \to B + S$ $\vec{k}[S] \ll \frac{\overleftarrow{k}_{A_1}}{K_{A_2} p_{A_2}}; \frac{\overleftarrow{k}_{A_2}}{K_{A_1} p_{A_1}}; \overleftarrow{k}_B \ll \frac{\vec{k}[S] K_{A_1} K_{A_2} p_{A_1} p_{A_2}}{1 + K_{A_1} p_{A_1} + K_{A_2} p_{A_2} + K}$ $[S]_0/[S]^2 \approx \frac{\vec{k}}{\overleftarrow{k}_B} K_{A_1} K_{A_2} p_{A_1} p_{A_2}$
II/b_3	$A_1 + 2\,S \rightleftarrows 2\,(^1/_2\,A_1)_a$ (1) $J_1 = A_s(\vec{k}_{A_1} p_{A_1}[S]^2 - \overleftarrow{k}_{A_2}[^1/_2\,A_1]_a^2)$ $2\,(^1/_2\,A_1)_a + A_2 \to (B)_a + S$ (2) $J_2 = A_s \vec{k} p_{A_2}[^1/_2\,A_1]_a^2$ $(B)_a \rightleftarrows B + S$ (3) $J_3 = A_s(\overleftarrow{k}_B [B]_a - \vec{k}_B p_B[S])$ $K_{A_1} \equiv \frac{\vec{k}_{A_1}}{\overleftarrow{k}_{A_1}}; K_B \equiv \frac{\vec{k}_B}{\overleftarrow{k}_B}$ $[S]_0 = [S] + [^1/_2\,A_1]_a + [B]_a$	(1) $A_1 + 2\,S \to 2\,(^1/_2\,A_1)_a$ $\vec{k}_{A_1} p_{A_1} \ll \vec{k} p_{A_2}, \frac{\vec{k}_B p_B}{[S]}$ $[S]_0/[S] \approx 1 + K_B p_B$ (2) $2\,(^1/_1\,A_1) + A_2 \to (B)_a + S$ $\vec{k} p_{A_2} \ll \overleftarrow{k}_{A_1}, \frac{\vec{k}_B p_B}{[S] K_{A_1} p_{A_1}}$ $[S]_0/[S] \approx 1 + \sqrt{K_{A_1} p_{A_1}} + K_B p_B$ (3) $(B)_a \to B + S$ $\vec{k} p_{A_2} \ll \overleftarrow{k}_{A_1}; \overleftarrow{k}_B \ll \frac{\vec{k}[S] K_{A_1} p_{A_1} p_{A_2}}{1 + \sqrt{K_{A_1} p_{A_1}} + K_B p_B}$ $[S]_0/[S]^2 \approx \frac{\vec{k}}{\overleftarrow{k}_B} K_{A_1} p_{A_1} p_{A_2}$

surface area of catalyst; k = rate constant; $[S]$ = surface concentration of active constant; J = reaction rate

Kinetic equation for the catalytic reaction	Type number
$J \approx A_S[S]_0 \vec{k}_{A_1} \frac{p_{A_1}}{1 + K_B p_B}$	II/b_1 — 1
$J \approx A_S[S]_0 K_{A_1} \vec{k} \frac{p_{A_1} p_{A_2}}{1 + K_{A_1} p_{A_1} + K_B p_B}$	II/b_1 — 2
$J \approx A_S[S]_0 \overleftarrow{k}_B$	II/b_1 — 3
$J \approx A_S [S]_0 \vec{k}_{A_1} \frac{p_{A_1}}{1 + K_{A_2} p_{A_2} + K_B p_B}$	II/b_2 — 1
$J \approx A_S [S]_0 \vec{k}_{A_2} \frac{p_{A_2}}{1 + K_{A_1} p_{A_1} + K_B p_B}$	II/b_2 — 2
$J \approx A_S[S]_0^2 K_{A_1} K_{A_2} \vec{k} \frac{p_{A_1} p_{A_2}}{(1 + K_{A_1} p_{A_1} + K_{A_2} p_{A_2} + K_B p_B)^2}$	II/b_2 — 3
$J \approx A_S[S]_0 \overleftarrow{k}_B$	II/b_2 — 4
$J \approx A_S [S]_0^2 \vec{k}_{A_1} \frac{p_{A_1}}{(1 + K_B p_B)^2}$	II/b_3 — 1
$J \approx A_S[S]_0^2 K_{A_1} \vec{k} \frac{p_{A_1} p_{A_2}}{(1 + \sqrt{K_{A_1} p_{A_1}} + K_B p_B)^2}$	II/b_3 — 2
$J \approx A_S[S]_0 \overleftarrow{k}_B$	II/b_3 — 3

TABLE IV.3

Comparison of the kinetic and rate equations for the catalytic reaction $A \rightarrow B$

Serial number	Mathematical form of rate equation		Kinetic equation	Type number	Region of validity	k	a	b	c
1.	$J = \text{const.}$	$J = a$	$J = k$	$\mathrm{I}/a - 2$	$K_A p_A \ll 1 + K_B p_B$	$A_s[S]_0 \vec{k}$	k	0	0
				$\mathrm{I}/a - 3$	—	$A_s[S]_0 \overleftarrow{k}_B$			
2.	$J = f(p_A)$	$\frac{p_A}{J} = a + bp_A$	$J = kp_A$	$\mathrm{I}/a - 1$	$1 \ll K_B p_B$	$A_s[S]_0 \vec{k}_A$	$\frac{1}{k}$	0	0
				$\mathrm{I}/a - 2$	$1 \ll K_A p_A + K_B p_B$	$A_s[S]_0 K_A \vec{k}$			
			$J = k \frac{p_A}{1 + K_A p_A}$		$1 \sim K_A p_A \gg K_B p_B$		$\frac{1}{k}$	$\frac{K_A}{k}$	0
3.	$J = f(p_A, p_B)$	$\frac{p_A}{J} = a + bp_A + cp_B$	$J = k \frac{p_A}{p_B}$	$\mathrm{I}/a - 1$	$K_B p_B \gg 1$	$A_s[S]_0 \frac{1}{K_B} \vec{k}_A$	0	0	$\frac{1}{k}$
				$\mathrm{I}/a - 2$	$K_B p_B \gg 1 + K_A p_A$	$A_s[S]_0 \frac{K_A}{K_B} \vec{k}$			
			$J = k \frac{p_A}{1 + k_B p_B}$	$\mathrm{I}/a - 1$	$1 \sim K_B p_B$	$A_s[S]_0 \vec{k}_A$	$\frac{1}{k}$	0	$\frac{K_B}{k}$
					$1 \sim K_B p_B \gg K_A p_A$				
			$J = k \frac{p_A}{K_A p_A + K_B p_B}$	$\mathrm{I}/a - 2$	$K_A p_A \sim K_B p_B \gg 1$	$A_s[S]_0 K_A \vec{k}$	0	$\frac{K_A}{k}$	$\frac{K_B}{k}$
			$J = k \frac{p_A}{1 + K_A p_A + K_B p_B}$		$1 \sim K_A p_A \sim K_B p_B$		$\frac{1}{k}$	$\frac{K_A}{k}$	$\frac{K_B}{k}$

$$1.\ J = \text{const.}$$

$$2.\ J = f(p_A)$$

$$3.\ J = f(p_A,\ p_B)$$

i.e. (1) the rate is independent of the partial pressures of the components, (2) it depends on the partial pressure of the starting component only, and (3) it is affected by the partial pressures of both components. The first thing to do is to choose the kinetic equation governing the experimental data. The experimental points or the rate values obtained by calculation from the former are then plotted using the corresponding transformation (Table IV.3, column 3).

If the first extreme case of the rate equation is valid ($J = a$), then according to the fifth column of Table IV.3 the kinetic equation may correspond to classes I/a–2 or I/a–3. Hence, the constant ($a = k$) may have two different meanings (Table IV.3, column 7).

If the rate equation referring to the second extreme case has been chosen, then further selection should be made, depending on the value of the slope of the function $p_A/J = f(p_A)$. If this is zero ($b = 0$), the kinetic equation may correspond to categories I/a–1 or I/a–2. Accordingly, the calculated constant ($a = 1/k$) may have different meanings ($A_s[S]_0\vec{k}_A$ or $A_s[S]_0K_A\vec{k}$). However, if the slope of the function $p_A/J = f(p_A)$ is different from zero then the kinetic equation relates to the case I/a–2 only. Thus, the meanings of the constants estimated from the slope (a) and the intercept (b) of the straight line are clear-cut.

Finally, if we are dealing with the third of the extreme cases, the constants of the kinetic and rate equations can be estimated in the following way. Using the rate curves determined for different initial values of p_A and p_B and the transformation in Table IV.3, the functions $J(p_A)$ and $J(p_B)$ referring to constant p_B or p_A values are plotted. If the plot is linear then four different classes can be chosen. This selection is based on the values of the constants in the rate equation (see the notation in Table IV.3):

(1) $a \approx b \approx 0$, $c \neq 0$. Classes I/a–1 and I/a–2. Accordingly, the physical meaning of the constant k is:

$$k = A_s[S]_0\vec{k}_A/K_B \text{ or } k = A_s[S]_0\vec{k}K_A/K_B.$$

(2) $b \approx 0$, $a \neq 0$, $c \approx 0$, Again classes I/a–1 and I/a–2 are valid, with the same consequences for the physical meaning of constant k.

(3) $a \approx 0$, $b \neq 0$, $c \neq 0$ and

(4) $a \neq 0$, $b \neq 0$, $c \neq 0$ correspond to class I/a–2, with the only difference being that in the former case the constants of the kinetic equations (k, K_A and K_A) cannot be determined separately, whereas in the latter case all three constants can be estimated from those of the rate equations: a, b and c.

TABLE

Comparison of the kinetic and rate equations

Serial No.	Mathematical form of rate equation		Kinetic equation	Type number
1.	$J = \text{const.}$	$J = a$	$J = k$	$II/a_1 - 3$
				$II/a_2 - 4$
				$II/a_3 - 3$
				$II/a_2 - 3$
				$II/a_1 - 2$
2.	$J = f(p_A)$	$\frac{p_A}{J} = a + b\,p_A$	$J = k\,p_A$	$II/a_1 - 1$
				$II/a_2 - 1$
				$II/a_3 - 1$
				$II/a_1 - 2$
				$II/a_2 - 2$
				$II/a_3 - 2$
			$J = k \frac{p_A}{1 + K_A p_A}$	$II/a_1 - 2$
		$\sqrt{\frac{p_A}{J}} = a + b\,p_A$	$J = k \frac{1}{p_A}$	$II/a_2 - 2$
				$II/a_3 - 2$
			$J = k \frac{p_A}{(1 + K_A p_A)^2}$	$II/a_2 - 2$
				$II/a_3 - 2$

for the catalytic reaction $A \rightarrow B_1 + B_2$

Region of validity	k	a	b	c	d
—					
—	$A_S[S]_0 \overleftarrow{k}_{B_1}$				
—		k	0	0	0
—	$A_S[S]_0 \overrightarrow{k}_{B_2}$				
$K_A p_A \gg 1 + K_{B_1} p_{B_1}$	$A_S[S]_0 \overrightarrow{k}$				
$1 \gg K_{B_1} p_{B_1}$					
$1 \gg K_{B_1} p_{B_1} + K_{B_2} p_{B_2}$	$A_S[S]_0 \overrightarrow{k}_A$				
$1 \gg \sqrt{K_{B_1} p_{B_1}}$					
$1 \gg K_A p_A + K_{B_1} p_{B_1}$	$A_S[S]_0 K_A \overrightarrow{k}$	$\frac{1}{k}$	0	0	0
$1 \gg K_A p_A + K_{B_1} p_{B_1} + K_{B_2} p_{B_2}$	$A_S[S]_0^2 K_A \overrightarrow{k}$				
$1 \gg K_A p_A + \sqrt{K_{B_1} p_{B_1}}$					
$1 \sim K_A p_A \gg K_{B_1} p_{B_1}$	$A_S[S]_0 K_A \overrightarrow{k}$	$\frac{1}{k}$	$\frac{K_A}{k}$	0	0
$K_A p_A \gg 1 + K_{B_1} p_{B_1} + K_{B_2} p_{B_2}$	$A_S[S]_0^2 \frac{1}{K_A} \overrightarrow{k}$	0	$\frac{1}{\sqrt{k}}$	0	0
$K_A p_A \ll 1 + \sqrt{K_{B_1} p_{B_1}}$					
$1 \sim K_A p_A \gg K_{B_1} p_{B_1} + K_{B_2} p_{B_2}$	$A_S[S]_0^2 K_A \overrightarrow{k}$	$\frac{1}{\sqrt{k}}$	$\frac{K_A}{\sqrt{k}}$	0	0
$1 \sim K_A p_A \gg \sqrt{K_{B_1} p_{B_1}}$					

TABLE

Comparison of the kinetic and rate equations

Serial No.	Mathematical form of rate equation		Kinetic equation	Type number
3.	$J=f(p_A, p_{B_1})$	$\frac{p_A}{J} = a + bp_A + cp_{B_1}$	$J = k\frac{p_A}{p_{B_1}}$	II/a_1 — 1
				II/a_2 — 1
				II/a_1 — 2
				II/a_3 — 2
			$J = k\frac{p_A}{1 + K_{B_1} p_{B_1}}$	II/a_1 — 1
				II/a_2 — 1
				II/a_1 — 2
			$J = k\frac{p_A}{K_A p_A + K_{B_1} p_{B_1}}$	II/a_1 — 2
			$J = k\frac{p_A}{1 + K_A p_A + K_{B_1} p_{B_1}}$	
		$\frac{p_A}{J} = a + c\sqrt{p_{B_1}}$	$J = k\frac{p_A}{\sqrt{p_{B_1}}}$	II/a_3 — 1
			$J = k\frac{p_A}{1 + \sqrt{K_{B_1} p_{B_1}}}$	
		$\sqrt{\frac{p_A}{J}} = a + bp_A + cp_{B_1}$	$J = k\frac{p_A}{(p_{B_1})^2}$	II/a_2 — 2
			$J = k\frac{p_A}{(1 + K_{B_1} p_{B_1})^2}$	
			$J = k\frac{p_A}{(K_A p_A + K_{B_1} p_{B_1})^2}$	
			$J = k\frac{p_A}{(1 + K_A p_A + K_{B_1} p_{B_1})^2}$	

for the catalytic reaction $A \rightarrow B_1 + B_2$

<table>
<tr><th>Region of validity</th><th>k</th><th>a</th><th>b</th><th>c</th><th>d</th></tr>
<tr><td>$Kp_{B_1} \gg 1$</td><td rowspan="2">$A_S [S]_0 \frac{1}{K_{B_1}} \vec{k}_A$</td><td rowspan="4">0</td><td rowspan="4">0</td><td rowspan="4">$\frac{1}{k}$</td><td rowspan="4">0</td></tr>
<tr><td>$K_{B_1} p_{B_1} \gg 1 + K_{B_2} p_{B_2}$</td></tr>
<tr><td>$K_{B_1} p_{B_1} \gg 1 + K_A p_A$</td><td>$A_S[S[_0 \frac{K_A}{K_{B_1}} \vec{k}$</td></tr>
<tr><td>$\sqrt{K_{B_1} p_{B_1}} \gg 1 + K_A p_A$</td><td>$A_S[S]_0^2 \frac{K_A}{K_{B_1}} \vec{k}$</td></tr>
<tr><td>$1 \sim K_{B_1} p_{B_1}$</td><td rowspan="2">$A_S[S]_0 \vec{k}_A$</td><td rowspan="3">$\frac{1}{k}$</td><td rowspan="3">0</td><td rowspan="3">$\frac{K_{B_1}}{k}$</td><td rowspan="3">0</td></tr>
<tr><td>$1 \sim K_{B_1} p_{B_1} \gg K_{B_2} p_{B_2}$</td></tr>
<tr><td>$1 \sim K_{B_1} p_{B_1} \gg K_A p_A$</td><td>$A_S[S]_0 K_A \vec{k}$</td></tr>
<tr><td>$K_A p_A \sim K_{B_1} p_{B_1} \gg 1$</td><td rowspan="2">$A_S[S]_0 K_A \vec{k}$</td><td>0</td><td>$\frac{K_A}{k}$</td><td>$\frac{K_{B_1}}{k}$</td><td>0</td></tr>
<tr><td>$1 \sim K_A p_A \sim K_{B_1} p_{B_1}$</td><td>$\frac{1}{k}$</td><td>$\frac{K_A}{k}$</td><td>$\frac{K_{B_1}}{k}$</td><td>0</td></tr>
<tr><td>$\sqrt{K_{B_1} p_{B_1}} \gg 1$</td><td>$A_S[S]_0 \frac{1}{\sqrt{K_{B_1}}} \vec{k}_A$</td><td>0</td><td>0</td><td>$\frac{1}{k}$</td><td>0</td></tr>
<tr><td>$1 \sim \sqrt{K_{B_1} p_{B_1}}$</td><td>$A_S[S]_0 \vec{k}_A$</td><td>$\frac{1}{k}$</td><td>0</td><td>$\frac{\sqrt{K_{B_1}}}{k}$</td><td>0</td></tr>
<tr><td>$K_{B_1} p_{B_1} \gg 1 + K_A p_A + K_{B_3} p_{B_3}$</td><td>$A_S[S]_0^2 \frac{K_A}{(K_B)^2} \vec{k}$</td><td>0</td><td>0</td><td>$\frac{1}{\sqrt{k}}$</td><td>0</td></tr>
<tr><td>$1 \sim K_{B_1} p_{B_1} \gg K_A p_A + K_{B_2} p_{B_2}$</td><td rowspan="3">$A_S[S]_0^2 K_A \vec{k}$</td><td>$\frac{1}{\sqrt{k}}$</td><td>0</td><td>$\frac{K_{B_1}}{\sqrt{k}}$</td><td>0</td></tr>
<tr><td>$K_A p_A \sim K_{B_1} p_{B_1} \gg 1 + K_{B_2} p_{B_2}$</td><td>0</td><td>$\frac{K_A}{\sqrt{k}}$</td><td>$\frac{K_{B_1}}{\sqrt{k}}$</td><td>0</td></tr>
<tr><td>$1 \sim K_A p_A \sim K_{B_1} p_{B_1} \gg K_{B_2} p_{B_2}$</td><td>$\frac{1}{\sqrt{k}}$</td><td>$\frac{K_A}{\sqrt{k}}$</td><td>$\frac{K_{B_1}}{\sqrt{k}}$</td><td>0</td></tr>
</table>

TABLE

Comparison of the kinetic and rate equations

Serial No.	Mathematical form of rate equation		Kinetic equation	Type number
3.	$J = f(p_A, p_{B_1})$	$\sqrt{\frac{p_A}{J}} = a + bp_A + c\sqrt{p_{B_1}}$	$J = k\frac{p_A}{(1 + \sqrt{K_{B_1}p_{B_1}})^2}$	II/a_3 — 2
			$J = k\frac{p_A}{(K_A p_A + \sqrt{K_{B_1} p_{B_1}})^2}$	
			$J = k\frac{p_A}{(1 + K_A p_A + \sqrt{K_{B_1}p_{B_1}})^2}$	
4.	$J = f(p_A, p_{B_2})$	$\frac{p_A}{J} = a + dp_{B_2}$	$J = k\frac{p_A}{p_{B_2}}$	II_1a_2 — 1
			$J = k\frac{p_A}{1 + K_{B_2} p_{B_2}}$	
		$\sqrt{\frac{p_A}{J}} = a + bp_A + dp_{B_2}$	$J = k\frac{p_A}{(p_{B_2})^2}$	II/a_2 — 2
			$J = k\frac{p_A}{(1 + K_{B_2} p_{B_2})^2}$	
			$J = k\frac{p_A}{(K_A p_A + K_{B_2} p_{B_2})^2}$	
			$J = k\frac{p_A}{(1 + K_A p_A + K_{B_2} p_{B_2})^2}$	
5.	$J = f(p_A, p_{B_1}, p_{B_2})$	$\frac{p_A}{J} = a + cp_{B_1} + dp_{B_2}$	$J = k\frac{p_A}{K_{B_1} p_{B_1} + K_{B_2} p_{B_2}}$	II/a_2 — 1
			$J = k\frac{p_A}{1 + K_{B_1} p_{B_1} + K_{B_2} p_{B_2}}$	
		$\sqrt{\frac{p_A}{J}} = a + bp_A + cp_{B_1} + dp_{B_2}$	$J = k\frac{p_A}{(K_{B_1}p_{B_1} + K_{B_2}p_{B_2})^2}$	II/a_2 — 2
			$J = k\frac{p_A}{(1 + K_{B_1}p_{B_1} + K_{B_2} p_{B_2})^2}$	
			$J = k\frac{p_A}{(K_A p_A + K_{B_1} p_{B_1} + K_{B_2} p_{B_2})^2}$	
			$J = k\frac{p_A}{(1 + K_A p_A + K_{B_1} p_{B_1} + K_{B_2} p_{B_2})^2}$	

for the catalytic reaction $A \rightarrow B_1 + B_2$

Region of validity	k	a	b	c	d
$1 \sim \sqrt{K_{B_1} p_{B_1}} \gg K_A p_A$		$\frac{1}{\sqrt{k}}$	0	$\frac{\sqrt{K_{B_1}}}{\sqrt{k}}$	0
$K_A p_A \sim \sqrt{K_{B_1} p_{B_1}} \ll 1$	$A_S[S]_0^2 K_A \vec{k}$	0	$\frac{K_A}{\sqrt{k}}$	$\frac{\sqrt{K_{B_1}}}{\sqrt{k}}$	0
$1 \sim K_A p_A \sim \sqrt{K_{B_1} p_{B_1}}$		$\frac{1}{\sqrt{k}}$	$\frac{K_A}{\sqrt{k}}$	$\frac{\sqrt{K_{B_1}}}{\sqrt{k}}$	0
$K_{B_2} p_{B_2} \gg 1 + K_{B_1} p_{B_1}$	$A_S[S]_0 \frac{1}{K_{B_2}} \vec{k}_A$	0	0	0	$\frac{1}{k}$
$1 \sim K_{B_2} p_{B_2} \gg K_{B_1} p_{B_1}$	$A_S[S]_0 \vec{k}_A$	$\frac{1}{k}$	0	0	$\frac{K_{B_2}}{k}$
$K_{B_2} p_{B_2} \gg + 1 K_A p_A + K_{B_1} p_{B_1}$	$A_S[S]_0^2 \frac{K_A}{(K_{B_2})^2} \vec{k}$	0	0	0	$\frac{1}{\sqrt{k}}$
$1 \sim K_{B_2} p_{B_2} \gg K_A p_A + K_{B_1} p_{B_1}$		$\frac{1}{\sqrt{k}}$	0	0	$\frac{K_{B_2}}{\sqrt{k}}$
$K_A p_A \sim K_{B_2} p_{B_2} \gg 1 \sim K_{B_1} p_{B_1}$	$A_S[S]_0^2 K_A \vec{k}$	0	$\frac{K_A}{\sqrt{k}}$	0	$\frac{K_{B_2}}{\sqrt{k}}$
$1 \sim K_A p_A \sim K_{B_2} p_{B_2} \gg K_{B_1} p_{B_1}$		$\frac{1}{\sqrt{k}}$	$\frac{K_A}{\sqrt{k}}$	0	$\frac{K_{B_2}}{\sqrt{k}}$
$K_{B_1} p_{B_1} \sim K_{B_2} p_{B_2} \gg 1$	$A_S[S]_0 \vec{k}_A$	0	0	$\frac{K_{B_1}}{k}$	$\frac{K_{B_2}}{k}$
$1 \sim K_{B_1} p_{B_1} \sim K_{B_2} p_{B_2}$		$\frac{1}{k}$	0	$\frac{K_{B_1}}{\sqrt{k}}$	$\frac{K_{B_2}}{k}$
$K_{B_1} p_{B_1} \sim K_{B_2} p_{B_2} \gg 1 K_A p_A$		0	0	$\frac{K_{B_1}}{\sqrt{k}}$	$\frac{K_{B_2}}{\sqrt{k}}$
$1 \sim K_{B_1} p_{B_1} \sim K_{B_2} p_{B_2} \gg K_A p_A$	$A_S[S]_0^2 K_A \vec{k}$	$\frac{1}{\sqrt{k}}$	0	$\frac{K_{B_1}}{\sqrt{k}}$	$\frac{K_{B_2}}{\sqrt{k}}$
$K_A p_A \sim K_{B_1} p_{B_1} \sim K_{B_2} p_{B_2} \gg 1$		0	$\frac{K_A}{\sqrt{k}}$	$\frac{K_{B_1}}{\sqrt{k}}$	$\frac{K_{B_2}}{\sqrt{k}}$
$1 \sim K_A p_A \sim K_{B_1} p_{B_1} \sim K_{B_2} p_{B_2}$		$\frac{1}{\sqrt{k}}$	$\frac{K_A}{\sqrt{k}}$	$\frac{K_{B_1}}{\sqrt{k}}$	$\frac{K_{B_2}}{\sqrt{k}}$

Comparison of the kinetic and rate equations

Serial No.	Mathematical form of rate equation		Kinetic equation	Type number
				II/b_1 — 3
1.	J = const.	$J = a$	$J = k$	II/b_2 — 4
				II/b_3 — 3
				II/b_1 — 1
2.	$J = f(p_{A_1})$	$J = bp_{A_1}$	$J = kp_{A_1}$	II/b_2 — 1
				II/b_3 — 1
				II/b_1 — 2
3.	$J = f(p_{A_2})$	$J = cp_{A_2}$	$J = kp_{A_2}$	II/b_2 — 2
				II/b_3 — 2
				II/b_2 — 1
		$\frac{p_{A_1}}{J} = a + bp_{A_1}$	$J = k\frac{p_{A_1}}{p_{A_2}}$	II/b_2 — 3
			$J = k\frac{p_{A_1}}{1 + K_{A_2}p_{A_2}}$	II/b_2 — 1
4.	$J=f(p_{A_1}, p_{A_2})$			II/b_2 — 2
		$\frac{p_{A_2}}{J} = a + cp_{A_1}$	$J = k\frac{p_{A_2}}{p_{A_1}}$	II/b_2 — 3
			$J = k\frac{p_{A_2}}{1 + K_{A_1}p_{A_1}}$	II/b_2 — 2

for the catalytic reaction $A_1 + A_2 \rightarrow B$

Region of validity	k	a	b	c	d
—					
—	$A_S[S]_0 \overleftarrow{k}_B$	k	0	0	0
—					
$1 \gg K_B p_B$					
$1 \gg K_{A_2} p_{A_2} + K_B p_B$	$A_S]S]_0 \overrightarrow{k}_{A_1}$	0	k	0	0
$1 \gg K_B p_B$					
$K_{A_1} p_{A_1} \gg 1 + K_B p_B$	$A[S]_0 \overrightarrow{k}$				
$1 \gg K_{A_1} p_{A_1} + K_B p_B$	$A_S[S]_0 \overrightarrow{k}_{A_2}$	0	0	k	0
$\sqrt{K_{A_1} p_{A_1}} \gg 1 + K_B p_B$	$A_S[S]_0^2 \overrightarrow{k}$				
$K_{A_2} p_{A_2} \gg 1 + K_B p_B$	$A_S[S]_0 \frac{1}{K_{A_2}} \overrightarrow{k}_{A_1}$	0	0	$\frac{1}{k}$	0
$K_{A_2} p_{A_2} \gg 1 + K_{A_1} p_{A_1} + K_B p_B$	$A_S[S]_0^2 \frac{K_{A_1}}{K_{A_2}} \overrightarrow{k}$				
$1 \sim K_{A_2} p_{A_2} \gg K_B p_B$	$A_S[S]_0 \overrightarrow{k}_{A_1}$	$\frac{1}{k}$	0	$\frac{K_{A_2}}{k}$	0
$K_{A_1} p_{A_1} \gg 1 + K_B p_B$	$A_S[S]_0 \frac{1}{K_{A_1}} \overrightarrow{k}_{A_2}$	0	$\frac{1}{k}$	0	0
$K_{A_1} p_{A_1} \gg 1 + K_{A_2} p_{A_2} + K_B p_B$	$A_S[S]_0^2 \frac{K_{A_1}}{K_{A_1}} \overrightarrow{k}$				
$1 \sim K_{A_1} p_{A_1} \gg K_B p_B$	$A_S[S]_0 \overrightarrow{k}_{A_2}$	$\frac{1}{k}$	$\frac{K_{A_1}}{k}$	0	0

TABLE

Comparison of the kinetic and rate equations

Serial No.	Mathematical form of rate equation		Kinetic equation	Type number
4.	$J=f(p_{A_1}, p_{A_2})$	$\frac{p_{A_1} p_{A_2}}{J} = a + b p_{A_1}$	$J = k p_{A_1} p_{A_2}$	$II/b_1 - 2$
				$II/b_3 - 2$
				$II/b_2 - 3$
			$J = k \frac{p_{A_1} p_{A_2}}{1 + K_{A_1} p_{A_1}}$	$II/b_1 - 2$
		$\sqrt{\frac{p_{A_1} p_{A_2}}{J}} = a + b p_{A_1} + c p_{A_2}$	$J = k \frac{p_{A_1} p_{A_2}}{(1 + K_{A_1} p_{A_1})^2}$	$II/b_2 - 3$
			$J = k \frac{p_{A_1} p_{A_2}}{(1 + K_{A_2} p_{A_2})^2}$	
			$J = k \frac{p_{A_1} p_{A_2}}{(K_{A_1} p_{A_1} + K_{A_2} p_{A_2})^2}$	
			$J = k \frac{p_{A_1} p_{A_2}}{(1 + K_{A_1} p_{A_1} + K_{A_2} p_{A_2})^2}$	
		$\sqrt{\frac{p_{A_1} p_{A_2}}{J}} = a + b\sqrt{p_{A_1}}$	$J = k \frac{p_{A_1} p_{A_2}}{(1 + \sqrt{K_{A_1} p_{A_1}})^2}$	$II/b_3 - 2$
5.	$J=f(p_A p_{As})$	$\frac{p_{A_1}}{J} = a + d p_B$	$J = k \frac{p_{A_1}}{p_B}$	$II/b_1 - 1$
				$II/b_2 - 1$
			$J = k \frac{p_{A_1}}{1 + K_B p_B}$	$II/b_1 - 1$
				$II/b_2 - 1$
		$\sqrt{\frac{p_{A_1}}{J}} = a + d p_B$	$J = k \frac{p_{A_1}}{(p_B)^2}$	$II/b_3 - 1$
			$J = k \frac{p_{A_1}}{(1 + K_B p_B)^2}$	

for the catalyzic reaction $A_1 + A_2 \rightarrow B$

Region of validity	k	a	b	c	d
$1 \gg K_{A_1}p_{A_1} + K_B\, p_B$	$A_S[S]_0\, K_{A_1}\, \vec{k}$				
$1 \gg \sqrt{K_{A_1}p_{A_1}} + K_B\, p_B$	$A_S[S]_0^2\, K_{A_1}\, \vec{k}$	$\frac{1}{k}$	0	0	0
$1 \gg K_{A_1}\, p_{A_1} + K_{A_2}\, p_{A_2} + K_B\, p_B$	$A_S[S]_0^2 K_{A_1}\, K_{A_2}\, \vec{k}$				
$1 \sim K_{A_1}\, p_{A_1} \gg K_B\, p_B$	$A_S[S]_0\, K_{A_1}\, \vec{k}$	$\frac{1}{k}$	$\frac{K_{A_1}}{k}$	0	0
$1 \sim K_{A_1}\, p_{A_1} \gg K_{A_2}\, p_{A_2} + K_B\, p_B$		$\frac{1}{\sqrt{k}}$	$\frac{K_{A_1}}{\sqrt{k}}$	0	0
$1 \sim K_{A_2}\, p_{A_2} \gg K_{A_1}\, p_{A_1} + K_B\, p_B$	$A_S[S]_0^2\, K_{A_1}\, K_{A_2}\, \vec{k}$	$\frac{1}{\sqrt{k}}$	0	$\frac{K_{A_2}}{\sqrt{k}}$	0
$K_{A_1}\, p_{A_1} \sim K_{A_2}\, p_{A_2} \gg 1 + K_B\, p_B$		0	$\frac{K_{A_1}}{\sqrt{k}}$	$\frac{K_{A_2}}{\sqrt{k}}$	0
$1 \sim K_{A_1}\, p_{A_1} \sim K_{A_2}\, p_{A_2} \gg K_B\, p_B$		$\frac{1}{\sqrt{k}}$	$\frac{K_{A_1}}{\sqrt{k}}$	$\frac{K_{A_2}}{\sqrt{k}}$	0
$1 \sim \sqrt{K_{A_1}\, p_{A_1}} \gg K_B\, p_B$	$A_S[S]_0^2\, K_{A_1}\, \vec{k}$	$\frac{1}{\sqrt{k}}$	$\frac{\sqrt{K_{A_1}}}{\sqrt{k}}$	0	0
$K_B\, p_B \gg 1$	$A_S[S]_0 \frac{1}{K_B}\, \vec{k}_{A_1}$	0	0	0	$\frac{1}{k}$
$K_B\, p_B \gg 1 + K_{A_2}\, p_{A_2}$					
$1 \sim K_B\, p_B$	$A_S[S]_0\, \vec{k}_{A_1}$	$\frac{1}{k}$	0	0	$\frac{K_B}{k}$
$1 \sim K_B\, p_B \gg K_{A_2}\, p_{A_2}$					
$K_B p_B \gg 1$	$A_S[S]_0^2\, \frac{1}{(K_B)^2}\, \vec{k}_{A_1}$	0	0	0	$\frac{1}{\sqrt{k}}$
$1 \sim K_B\, p_B$	$A_S[S]_0^2\, \vec{k}_{A_1}$	$\frac{1}{\sqrt{k}}$	0	0	$\frac{K_B}{\sqrt{k}}$

TABLE

Comparison of the kinetic and rate equations

Serial No.	Mathematical form of rate equation		Kinetic equation	Type number
6.	$J = f(p_{A_2}, p_B)$	$\frac{p_{A_2}}{J} = a + dp_B$	$J = k \frac{p_{A_2}}{p_B}$	II/b_2 — 2
			$J = k \frac{p_{A_2}}{1 + K_B p_B}$	
7.	$J = f(p_{A_1}, p_{A_2}, p_B)$	$\frac{p_{A_1}}{J} = a + cp_{A_2} + dp_B$	$J = k \frac{p_{A_1}}{K_{A_2} p_{A_2} + K_B p_B}$	II/b_2 — 1
			$J = k \frac{p_{A_1}}{1 + K_{A_2} p_{A_2} + K_B p_B}$	
		$\frac{p_{A_2}}{J} = a + bp_{A_1} + dp_B$	$J = k \frac{p_{A_2}}{K_{A_1} p_{A_1} + K_B p_B}$	II/b_2 — 2
			$J = k \frac{p_{A_2}}{1 + K_{A_1} p_{A_1} + K_B p_B}$	
		$\frac{p_{A_1} p_{A_2}}{J} = a + bp_{A_1} + dp_B$	$J = k \frac{p_{A_1} p_{A_2}}{p_B}$	II/b_2 — 2
			$J = k \frac{p_{A_1} p_{A_2}}{1 + K_B p_B}$	
			$J = k \frac{p_{A_1} p_{A_2}}{K_{A_1} p_{A_1} + K_B p_B}$	
			$J = k \frac{p_{A_1} p_{A_2}}{1 + K_{A_1} p_{A_1} + K_B p_B}$	

Inspection of the computation procedures presented above reveals that even for such simple reactions (and assuming a simple reaction mechanism) the relationship between the kinetic and the rate equations can be found only by tedious work requiring data from a number of carefully planned experiments. Another conclusion to be drawn is that the knowledge of the mathematical form of the kinetic or rate equation is not always sufficient to reveal the unambiguous physical meaning of the constants, and to establish the rate-controlling step and consequently the mechanism of the reaction.

The above is valid to an even greater extent for complex reactions. The physical interpretation of the constants of kinetic equations, and the confirmation of the mechanism on the basis of kinetic equations should therefore be made with caution. It can be seen from Table IV.3 that, for example, the constant k of the kinetic equation (if it can be evaluated from the rate equations) may involve not only the product $A_s[S]_0$, but also the for the catalytic reaction $A_1 + A_2 \rightarrow B$

Region of validity	k	a	b	c	d
$K_B p_B \gg 1 + K_{A_1} p_{A_1}$	$A_S[S]_0 \frac{1}{K_B} \vec{k}_{A_2}$	0	0	0	$\frac{1}{k}$
$1 \sim K_B p_B \gg K_{A_1} p_{A_1}$	$A_S[S]_0 \vec{k}_{A_2}$	$\frac{1}{k}$	0	0	$\frac{K_B}{k}$
$K_{A_2} p_{A_2} \sim K_B p_B \gg 1$	$A_S[S]_0 \vec{k}_{A_1}$	0	0	$\frac{K_{A_2}}{k}$	$\frac{K_B}{k}$
$1 \sim K_{A_2} p_{A_2} \sim K_B p_B$		$\frac{1}{k}$	0	$\frac{K_{A_2}}{k}$	$\frac{K_B}{k}$
$K_{A_1} p_{A_1} \sim K_B p_B \gg 1$	$A_S[S]_0 \vec{k}_{A_2}$	0	$\frac{K_{A_1}}{k}$	0	$\frac{K_B}{k}$
$1 \sim K_{A_1} p_{A_1} \sim K_B p_B$		$\frac{1}{k}$	$\frac{K_{A_1}}{k}$	0	$\frac{K_B}{k}$
$K_B p_B \gg 1 + K_{A_1} p_{A_1}$	$A_S[S]_0 \frac{K_{A_1}}{k} \vec{k}$	0	0	0	$\frac{1}{k}$
$1 \sim K_B p_B \gg K_{A_1} p_{A_1}$	$A_S[S]_0 K_{A_1} \vec{k}$	$\frac{1}{k}$	0	0	$\frac{K_B}{k}$
$K_{A_1} p_{A_1} \sim K_B p_B \gg 1$		0	$\frac{K_{A_1}}{k}$	0	$\frac{K_B}{k}$
$1 \sim K_{A_1} p_{A_1} \sim K_B p_B$		$\frac{1}{k}$	$\frac{K_{A_1}}{k}$	0	$\frac{K_B}{k}$

rate constant for the adsorption of component A, the rate constant for the desorption of component B, the rate constant of the surface reaction, and the same quantities multiplied by the adsorption equilibrium constants.

If the mechanism of the reaction cannot be confirmed unequivocally by using the equation, therefore, it is reasonable to regard the constant k of the kinetic equation as an apparent or overall constant. Accordingly, the activation energy ($E^{\ddagger}$) estimated from the Arrhenius plot ($\ln k = \ln k_0 - E^{\ddagger}/RT$) of the temperature-dependence of constant k is called the resulting activation energy, since it may include activation energy of desorption or adsorption, or the algebraic sum of the activation energy of the surface reaction and adsorption heats. Consequently, when the mechanism is studied, it is of primary importance to elucidate whether the kinetic equation corresponds to one mechanism only. If possible, the mechanism should also be confirmed by non-kinetic methods.

TABLE

Comparison of the kinetic and rate equations

Serial No.	Mathematical form of rate equation		Kinetic equation	Type number
7.	$J = f(p_{A_1}, p_{A_2}, p_B)$		$J = k \frac{p_{A_1} p_{A_2}}{(p_B)^2}$	II/b_2 — 3 II/b_3 — 2
			$J = k \frac{p_{A_1} p_{A_2}}{(1 + K_B p_B)^2}$	II/b_2 — 3 II/b_3 — 2
		$\sqrt{\frac{p_{A_1} p_{A_2}}{J}} = a + bp_{A_1} + cp_{A_2} + dp_B$	$J = k \frac{p_{A_1} p_{A_2}}{(K_{A_1} p_{A_1} + K_B p_B)^2}$	II/b_2 — 3
			$J = k \frac{p_{A_1} p_{A_2}}{(1 + K_{A_1} p_{A_1} + K_B p_B)^2}$	
			$J = k \frac{p_{A_1} p_{A_2}}{(K_{A_2} p_{A_2} + K_B p_B)^2}$	
			$J = k \frac{p_{A_1} p_{A_2}}{(1 + K_{A_2} p_{A_2} + K_B p_B)^2}$	
			$J = k \frac{p_{A_1} p_{A_2}}{(K_{A_1} p_{A_1} + K_{A_2} p_{A_2} + K_B p_A)^2}$	
			$J = k \frac{p_{A_1} p_{A_2}}{(1 + K_{A_1} p_{A_1} + K_{A_2} p_{A_2} + K_B p_B)^2}$	
		$\sqrt{\frac{p_{A_1} p_{A_2}}{J}} = a + b\sqrt{p_{A_1}} + dp_B$	$J = k \frac{p_{A_1} p_{A_2}}{(\sqrt{K_{A_1} p_{A_1}} + K_B p_B)^2}$	II/b_3 — 2
			$J = k \frac{p_{A_1} p_{A_2}}{(1 + \sqrt{K_{A_1} p_{A_1}} + K_B p_B)^2}$	

PARAGRAPH 2.2.2. DETERMINATION OF THE KINETIC EQUATION FOR THE CATALYTIC REACTION $A \rightarrow B_1 + B_2$

In this case, if the mechanisms shown in Table IV.2/2 are assumed, nine extreme classes can be distinguished. As regards the mathematical form of the rate equation six extreme cases can be found which divide further into 11 or 30 subcases. These are collected in Tables IV.4/1, IV/.4/2 and IV.4/3. The appropriate rate and kinetic equations are found by the techniques presented for the reaction $A \rightarrow B$, with the only difference being that the procedure is significantly longer owing to the greater number of subcases.

IV.5/4

for the catalytic reaction $A_1 + A_2 \rightarrow B$

Region of validity	k	a	b	c	d
$K_B p_B \gg 1 + K_{A_1} p_{A_1} + K_{A_2} p_{A_2}$	$A_s[S]_0^2 \frac{K_{A_1} K_{A_2} \vec{k}}{(K_B)^2}$	0	0	0	$\frac{1}{\sqrt{k}}$
$K_B p_B \gg 1 + \sqrt{K_{A_1} p_{A_1}}$	$A_s[S]_0^2 \frac{K_{A_1}}{(K_B)^2} \vec{k}$				
$1 \sim K_B p_B \gg K_{A_1} p_{A_1} + K_{A_2} p_{A_2}$	$A_s[S]_0^2 K_{A_1} K_{A_2} \vec{k}$	$\frac{1}{\sqrt{k}}$	0	0	$\frac{K_B}{\sqrt{k}}$
$1 \sim K_B p_B \gg \sqrt{K_{A_1} p_{A_1}}$	$A_s[S]_0^2 K_{A_1} \vec{k}$				
$K_{A_1} p_{A_1} \sim K_B p_B \gg 1 + K_{A_2} p_{A_2}$	$A_s[S]_0^2 K_{A_1} K_{A_2} \vec{k}$	0	$\frac{K_{A_1}}{\sqrt{k}}$	0	$\frac{K_B}{\sqrt{k}}$
$1 \sim K_{A_1} p_{A_1} \sim K_B p_B \gg K_{A_2} p_{A_2}$		$\frac{1}{\sqrt{k}}$	$\frac{K_{A_1}}{\sqrt{k}}$	0	$\frac{K_B}{\sqrt{k}}$
$K_{A_2} p_{A_2} \sim K_B p_B \gg 1 + K_{A_1} p_{A_1}$		0	0	$\frac{K_{A_2}}{\sqrt{k}}$	$\frac{K_B}{\sqrt{k}}$
$1 \sim K_{A_2} p_{A_2} \sim K_B p_B \gg K_{A_1} p_{A_1}$		$\frac{1}{\sqrt{k}}$	0	$\frac{K_{A_2}}{\sqrt{k}}$	$\frac{K_B}{\sqrt{k}}$
$K_{A_1} p_{A_1} \sim K_{A_2} p_{A_2} \sim K_B p_B \gg 1$		0	$\frac{K_{A_1}}{\sqrt{k}}$	$\frac{K_{A_2}}{\sqrt{k}}$	$\frac{K_B}{\sqrt{k}}$
$1 \sim K_{A_1} p_{A_1} \sim K_{A_2} p_{A_2} \sim K_B p_B$		$\frac{1}{\sqrt{k}}$	$\frac{K_{A_1}}{\sqrt{k}}$	$\frac{K_{A_2}}{\sqrt{k}}$	$\frac{K_B}{\sqrt{k}}$
$\sqrt{K_{A_1} p_{A_1}} \sim K_B p_B \gg 1$	$A_s[S]_0^2 K_{A_1} \vec{k}$	0	$\frac{\sqrt{K_{A_1}}}{\sqrt{k}}$	0	$\frac{K_B}{\sqrt{k}}$
$1 \sim \sqrt{K_{A_1} p_{A_1}} \sim K_B p_B$		$\frac{1}{\sqrt{k}}$	$\frac{\sqrt{K_{A_1}}}{\sqrt{k}}$	0	$\frac{K_B}{\sqrt{k}}$

PARAGRAPH 2.2.3. DETERMINATION OF THE KINETIC EQUATION FOR THE CATALYTIC REACTION $A_1 + A_2 \rightarrow B$

Similarly as for the reactions discussed previously, Table IV.5/1–IV.5/4 have been compiled on the basis of Table IV.2/3. The Tables contain the rate equations, their linearized forms, the subcases, the relationships between the rate and kinetic equations and the possible physical meanings of the constants of the rate equations. It can be seen from these Tables that there are 7 extreme cases which can be divided into 16 or 38 subcases. The determination of the kinetic and rate equations is analogous in principle to the methods outlined above.

SECTION 2.3. EXAMPLES OF THE DETERMINATION OF THE RATE AND KINETIC EQUATIONS IN VARIOUS LABORATORY REACTORS

PARAGRAPH 2.3.1. STATIC REACTORS

Szabó et al. [8] studied the catalytic decomposition of N_2O in a static reactor on a series of CuO catalysts differing in their method of pretreatment. The reaction obeys the following stoichiometric equation:

$$N_2O = 1/2\ O_2 + N_2;$$

$$A \rightarrow B_1 + B_2 \qquad \text{(IV.96)}$$

i.e. the stoichiometric numbers are: $\nu_A = -1$; $\nu_{B_1} = 1/2$; $\nu_{B_2} = 1$. Thus, using Eq. (IV.82) the reaction rate is:

$$J = \frac{2}{R}\left(\frac{v_r}{T_r} + \frac{v_m}{T_m} + \frac{v_d}{T_d}\right)\frac{dp}{dt} \qquad \text{(IV.97)}$$

Let us introduce the following notation:

$$\alpha \equiv \frac{2}{R}\left(\frac{v_r}{T_r} + \frac{v_m}{T_m} + \frac{v_d}{T_d}\right); \quad P \equiv p - p^0 \qquad \text{(IV.98)}$$

where p^0 is the pressure of the reaction mixture at time $t = 0$. With this notation Eq. (IV.97) is written:

$$J = \alpha\frac{dP}{dt} \qquad \text{(IV.99)}$$

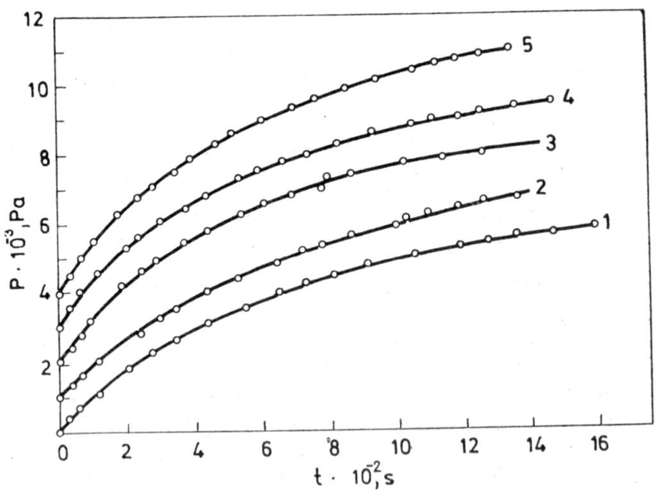

Fig. IV.14. Variation of the total pressure as a function of time during the decomposition of N_2O a CuO catalyst

Results obtained using the same reactor and catalyst at a given temperature are shown in Table IV.6 and Fig. IV.14. (The origins of the ordinates for the different curves are successively shifted by 10^3 Pa.) The measure-

TABLE IV.6

Catalytic decomposition of N_2O on a CuO catalyst in a static reactor

$v_r = 40\ cm^3$; $v_m = 35\ cm^3$; $v_d = 5\ cm^3$

$t_r = 573\ °C$; $t_m = 27\ °C$; $t_d = 27-573\ °C$

No. of exp.	1		2		3		4		5	
p_A^0, 10^3Pa	13.2		13.4		13.0		14.8		16.0	
$p_{B_1}^0$, 10^3Pa	10.9		14.8		0.0		7.1		3.5	
$p_{B_2}^0$, 10^3Pa	21.7		29.6		0.0		14.2		7.1	
	t, s	P, 10^3Pa	t, s	P, 10^3Pa	t, s	P, 10^3Pa	t, s	P, 10^3Pa	t, s	P, 10^3Pa
	0.0	0.00	0.0	0.00	0.0	0.00	0.0	0.00	0.0	0.00
	28.5	0.33	27.1	0.33	32.6	0.40	35.7	0.60	28.8	0.53
	61.6	0.73	62.3	0.67	63.1	0.80	67.8	1.00	70.2	1.07
	115.1	1.07	111.1	1.00	97.1	1.20	112.1	1.54	103.8	1.47
	209.5	1.80	231.9	1.80	190.2	2.14	209.3	2.34	185.7	2.27
	274.7	2.20	296.0	2.20	238.6	2.54	246.6	2.60	233.4	2.67
	343.2	2.60	346.0	2.47	291.4	2.94	306.9	3.00	281.9	3.07
	443.6	3.14	443.7	3.00	370.5	3.47	377.4	3.40	342.2	3.47
	548.6	3.54	523.8	3.40	449.7	3.87	446.2	3.80	400.2	3.87
	651.4	3.94	641.0	3.80	540.6	4.27	532.7	4.20	463.7	4.27
	729.7	4.20	714.8	4.14	612.6	4.54	596.7	4.46	513.0	4.54
	813.5	4.47	781.1	4.34	684.1	4.80	665.2	4.74	602.8	4.94
	913.2	4.73	879.9	4.60	784.6	5.06	741.6	5.00	694.3	5.33
	1053.5	5.00	997.1	4.87	794.8	5.34	827.2	5.26	765.1	5.60
	1184.2	5.26	1023.6	5.13	868.6	5.46	928.9	5.53	842.5	5.86
	1267.1	5.40	1096.4	5.26	1027.0	5.73	1045.6	5.80	939.8	6.13
	1355.7	5.53	1183.5	5.40	1132.8	5.86	1107.7	5.93	1051.3	6.40
	1471.2	5.67	1261.2	5.53	1253.4	6.00	1179.0	6.06	1110.2	6.53
	1589.2	5.80	1361.8	5.66			1247.4	6.20	1173.9	6.66
							1354.8	6.33	1251.6	6.80
							1458.0	6.46	1334.8	6.93

ments were reproducible and no ageing of the catalyst was observed. Consequently, the entire P vs. t curve can be used to estimate the rate.

The determination of the reaction rate requires a knowledge of the values of dP/dt. The character of the P vs. t curve renders possible the use of the following good approximation:

$$\frac{dP}{dt} \approx \frac{\Delta P}{\Delta t} \tag{IV.100}$$

The values of $\Delta P/\Delta t$ are obtained by reading the ordinates for identical intervals from the curves in Fig. IV.14. Since the error in reading off the P values is equal to ± 50 Pa the time intervals are chosen such that $\Delta P > > 50$ Pa. Part A of Table IV.7 shows the values of P which were selected as satisfying the above requirement. Values of $\Delta P/\Delta t$ are calculated by using the following equation

$$\frac{\Delta P}{\Delta t}(t_i) = \frac{P(t_{i+1}) - P(t_{i-1})}{t_{i+1} - t_{i-1}} \tag{IV.101}$$

The results are presented in part B of Table IV.7.

TABLE IV.7

Catalytic decomposition of N_2O. Illustration of the calculation procedure described in the text.

t, s	A					B				
	P, 10^3Pa					$\Delta P/\Delta t$, Pa s^{-1}				
	1	2	3	4	5	1	2	3	4	5
0	0.13	0.13	0.00	0.13	0.13	—	—	—	—	—
50	0.60	0.53	0.67	0.73	0.80	8.66	8.00	10.66	12.00	12.66
100	1.00	0.93	1.20	1.33	1.33	8.00	8.00	10.66	10.66	11.32
150	1.40	1.33	1.73	1.80	1.93	7.33	7.33	10.00	8.66	10.00
200	1.73	1.66	2.20	2.20	2.40	6.66	6.66	8.66	8.00	8.66
250	2.07	2.00	2.60	2.60	2.80	6.00	6.00	8.00	7.33	8.00
300	2.33	2.27	3.00	2.93	3.27	5.60	5.33	6.93	6.66	7.66
400	2.87	2.73	3.60	3.53	3.93	4.93	4.66	5.60	5.33	6.00
500	3.33	3.20	4.07	4.00	4.40	4.33	4.26	4.67	4.67	5.33
600	3.73	3.60	4.53	4.47	4.93	3.66	4.00	4.00	4.33	4.66
700	4.06	4.00	4.86	4.86	5.33	3.33	3.67	3.33	3.66	3.66
800	4.40	4.33	5.20	5.20	5.66	2.80	3.33	2.80	3.00	3.00
1000	4.86	4.93	5.66	5.66	6.27	2.13	2.67	1.87	2.20	2.53
1200	5.26	5.40	5.93	6.06	6.66	1.86	2.00	1.33	1.80	2.00
1400	5.60	5.73	6.20	6.40	7.06	—	—	—	—	—

The value of P at $t = 0$ is by definition zero. Any deviations from this value are due to uncertainties in the determination of the time $t = 0$.

The data in parts A and B of Table IV.7 are used to plot $\Delta P/\Delta t$ against P. These curves are shown in Fig. IV.15 (the origin of the ordinates for the different curves are successively shifted by 5 Pa s^{-1}).

The direct experimental data are plotted in Fig. IV.14. The abscissa and the ordinate are scaled to conform with the accuracy of the experiments. Since the accuracy in reading off the pressure is ± 50 Pa, 1 mm on the graph-paper is taken to correspond to 100 Pa. In this way, 50 Pa is represented by 0.5 mm, i.e. the interval which can still be estimated on the paper. On the other hand, the scale of the time axis (abscissa) is independent of the accuracy with which the time is measured. From the data in Table IV.7 we have

$$\left(\frac{\Delta P}{\Delta t}\right)_{max} \approx 10 \text{ Pa s}^{-1}$$

The errors in ΔP are those in P, i.e. ± 50 Pa. Consequently, the required accuracy in measuring the reaction time is

$$\Delta t \sim \frac{\pm 50}{10} = \pm 5 \text{ s}$$

and the scale of the abscissa in Fig. IV.14 is chosen so that 1 mm corresponds to 10 s. The scales in Fig. IV.15 were chosen by similar considerations. However, the data required to draw these curves are calculated from the curves in Fig. IV.14. The accuracy of reading off the coordinates in Fig. IV.14 must, therefore, be taken into account in this latter case.

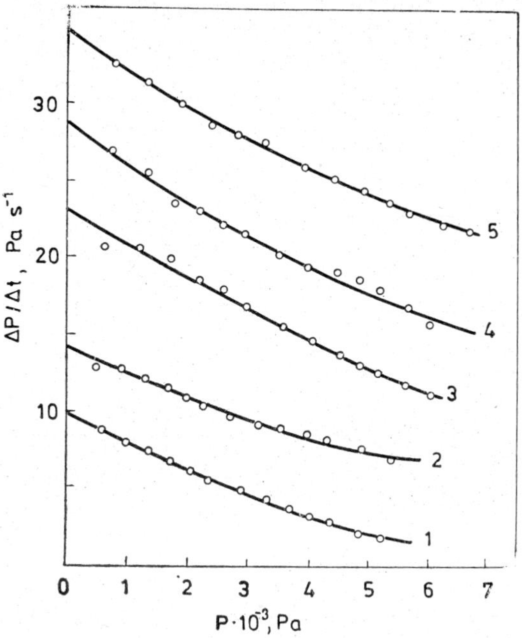

Fig. IV.15. Catalytic decomposition of N_2O. Illustration of the calculation procedure described in the text

A knowledge of the function

$$J = f(p_A, p_{B_1}, p_{B_2}) \qquad \text{(IV.102)}$$

is necessary to determine the rate equation.

According to the law of the conservation of matter the following relationships can be formulated from Eqs (IV.84), (IV.96) and (IV.98):

$$p_A = p_A^0 - 2\,P;\; p_{B_1} = p_{B_1}^0 + P;\; p_{B_2} = p_{B_2}^0 + 2\,P \qquad \text{(IV.103)}$$

where p_A^0, $p_{B_1}^0$ and $p_{B_2}^0$ are the initial partial pressures ($t = 0$) of the components.

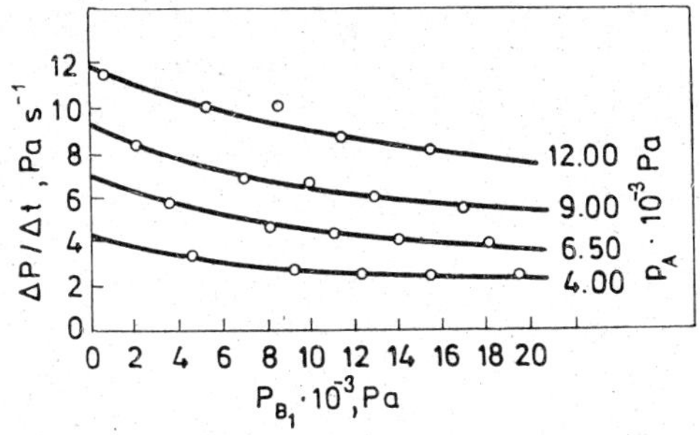

Fig. IV.16. Catalytic decomposition of N_2O. Illustration of the calculation procedure described in the text

The experimental data indicate that the reaction rate is not affected by the partial pressure of N_2, and therefore Eq. (IV.102) can be modified:

$$J = f(p_A, p_{B_1}) \qquad \text{(IV.104)}$$

Using Eq. (IV.103), Table IV.6 and Fig. IV.15, Table IV.8 can be compiled and the functions

TABLE IV.8

Catalytic decomposition of N_2O. Illustration of the calculation procedure described in the text.

p_A, 10^3 Pa	P, 10^3 Pa					$\frac{\Delta P}{\Delta t}$, Pa s^{-1}					p_{B_1} 10^3 Pa				
	1	2	3	4	5	1	2	3	4	5	1	2	3	4	5
4.00	4.60	4.70	4.70	5.40	6.00	2.5	2.6	3.4	2.2	2.3	15.50	19.50	4.70	12.50	9.50
6.50	3.35	3.45	3.45	4.15	4.75	4.0	3.9	5.8	4.3	4.7	14.25	18.25	3.45	11.25	8.25
9.00	2.10	2.20	2.20	2.90	3.50	6.0	5.6	8.4	6.6	6.3	13.00	17.00	2.20	10.00	7.00
12.00	0.60	0.70	0.70	1.40	2.00	8.6	8.0	11.5	10.0	9.3	11.50	15.50	0.70	8.50	5.50

TABLE IV.9

Catalytic decomposition of N_2O. Illustration of the calculation procedure described in the text.

p_{B_1} 10^3 Pa	P, 10^3 Pa					$\frac{\Delta P}{\Delta t}$, Pa s^{-1}					P_A, 10^3 Pa				
	1	2	3	4	5	1	2	3	4	5	1	2	3	4	5
20.00	—	5.20	—	—	—	—	2.1	—	—	—	—	2.93	—	—	—
18.65	—	3.87	—	—	—	—	3.6	—	—	—	—	5.60	—	—	—
17.34	—	2.53	—	—	—	—	5.3	—	—	—	—	8.27	—	—	—
16.00	5.13	1.20	—	—	—	2.0	7.2	—	—	—	2.93	10.92	—	—	—
14.66	3.80	—	—	—	—	3.5	—	—	—	—	5.60	—	—	—	—
13.33	2.47	—	—	6.20	—	5.5	—	—	1.9	—	8.26	—	—	2.40	—
12.00	1.13	—	—	4.87	—	7.5	—	—	3.5	—	10.93	—	—	5.07	—
10.66	—	—	—	3.53	—	—	—	—	5.3	—	—	—	—	7.73	—
9.33	—	—	—	2.20	5.80	—	—	—	7.9	3.2	—	—	—	10.40	4.40
8.00	—	—	—	0.87	4.46	—	—	—	10.7	5.2	—	—	—	13.07	7.06
6.66	—	—	—	—	3.13	—	—	—	—	7.3	—	—	—	—	9.73
5.33	—	—	5.33	—	1.80	—	—	2.5	—	10.1	—	—	2.67	—	12.40
4.00	—	—	4.00	—	0.47	—	—	4.7	—	13.3	—	—	5.34	—	15.07
2.67	—	—	2.67	—	—	—	—	7.3	—	—	—	—	8.00	—	—
1.33	—	—	1.33	—	—	—	—	10.8	—	—	—	—	10.66	—	—
0.00	—	—	0.00	—	—	—	—	15.2	—	—	—	—	13.33	—	—

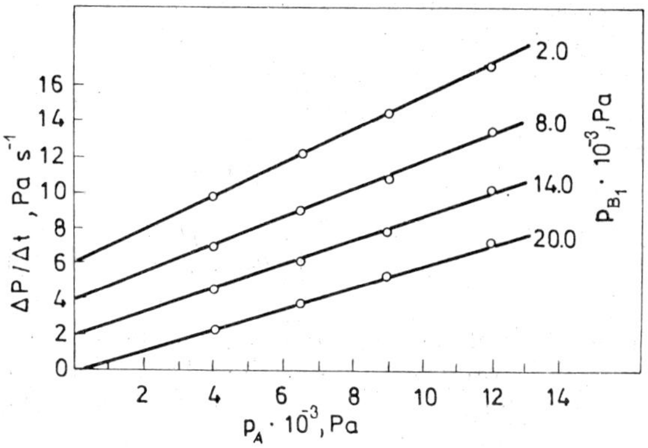

Fig. IV.17. Catalytic decomposition of N_2O. Illustration of the calculation procedure described in the text

$\Delta P/\Delta t = f(p_{B_1})$ can be plotted for different values of $p_A =$ const. These curves are to be seen in Fig. IV.16. Further, the curves representing the functions $\Delta P/\Delta t = f(p_A)$ (p_{B_1} = const.) can be drawn (Fig. IV.17). The origins of the ordinates for the different curves are successively shifted by 2 Pa s^{-1} in both figures.

It is seen in Figs IV.16 and IV.17 that $\Delta P/\Delta t$ is proportional to p_A, and inversely proportional to p_{B_1}.

In accordance with Eq. (IV.96), the catalytic reaction studied belongs to the following reaction type

$$A \rightarrow B_1 + B_2$$

Rate equation (IV.104) corresponds to the third case in Table IV.4. Since $\Delta P/\Delta t$ is proportional to p_A, selection is necessary from four "classes". This is facilitated by plotting the functions

$$p_A \Big/ \frac{\Delta P}{\Delta t} = f(p_{B_1}),\ p_A \Big/ \frac{\Delta P}{\Delta t} = $$

$$= f(\sqrt{p_{B_1}}),\ \sqrt{p_A \Big/ \frac{\Delta P}{\Delta t}} = f(p_{B_1})$$

and $\sqrt{p_A \Big/ \frac{\Delta P}{\Delta t}} = f(\sqrt{p_{B_1}})$. For this purpose, Table IV.9 is first compiled using Eq. (IV.103), Table IV.6 and Fig. IV.15. Table IV.10 is then pre-

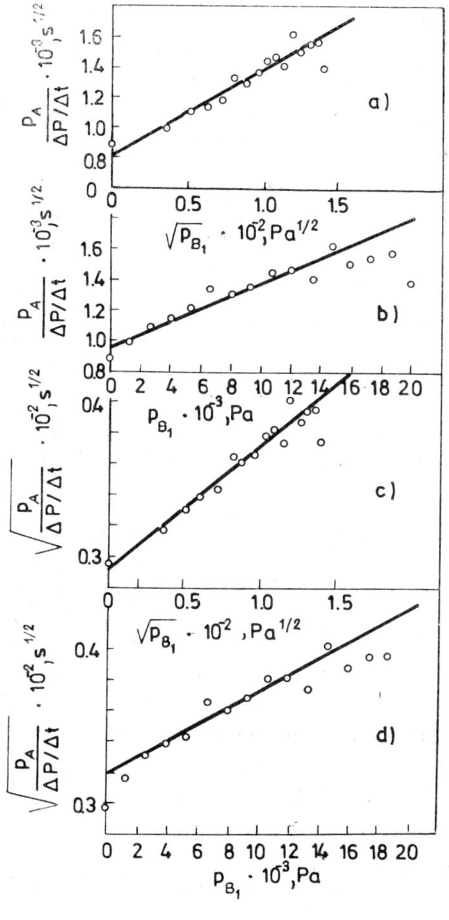

Fig. IV.18. Catalytic decomposition of N_2O. Illustration of the calculation procedure described in the text pared by appropriate transformation of the data in Table IV.9. The resulting plots are shown in Fig. IV.18. Since linear relations are sought, only the following rate equations can be chosen:

$$p_A \Big/ \frac{\Delta P}{\Delta t} = a^{\mathrm{I}} + c^{\mathrm{I}} \sqrt{p_{B_1}} \qquad \text{(IV.105)}$$

$$\sqrt{p_A \Big/ \frac{\Delta P}{\Delta t}} = a^{\mathrm{II}} + c^{\mathrm{II}} \sqrt{p_{B_1}} \qquad \text{(IV.106)}$$

The scatter of the experimental points renders a choice between these cases impossible on the basis of Figs IV.18a and IV.18c. The method of least squares offers a solution to the problem. The constants in Eqs (IV.105)

TABLE IV.10

Catalytic decomposition of N_2O
Illustration of the calculation procedure described in the text

p_{B_1}, 10^3 Pa	$\sqrt{p_{B_1}}$, 10^2 Pa$^{1/2}$	$\frac{p_A}{\Delta P/\Delta t}$, 10^3 s	$\sqrt{\frac{p_A}{\Delta P/\Delta t}}$, 10^2 s$^{1/2}$
0.00	0.000	0.88	0.296
1.33	0.365	0.99	0.316
2.67	0.516	1.09	0.330
4.00	0.632	1.14	0.338
5.33	0.730	1.18	0.344
6.66	0.816	1.33	0.364
8.00	0.894	1.30	0.360
9.33	0.966	1.35	0.367
10.66	1.033	1.45	0.380
12.00	1.095	1.46	0.382
13.33	1.155	1.40	0.374
14.66	1.210	1.61	0.401
16.00	1.265	1.50	0.387
17.34	1.316	1.55	0.394
18.65	1.365	1.56	0.395
20.00	1.415	1.38	0.372

and (IV.106) are estimated using this method. The variances of the values of $p_A \Big/ \frac{\Delta P}{\Delta t}$ calculated for both equations are then compared.

Both Eqs (IV.105) and (IV.106) are of the following form:

$$y = A + Bx$$

Thus, according to the least squares principle we have:

$$\frac{\partial}{\partial A} \sum_{i=1}^{n} (y_i - A - Bx_i)^2 = 0,$$

$$\frac{\partial}{\partial B}\sum_{i=1}^{n}(y_i - A - Bx_i)^2 = 0$$

The values of A and B are obtained by solving the following system of equations:

$$\sum_{i=1}^{n} y_i = nA + B\sum_{i=1}^{n} x_i$$

$$\sum_{i=1}^{n} x_i y_i = A\sum_{i=1}^{n} x_i + B\sum_{i=1}^{n} x_i^2$$

where n is the number of experimental data.

The computation is illustrated in Table IV.11, from which it can be seen that Eq. (IV.105) gives rise to a smaller variance (although the difference is too small to make a definite decision), and consequently this is accepted as being valid. This result, together with Eqs (IV.99) and (IV.100) and Table IV.4/2, leads to the conclusion that the catalytic reaction studied is of the type II/a_3–1. The kinetic equation of this type (see Table IV.2/2) is:

$$J = A_s[S]_0 \vec{k}_A \frac{p_A}{1 + \sqrt{K_{B_1} p_{B_1}}} \qquad \text{(IV.107)}$$

and the adsorption of component A (N_2O) is the rate-controlling step. Eq. (IV.107) can formally be transformed as follows:

$$J = k \frac{p_A}{1 + \sqrt{K_{B_1} p_{B_1}}} \qquad \text{(IV.108)}$$

Comparing Eq. (IV.105) with Eq. (IV.108), and considering Eqs (IV.99) and (IV.100), we have

$$k = \frac{\alpha}{a^{\mathrm{I}}}; \quad K_{B_1} = \left(\frac{c^{\mathrm{I}}}{a^{\mathrm{I}}}\right)^2 \qquad \text{(IV.109)}$$

From the data in Table IV.11:

$$a^{\mathrm{I}} = 8.56\ 10^2\ \mathrm{s};$$

$$c^{\mathrm{I}} = 5.06\ \mathrm{s\ Pa^{-1/2}} \qquad \text{(IV.110)}$$

The average temperature in the connecting tubes $T_d = 300$ °K (see Table IV.6), and thus from Eq. (IV.98):

$$\alpha = \frac{2}{R}\left(\frac{40}{846.13} + \frac{35}{300.16} + \frac{5}{573.16}\right), \qquad \text{(IV.111)}$$

$$R = 8.31\ 10^6\ \mathrm{cm^3\ Pa\ mole^{-1}\ K^{-1}}$$

and

$$\alpha = \frac{2}{8.31\ 10^6}(0.0473 + 0.1168 + 0.0087) = 4.16\ 10^{-7}\ \mathrm{mole\ Pa^{-1}} \qquad \text{(IV.112)}$$

Catalytic decomposition of N_2O. Illustration of the calculation

i	$10^{-2} x_i \equiv$ $\equiv 10^{-2} \sqrt{p_{B_1}}$	$10^{-4} x_i^2 \equiv$ $\equiv 10^{-4}\, p_{B_1}$	$10^{-2} y_i \equiv$ $\equiv 10^{-2} \times$ $\times \frac{p_A}{\Delta P/\Delta t}$	$x_i y_i\, 10^{-4}$	$10^{-2}(y_i)_c \equiv$ $\equiv 10^{-2}\left(\frac{p_A}{\Delta P/\Delta t}\right)_c$	$10^{-2} \Delta_i \equiv$ $\equiv 10^{-2}[y_i - (y_i)_c]$	$10^{-4} \Delta_i^2$
1	0.000	0.000	8.80	0.000	8.56	+0.24	0.058
2	0.365	0.133	9.90	3.614	10.41	−0.51	0.260
3	0.516	0.267	10.90	5.624	11.17	−0.27	0.073
4	0.632	0.400	11.40	7.205	11.76	−0.36	0.130
5	0.730	0.533	11.80	8.614	12.25	−0.45	0.202
6	0.816	0.666	13.30	10.853	12.69	+0.61	0.372
7	0.894	0.800	13.00	11.622	13.08	−0.08	0.006
8	0.966	0.933	13.50	13.041	13.45	+0.05	0.003
9	1.033	1.066	14.50	14.978	13.79	+0.71	0.504
10	1.095	1.200	14.60	15.987	14.10	+0.50	0.250
11	1.155	1.333	14.00	16.170	14.40	−0.40	0.160
12	1.210	1.466	16.10	19.481	14.68	+1.42	2.019
13	1.265	1.600	15.00	18.975	14.96	+0.04	0.002
14	1.316	1.734	15.50	20.398	15.22	+0.28	0.078
15	1.365	1.865	15.60	21.294	15.47	+0.13	0.017
16	1.415	2.000	13.80	19.527	15.72	−1.92	3.685
1–16	14.773	15.996	211.70	207.383			7.819

$$10^{-2} \frac{p_A}{\Delta P/\Delta t} = 8.56 + 5.06\, 10^{-2} \sqrt{p_{B_1}}\ ;\quad 10^{-4} \frac{\Sigma \Delta_i^2}{n} = 0.488$$

If only the mechanisms presented in Section 1.5 are considered then Eqs (IV.109), (IV.110) and (IV.111) give the following values:

$$k = 4.84\ 10^{-10}\ \text{mole Pa}^{-1}\ \text{s}^{-1};\qquad K_{B_1} = 3.49\ 10^{-5}\ \text{Pa}^{-1} \tag{IV.113}$$

PARAGRAPH 2.3.2. GRADIENTLESS RECIRCULATION REACTOR

Móger [9] has examined the catalytic hydrogenation of cyclohexene in a gradientless recirculation reactor, using platinized platinum as the catalyst.

The stoichiometric equation of the reaction is:

$$H_2 + C_6H_{10} = C_6H_{12};\qquad A_1 + A_2 \rightarrow B \tag{IV.114}$$

and consequently the stoichiometric numbers are: $\nu_{A_1} = -1$; $\nu_{A_2} = -1$; $\nu_B = 1$.

The rate equation can be expressed on the basis of Eq. (IV.82):

procedure described in the text (c: calculated values)

$10^{-1} y'_i \equiv \equiv 10^{-1} \sqrt{\frac{p_A}{\Delta P/\Delta t}}$	$x_i y'_i 10^{-3}$	$10^{-1}(y'_i)_c$	$10^{-2}(y'^2_{ic})]$	$10^{-2} \Delta'_i \equiv \equiv 10^{-2}[y'_i - (y'_i)^2_c]$	$10^{-4}(\Delta'_i)^2$
2.96	0.000	2.96	8.75	+0.05	0.003
3.16	1.153	3.22	10.38	—0.48	0.230
3.30	1.703	3.33	11.10	—0.20	0.040
3.38	2.136	3.41	11.62	—0.22	0.048
3.44	2.511	3.49	12.19	—0.39	0.152
3.64	2.970	3.55	12.60	+0.70	0.490
3.60	3.218	3.60	12.96	+0.04	0.002
3.67	3.545	3.66	13.40	+0.10	0.010
3.80	3.925	3.70	13.70	+0.80	0.640
3.82	4.183	3.75	14.08	+0.52	0.270
3.74	4.320	3.79	14.37	—0.37	0.137
4.01	4.852	3.83	14.68	+1.42	2.020
3.87	4.896	3.87	14.98	+0.02	0.000
3.94	5.185	3.91	15.30	+0.20	0.040
3.95	5.392	3.94	15.52	+0.08	0.006
3.72	5.264	3.98	15.85	—2.05	4.200
58.00	55.254				8.288

$$10^{-1} \sqrt{\frac{p_A}{\Delta P/\Delta t}} = 2.958 + 0.722\ 10^{-2} \sqrt{p_{B_1}}\ ;\ 10^{-4} \frac{\Sigma(\Delta'_i)^2}{n} = 0.518$$

$$J = -\frac{1}{R}\left(\frac{v_r}{T_r} + \frac{v_m}{T_m} + \frac{v_d}{T_d}\right)\frac{dp}{dt} \tag{IV.115}$$

Introducing the following notations

$$\alpha \equiv \frac{1}{R}\left(\frac{v_r}{T_r} + \frac{v_m}{T_m} + \frac{v_d}{T_d}\right);\ P \equiv p_0 - p \tag{IV.116}$$

Eq. (IV.115) can be written:

$$J = \alpha \frac{dP}{dt} \tag{IV.117}$$

The experimental data are shown in Table IV.12 and Fig. IV.19. (The origins of the ordinates for the different curves are successively shifted by 10^3 Pa.)

Ageing of the catalyst was observed during the experiments. For the experiments to be reproducible, the regeneration of the catalyst was necessary. Accordingly, the P vs. t curve is not suitable for the determination of the rate equation. Instead, values extrapolated to $t = 0$, $(dP/dt)_0$,

TABLE IV.12

Hydrogenation of cyclohexene on a Pt catalyst in a gradientless recirculation reactor

$v_r = 50\ cm^3,\ v_m = 10\ cm^3,\ v_b = 100\ cm^3$

$t_r = 50\ °C,\ t_m = 20\ °C,\quad t_b = 20–50\ °C$

	1		2		3		4		5		6		7		8	
$p^0_{A_1}$, 10^3 Pa	2.88		5.46		7.94		13.33		26.52		5.40		10.80		16.00	
$p^0_{A_2}$, 10^3 Pa	2.67		2.67		2.67		2.67		2.67		5.34		5.34		5.34	
	t, s	P, 10^3 Pa	t, s	P, 10^3 Pa	t, s	P, 10^3 Pa	t, s	P, 10^3 Pa	t, s	P, 10^3 Pa	t, s	P, 10^3 Pa	t, s	P, 10^3 Pa	t, s	P, 10^3 Pa
	0	0	0	0	0	0	0	0	0	0	0	0	0	0	0	0
	15	0.09	18	0.13	25	0.20	7	0.13	16	0.13	5	0.07	10	0.13	15	0.13
	40	0.23	40	0.27	40	0.33	20	0.27	30	0.27	18	0.20	18	0.27	25	0.27
	78	0.36	65	0.40	60	0.47	35	0.40	46	0.40	45	0.47	40	0.53	40	0.40
	110	0.49	85	0.53	85	0.60	57	0.53	60	0.53	58	0.60	63	0.80	55	0.53
	225	0.89	115	0.67	102	0.73	76	0.67	78	0.67	73	0.73	105	1.07	70	0.80
	263	1.03	150	0.80	125	0.87	100	0.80	130	0.93	100	0.87	123	1.33	105	1.07
	308	1.16	190	0.93	180	1.13	120	0.93	180	1.20	120	1.00	140	1.47	125	1.33
	355	1.29	220	1.07	250	1.40	147	1.07	255	1.47	140	1.13	180	1.73	160	1.60
	420	1.43	258	1.20	290	1.53	178	1.20	345	1.80	162	1.27	222	2.00	190	1.87
	488	1.56	300	1.33	395	1.80	210	1.33	495	2.00	205	1.67	275	2.27	245	2.27
	565	1.69	355	1.47	515	2.07	240	1.47	590	2.13	252	1.93	320	2.53	320	2.67
	638	1.83	408	1.60			327	1.73			305	2.20	385	2.80	390	3.07
			515	1.87			440	2.00			360	2.47	415	2.93		
			760	2.13			620	2.27			420	2.73	510	3.20		
											495	3.00				
											690	3.27				
											725	3.53				

should be used. The extrapolation is best based on the mathematical form of the function $P = f(t)$. The curves in Fig. IV.19 can be represented by the exponential relationship:

$$P = a[1 - \exp(-bt)] \qquad \text{(IV.118)}$$

Hence

$$\left(\frac{dP}{dt}\right)_0 = ab \qquad \text{(IV.119)}$$

The constants in Eq. (IV.118) can be determined as follows: Denote the pressures measured at equal time intervals by P_i and P_{i+1}. The following expression results from Eq. (IV.118):

$$P_i = a[1 - \exp(b\Delta t)] + \exp(b\Delta t)P_{i+1} \qquad \text{(IV.120)}$$

Thus, if the pressure values measured at equal intervals are plotted as functions of each other in the above form a straight line is obtained whose slope and intercept are used to calculate the constants a and b. In practice the ordinates of the curves (Fig. IV.19) are read off at equal intervals ($\Delta t = 50$ s) of the abscissa. These pressure values are collected in Table IV.14. They are next plotted according to Eq. (IV.120) (see Fig. IV.20; the origins of the abscissa for the different curves are successively shifted by 10^3 Pa). The slopes and intercepts of these lines are then determined. The next step is to calculate the constants a and b and, consequently, the value of $(dP/dt)_0$. The results of the calculations are also shown in Table IV.13.

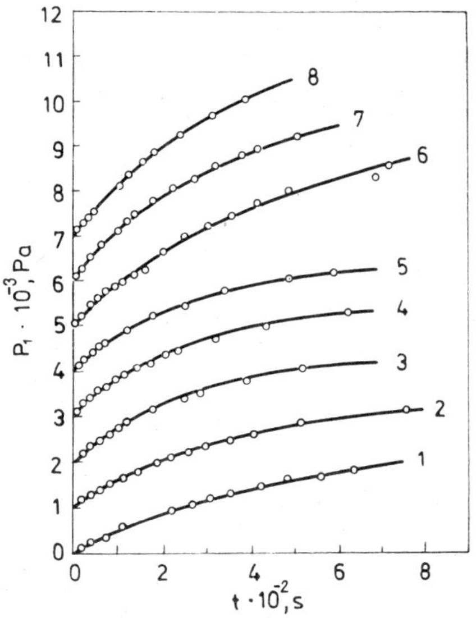

Fig. IV.19. Dependence of total pressure on time for the catalytic hydrogenation of cyclohexene on a platinized platinum catalyst

The experiments revealed that the pressure of cyclohexane did not affect the reaction rate, i.e. the rate equation of the hydrogenation is of the following general form:

$$J = f(p_{A_1}, p_{A_2}) \qquad \text{(IV.121)}$$

According to Eq. (IV.114), the catalytic hydrogenation of cyclohexene is therefore of the type:

$$A_1 + A_2 \rightarrow B$$

The kinetic and rate equation for this type of reaction are collected in Table IV.5. The general form of the rate equation Eq. (IV.121) may correspond to any of the following linearized expressions (see Tables IV.5/1 and IV.5/2):

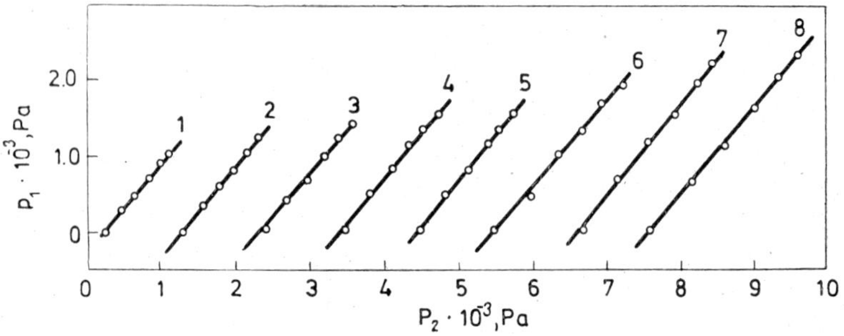

Fig. IV.20. Hydrogenation of cyclohexene. Illustration of the calculation procedure described in the text

$$\frac{p^0_{A_1}}{\left(\frac{\mathrm{d}P}{\mathrm{d}t}\right)_0} = a^{\mathrm{I}} + c^{\mathrm{I}} p^0_{A_2} \qquad \text{(IV.123)}$$

$$\frac{p^0_{A_2}}{\left(\frac{\mathrm{d}P}{\mathrm{d}t}\right)_0} = a^{\mathrm{II}} + b_{\mathrm{II}} p^0_{A_1} \qquad \text{(IV.124)}$$

TABLE IV.13

Hydrogenation of cyclohexene
Illustration of the calculation procedure described in the text

t s	P (calculated), 10^3 Pa							
	1	2	3	4	5	6	7	8
0	0.00	0.00	0.00	0.00	0.00	0.00	0.00	0.00
50	0.27	0.33	0.40	0.47	0.47	0.47	0.67	0.60
100	0.47	0.60	0.67	0.80	0.80	1.00	1.13	1.13
150	0.67	0.80	1.00	1.13	1.13	1.33	1.53	1.60
200	0.87	1.00	1.20	1.33	1.33	1.67	1.93	2.00
250	1.00	1.20	1.40	1.53	1.53	1.93	2.20	2.33
300	1.13	1.33	1.60	1.73	1.73	2.20	2.46	2.60
$e^{b\Delta t}$	1.165	1.200	1.130	1.190	1.240	1.200	1.200	1.220
$a(1 - e^{b\Delta t})$	−287	−367	−373	−453	−520	−600	−759	−746
$a\ 10^{-3}$	1.75	1.82	2.87	2.39	2.16	3.00	3.80	3.40
$b\ 10^3$	3.04	3.68	2.49	3.50	4.34	3.68	3.68	4.02
$\left(\frac{\mathrm{d}P}{\mathrm{d}t}\right)_0$	5.33	6.80	7.19	8.39	9.46	10.93	13.86	13.72

$$\frac{p^0_{A_1} p^0_{A_2}}{\left(\dfrac{\mathrm{d}P}{\mathrm{d}t}\right)_0} = a^{\mathrm{III}} + b^{\mathrm{III}} p^0_{A_1} \qquad \text{(IV.125)}$$

$$\sqrt{\frac{p^0_{A_1} p^0_{A_2}}{\left(\dfrac{\mathrm{d}P}{\mathrm{d}t}\right)_0}} = a^{\mathrm{IV}} + b^{\mathrm{IV}} p^0_{A_1} + c^{\mathrm{IV}} p^0_{A_2} \qquad \text{(IV.126)}$$

$$\sqrt{\frac{p^0_{A_1} p^0_{A_2}}{\left(\dfrac{\mathrm{d}P}{\mathrm{d}t}\right)_0}} = a^{\mathrm{V}} + b^{\mathrm{V}} \sqrt{p^0_{A_1}} \qquad \text{(IV.}$$

To check the applicability of the above equations Table IV.14 has been compiled using the data in Tables IV.12 and IV.13. It is seen immediately that neither Eq. (IV.123) $\left[\text{since } p^0_{A_1} \Big/ \left(\dfrac{\mathrm{d}P}{\mathrm{d}t}\right)_0 \text{ depends on } p_{A_1}\right]$ nor Eq. (IV.124) $\left[p^0_{A_2} \Big/ \left(\dfrac{\mathrm{d}P}{\mathrm{d}t}\right)_0 \text{ decreases with increasing values of } p^0_{A_1}\right]$ is valid. Further

TABLE IV.14

Hydrogenation of cyclohexene
Illustration of the calculation procedure described in the text.

No. of experiment	1	2	3	4	5	6	7	8
p^0_{A}, 10^3 Pa	2.89	5.46	7.93	13.33	26.51	5.40	10.79	15.99
p^0_{A}, 10^3 Pa	2.66	2.66	2.66	2.66	2.66	5.33	5.33	5.33
$\left(\frac{\mathrm{d}P}{\mathrm{d}t}\right)_0$, Pa s^{-1}	5.33	6.80	7.20	8.40	9.46	10.92	13.86	13.73
$p^0_{A_1} \Big/ \left(\frac{\mathrm{d}P}{\mathrm{d}t}\right)_0$, 10^2 s	5.5	8.1	11.1	16.0	28.2	4.9	7.8	11.7
$p^0_{A_2} \Big/ \left(\frac{\mathrm{d}P}{\mathrm{d}t}\right)_0$, 10^2 s	5.1	4.0	3.7	3.2	2.8	4.8	3.8	3.9
$p^0_{A_1} p^0_{A_2} \Big/ \left(\frac{\mathrm{d}P}{\mathrm{d}t}\right)_0$, 10^6 Pa s	1.5	2.1	3.1	4.3	7.5	2.5	4.1	6.1
$\sqrt{p^0_{A_1} p^0_{A_2} \Big/ \left(\frac{\mathrm{d}P}{\mathrm{d}t}\right)_0}$, 10^3 $Pa^{1/2}$ $s^{1/2}$	1.2	1.4	1.8	2.1	2.7	1.6	2.0	2.5
$\sqrt{p^0_{A_1}}$, 10^2 $Pa^{1/2}$	0.54	0.74	0.89	1.16	1.63	0.73	1.04	1.26

selection can be performed by plotting the experimental data using the respective transformed variables whose values are given in Table IV.14. Inspection of the curves shown in Fig. IV.21 leads to the conclusion that, owing to the dispersion of the experimental data, Eqs (IV.125), (IV.126) and (IV.127) may be equally acceptable. The most adequate equation is again chosen by using the method of least squares, as shown for the decomposition of N_2O. (The method of least squares should be handled with due care. It is a common occurrence for the variance, calculated for a given approximation, to appear to be the least, whilst the deviations exhibit a definite tendency. In such a case the variance cannot be used to decide the validity of an equation. Besides, the method of least squares operates with the ordinate errors only; this is not correct in most cases.) The variances resulting from the above calculations are:

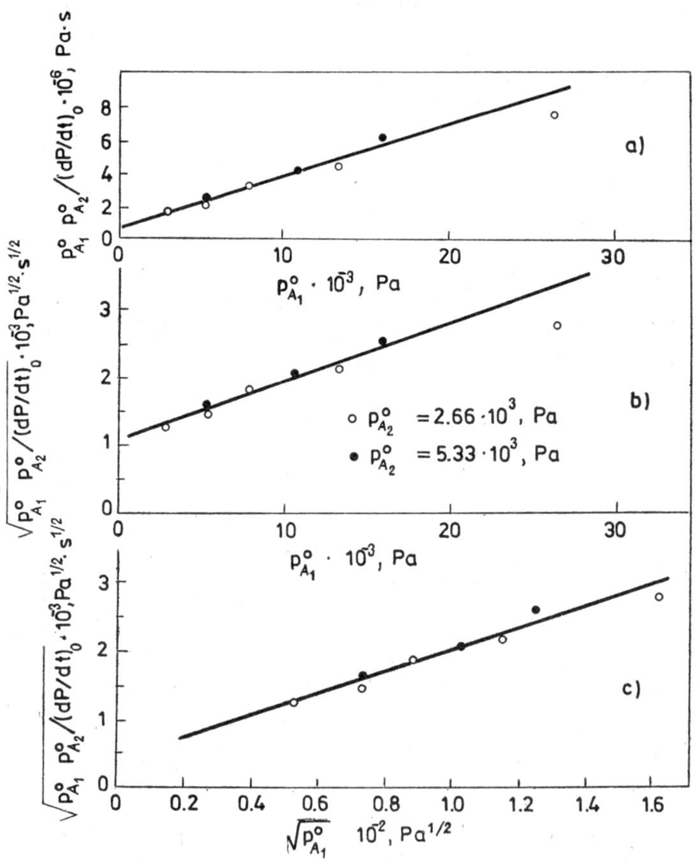

Fig. IV.21. Hydrogenation of cyclohexene. Illustration of the calculation procedure described in the text

$$\text{for Eq. (IV.125)}\ \frac{\Sigma\Delta_i^2}{n} = 1.77\ \ 10^{11}\ \text{Pa}^2\ \text{s}^2$$

$$\text{for Eq. (IV.126)}\ \frac{\Sigma\Delta_i^2}{n} = 3.90\ \ 10^{11}\ \text{Pa}^2\ \text{s}^2 \qquad \text{(IV.128)}$$

$$\text{for Eq. (IV.127)}\ \frac{\Sigma\Delta_i^2}{n} = 1.42\ \ 10^{11}\ \text{Pa}^2\ \text{s}^2$$

i.e. the experimental data are most precisely described by Eq. (IV.127). The constants of Eq. (IV.127) are accordingly calculated as:

$$a^{\text{V}} = 4.27\ \ 10^2\ \text{Pa}^{1/2}\ \text{s}^{1/2}$$
$$b^{\text{V}} = 14.8\ \text{s}^{1/2} \qquad \text{(IV.129)}$$

According to Table IV.5/1, Eq. (IV.127) corresponds to the following kinetic equation:

$$J = k\frac{p_{A_1}p_{A_2}}{(1+\sqrt{K_{A_1}p_{A_1}})^2} \qquad \text{(IV.130)}$$

(naturally, only the mechanisms presented in Section 1.5 are considered). Eq. (IV.130) represent type II/b_3–2. The characteristics of this type of reaction can be noted from Table IV.2/3: dissociative adsorption of component A_1 (hydrogen) takes place; component A_2 (cyclohexene) is not absorbed; and the reaction rate is determined by the surface reaction.

Using Eqs (IV.117), (IV.127), (IV.129) and (IV.130) the following constants can be determined:

$$k = \frac{\alpha\ 10^{-4}}{4.27^2}\ \text{mole Pa}^{-2}\ \text{s}^{-1};$$

$$K_{A_1} = \left(\frac{14.8}{4.27}\right)^2 10^{-4}\ \text{Pa}^{-1}$$

The value of α is obtained using the data in Table IV.12 and Eq. (IV.116) ($t_{\text{d}} = 35\ °\text{C}$):

$$\alpha = 6.18\ \ 10^{-8}\ \text{mole Pa}^{-1}$$

hence

$$k = 3.28\ \ 10^{-13}\ \text{mole Pa}^{-2}\ \text{s}^{-1}$$
$$K_{A_1} = 1.2\ \ 10^{-13}\ \text{Pa}^{-1} \qquad \text{(IV.131)}$$

The constant k in the kinetic equation (IV.130) includes the following terms (see Table IV.5/2):

$$k = A_{\text{s}}[S]_0^2\vec{k}K_{A_1} \qquad \text{(IV.132)}$$

From Eqs (IV.131) and (IV.132):

$$A_{\text{s}}[S]_0^2\,\vec{k} = \frac{k}{K_{A_1}} = 2.73\ \ 10^{-10}\ \text{mole Pa}^{-1}\ \text{s}^{-1} \qquad \text{(IV.133)}$$

Kiperman [7] studied the hydrogenation of ethylene on a nickel catalyst in a gradientless stirred flow reactor. The reaction mixture diluted with nitrogen was introduced into the reactor at a pressure of 1 atm with a constant flow velocity. The mixture leaving the reactor was cooled to condense ethylene and ethane, and the [hydrogen] : [nitrogen] ratio was measured in the residual gas. This ratio was used to evaluate the composition of the reaction mixture. The results are presented in Table IV.15. The reaction obeys the following stoichiometric equation.

$$H_2 + C_2H_4 = C_2H_6; \quad A_1 + A_2 \rightarrow B \qquad \text{(IV.134)}$$

TABLE IV.15

Hydrogenation of ethylene on a Ni catalyst in a gradientless stirred flow reactor

$m_S = 0.1$ g $\quad t_r = 77.0$ °C $\quad p = 1.012\ 10^5$ Pa

$A_1 \equiv H_2 \quad A_2 \equiv C_2H_4 \quad B \equiv C_2H_6 \quad 0 \equiv N_2$

No	F' 10^3 cm^3 s^{-1} (20.0 °C)	$(p_{A_1})_0$ 10^4 Pa	$(p_{A_2})_0$ 10^4 Pa	$(p_B)_0$, 10 Pa	$(p_0)_0$, 10^4 Pas	$(p_{A_1})_1$, 10^4 Pa	$(p_{A_2})_1$, 10^4 Pa	(p_{B1}), 10^4 Pa	$(p_0)_1$, 10^4 Pa
1	4.86	2.09	3.16	—	4.88	1.03	2.25	1.33	5.52
2	12.52	2.09	3.16	—	4.88	1.56	2.70	0.67	5.20
3	21.40	2.09	3.16	—	4.88	1.65	2.77	0.55	5.16
4	36.80	2.09	3.16	—	4.88	1.87	2.96	0.29	5.00
5	42.80	2.09	3.16	—	4.88	1.88	2.98	0.25	5.01
6	55.20	2.09	3.16	—	4.88	1.95	3.03	0.18	4.97
7	21.45	2.13	3.81	—	4.18	1.76	3.52	0.47	4.37
8	22.45	2.52	3.54	—	4.06	2.08	3.16	0.57	4.30
9	24.95	3.34	3.27	—	3.52	2.91	2.83	0.64	3.74
10	30.40	4.72	2.74	—	2.66	4.28	2.14	0.82	2.88
11	33.60	5.33	2.05	—	2.74	4.85	1.28	0.96	3.03

Hence, in agreement with Eqs (IV.85) and (IV.86), the reaction rate is expressed by:

$$J = F[(y_B)_1 - (y_B)_0] \qquad \text{(IV.135)}$$

The flow (F) of the reaction mixture and the molar concentrations of the components referred to unit mass of the reaction mixture (y) can be calculated from the experimental data. The computations are outlined below:

The molecular weights of the components (M) are:

$$\begin{aligned} M_{A_1}\ (H_2) &= 2.02 \text{ g mole}^{-1} \\ M_{A_2}\ (C_2H_4) &= 28.11 \text{ g mole}^{-1} \\ M_O\ (N_2) &= 28.02 \text{ g mole}^{-1} \\ M_B\ (C_2H_6) &= 30.13 \text{ g mole}^{-1} \end{aligned} \qquad \text{(IV.136)}$$

Regarding the reaction mixture as an ideal gas and applying the gas law we have:

$$p_i v = \frac{m_i}{M_i} RT;$$
$$m_i = p_i M_i \frac{v}{RT} \qquad \text{(IV.137)}$$

and

$$F \equiv \frac{\mathrm{d}m_i}{\mathrm{d}t} = \frac{\mathrm{d} \sum_i m_i}{\mathrm{d}t} = [(p_{A_1})_0 M_{A_1} + (p_{A_2})_0 M_{A_2} + (p_O)_0 M_O] \frac{F'}{RT} \qquad \text{(IV.138)}$$

where

$$\frac{\mathrm{d}v}{\mathrm{d}t} = F'$$

The molar concentration referred to unit mass of reaction mixture for any component is obtained using Eq. (IV.137):

$$y_i = \frac{m_i/M_i}{\sum_i m_i} = \frac{p_i}{\sum_i p_i M_i} \qquad \text{(IV.139)}$$

i.e.

$$(y_B)_1 = \frac{(p_B)_1}{(p_{A_1})_1 M_{A_1} + (p_{A_2})_1 M_{A_2} + (p_B)_1 M_B + (p_O)_1 M_O} \qquad \text{(IV.140)}$$

The results of applying Eqs (IV. 135), (IV.138) and (IV.140) to the experimental data are collected in Table IV.16.

TABLE IV.16

Hydrogenation of ethylene
Illustration of the calculation procedure described in the text

No. of experiment	F, $g_g\ s^{-1}$	$(y_B)_1$, 10^{-2} mole g_g^{-1}	J, 10^{-2} mole s^{-1}	$\frac{J}{(p_{A_1})_1}$, 10^{-6} mole $Pa^{-1}\ s^{-1}$
1	4.58	0.514	2.35	2.29
2	11.80	0.272	3.21	2.06
3	20.18	0.227	4.59	2.78
4	34.75	0.124	4.30	2.25
5	40.30	0.107	4.32	2.30
6	52.05	0.077	4.01	2.06
7	20.18	0.198	3.98	2.27
8	20.18	0.249	5.02	2.41
9	20.18	0.306	6.16	2.12
10	20.18	0.470	9.47	2.21
11	20.18	0.603	12.19	2.36
				25.32

These calculations reveal (see Table IV.16) that the experimental data can be described by the following kinetic equation:

$$J = k\, p_{A_1} \tag{IV.141}$$

According to Eq. (IV.134) the hydrogenation of ethylene is classified as a reaction of type II/b, i.e.

$$A_1 + A_2 \rightarrow B$$

With respect to the mechanism, this reaction belongs to groups II/b_1–1, II/b_2–1 and II/b_3–1, i.e. hydrogen adsorption is the rate-controlling step (considering only the mechanisms discussed in Section 1.5):

$$k = A_s[S]_0\, \vec{k}_{A_1} = 2.30 \ 10^{-6} \ \text{mole Pa}^{-1}\text{s}^{-1} \tag{IV.142}$$

PARAGRAPH 2.3.4. DYNAMIC INTEGRAL REACTOR

Kalló et al. [10] have studied the catalytic dehydration of diethyl ether on an Al_2O_3 catalyst in a dynamic integral reactor.

The stoichiometric equation of the reaction is:

$$(C_2H_5)_2O = 2\ C_2H_4 + H_2O;$$

$$A \rightarrow B_1 + B_2 \tag{IV.143}$$

Liquid ether was continuously fed into the reactor. The evaporated ether established a flow of constant mass velocity in the reactor. The gas-vapour mixture leaving the reactor at 1 atm was passed through a cooler. The amount of ethylene formed was measured using a soap-film flow meter. The experimental results are shown in Table IV.17. The rate equation is expressed according to Eqs (IV.92) and (IV.143) as follows:

$$\frac{1}{S} = -\int_{(y_A)_0}^{(y_A)_1} \frac{dy_A}{\bar{j}} = \frac{1}{2}\int_{(y_{B_1})_0}^{(y_{B_1})_1} \frac{dy_{B_1}}{\bar{j}} \tag{IV.144}$$

TABLE IV.17

Catalytic dehydration of diethyl ether on an Al_2O_3 catalyst in a dynamic integral reactor

$m_s = 4.12$ g; $t_r = 348$ °C; $P = 1.012\ 10^5$ Pa

$A \equiv (C_2H_5)_2O;$ $B_1 \equiv C_2H_4;$ $B_2 \equiv H_2O$

No. of experiment	1	2	3	4	5
$F_A''\ 10^{-3}\ \text{cm}^{-3}\ \text{s}^{-1}$ (liquid ether at 25 °C)	4.99	6.38	9.16	18.61	70.50
$F_{B_1}'\ (\text{cm}^3\ \text{s}^{-1})$ (in gaseous state at 25 °C)	1.855	2.100	2.370	2.980	2.320

The density of the pure liquid ether (at 25 °C) introduced into the reactor is:

$$\varrho_A = 0.712 \text{ g cm}^{-3} \tag{IV.145}$$

Using this value and those of F''_A taken from Table IV.17, the mass flow of the feed, F, and (if m_s is known) the space velocity, S, can be calculated:

$$F = F''_A \varrho_A; \; S \equiv F/m_s \tag{IV.146}$$

Given the molecular weights

$$M_A = 74.08 \text{ g mole}^{-1}; \quad M_{B_1} = 28.03 \text{ g mole}^{-1}; \quad M_{B_2} = 18.02 \text{ g mole}^{-1} \tag{IV.147}$$

y_A, y_{B_1} and y_{B_2} can be calculated in the following way.

It was mentioned above that only ether was fed into the reactor, so that:

$$(y_A)_0 = \frac{1}{M_A}; \quad (y_{B_1})_0 = 0; \quad (y_{B_2})_0 = 0 \tag{IV.148}$$

From Eqs (IV.144) and (IV.148) the material balance is:

$$(y_A)_0 - (y_A)_1 = \frac{1}{2}(y_{B_1})_1 = (y_{B_2})_1 \tag{IV.149}$$

$(y_{B_1})_1$ is given by the space velocity of the evolved ethylene, applying the perfect gas law:

$$(y_{B_1})_1 = \frac{F'_{B_1}\, 273.16}{298.16 \;\; 22410\, F} \tag{IV.150}$$

Performing the calculations as indicated above, the results shown in Table IV.18 are obtained.

TABLE IV.18

Catalytic dehydration of diethyl ether
Illustration of the calculation procedure described in the text

No. of experiment	1	2	3	4	5
F, 10^{-3} g_g s^{-1}	3.55	4.54	6.52	13.24	50.10
S, 10^{-3} g_g g_s^{-1} s^{-1}	0.86	1.10	1.58	3.22	12.18
$(y_A)_0$, 10^{-2} mole g_g^{-1}	1.35	1.35	1.35	1.35	1.35
$(y_A)1$, 10^{-2} mole g_g^{-1}	0.283	0.405	0.608	0.891	1.255
$\frac{1}{S}$, 10^3 g_s s g_g^{-1}	1.162	0.910	0.633	0.310	0.083

According to Eq. (IV.143) the dehydration of diethyl ether is of the type:

$$A \rightarrow B_1 + B_2$$

In this series of experiments the starting substance was always pure ether, i.e. the initial concentrations of ether, A, ethylene, B_1, and water, B_2 were not varied. Thus, the roles of the concentrations of the individual components in the kinetic equation cannot be elucidated. This is because the interrelations of the partial pressures of the components are fixed by the stoichiometric equation.

According to the definition of the partial pressures:

$$p_A = \frac{y_A}{y_A + y_{B_1} + y_{B_2}} P; \quad p_{B_1} = \frac{y_{B_1}}{y_A + y_{B_1} + y_{B_2}} P; \quad p_{B_2} = \frac{y_{B_2}}{y_A + y_{B_1} + y_{B_2}} P \tag{IV.151}$$

Taking into account Eq. (IV.149), Eqs (IV.151) become:

$$p_A = \frac{y_A}{3(y_A)_0 - 2\,y_A} P; \quad p_{B_1} = \frac{2\,(y_A)_0 - 2\,y_A}{3\,(y_A)_0 - 2\,y_A} P; \quad p_{B_2} = \frac{(y_A)_0 - y_A}{3\,(y_A)_0 - 2\,y_A} P$$

Hence:

$$\frac{p_A}{P} = \frac{y_A}{3(y_A)_0 - 2y_A}; \quad \frac{p_{B_1}}{P} = \frac{2}{3}\left(1 - \frac{p_A}{P}\right); \quad \frac{p_{B_2}}{P} = \frac{1}{3}\left(1 - \frac{p_A}{P}\right) \tag{IV.152}$$

The rate equations to be considered are (see Tables IV.4/1, IV.4/2 and IV.4/3):

$$J = \frac{p_A}{a^{\mathrm{I}} + b^{\mathrm{I}} p_A + c^{\mathrm{I}} p_{B_1}}; \quad \frac{p_A}{a^{\mathrm{I}} + b^{\mathrm{I}} p_A + d^{\mathrm{I}} p_{B_2}} \tag{IV.153a}$$

$$J = \frac{p_A}{(a^{\mathrm{II}} + b^{\mathrm{II}} p_A + c^{\mathrm{II}} p_{B_1} + d^{\mathrm{II}} p_{B_2})^2} \tag{IV.153b}$$

$$J = \frac{p_A}{a^{III} + c^{III}\sqrt{p_{B_1}}} \qquad \text{(IV.153c)}$$

$$J = \frac{p_A}{(a^{IV} + b^{IV}p_A + c^{IV}\sqrt{p_{B_1}})^2} \qquad \text{(IV.153d)}$$

On the basis of Eqs (IV. 152), the equations can be written in the following forms:

$$J = \frac{p_A/P}{\left(\frac{a^I}{P} + \frac{2}{3}c^I\right) + \left(b^I - \frac{2}{3}c^I\right)\frac{p_A}{P}}$$

$$J = \frac{p_A/P}{\left(\frac{a^I}{P} + \frac{1}{3}d^I\right) + \left(b^I - \frac{1}{3}d^I\right)\frac{p_A}{P}}$$

$$J = \frac{p_A/P}{P\left[\left(\frac{a^{II}}{P} + \frac{3}{2}c^{II} + \frac{1}{3}d^{II}\right) + \left(b^{II} - \frac{2}{3}d^{II} - \frac{1}{3}d^{II}\right)p_A/P\right]^2}$$

$$J = \frac{p_A/P}{a^{III}/P + \sqrt{\frac{2}{3}\frac{c^{III}}{\sqrt{P}}}\sqrt{1 - \frac{p_A}{P}}}$$

$$J = \frac{p_A/P}{P\left[\left(\frac{a^{IV}}{P}\right) + b^{IV}\frac{p_A}{P} + \sqrt{\frac{2}{3}\frac{c^{IV}}{\sqrt{P}}}\sqrt{1 - \frac{p_A}{P}}\right]^2}$$

Introducing the notation

$$x = \frac{p_A}{P} \qquad \text{(IV.154)}$$

and linearizing the above expressions, we obtain:

$$\frac{x}{J} = \left(\frac{a^I}{P} + \frac{2}{3}c^I\right) + \left(b^I - \frac{2}{3}c^I\right)\mathrm{x}\,;$$

$$\frac{x}{J} = \left(\frac{a^I}{P} + \frac{1}{3}d^I\right) + \left(b^I - \frac{3}{1}d^I\right)x\,; \qquad \text{(IV.155a)}$$

$$\frac{x}{J} = \alpha^I + \beta^I x$$

$$\sqrt{\frac{x}{J}} = \sqrt{P}\left[\left(\frac{a^{II}}{P} + \frac{2}{3}c^{II} + \frac{1}{3}d^{II}\right) + \left(b^{II} - \frac{2}{3}c^{II} - \frac{1}{3}d^{II}\right)x\right]$$

$$\sqrt{\frac{x}{J}} = \alpha^{II} + \beta^{II}x \qquad \text{(IV.155b)}$$

$$\frac{x}{J} = \frac{a^{III}}{P} + \sqrt{\frac{2}{3}\frac{c^{III}}{\sqrt{P}}}\sqrt{1-x}\,;$$

$$\frac{x}{J} = \alpha^{III} + \gamma^{III}\sqrt{1-x} \quad \text{(IV.155c)}$$

$$\sqrt{\frac{x}{J}} = \sqrt{P}\left[\frac{a^{IV}}{P} + b^{IV}x + \sqrt{\frac{2}{3}\frac{c^{IV}}{P}}\sqrt{1-x}\right]$$

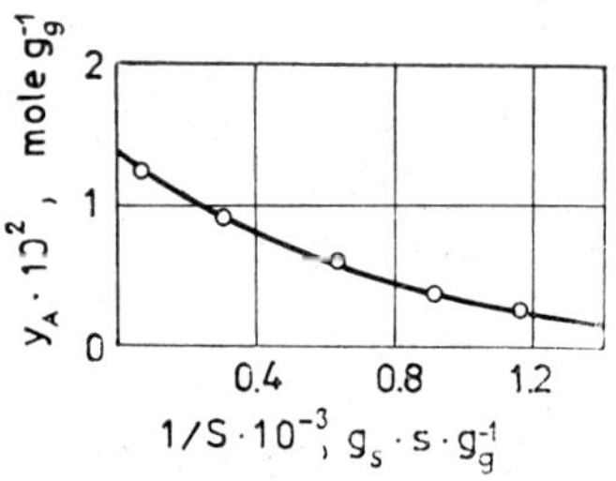

Fig. IV.22. Variation of diethyl ether content of the product as a function of reciprocal space velocity during the dehydration of diethyl ether on an Al_2O_3 catalyst

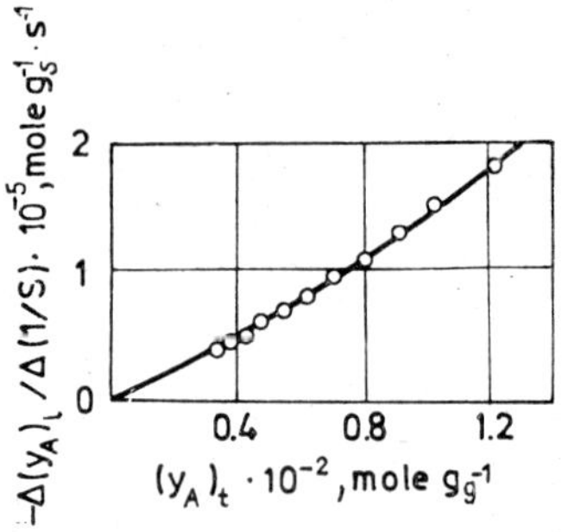

Fig. IV.23. Catalytic dehydration of diethyl ether. Illustration of the calculation procedure described in the text

$$\sqrt{\frac{x}{J}} = \alpha^{IV} + \beta^{IV}x + \gamma^{IV}\sqrt{1-x} \quad \text{(IV.155d)}$$

The reaction rate is determined in accordance with Eq. (IV.91):

$$\bar{j} = -\frac{d(y_A)_1}{d(1/S)} \approx -\frac{\Delta(y_A)_1}{\Delta(1/S)} \quad \text{(IV.156)}$$

$(y_A)_1$ is plotted against $1/S$ using the data in Table IV.18 (Fig. IV.22). The ordinates $(y_A)_1$ of the curves obtained are read off for identical $\Delta(1/S)$ values and the quotients $-\Delta(y_A)_1/\Delta(1/S)$ are calculated:

$$\frac{\Delta(y_A)_1}{\Delta(1/S)}(1/S)_i = \frac{(y_A)_1\,(1/S)_{i+1} - (y_A)_1\,(1/S)_{i-1}}{(1/S)_{i+1} - (1/S)_{i-1}} \quad \text{(IV.157)}$$

These difference quotients are plotted against $(y_A)_1$ (Fig. IV.23), and the function $\bar{j}(y_A)$ is obtained [see Eq. (IV.156)]. Using the known values of $\bar{j}(y_A)$ and m_s, the function $J(y_A)$, too, can be calculated. The results of the calculations are collected in Table IV.19.

The experimental data transformed according to Eqs (IV.155a)–(IV.155d) are shown in Table IV.19 and Fig. IV.24. It is seen in Fig IV.24a that the experimental data fall on a straight line if the transformation defined by Eq. (IV.155a) is used, i.e. this equation can be regarded as being correct.

TABLE IV.19

Catalytic dehydration of diethyl ether. Illustration of the calculation procedure described in the text

$\frac{1}{S}$, 10^3 g_g^{-1} s g_s	0.0	0.1	0.2	0.3	0.4	0.5	0.6	0.7	0.8	0.9	1.0	1.1	1.2
$(y_A)_1$, 10^{-2} mole g_g^{-1}	1.42	1.21	1.06	0.91	0.80	0.70	0.61	0.54	0.47	0.42	0.37	0.33	0.29
$-\frac{\Delta (y_A)_1}{\Delta \frac{1}{S}}$, 10^{-5} mole g_s^{-1} s^{-1}	—	1.80	1.50	1.30	1.05	0.95	0.80	0.70	0.60	0.50	0.45	0.40	—
$(y_A)_1$, 10^{-2} mole g_g^{-1}	0.20	0.30	0.40	0.50	0.60	0.70	0.80	0.90	1.00	1.10	1.20	1.30	—
$\bar{j}$, 10^{-5} mole g_s^{-1} s^{-1}	0.25	0.37	0.50	0.64	0.77	0.92	1.08	1.24	1.40	1.57	1.76	1.98	—
J, 10^{-5} mole s^{-1}	1.02	1.53	2.05	2.64	3.18	3.80	4.45	5.10	5.76	6.46	7.25	8.15	—
$[3(y_A)_0 - 2y_A]$, 10^{-2} mole g_g^{-1}	3.65	3.45	3.25	3.05	2.85	2.65	2.45	2.25	2.05	1.85	1.65	1.45	—
$x \equiv p_A P\, 10^2 \equiv \frac{y_A}{3(y_A)_0 - 2y_A}\, 10^2$	5.64	8.70	12.30	16.40	21.00	26.40	32.70	40.00	48.80	59.50	72.90	89.70	—
$\frac{x}{J}$, 10^3 s $mole^{-1}$	5.51	5.69	6.00	6.20	6.60	6.95	7.35	7.85	8.46	9.20	10.05	11.00	—
$\sqrt{\frac{x}{J}}$, 10 $s^{1/2}$ $mole^{1/2}$	7.43	7.54	7.75	7.86	8.13	8.34	8.56	8.85	9.20	9.59	10.00	10.49	—
$10(1 - x)$	9.44	9.13	8.77	8.36	7.90	7.36	6.73	6.00	5.12	4.05	2.71	1.03	—
$10\sqrt{1 - x}$	9.72	9.56	9.35	9.14	8.89	8.58	8.21	7.75	7.16	6.36	5.20	3.21	—
$108 - \sqrt{\frac{x}{J}}$, $s^{1/2}$ $mole^{-1/2}$	33.7	32.6	30.5	29.4	26.7	24.6	22.4	19.5	16.0	12.1	8.00	3.1	—
$\frac{108 - \sqrt{x/J}}{\sqrt{1 - x}}$, $s^{1/2}$ $mole^{-1/2}$	34.7	34.1	32.9	3.22	3.10	28.7	27.2	25.2	22.3	19.0	15.4	9.7	—

There is a strong deviation from linearity in Fig. IV.24b, and consequently Eq. (IV.155b) is not valid. The same holds for Eq. (IV.155c), since a straight line cannot be drawn for this set of points (Fig. IV.24c) and the curve de-

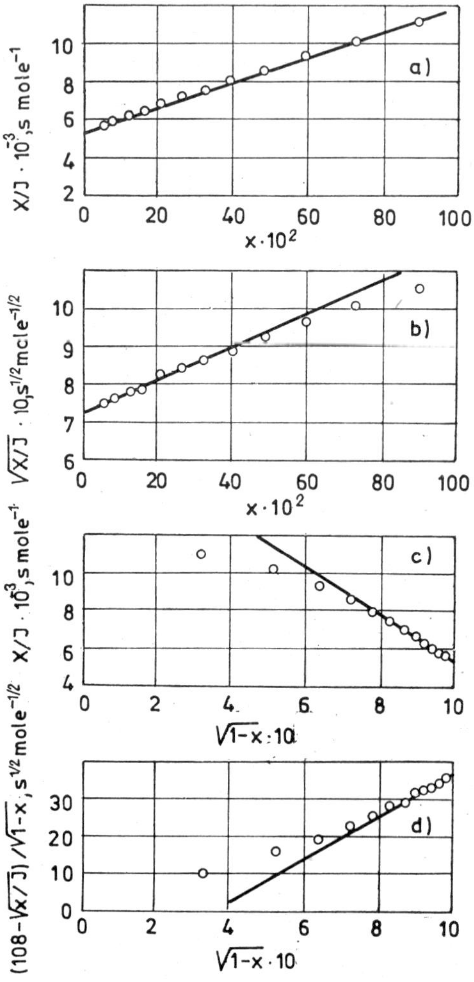

Fig. IV.24. Catalytic dehydration of diethyl ether. Illustration of the calculation procedure described in the text

creases with increasing values of $\sqrt{1-x}$, whereas according to Eq. (IV.155c) an increasing curve is expected, for c^{III}/P may assume only positive values.

The validity of Eq. (IV.155d) can be checked as follows. The value of $x/J = 11.7\ 10^3$ s mole^{-1} at $x = 1$ (Fig. IV.24a). Hence $(\sqrt{x/J})_{x=1} = 1.08\ 10^2$ s$^{1/2}$ mole$^{-1/2}$. From Eq (IV.155d):

$$\sqrt{\frac{1}{J}} - \sqrt{\frac{x}{J}} = \beta^{IV}(1-x) - \gamma^{IV}\sqrt{1-x}$$

i.e. considering the numerical value of $\sqrt{\frac{1}{J}}$:

$$\frac{108 - \sqrt{\frac{x}{J}}}{\sqrt{1-x}} = -\gamma^{IV} + \beta^{IV}\sqrt{1-x} \tag{IV.158}$$

It is seen in Fig. IV.24d that the experimental points plotted according to Eq. (IV.158) do not obey a linear function, i.e. Eq. (IV.155d) cannot be applied.

From Fig. IV.24a the constants of Eq. (IV.155a) are:

$$\begin{aligned} \alpha^{I} &= 5.1\ 10^{3}\ \text{s mole}^{-1} \\ \beta^{I} &= 6.6\ 10^{3}\ \text{s mole}^{-1} \end{aligned} \tag{IV.159}$$

According to Eq. (IV.155a):

$$\alpha^{I} \equiv \left(\frac{a^{I}}{P} + \frac{2}{3}c^{I}\right); \quad \beta^{I} \equiv \left(b_{1} - \frac{2}{3}c^{I}\right)$$

or

$$\alpha^{I} \equiv \left(\frac{a^{I}}{P} + \frac{1}{3}d^{I}\right); \quad \beta^{I} \equiv \left(b^{I} - \frac{1}{3}d^{I}\right) \tag{IV.160}$$

The given series of experiments does not allow the separation of the individual constants. The information that can be gained from Eqs (IV.159) and (IV.160) is:

$$\frac{a^{I}}{P} + \frac{2}{3}c^{I} > 0; \quad \frac{2}{3}c^{I} < b^{I} \neq 0$$

and

$$\frac{a_{I}}{p_{B}} + \frac{1}{3}d^{I} > 0; \quad \frac{1}{3}d^{I} < b^{I} \neq 0 \tag{IV.161}$$

This type of rate equation is expected for reactions II/a_1–2 (see Table IV.4), in which the rate is controlled by the surface reaction and only one of the products is adsorbed (see Table IV.2/2). The kinetic equations corresponding to the valid rate equation are:

$$J = k\frac{p_A}{1 + K_A p_A + K_{B_1} p_{B_1}}$$

and

$$J = k\frac{p_A}{1 + K_A p_A + K_{B_2} p_{B_2}} \tag{IV.162}$$

From the comparison of Eqs (IV.153a), (IV.155a) and (IV160)–(IV.162) the following expressions can be formulated:

$$\alpha^{\mathrm{I}} = \frac{1}{k}\left(\frac{1}{P} + \frac{2}{3}K_{B_1}\right); \quad \beta^{\mathrm{I}} = \frac{1}{k}\left(K_A - \frac{2}{3}K_{B_1}\right) \tag{IV.163a}$$

or

$$\alpha^{\mathrm{I}} = \frac{1}{k}\frac{1}{P}; \quad \beta^{\mathrm{I}} = \frac{1}{k}K_A; \quad \text{if} \quad K_{B_1} \approx 0 \tag{IV.163b}$$

If component B_2 is adsorbed, then:

$$\alpha^{\mathrm{I}} = \frac{1}{k}\left(\frac{1}{P} + \frac{1}{3}K_{B_2}\right); \quad \beta^{\mathrm{I}} = \frac{1}{k}\left(K_A - \frac{1}{3}K_{B_2}\right) \tag{IV.164a}$$

or

$$\alpha^{\mathrm{I}} = \frac{1}{k}\frac{1}{P}; \quad \beta^{\mathrm{I}} = \frac{1}{k}K_A; \quad \text{if} \quad K_{B_2} \approx 0 \tag{IV.164b}$$

where

$$k = A_s[S]_0 \vec{k} K_A \tag{IV.165}$$

Since the proportions of the components are strictly determined by the stoichiometric numbers, a choice between kinetic equations (IV.163a)–(IV.163d) is impossible on the basis of these measurements. Accordingly, the physical meaning of the constants which can be calculated are ambiguous. If Eq. (IV.163b) or (IV.164b) is valid, then:

$$k = \frac{1}{Pd^{\mathrm{I}}}; \quad K_A = \frac{\beta^{\mathrm{I}}}{\alpha^{\mathrm{I}}}\frac{1}{P} \tag{IV.166}$$

on the condition that either $K_{B_1} \approx 0$ or $K_{B_2} \approx 0$
i.e. both k and K_A can be separately calculated from the experimentally determined α_0 and β_0. However, if Eq. (IV.163a) or (IV.164a) is valid, then none of the constants k, K_A, K_{B_1} and K_{B_2} can be separately estimated. (Naturally, it must be assumed that only the mechanisms discussed in Section 1.5 are possible.)

PARAGRAPH 2.3.5. DYNAMIC DIFFERENTIAL REACTOR

Tétényi et al. [11] studied the catalytic dehydrogenation of cyclohexane on a nickel catalyst in a dynamic differential reactor. Liquid cyclohexane was introduced into the reactor at a constant velocity. The resulting vapour ensured constant mass flow. Argon was used as the inert gas. The product was cooled to condense benzene and cyclohexane, and the volume of gases (H_2 or H_2+Ar) was measured using a soap-film flow meter. The total pressure was 1.012 10^5 Pa in each experi ent. The stoichiometric equation of the reaction is:

$$C_6H_{12} = 3H_2 + C_6H_6; \qquad A \rightarrow B_1 + B_2 \tag{IV.167}$$

The experimental results are contained in Table IV.20. The rate equation is expressed on the a basis of Eq. (IV.95) and (IV.167):

$$J = \frac{1}{3} \frac{\Delta n_{B_1}}{\Delta t} \tag{IV.168}$$

The values of the rate calculated using this equation are also shown in Table IV.20. The partial pressures of the components can be calculated as follows:

$$p_A = \frac{(F')_A}{(F')_0 + (F')_A + (F')_{B_1}} P \tag{IV.169}$$

$$p_{B_1} = \frac{(F')_{B_1}}{(F')_0 + (F')_A + (F')_{B_1}}$$

the results of the calculations are given in Table IV.20.

Dehydrogenation of cyclohexane belongs to the reaction type

$$A \rightarrow B_1 + B_2 \tag{IV.170}$$

similarly to the dehydration of ether. According to Table IV.4, the rate equations to be considered are as follows (the mechanisms are confined to those discussed in Section 1.5):

$$\frac{p_A}{J} = a^{\mathrm{I}} + b^{\mathrm{I}} p_A + c^{\mathrm{I}} p_{B_1} \tag{IV.171a}$$

$$\sqrt{\frac{p_A}{J}} = a^{\mathrm{II}} + b^{\mathrm{II}} p_A + c^{\mathrm{II}} p_{B_1} \tag{IV.171b}$$

$$\frac{p_A}{J} = a^{\mathrm{III}} + c^{\mathrm{III}} \sqrt{p_{B_1}} \tag{IV.171c}$$

$$\sqrt{\frac{p_A}{J}} = a^{\mathrm{IV}} + b^{\mathrm{IV}} p_A + c^{\mathrm{IV}} \sqrt{p_{B_1}} \tag{IV.171d}$$

The results of the calculations necessary to check the above equations are also collected in Table IV.20.

This Table indicates that Eq. (IV.171a) is not valid. In experiments 1–5, $p_{B_1} = 0$. Consequently, according to Eq. (IV.171c) p_A/J should remain constant. This is not consistent with the experimental results. If $p_{B_1} =$ $=$ const., Eqs (IV.171a), (171b) and (IV.171d) assume the following forms:

$$\frac{p_A}{J} = a^{\mathrm{I}} + b^{\mathrm{I}} p_A \tag{IV.172a}$$

$$\sqrt{\frac{p_A}{J}} = a^{\mathrm{II}} + b^{\mathrm{II}} p_A ; \qquad \sqrt{\frac{p_A}{J}} = a^{\mathrm{IV}} + b^{\mathrm{II}} p_A \tag{IV.172b}$$

Consequently, depending upon whether Eq. (IV.171a) or one of Eqs (IV.171b) and (IV.171d) is valid, either p_A/J or $\sqrt{p_A/J}$ is a linear function of p_A.

The experimental data plotted according to Eqs (IV.172a) and (IV.172b) are shown in Fig. IV.25. It is seen that a choice between these two equations is almost impossible. Since the number of experimental data is rather small,

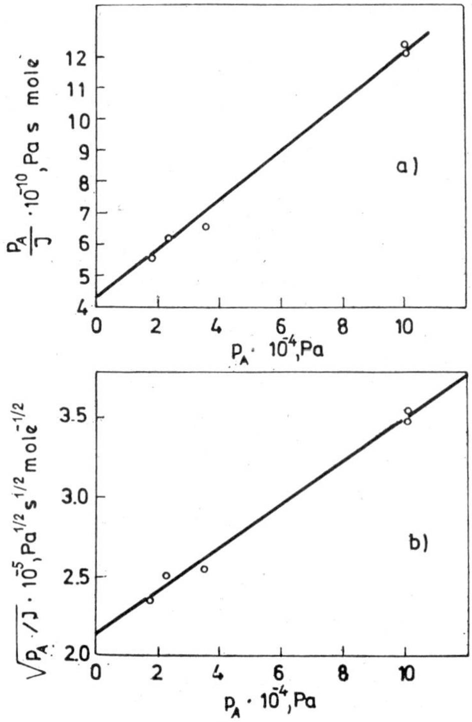

Fig. IV.25. Catalytic dehydrogenation of cyclohexane. Illustration of the calculation procedure described in the text

TABLE

Catalytic dehydrogenation of cyclohexane on a

$t_r = 298$ °C, $P = 1.012 \ 10^5$ Pa, $m_s = 10.0$ g

No. of experiment	F'_A 10^{-5} mole s^{-1}	F'_0 10^{-5} mole s^{-1}	F'_{B1} 10^{-5} mole s^{-1}	$\frac{\Delta n_{B1}}{\Delta t}$ 10^{-7} mole s^{-1}	J 10^{-7} mole s^{-1}
1	0.87	0.00	—	25.23	8.41
2	0.87	1.62	—	16.32	5.44
3	0.87	2.79	—	11.61	3.38
4	0.87	3.90	—	10.05	3.35
5	0.87	0.00	—	24.33	8.11
6	0.87	0.00	1.59	11.16	3.72
7	0.87	0.00	2.62	8.49	2.83
8	0.87	0.00	3.94	6.24	2.08

the variance criterion is also of limited use. The estimation of the constants in Eqs (IV.172a) and (IV.172b) by the method of least squares results in the following expressions:

$$\frac{p_A}{J} = 3.99\ 10^{10} + 8.11\ 10^5\, p_A \qquad \text{(IV.173a)}$$

$$\sqrt{\frac{p_A}{J}} = 2.11\ 10^5 + 1.37\, p_A \qquad \text{(IV.173b)}$$

The variance is

$$\bar{\Delta} = \left[\left(\frac{p_A}{J}\right)_c - \left(\frac{p_A}{J}\right)_m\right]^2 \frac{1}{n} \qquad \text{(IV.174)}$$

where the subscripts c and m refer to calculated and measured values, respectively, and n is the number of experimental data.

The variances calculated from Eqs (IV.173a) and (IV.173b) are:

(IV.173a) $\bar{\Delta}^2 = 6.6\ 10^{18}\ \mathrm{Pa}^2\ \mathrm{s}^2\ \mathrm{mole}^{-2}$ (IV.175)

(IV.173b) $\bar{\Delta}^2 = 4.8\ 10^{18}\ \mathrm{Pa}^2 . \mathrm{s}^2\ \mathrm{mole}^{-2}$

i. e. the difference between them is so small that a decision in favour of one, or other, equation is impossible. Both are therefore considered valid, temporarily. In the series of experiments 6, 7 and 8, p_{B_1} also varies. Since the values of a and b are known from Eqs (IV.173a) and (IV.173b), Eqs (IV.171a), (IV.171b) and (IV.171d) can be rewritten as follows:

$$\mu^{\mathrm{I}} \equiv \frac{1}{p_{B_1}}\left(\frac{p_A}{J} - 3.99\ 10^{10} - 8.11\ 10^5\, p_A\right) = c^{\mathrm{I}} \qquad \text{(IV.176a)}$$

$$\mu^{\mathrm{II}} \equiv \frac{1}{p_{B_1}}\left(\sqrt{\frac{p_A}{J}} - 2.11\ 10^5 - 1.37\, p_A\right) = c^{\mathrm{II}} \qquad \text{(IV.176b)}$$

IV.20

Ni catalyst in a dynamic differential reactor

$A \equiv C_6H_{12}$, $B_1 \equiv H_2$, $B_2 \equiv C_6H_6$, $O \equiv Ar$

No. of experiment	p_A 10^4 Pa	p_{B_1} 10^4 Pa	$\frac{p_A}{J}$ 10^{10} Pa s mole^{-1}	$\sqrt{\frac{p_A}{J}}$ 10^{10} Pa$^{1/2}$ s$^{1/2}$ mole$^{1/2}$	$\sqrt{p_{B_1}}$ 10^2 Pa$^{1/2}$
1	10.12	—	12.04	3.47	—
2	3.55	—	6.52	2.55	—
3	2.41	—	6.24	2.50	—
4	1.85	—	5.54	2.35	—
5	10.12	—	12.45	3.53	—
6	3.72	6.54	9.65	3.11	2.56
7	2.52	7.61	8.90	2.98	2.76
8	1.83	8.31	8.77	2.96	2.88

$$\mu^{IV} \equiv \frac{1}{\sqrt{p_{B_1}}}\left(\sqrt{\frac{p_A}{J}} - 2.11\ 10^5 - 1.37\ p_A\right) = c^{IV} \qquad \text{(IV.176d)}$$

The results of calculations based upon the above equations are given in Table IV.21. It is seen from this Table that the available experimental data are not sufficient to decide which of Eqs (IV.176a), (IV.176b) and (IV.176d) is best.

TABLE IV.21

Catalytic dehydrogenation of cyclohexane
Illustration of the calculation procedure described in the text.

p_{B_1}, 10^4 Pa	$\sqrt{p_{B_1}}$, 10^2 Pa$^{1/2}$	c^I 10^5 s mole^{-1}	c^{II} $x^{1/2}$ Pa$^{-1/2}$ mole$^{-1/2}$	c^{IV} 10^2 s$^{1/2}$ mole$^{-1/2}$
6.54	2.56	4.216	0.772	1.973
7.61	2.76	3.783	0.689	1.900
8.31	2.88	3.981	0.713	2.064
	average	3.993	0.726	1.979

Using Eqs (IV.173a) and (IV.173b) and the mean values given in Table IV.21, Eqs (IV.171a), (171b) and (IV.171d) are written in the following forms:

$$\frac{p_A}{J} = 3.99\ 10^{10} + 8.11\ 10^5\ p_A + 3.993\ 10^5\ p_{B_1} \qquad \text{(IV.177a)}$$

$$\frac{p_A}{J} = (2.11\ 10^5 + 1.37\ p_A + 0.726\ p_{B_1})^2 \qquad \text{(IV.177b)}$$

$$\frac{p_A}{J} = (2.11\ 10^5 + 1.37\ p_A + 1.979\ 10^2 \sqrt{p_{B_1}})^2 \qquad \text{(IV.177d)}$$

Values of $\bar{\Delta}^2$ can be estimated from the above equations:

$$\begin{aligned} \bar{\Delta}^2\,(177a) &= 4.26\ 10^{18}\ \text{Pa}^2\ \text{s}^2\ \text{mole}^{-2} \\ \bar{\Delta}^2\,(177b) &= 3.38\ 10^{18}\ \text{Pa}^2\ \text{s}^2\ \text{mole}^{-2} \\ \bar{\Delta}^2\,(177d) &= 3.20\ 10^{18}\ \text{Pa}^2\ \text{s}^2\ \text{mole}^{-2} \end{aligned} \qquad \text{(IV.178)}$$

From a comparison of the variances in Eq. (IV.178), Eq. (IV.177d), i.e. rate equation (IV.171d), is the most probable. Owing to insufficient experimental data and the lack of any marked difference between the variances, however, the mechanism cannot be considered certain. If the variances, $\bar{\Delta}^2$, are chosen as criteria (which is somewhat dubious), the most probable rate equation for the catalytic dehydrogenation of cyclohexane is Eq. (IV.171d), i.e.

$$J = \frac{p_A}{(a + bp_A + c\sqrt{p_{B_1}})^2} \qquad \text{(IV.179)}$$

This rate equation is expected for reactions of type II/a_3–2 (See Table IV.4/3) and the corresponding kinetic equation is:

$$J = k\frac{p_A}{(1 + K_A p_A + \sqrt{K_{B_1} p_{B_1}})^2} \qquad \text{(IV.180)}$$

where

$$k = A_s[S]_0^2 \vec{k} K_A$$

The following conclusions can be drawn from Table IV.2/2: in such a catalytic dehydrogenation of cyclohexane on a Ni catalyst the rate-controlling step is the surface reaction, and the dissociative adsorption of hydrogen takes place on the catalyst surface. From Eqs (IV.177d) and (IV.180) we have:

$$k = 2.25\ 10^{-1}\ \text{mole s}^{-1}\ \text{Pa}^{-1};$$

$$K_A = 6.45\ 10^{-6}\ \text{Pa}^{-1};$$

$$A_s[S]_0^2\vec{k} = 3.49\ 10^{-5}\ \text{mole s}^{-1} \qquad \text{(IV.181)}$$

$$K_{B_1} = 0.876\ 10^{-6}\ \text{Pa}^{-1}$$

The main objective in giving the above examples was to illustrate the treatment of experimental data obtained by the use of the various reactors. The method involved the examination of the validity of the kinetic equations which can be derived from theoretically possible mechanisms. Neither the mechanisms assumed nor the kinetic equations collected in Tables IV.3, 4 and 5 are exhaustive. Accordingly, none of the examples was intended to confirm a mechanism. The only aim was to show the elucidation of possible mechanisms on the basis of experimental data. The mechanisms found in this way should not be regarded as proved, since the kinetic data alone, particularly those determined at a single temperature, are not suitable for this purpose; a more extensive study, including, for example, energy factors, is required. The above examples show that, in several cases, not even the possible mechanism of reaction can be found from kinetic data alone; this is explained by the limitation of experimental accuracy. At present the number of heterogeneous catalytic reactions whose mechanisms have been unequivocally proved is very small.

NOTATIONS AND DIMENSIONS

l	length
n	mole
m_g	mass (gas)
m_s	mass (catalyst)
t	time
T	temperature
a, b, c, d	constants of the rate equation
A_i	reactant
B_j	product
$(A_i)_a, (B_j)_a$	adsorbed components
$[A_i]_a, [B_j]_a$	surface concentrations of adsorbed components, $n\ l^{-2}$
A_s	surface of catalyst
c	volume concentration, $n\ l^{-3}$
D	diffusion coefficient, $l^2\ t^{-1}$
$D^+ \equiv D_{eff}$	diffusion coefficient referring to porous grain
D'	overall diffusion coefficient in equations of type Fick–I, describing convective and diffusion mass transport simultaneously
F	mas velocity of gas flow $m_g\ t^{-1}$
G	mass velocity of gas flow in pore phase
$\bar{j}$	specific rate of reaction
J	reaction rate, $n\ t^{-1}$
$\vec{J}\overleftarrow{J}$	rate of elementary chemical reaction, $n\ t^{-1}$
J_{A_i}, J_{B_j}	rates of adsorption of components $n\ t^{-1}$
k	rate constant
$\vec{k}\overleftarrow{k}$	rate constant of an elementary chemical reaction taking place on the surface
$\vec{k}_{A_i}, \vec{k}_{B_j}$	rate constants for adsorption processes of components
$\overleftarrow{k}_{A_i}, \overleftarrow{k}_{B_j}$	rate constants for desorption processes of components
K	equilibrium constant of chemical reaction
K_{A_i}, K_{B_j}	adsorption equilibrium constants for components
l_p	pore length, l
m_{A_i}, m_{B_j}	masses of components m_g
m_s	mass of catalyst, m_s
n_{A_i}, n_{B_j}	amounts of components, n
$\dot{N}$	component transport, $n\ l^{-2}\ t^{-1}$
O	inert gas
p_{A_i}, p_{B_j}	partial pressures of components, $P\ a$, kg $m^{-1}\ s^{-2}$

P	pressure in the bulk phase
r_p	pore radius, l
R	gas constant
R_f	flow resistance, $m_g^{-1}\ l^3\ t$
$\{S$	space velocity of gas flow, $m_g\ m_s^{-1}\ t^{-1}$
S	selectivity
(S)	free active area of catalyst surface
$[S]$	concentration of free active sites of catalyst, $n\ l^{-2}$
$[S]_0$	total concentration of active sites of catalyst
t	reaction time, t
T	temperature, T
v	gas volume, l^3
y_{A_i}, y_{B_j}	component concentrations, $n\ m_g^{-1}$
ν_{A_i}, ν_{B_j}	stoichiometric numbers for components
ϱ	density, $m_g\ l^{-3}$

REFERENCES TO PART IV

1. Bokhoven, C. et al.: J. Phys. Chem. **58,** 471 (1954)
2. Fejes, P. and Kalló, D.: Acta Chim. Acad. Sci. Hung. **39,** 213 (1963)
3. Wheeler, A.: Adv. Cat. **3,** 250 (1951)
4. Wicke, E. and Brötz, W.: Chem. Ing. Techn. **21,** 219 (1949)
5. Thiele, E. W.: Ind. Eng. Chem. **31,** 216 (1939)
6. Weisz, P. B. and Swegler, E. W.: J. Phys. Chem. **59,** 823 (1955)
7. Kiperman, S. A. and Kaplan, G. I.: Kinetika i Kataliz **5,** 888 (1964)
8. Batta, I., Solymosi, F. and Szabó, Z.: J. Catalysis **1,** 103 (1962)
9. Móger, D.: Izuchenie kinetiki i mehanizma gidrirovaniya i samogidrirovaniya ciklogeksena na Pt-katalizatore s pomoshchu deuteriya metodom gazovoy hromatografii i mass-spektrometrii. Thesis, Moscow, 1963
10. Kalló, D.: Private communication, 1959
11. Tétényi, P., Király, J. and Babernics, L.: Acta Chim. Acad. Sci. Hung. **29,** 35 (1961)

HANDBOOKS

1. Advances in Catalysis. Vols. 1, 2, etc. Academic Press Inc., New York, from 1949
2. Emmett, P. H.: Catalysis. Vols. 1, 2, etc. Reinhold Publ. Co., New York, from 1952
3. Hougen, O. A., Watson, K. M.: Chemical Process Principles, J. Wiley and Sons Inc., New York, 1950
4. Schwab, G. M.: Handbuch der Katalyse. Vols. 1, 2, etc. Springer Verlag, Wien, from 1941